VIRAL VECTORS

VIRAL VECTORS

Gene Therapy and Neuroscience Applications

Edited by

Michael G. Kaplitt, MD, PhD
Division of Neurosurgery
Department of Surgery and Department of Neurology
New York Hospital–Cornell University Medical College
and Laboratory of Neurobiology and Behavior
The Rockefeller University
New York, New York

Arthur D. Loewy, PhD
Department of Anatomy and Neurobiology
School of Medicine
Washington University
St. Louis, Missouri

ACADEMIC PRESS

San Diego New York Boston
London Sydney Tokyo Toronto

Cover photograph: Photomicrographs of hippocampus demonstrating ßgal expression from a HSV vector (See Chapter 9, Figure 2).

This book is printed on acid-free paper. ♾

Academic Press, Inc.
A Division of Harcourt Brace & Company
525 B Street, Suite 1900, San Diego, California 92101-4495

United Kingdom Edition published by
Academic Press Limited
24-28 Oval Road, London NW1 7DX

Library of Congress Cataloging-in-Publication Data

Viral vectors : tools for study and genetic manipulation of the nervous system / edited by Michael G. Kaplitt, Arthur D. Loewy.
p. cm.
Includes bibliographical references and index.
ISBN 0-12-397570-0 (case). -- ISBN 0-12-397571-9 (pbk)
1. Central nervous system--Diseases--Gene therapy. 2. Genetic vectors. 3. Viruses. I. Kaplitt, Michael G. II. Loewy, Arthur D.
[DNLM: 1. Genetic Vectors. 2. Central Nervous System Diseases--genetics. 3. Central nervous System Diseases--therapy. 4. Central Nervous System Diseases--physiology. 5. Genes, Viral--genetics. 6. Gene Expression Regulation, Viral. 7. Gene Therapy. QH 442.2 V813 1995]
RC361.V55 1995
616.8--dc20
DNLM/DLC
for Library of Congress 94-49185
CIP

PRINTED IN THE UNITED STATES OF AMERICA
95 96 97 98 99 00 EB 9 8 7 6 5 4 3 2 1

Contents

CHAPTER 6

HSV Vectors in the Mammalian Brain: Molecular Analysis of Neuronal Physiology, Generation of Genetic Models, and Gene Therapy of Neurological Disorders

Matthew J. During

CHAPTER 7

The Use of Defective Herpes Simplex Virus Amplicon Vectors to Modify Neural Cells and the Nervous System

Howard J. Federoff

CHAPTER 8

Viral Vector-Mediated Gene Transfer in the Nervous System: Application to the Reconstruction of Neural Circuits in the Injured Mammalian Brain

J. Verhaagen, W. T. J. M. C. Hermens, A. J. G. D. Holtmaat, A. B. Oestreicher, W. H. Gispen, and M. G. Kaplitt

CHAPTER 9

The Use of Herpes Simplex Virus Vectors for Protection from Necrotic Neuron Death

Dora Y. Ho, Matthew S. Lawrence, Timothy J. Meier, Sheri L. Fink, Rajesh Dash, Tippi C. Saydam, and Robert M. Sapolsky

CHAPTER 10

In Vivo Promoter Analysis in the Adult Central Nervous System Using Viral Vectors

Jun Yin, Michael G. Kaplitt, and Donald W. Pfaff

CHAPTER 11

Adenoviral-Mediated Gene Transfer: Potential Therapeutic Applications

Beverly L. Davidson and Blake J. Roessler

CHAPTER 12

Transfer and Expression of Potentially Therapeutic Genes into the Mammalian Central Nervous System in Vivo Using Adeno-Associated Viral Vectors

Michael G. Kaplitt and Matthew J. During

CHAPTER 13

Genetic Modification of Cells with Retrovirus Vectors for Grafting into the Central Nervous System

Un Jung Kang

CHAPTER 20

Molecular Properties of Alphaherpesviruses Used in Transneuronal Pathway Tracing

Thomas C. Mettenleiter

CHAPTER 21

Molecular Analysis of Rabies and Pseudorabies Neurotropism

Patrice Coulon and Anne Flamand

Contributors

Numbers in parentheses indicate the pages on which the authors' contributions begin.

Rajat Bannerji (75) Department of Molecular Biology, Memorial–Sloan Kettering Cancer Center, New York, New York 10021.

Jeffrey S. Bartlett (55) Gene Therapy Center, University of North Carolina, Chapel Hill, North Carolina 27599.

Mary Ann Bender (1) Department of Molecular Genetics and Biochemistry, School of Medicine, University of Pittsburgh, Pittsburgh, Pennsylvania 15261.

Xandra O. Breakefield (239) Department of Surgery and Molecular Neurogenetics Laboratory, Harvard Medical School, Massachusetts General Hospital, Charlestown, Massachusetts 02129.

J. Patrick Card (319) Department of Neuroscience, University of Pittsburgh, Pittsburgh, Pennsylvania 15260.

E. Antonio Chiocca (239) Department of Surgery and Molecular Neurogenetics Laboratory, Harvard Medical School, Massachusetts General Hospital, Charlestown, Massachusetts 02129.

Patrice Coulon (395) Laboratoire de Génétique des Virus, CNRS, 91198 Gif sur Yvette, France.

Josep Dalmau (275) Department of Neurology and the Cotzias Laboratory of Neuro-Oncology, Cornell University Medical College, New York, New York 10021.

Rajesh Dash (133) Department of Biological Sciences, Stanford University, Stanford, California 94305.

Beverly L. Davidson (173) Department of Internal Medicine, University of Iowa, Iowa City, Iowa 52242.

Neal DeLuca (1) Department of Molecular Genetics and Biochemistry, School of Medicine, University of Pittsburgh, Pittsburgh, Pennsylvania 15261.

Matthew J. During (89, 193) Department of Neurosurgery, Yale University School of Medicine, New Haven, Connecticut 06520.

Howard J. Federoff (109) Departments of Medicine and Neuroscience, Albert Einstein College of Medicine, Bronx, New York 10461.

David J. Fink (1) Department of Neurology, University of Michigan Medical School, Ann Arbor, Michigan 48109.

Sheri L. Fink (133) Department of Biological Sciences, Stanford University, Stanford, California 94305.

Anne Flamand (395) Laboratoire de Génétique des Virus, CNRS, 91198 Gif sur Yvette Cedex, France.

Niza Frenkel (25) Department of Cell Research and Immunology, Tel Aviv University, Tel Aviv, Israel.

W. H. Gispen (119) Department of Pharmacology, Rudolf Magnus Institute for Neuroscience, 3508 TA Utrecht, The Netherlands.

Joseph C. Glorioso (1) Department of Molecular Genetics and Biochemistry, School of Medicine, University of Pittsburgh, Pittsburgh, Pennsylvania, 15261.

William F. Goins (1) Department of Molecular Genetics and Biochemistry, School of Medicine, University of Pittsburgh, Pittsburgh, Pennsylvania 15261.

W. T. J. M. C. Hermens (119) Netherlands Institute for Brain Research, 1105 AZ Amsterdam, The Netherlands, and Rudolph Magnus Institute for Neuroscience, Department of Pharmacology, Stratenum, The Netherlands.

Dora Y. Ho (133) Department of Biological Sciences, Stanford University, Stanford, California 94305.

A. J. G. D. Holtmaat (119) Netherlands Institute for Brain Research, 1105 AZ Amsterdam, The Netherlands, Rudolph Magnus Institute for Neuroscience, Department of Pharmacology, Stratenum, The Netherlands.

William Hunter (259) Molecular Neurosurgery Laboratory and Georgetown Brain Tumor Center, Department of Neurosurgery, Georgetown University Medical Center, Washington, District of Columbia 20007.

Un Jung Kang (211) Department of Neurology, The University of Chicago, Chicago, Illinois 60637.

Michael G. Kaplitt (119, 157, 193) Laboratory of Neurobiology and Behavior, The Rockefeller University, New York, New York 10021.

Ann D. Kwong (25) Schering Plough Research Institute, Kenilworth, New Jersey 07033.

Matthew S. Lawrence (133) Department of Biological Sciences, Stanford University, Stanford, California 94305.

Arthur D. Loewy (349) Department of Anatomy and Neurobiology, Washington University School of Medicine, St. Louis, Missouri 63110.

Marla B. Luskin (411) Department of Anatomy and Cell Biology, Emory University School of Medicine, Atlanta, Georgia 30322.

Robert Martuza (259) Molecular Neurosurgery Laboratory and Georgetown Brain Tumor Center, Department of Neurosurgery, Georgetown University Medical Center, Washington, District of Columbia 20007.

Kieran W. McDermott (411) Department of Anatomy, University College, Cork, Ireland.

Timothy J. Meier (133) Department of Biological Sciences, Stanford University, Stanford, California 94305.

Thomas C. Mettenleiter (367) Federal Research Centre for Virus Diseases of Animals, Institute of Molecular and Cellular Virology, D-17498 Insel Riems, Germany.

A. B. Oestreicher (119) Rudolf Magnus Institute for Neuroscience, Department of Pharmacology, 3508 TA Utrecht, The Netherlands.

Donald W. Pfaff (157) Laboratory of Neurobiology and Behavior, The Rockefeller University, New York, New York 10021.

Samuel Rabkin (259) Molecular Neurosurgery Laboratory and Georgetown Brain Tumor Center, Department of Neurosurgery, Georgetown University Medical Center, Washington, District of Columbia 20007.

Blake J. Roessler (173) Department of Internal Medicine, University of Michigan, Ann Arbor, Michigan 48109.

Myrna R. Rosenfeld (275) Department of Neurology and the Cotzias Laboratory of Neuro-Oncology, Cornell University Medical College, New York, New York 10021.

Richard J. Samulski (55) Gene Therapy Center, University of North Carolina, Chapel Hill, North Carolina 27599.

Robert M. Sapolsky (133) Department of Biological Sciences, Stanford University, Stanford, California 94305.

Tippi C. Saydam (133) Department of Biological Sciences, Stanford University, Stanford, California 94305.

Thomas Shenk (43) Department of Molecular Biology, Howard Hughes Medical Institute, Princeton University, Princeton, New Jersey 08544.

Evan Y. Snyder (435) Department of Neurology and Pediatrics, Harvard Medical School, Children's Hospital, Boston, Massachusetts 02115.

Takashi Tamiya (239) Department of Surgery and Molecular Neurogenetics Laboratory, Harvard Medical School, Massachusetts General Hospital, Charlestown, Massachusetts 02129.

Gabriella Ugolini (293) Laboratoire de Génétique des Virus, CNRS, F-91198 Gif-Sur-Yvette, France.

J. Verhaagen (119) Netherlands Institute for Brain Research, 1105 AZ Amsterdam, The Netherlands, Rudolph Magnus Institute for Neuroscience, Department of Pharmacology, Stratenum, The Netherlands.

Jan J. Verschuuren (275) Department of Neurology and the Cotzias Laboratory of Neuro-Oncology, Cornell University Medical College, New York, New York 10021.

Ming X. Wei (239) Department of Surgery and Molecular Neurogenetics Laboratory, Harvard Medical School, Massachusetts General Hospital, Charlestown, Massachusetts 02129.

Jun Yin (157) The Rockefeller University, New York, New York 10021.

Preface

Genetic manipulation of the mammalian central nervous system has progressed from a theoretical concept to an explosive field with remarkable speed over the past several years. It is not surprising that there would be great interest in such an endeavor, but few could have predicted the technical successes that are now permitting previously unimaginable neurobiological explorations. Gene transfer has long been a mainstay of molecular and cellular biology. The ability to either express a novel gene product or overexpress an endogenous protein within a mammalian cell in tissue culture has yielded a wealth of information about basic cellular and subcellular processes. This concept was extended to animal research with the advent of transgenic mice. Although this technology has been extremely powerful, it is very time consuming and results in genetic alteration of every cell in the resulting animal. Given the complexities of the mammalian central nervous system, neurobiologists began to search for methods that would rapidly generate reagents for the selective manipulation of particular populations of cells. This search was intensified by those wishing to extend the gene therapy revolution to the treatment of nervous system disorders.

Viral vectors have become the vehicle of choice for genetic manipulation of the adult nervous system. This approach takes advantage of the inherent properties of viruses as evolutionary gene transfer agents. Viruses are essentially parasites that require the functions of a host cell in order to survive. They employ either DNA or RNA as the genetic material that encodes a limited set of virus-specific functions. Usually these include virus coat proteins and some replication and packaging machinery. The viral genome is encased in a lipid coat; in some instances this coat is enveloped in a second lipid layer, and viral proteins are embedded within this outer layer. These viral coat proteins mediate interactions with cellular receptors, thereby promoting highly efficient attachment to and penetration of the recipient cell. The result is entry of the viral genetic material into the host cell. Under lytic conditions, the virus genome will replicate and encode synthesis of structural proteins in order to generate progeny virus. This process often subverts and destroys the recipient cell. Certain viruses enter a lysogenic or latent state, which can result in retention of the viral genome within a surviving cell. In some instances, insertion of viral genetic material within a chromosome of the host cell can yield permanent alteration of the physiology of that cell. The identification of cellular homologs of viral oncogenes is but one example of the potential influence of viral-mediated gene transfer on the evolutionary development of higher eukaryotic organisms.

This volume is intended to provide a comprehensive overview of the current state of development and application of viral vectors to the study of nervous system function and treatment of neurologic disease. Although several systems and numerous applications are presented, all of the contributors share a common mission. Each group has attempted to exploit the benefits of naturally occurring viruses as highly efficient gene transfer agents while altering the viral genome in order to minimize adverse effects of viral infection upon the target cell. In recognition of the diversity of vectors and applications which may follow this theme, however, we have organized this book into four sections. Section I contains brief reviews from leading virologists describing the relevant biology of each virus currently in use as a vector. While the reader is encouraged to more fully explore the background of a particular approach, this section should adequately provide to those unfamiliar with one or more systems the basic information necessary to appreciate the ensuing chapters. Section II provides a broad view of several gene transfer applications that have utilized viral vectors. These range from purely basic studies intended to elucidate the function of a particular gene or regulatory signal to more clinically relevant experiments with clear implications for human gene therapy. Section III describes the work of several groups using viral vectors in novel ways to understand and treat tumors of the central nervous system. The contributors to Section IV use viral vectors in a somewhat different manner in order to study central nervous system development and to map previously unknown neuroanatomical relationships. These groups use specific vectors that can either tag a neuron during development or spread from one neuron to connecting cells in a controlled and defined manner.

Gene transfer into the adult mammalian nervous system is no longer an esoteric tool, but rather a growing field that is attracting the interest of a wide array of scientists. Viral vectors have become the principal vehicle for effective gene transfer. With the development of numerous viral vectors for different applications, researchers have entered into a debate to determine which vector is the most useful for neurobiologists. We will not attempt to offer an answer to this question, because it is our belief that this volume will demonstrate the value of all the systems. We hope that by synthesizing the collective work of the leading researchers in our field into a single volume, we will at once create a resource for state-of-the-art information and impress upon the remaining skeptics the importance of viral vector-mediated gene transfer. Since clinical trials of genetic therapy of brain tumors and certain nonneurological genetic diseases have already been initiated, the concepts contained within this volume should be of lasting value to both the experimental and the clinical neurobiologist, as well as to anyone who would use viral vectors outside the nervous system.

Michael G. Kaplitt
Arthur D. Loewy

Acknowledgments

A volume such as this is clearly the result of the efforts of more than just two people, and therefore we acknowledge the contributions of the following individuals. First we thank Dr. Donald Pfaff at The Rockefeller University for his advice and support during the early stages of developing this concept. Our publisher, Academic Press, has been an outstanding partner in this endeavor, and this was due in large part to the guidance and support of the Editorial Director for this project, Dr. Graham Lees. His enthusiasm for this concept, constant availability, and ease of manner all contributed to making this a very positive experience for everyone. This project would also have been difficult to complete without the invaluable aid of Dr. Lees' assistant, Karen Dempsey, who coordinated much of this project and could always be relied upon to provide assistance or information at a moments notice. We also thank Ken Fine and Monique Larson for seamlessly shepherding this book through the final stages of production and G. B. D. Smith, a freelance artist, for designing such a striking cover. Finally, we thank our wives, Melissa and Arleen, for their constant support and tolerance of phone calls and editorial obligations at all hours of the day and night.

CHAPTER

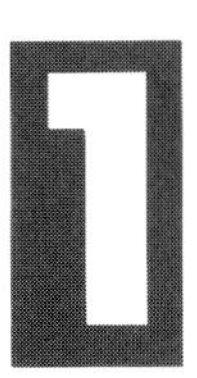

Herpes Simplex Virus as a Gene-Delivery Vector for the Central Nervous System

Joseph C. Glorioso
Mary Ann Bender
William F. Goins
Neal DeLuca
Department of Molecular Genetics and Biochemistry
University of Pittsburgh School of Medicine
Pittsburg, Pennsylvania

David J. Fink
Department of Neurology
University of Michigan Medical School
Ann Arbor, Michigan

I. Introduction

There are a wide variety of diseases which alter the biology of the brain and peripheral nervous system. These include tumors, immune pathological disorders, metabolic defects which often lead to abnormal brain development and mental retardation, and several common neurodegenerative syndromes such as Alzheimer's and Parkinson's disease. While some of these diseases are multifactorial, others arise from single gene defects. The rapid growth of new knowledge concerning the genetic basis of these conditions along with the increased understanding of the molecular and biochemical processes leading to brain pathology in specific disease states has opened up new opportunities for biologic therapy including gene transfer methods for synthesis of therapeutic products *in situ*. Gene therapy offers opportunities to transfer genes whose products would either ameliorate the disease condition or correct a genetic defect. However, with the limitations imposed by the blood–brain barrier on the transport of the replacement gene product to the central nervous system (CNS), it will be necessary in some cases to express this gene locally in the brain. Successful gene replacement therapy would thus provide endogenous production of the therapeutic product *in vivo*.

Most current gene therapy methodology involves *ex vivo* techniques which were developed using retroviral vectors. Retroviruses integrate their proviral genomes into the chromosome of the host cell providing a method to stably introduce genes into cells removed from the patient which can then be returned by cell transplantation procedures following gene transfer. Retroviruses carrying therapeutic genes are proficient tools for transferring genes to rapidly dividing neoplastic cells, fibroblasts and bone marrow stem cells (Anderson, 1992), and may be appropriate for gene delivery to the liver for hepatic enzyme disorders with associated nervous system damage (Grossman *et al.*, 1994). However, several drawbacks exist with this gene-transfer system. These include (i) the risk of cell transformation from integration of the retroviral provirus genome in a manner to activate cellular oncogenes, (ii) the inability to efficiently propagate many cell types *in vitro*, (iii) the failure of cellular implants to survive long term, and (iv) the difficulty in retrieving tissues for explantation which is particularly important in the case of the brain. Since retroviruses require dividing cells to enter the nucleus and integrate the therapeutic gene, they cannot infect postmitotic cells, such as neurons. This makes them unsuitable for gene transfer to brain. In addition, the implantation of cells into the cranial space requires the surgical creation of an open area within the brain to contain the transplant. This level of damage to normal neuronal tissue carries the risk of focal neurologic deficits.

A more direct *in vivo* approach is crucial for effective gene delivery to brain neuronal tissue. Both adenovirus (Akli *et al.*, 1993; Davidson *et al.*, 1993; Le Gal La Salle *et al.*, 1993) and herpesvirus-based (Andersen *et al.*, 1992; Fink *et al.*, 1992; Chiocca *et al.*, 1990; Glorioso *et al.*, 1992, 1994) vectors possess a wide host cell range and can efficiently infect nondividing cells. These vectors have each demonstrated their abilities to support expression of a foreign gene in the brain. However, it remains to be determined whether adenovirus vectors require low levels of viral gene expression in order to express the foreign gene and whether as a consequence, immune surveillance will result in rejection of the infected neurons. In contrast, herpes simplex virus (HSV) is particularly well suited for gene delivery to the brain because, unlike adenovirus, it is naturally neurotropic and has evolved a mechanism to remain latent and persist within neurons without expression of viral proteins or apparent brain damage. Genetic modifications of HSV would ensure that the virus could not express lytic functions and would endow it with the capacity to efficiently express transgenes during this latent phase. Appropriately engineered HSV vectors should be effective for gene therapy approaches to treat CNS disease.

II. Herpes Simplex Virus Vector Development

Our research efforts are directed toward the engineering of herpes simplex virus vectors as potential gene-transfer vehicles. We are moving toward an effective vector design by (i) capitalizing upon the latent state of the virus within neurons to serve as a scaffold to support foreign gene expression and by (ii) modifying the virus to remove the cytotoxic features of virus infection. This review will focus upon the biological features of HSV which are important in vector design and will highlight experiments which demonstrate foreign gene delivery by HSV. Although there is a body of literature dealing with the use of defective herpes vectors for packaging and delivery of plasmid amplicons as gene-transfer vehicles, we will limit this review to recombinant HSV vectors.

A. The Biology of Herpes Simplex Virus

The herpes simplex virion is composed of an interior icosahedral capsid composed of seven structural proteins within which is contained the viral DNA genome in association with core proteins. Surrounding the capsid is an amorphous matrix of proteins called the tegument (Roizman and Furlong, 1974) which consists of approximately 12 viral components (Roizman and Sears, 1993). The viral envelope is the outer shell of the virus particle and possesses at least 10 unique virus-encoded glycoproteins (See Fig. 1a). HSV infection initiates at epithelial or mucosal skin surfaces by the attachment and fusion of the viral envelope with cell surface membranes releasing the capsids into the cytoplasm. Once inside the cell, the capsid and tegument are transported to the nucleus where the viral DNA enters the nucleus through a nuclear pore (Batterson *et al.*, 1983). As outlined in Fig. 1, the replication of the virus leads to the production of progeny virus particles which are released from the infected cell. The infectious particles bind to and fuse with axon terminals innervating the site of primary infection. By retrograde transport, the viral capsids migrate to the cell body of the sensory ganglia where either a lytic or latent pattern of viral expression unfolds (Roizman and Sears, 1990). Although the lytic functions can be expressed, leading to further viral replication and cell death, typically the lytic cycle is interrupted and a latent state is established within the infected neurons. During latency, the viral lytic genes are silenced and a series of latency-specific RNAs are detected in the nuclei of latently infected neuronal cells (Fig. 1b). Reactivation from the latent state can occur due to exposure to various stimuli, including stress, trauma, or immunosuppressive drug therapies. Following reactivation, the lytic phase resumes with the production of viral progeny. Viral particles are transported either in an anterograde manner back to the primary site of infection with the development of visible lesions or, in rare cases, in a retrograde direction to the CNS, leading either to a latent infection or to a viral replication and spread with accompanying encephalitis.

B. Lytic Viral Infection

The viral genome is a linear double-stranded DNA molecule 152 kilobases (kb) in length. It is composed of unique long (U_L) and unique short (U_S)

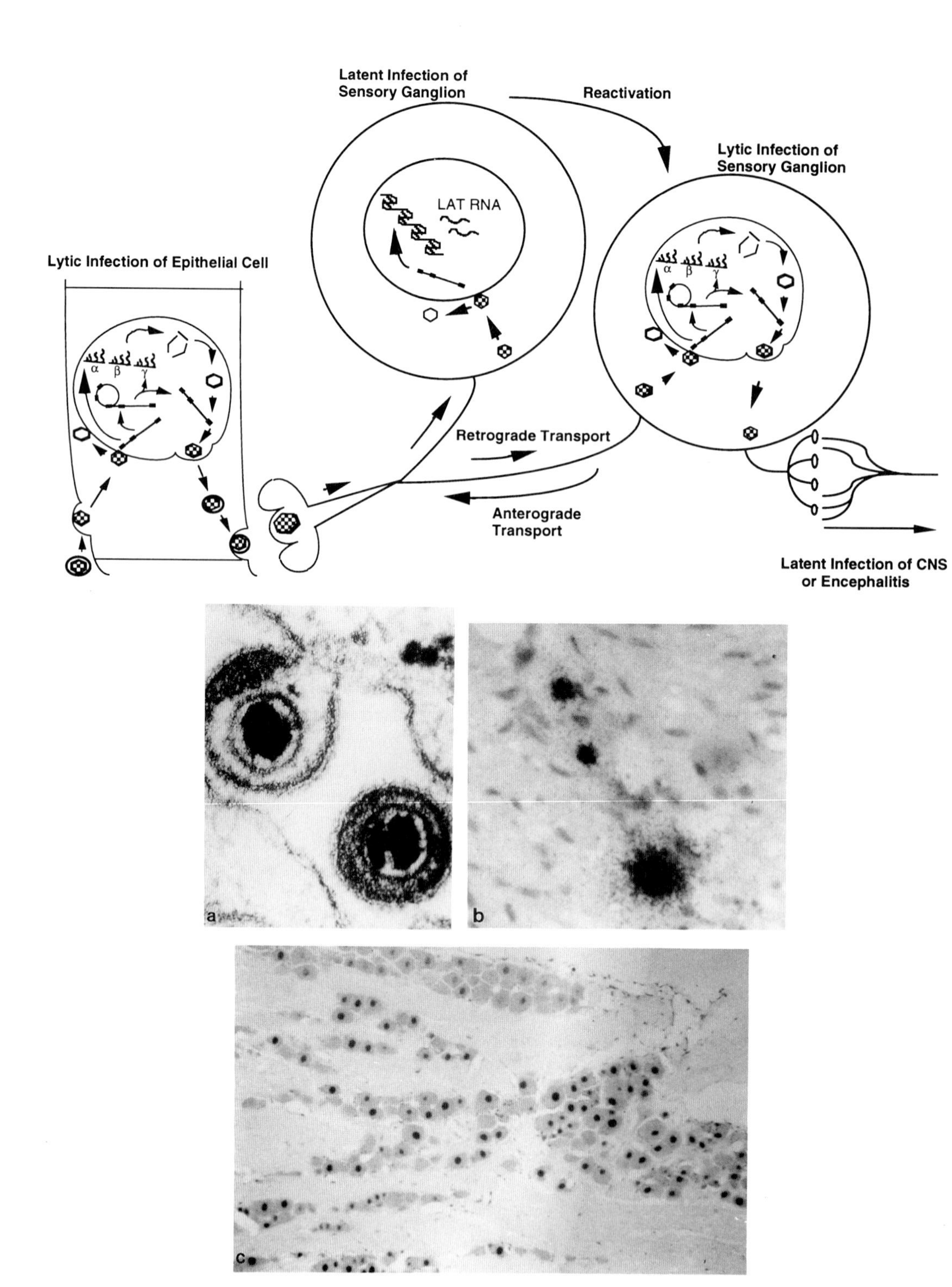

Latent Infection of Sensory Ganglion
Reactivation
LAT RNA
Lytic Infection of Sensory Ganglion
Lytic Infection of Epithelial Cell
α
β
γ
Retrograde Transport
Anterograde Transport
Latent Infection of CNS or Encephalitis
a
b
c

regions which are each flanked by inverted terminal repeats. HSV encodes at least 75 gene products which can be categorized as (i) nonstructural proteins which control gene expression and DNA replication or (ii) structural proteins which comprise the capsid and core protein components, tegument, and envelope glycoproteins. In Fig. 2, it can be seen that almost half of these viral functions is required for viral growth in cell culture (essential genes), while the other half encodes functions which contribute to productive infection in the neuronal host cell, but is not required for virus production in cultured cells (nonessential genes). Remarkably, the arrangement of these two sets of genes along the viral genome is such that large segments of tandem nonessential genes could potentially be removed without compromising viral growth. The right-hand end of the genome, for example, contains only three essential genes located within 35 kb of viral DNA (see Fig. 2). This organization would provide the opportunity to replace substantial segments of the viral genome with large continuous segments of foreign DNA.

During the lytic cycle, HSV genes are expressed in a coordinately regulated temporal pattern. This pattern follows a three step cascade (Fig. 3): the immediate early (IE or α), early (E or β), and late (L or γ) phases (Honess and Roizman, 1974). In addition, the late phase also requires viral DNA synthesis which involves E gene-encoded enzymatic functions. The regulated pattern of gene expression depends on both viral and cellular transcription factors. The viral

Figure 1 Life cycle of herpes simplex virus (HSV-1). When HSV encounters an epithehial or mucosal cell at the site of primary infection, the virus proceeds through the regulated cascade of lytic infection with the subsequent production of progeny virions. These particles can bind to and enter sensory neurons which innervate this region at which time the virus is transported by rapid retrograde flow along the axon to the nucleus of the neuron. There, the virus sets up one of two types of infection. In a latent infection, the viral DNA becomes associated with histone proteins, forming a chromatin structure, and expresses a set of latency-associated transcripts (LATs). Alternatively, the virus may set up a lytic infection and produce viral progeny. Latency may last for the lifetime of the host, but upon stimulation by such agents as stress or drugs, the lytic cycle may be reactivated. Progeny virus is transported either in an anterograde manner, returning to the site of the initial infection, or to neurons of the CNS. (a) Ultrastructure of the HSV virion. The dark inner core of the virus containing the viral DNA and core proteins is seen as a central opaque region within each virion. The core is encompassed by the symmetric icosahedral capsid. Amorphous tegument separates the capsid from the outer envelope of the virus which is acquired as the virus leaves the host cell. (b) Detection of LAT synthesis during latency. *In situ* hybridization techniques reveal that the LATs are present in approximately 5% of trigeminal ganglion (TG) neurons at 30 days postinfection. Analysis of TG slices shows those neurons which are positive for LAT expression, marked by the deposition of the darkly staining granules over the cells. (c) Detection of latently infected neurons within nerve tissue. Latent viral DNA can be detected in trigeminal ganglion neurons by *in situ* PCR at 30 days following infection with a wild-type virus. The extensive labeling of the TG neurons demonstrates the virus' ability to establish a latent state within a large number of neurons.

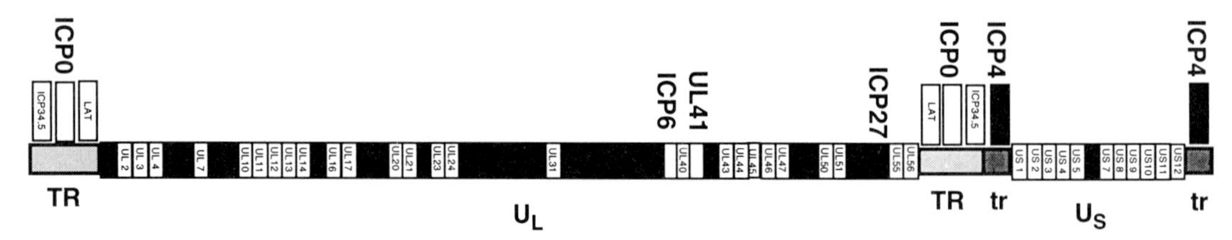

Figure 2 Map of the HSV genome. The 152,000-bp genome of HSV is organized into two regions, a unique long (U_L) and a unique short (U_S) segment, each flanked by terminal repeat regions (TR and tr). The 75 genes of HSV are classified into two groups; (i) those genes which are essential for growth in cell culture (black boxes) and (ii) those genes which are not essential for growth *in vitro,* but may provide increased viability in nondividing neuronal cell environments *in vivo* (clear boxes). Genes of interest which map to positions within the terminal repeats are placed above the shaded boxes of the repeat regions. Also listed in bold above the genome are the various viral genes which are candidates for inactivation to construct a viral vector which is not cytotoxic in cell culture.

factors are induced on infection by one of the HSV tegument proteins, VP16 (or αTIF, Vmw65), a product of the UL48 gene. VP16 enters the nucleus of the infected cell along with the viral genome, complexes with a cellular transcription factor (Oct 1) and boosts IE gene transcription by binding to specific sequences within the IE promoters. Upregulation of IE gene expression is important to the initiation of the viral lytic cycle *in vivo* since mutations which affect the transcriptional activating function of VP16 result in virus attenuation (Steiner *et al.*, 1990). VP16 boosts the transcription of five IE genes

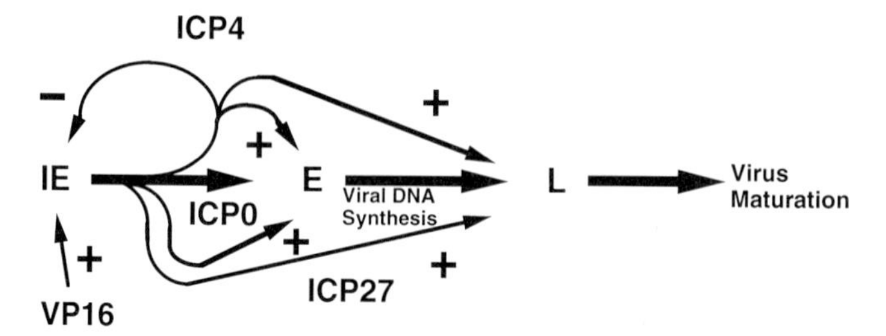

Figure 3 Transcriptional progression of the HSV lytic cycle. Three classes of HSV genes are expressed during the lytic cycle of viral growth, the immediate early (IE) or α, the early (E) or β, and the late (L) or γ genes. The IE gene products act as transactivators to stimulate expression of the early (E) and late (L) genes. Late (L) genes are generally expressed only after DNA replication has occurred. The viral gene products which regulate this cascade of transcription are depicted, as are the positive (+) or negative (−) effect they impose on gene expression. The VP16 protein, which is a component of the infecting virus' tegument, activates transcription of the IE genes. Three IE proteins (ICP4, ICP0, and ICP27) can activate transcription of both E and L genes. ICP4 is also able to inhibit the expression of the IE genes, including its own expression, once transcription of the E class of genes is initiated.

whose products represent infected cell proteins (ICP) 0, 4, 22, 27, and 47 (Cordingly *et al.*, 1983; Mackem and Roizman, 1982). Two IE genes, ICP4 and ICP27, are essential for viral replication (Dixon and Schaffer, 1980; Preston *et al.*, 1979; Sacks *et al.*, 1985; Watson and Clements, 1978) and are required for the expression of E and L genes. ICP4 can also autoregulate its own expression (Dixon and Schaffer, 1980) by repressing the activity of its promoter and the promoters of other IE genes during the later stages of viral replication (DeLuca and Schaffer, 1985; O'Hare and Hayward, 1985). The complex interplay among the IE gene products in regulating HSV gene expression illustrates their pivotal role in directing the lytic cycle.

C. Eliminating HSV Cytotoxicity

To establish an effective HSV vector system, it is necessary to remove virally induced cytotoxic functions, including those required for lytic replication, and to promote the establishment of latency without the possibility for reactivation. This can be accomplished by deleting IE genes, thereby preventing early and late gene expression. By deleting both copies of the ICP4 gene, for example, the virus can no longer initiate the lytic gene cascade. Rather, the virus expresses only the other four IE genes. Nevertheless, ICP4 deletion mutants remain fully competent to permanently enter into latency within neurons (Dobson *et al.*, 1990). The propagation of ICP4 deletion mutants in cell culture requires the use of cell lines which have been stably transduced with the ICP4 gene. The ICP4 gene product is provided in *trans* on infection with replication-defective ICP4$^-$ mutant viruses (DeLuca *et al.*, 1985) thereby providing a permissive environment for mutant virus propagation. One particular ICP4$^-$ mutant (designated d120) has been shown to be highly effective for gene transfer to nerve tissue and brain (Chiocca *et al.*, 1990; Dobson *et al.*, 1990) in the absence of demonstrable neuronal tissue damage. However, these ICP4 mutants are highly cytotoxic for most cell types in culture (Johnson *et al.*, 1992).

Additional steps are required to further reduce viral toxicity. Among the candidate genes are those which include nonessential structural components of the infectious particle. Although UV-irradiated virus itself is not substantially cytotoxic, indicating that the structural components of the virus are not a significant source of the cytotoxicity (Leiden *et al.*, 1980), the infecting virus carries within its tegument a virion host shutoff protein (vhs) (UL41 gene product, Fig. 2) which interferes with the synthesis of host cell proteins (Kwong and Frenkel, 1987) through the degradation of mRNA molecules (Oroskar and Read, 1989). However, disruption of the vhs function in the context of an ICP4$^-$ viral background still does not fully eliminate viral cytotoxicicty (Johnson *et al.*, 1992). Other viral gene candidates whose products may contribute to cytotoxicity include the four remaining IE genes which are overexpressed in ICP4$^-$ mutants. ICP27 and ICP0, in particular (Figs. 2 and 3), activate E and L viral gene expression and may be responsible for the low levels (1 to 5% of wild type) of viral mRNA and proteins detected in ICP4$^-$ mutant virus infections (DeLuca and Schaffer, 1988). The laboratory of Dr. DeLuca has recently engineered an ICP4$^-$ virus also deleted for the essential ICP27 gene. Moreover, he has deleted the vhs gene and the viral gene encoding the ICP6 protein, the large subunit of the viral ribonucleotide reductase. This multiply mutated virus is grown on a complementing cell line which has been biochemically transformed to express the ICP4 and ICP27 essential gene products upon virus infection. Not only is there a marked reduction in toxicity for infected cells in culture with this viral mutant, but insertion of the *lacZ* gene downstream of the ICP6 promoter enables expression of the reporter gene. Continuing efforts to develop additional mutiple mutants which also fail to express ICP0 should provide a fully noncytotoxic viral backbone into which foreign genes may be inserted and expressed without the expression and subsequent complications of lytic viral functions.

D. The Virus in Latency

During latency, the viral genome fails to support the productive lytic cycle, but rather becomes partially methylated (Dressler *et al.*, 1987) and associates with nucleosomal structures forming minichromosomes (Deshmane and Fraser, 1989). Latently infected nuclei may contain multiple copies of the viral genome, and recent work shows that a large majority of susceptible nuclei harbor latent virus (Ramakrishnan *et al.*, 1994) (Fig. 1c). In animal models, latency may be achieved experimentally in the peripheral nervous system through the inoculation of the skin (Cook and Stevens, 1973), cornea (Seiler and Schwab, 1984), or olfactory bulb (Stroop and Schaefer, 1987). Latency may also be established in specific foci of the CNS through stereotactic injections of virus suspensions into targeted brain regions (Bak *et al.*, 1977; McFarland *et al.*, 1986).

The pattern of transcription during latency (Hill, 1985) localizes to a small segment of the inverted repeat segments flanking the U_L region of the viral genome (Fig. 4A). The major latency-associated transcript (LAT) is 2 kb in length (Stevens *et al.*, 1987) and appears to arise as a stable intron spliced from a larger 8.77-kb poly-A$^+$ RNA (Farrell *et al.*, 1991). During latency, the 2-kb LAT is predominantly intranuclear, uncapped, and nonpolyadenylated (Devi-Rao *et al.*, 1991). No LAT-specific proteins have been detected during latency (Doerig *et al.*, 1991). Several minor species of 1.5 and 1.45 kb are also detected which may represent alternatively spliced or processed forms of the 2-kb species (Spivack and Fraser, 1987, 1988; Spivack *et al.*, 1991; Stevens, 1989; Wagner *et al.*, 1988).

The function of LAT in viral latency or reactivation is yet to be resolved although recent observations suggest several possibilities. *In vitro* studies have

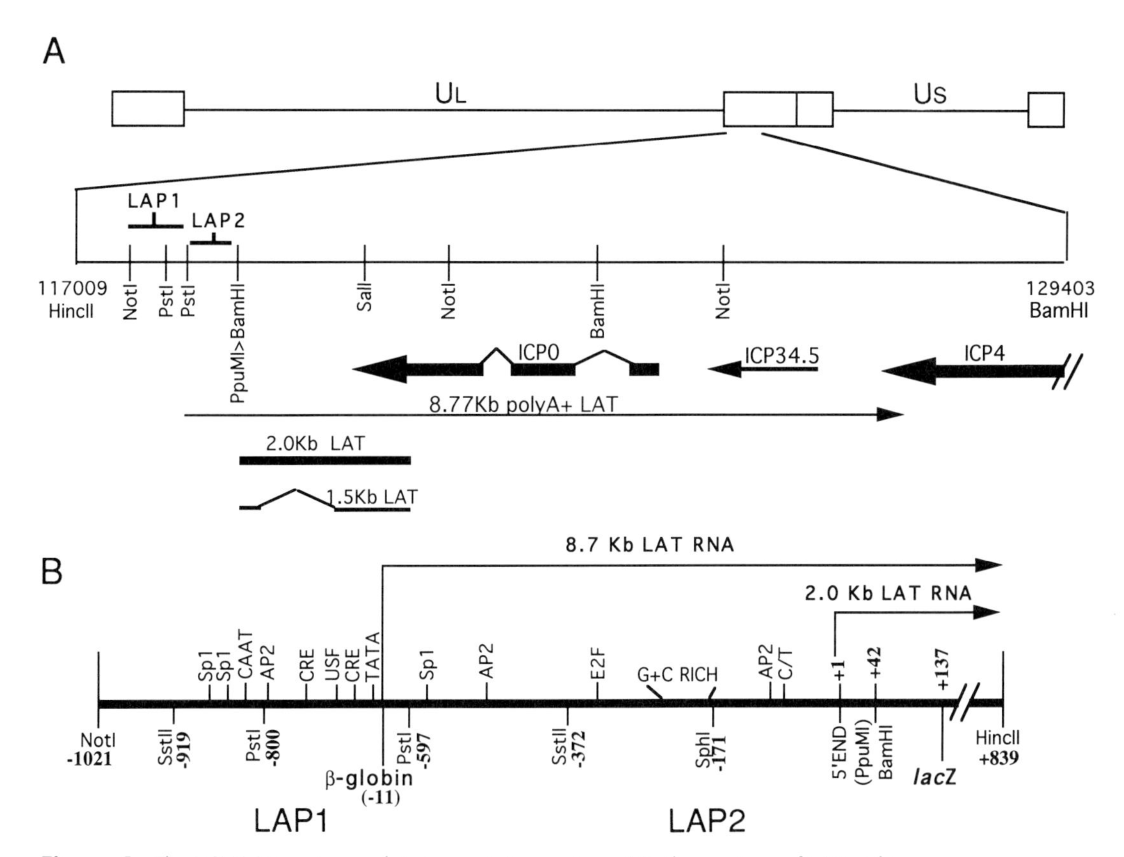

Figure 4 The HSV LAT region and its transcription pattern. (A) The position of LAT within the viral genome and the restriction map of the LAT region are outlined. Below the genomic map are the transcription products of the LAT region. The 8.7-kb polyadenylated LAT transcript is present during productive infection. The stable 2-kb LAT transcript, also detected in productive infection, is the major LAT transcript during latency, while a smaller 1.5-kb transcript is a spliced variant of the 2-kb LAT found exclusively during latency. (B) An expansion of the LAT promoter regions is depicted, displaying the *cis*-acting sites within LAP1 and LAP2. LAP1 is located over 600 bp upstream from the 5′ end of the stable 2-kb LAT. The site at which the β-globin reporter gene was inserted downstream of LAP1 in the LAP1-β-globin expression vector is depicted (Dobson *et al.*, 1989). LAP2 is located just upstream of the 5′ end of the 2-kb stable LAT (Goins *et al.*, 1994). The position of the β-galactosidase (*lacZ*) insert downstream of LAP2 is also given (Ho and Mocarski, 1989).

shown that LAT molecules associate with polyribosomes, implying a function in the regulation of mRNA translation into protein (N. Fraser, personal communication). Also, because LAT is transcribed from the DNA strand opposite ICP0 (Fig. 4A), it is possible that LAT may act in a manner to inhibit ICP0 mRNA transport, stability, or translation. The ability of the LAT intron to reduce IPC0 transactivation of a viral promoter in transient assays was demonstrated by Farrell *et al.* (1991). In view of this, it is surprising that many animal model studies of LAT-deletion viruses have shown that LAT is not an absolute requirement for the establishment and maintenance of latency (Hill *et al.*, 1990; Ho and Mocarski, 1989; Javier *et al.*, 1988; Leib *et al.*, 1989; Sedarati *et al.*, 1989; Steiner *et al.*, 1989; Stevens, 1989). Several reports demonstrate that while LAT^- viruses can efficiently establish latency within neurons (Hill *et al.*, 1990; Leib *et al.*, 1989; Sedarati *et al.*, 1989), they may differ in their ability to reactivate, displaying either wild-type (Block *et al.*, 1990; Javier *et al.*, 1988; Natarajan *et al.*, 1991; Sedarati *et al.*, 1989) or delayed reactivation kinetics (Hill *et al.*, 1990; Leib *et al.*, 1989; Sawtell and Thompson, 1992; Steiner *et al.*, 1989). These differences may reflect the complex interplay between virus and host cell factors leading to reactivation, highlighted by either the method of reactivation (Hill *et al.*, 1990) or the anatomic site of latency (Sawtell and Thompson, 1992). Moreover, the kinetics of reactivation should depend on the number of latent genomes established in the nervous system which were not quantitated in most cases.

The details of LAT gene expression are also very complex and require further elucidation. Two latency active promoters (LAPs) have been described which are also capable of functioning during lytic infection. These include LAP1 which is located at the 5′ end of the 8.7-kb transcript, but 700 bp upstream of the 5′ end of the 2-kb LAT (Batchelor and O-Hare, 1990, 1992; Devi-Rao *et al.*, 1991; Dobson *et al.*, 1989; Goins *et al.*, 1994; Javier *et al.*, 1988; Leib *et al.*, 1989; Zwaagstra *et al.*, 1989, 1990, 1991), and LAP2 just proximal to the 5′ end of the 2-kb LAT (Nicosia *et al.*, 1993; Goins *et al.*, 1994) (Fig. 4B). While LAP1 relies on a consensus TATA box to position the start of transcription, LAP2 is TATA-less, containing an initiator element at the transcription start site. LAP2 also displays a high level of sequence homology to the other TATA-less promoters of various housekeeping genes, including the human NGF (Sehgal *et al.*, 1988) and EGF (Ishii *et al.*, 1985) receptors and the protooncogenes PIM-1 (Meeker *et al.*, 1990), c-myc (Kolluri *et al.*, 1992), and murine c-Ki-ras (Hoffman *et al.*, 1987). LAP2 also shares over 75% homology within the first 100 bp to the TATA-less promoter for brain-derived neurotropic factor (Timmusk *et al.*, 1993) implying that LAP2 contains neuronally active elements.

Recent studies suggest that both LAP1 and LAP2 can independently serve as promoters. LAP1's activity was demonstrated by an HSV recombinant in which the β-globin gene was placed immediately downstream of the LAP1 TATA box (Fig. 4B). This virus directed the synthesis of β-globin mRNA in latently infected neurons of the peripheral nervous system (PNS) (Dobson *et al.*, 1989). As evidence for LAP2, Ho and Mocarski (1989) engineered a vector which contained the β-galactosidase gene inserted 137 bp downstream of the LAP2 promoter. This construct was also able to support the expression of the foreign gene in latently infected trigeminal neurons with the 5′ end of the chimeric mRNA mapping in proximity of the 2-kb LAT just 3′ of the LAP2 sequences. We have observed that a virus containing a 203-bp deletion of LAP1 sequences (Dobson *et al.*, 1989) is still able to drive LAT expression presumably from the LAP2 promoter during latency, although at a much lower level than that for LAP1 (Chen, Fink, and Glorioso, manuscript in preparation). Moreover, we have constructed recombinants in which LAP2 is capable of driving *lacZ* expression from the native locus (Chen, Fink, and Glorioso, manuscript in preparation) or the ectopic glycoprotein C gene locus (Goins *et al.*, 1994) in latently infected PNS and CNS neurons. It is as yet unclear whether LAP1 and LAP2 represent independent promoters or whether they function as a larger promoter complex. The LAPs are currently under study to determine the nature of their ability to drive transcription during the latent state, and to define promoter elements which may be able to maintain expression of strong (but normally silenced) promoters in a HSV gene-transfer vector.

III. Foreign Gene Expression by HSV Vectors

The adaptation of HSV as a tool to deliver genes to the nervous system is based on its natural nonpathogenic long-term persistence within neurons. To be effective for *in vivo* gene therapy, however, the virus must also provide correct levels of therapeutic gene expression. To meet this goal, several laboratories have engineered a variety of viral recombinants by (i) placing the foreign gene at different locations within the HSV genome under the direction of both viral and nonviral promoters and (ii) analyzing these recombinants for their ability to support transgene expression over time. Early work in the development of HSV-mediated foreign gene expression demonstrated the ability of recombinant viruses to use various HSV promoters to produce hepatitis B surface antigen (Shih *et al.*, 1984), the α and β globins (Panning and Smiley, 1989; Smiley *et al.*, 1987), and the HPRT enzyme (Palella *et al.*, 1988) from infected cells in culture. This expression was short lived, however, and exhibited early gene kinetics with a peak of transgene expression occurring at a few hours after infection. In the brain, similar kinetics and short-term expression were also observed when a replicating HPRT-expression vector was injected into the brains of mice (Palella *et al.*, 1989). HPRT-specific mRNA was detected at 3–5 days after inoculation, but the virus was lethal at high titers. A replication-compromised vector possessing the HSV late gene promoter for glycoprotein C driving a reporter gene resulted in transient expression following introduction into rat hippocampus (Fink *et al.*, 1992). Similarly, we introduced the *Escherichia coli lacZ* gene encoding β-galactosidase into rat hippocampal regions through stereotactic injection of a recombinant replication-defective

HSV vector mutant (d120). This vector was able to deliver the reporter gene under control of the strong human cytomegalovirus immediate-early (HCMV IE) promoter (Fig. 5A) and drive expression of high levels of β-galactosidase within hippocampal dentate gyrus neurons (Fig. 5B) in the absence of encephalitis or signs of neuronal damage. Again, reporter gene expression was transient, disappearing after 7 days. In addition, we have shown that a d120 ($ICP4^-$) vector with an expression cassette containing the Rous sarcoma virus LTR upstream of α-interferon failed to support long-term expression in rat hippocampus (Mester, Fink, and Glorioso, unpublished observation). Results from the laboratory of Breakefield (Andersen *et al.*, 1992; Chiocca *et al.*, 1990) confirm the pattern of short-lived transgene expression. Using a variety of replication- defective vectors containing β-galactosidase under the direction of several non-HSV promoters, Breakefield and co-workers observed that expression peaked early after inoculation and diminished greatly or disappeared by 2 weeks.

One explanation for this reduction in transgene expression over time is that the viral genome is lost from the infected neurons. Studies were conducted using a quantitative polymerase chain reaction (PCR) technique to determine the number of viral genomes retained within recombinant virus-infected CNS neurons at 2, 7, and 56 days after inoculation (Ramakrishnan *et al.*, 1994). This work revealed that approximately 10% of the inoculated viral genomes remained within a large number of neurons and were stable for the entire 8-week period. In addition, low levels of LAT RNA were detected by reverse transcriptase-PCR (RT-PCR), indicating reduced but continuous expression of LAT from the latent viral genomes. These data argue that a reduction of transgene expression over time is not due to the loss of the viral genomes, but rather to transient gene expression under control of a recombinant promoter. While some disease conditions may respond to a short burst of therapeutic gene expression, as in the production of a toxic product within a brain tumor mass, many therapeutic approaches will require continuous expression of a foreign gene from the viral genomes present in latently infected neurons.

A. Long-Term Expression from the LAT Promoters

The ability of HSV to maintain LAT expression during latency suggests that the LAPs may be adapted to express foreign genes from the viral vector within neurons. A viral recombinant with LAP1 driving expression of the rabbit β-globin gene (Dobson *et al.*, 1989) was successful in supporting long-term expression of β-globin mRNA for 3 weeks in latently infected PNS neurons. However, expression diminished by 20-fold once latency was established. Similarly, a β-glucuronidase cDNA clone was inserted just downstream of LAP1 sequences, and after inoculation by corneal scarification, positive enzymatic activity was detected at acute and latent (2–18 weeks) times in peripheral

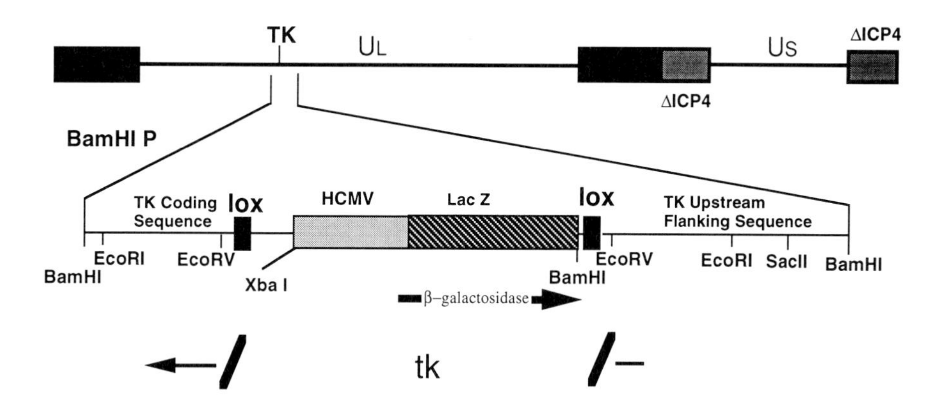

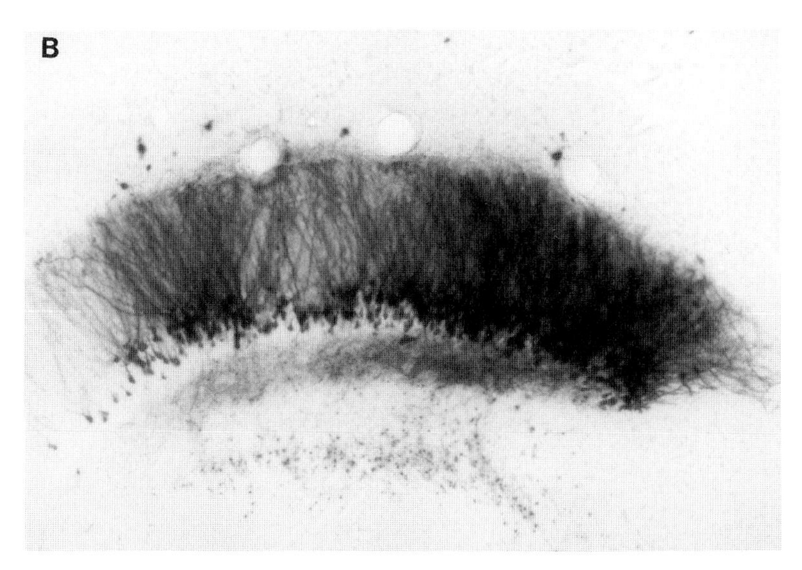

Figure 5 HSV vector-mediated expression of β-galactosidase from the HCMV IE promoter in rat CNS. (A) The recombinant virus d120 (ICP4$^-$), TK$^-$::HCMVIEp–*lacZ* with a gene cassette of the HCMV IE promoter driving the β-galactosidase gene (*lacZ*), was constructed by *cre–lox* recombination (Sauer *et al.,* 1987; Gage *et al.,* 1992) and contains a deletion of both copies of ICP4 (DeLuca *et al.,* 1985). (B) Stereotactic inoculation of 5 μl containing 5×10^7 pfu of this *lacZ* recombinant virus into rat hippocampus resulted in intense staining of granule cell neurons of the dentate gyrus at 2 days postinoculation. Although *lacZ* expression was transient, expression could not be detected in either the contralateral hippocampus or the other brain regions with projections to the hippocampal region. This confirms the inability of this recombinant viral vector to spread. In addition, we have been unable to detect any significant pathology resulting from this mutant even in long-term animals.

A

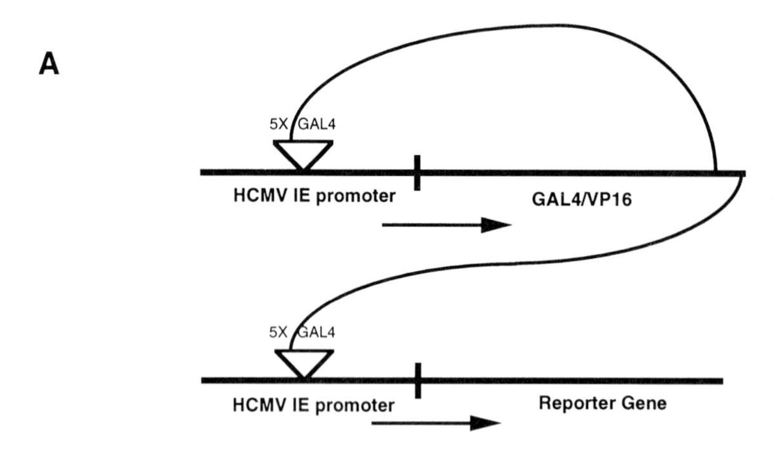

5X GAL4
HCMV IE promoter
GAL4/VP16
5X GAL4
HCMV IE promoter
Reporter Gene

B

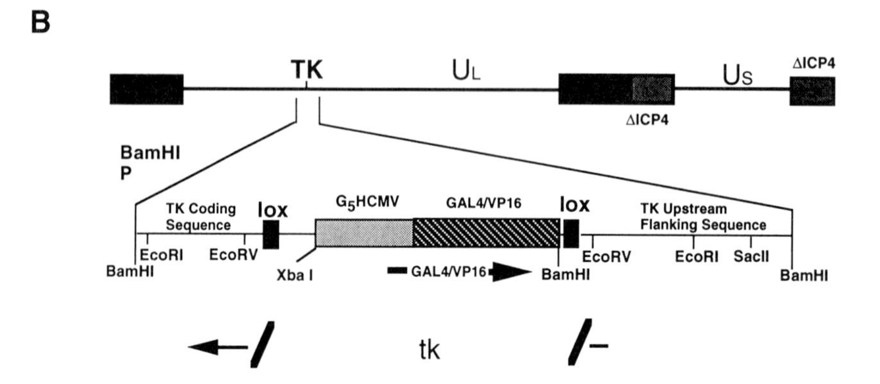

TK
UL
US
ΔICP4
ΔICP4
BamHI P
TK Coding Sequence
lox
G5HCMV
GAL4/VP16
lox
TK Upstream Flanking Sequence
BamHI
EcoRI
EcoRV
Xba I
GAL4/VP16
BamHI
EcoRV
EcoRI
SacII
BamHI
tk

C

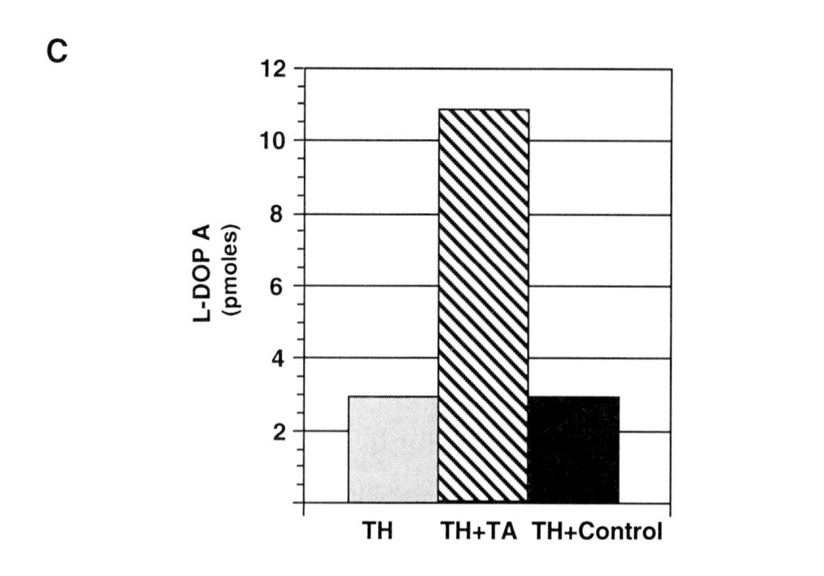

L-DOP A (pmoles)
12
10
8
6
4
2
TH
TH+TA
TH+Control

neurons of the trigeminal ganglia and also in brain stem, albeit at low levels due to a gradual reduction in the number of positively expressing cells (Wolfe *et al.*, 1992). In addition, recombinants containing *lacZ* or the nerve growth factor cDNA inserted downstream of LAP1 failed to produce transgene expression during latency, even using sensitive RT-PCR to detect the transgene messages (Margolis *et al.*, 1993). All of the these recombinants except the β-globin construct suffered deletions of the LAP2 sequences, implying that LAP2 may play a role in maintaining long-term expression in the PNS.

We have designed recombinant viruses in which the β-galactosidase gene was inserted downstream of LAP2 in either the ectopic glycoprotein C (gC) gene locus (Goins *et al.*, 1994) or in its native position in LAT. Both viruses were capable of supporting β-galactosidase expression as determined by immunohistochemistry, X-gal staining, and sensitive RT-PCR RNA detection assays in trigeminal ganglia for 42 to 300 days, consistent with latent infection. Moreover, we have recently observed that a replication-defective vector with the LAP2–β-galactosidase cassette in the gC locus could express β-galactosidase at 7 and 21 days postinoculation in the hippocampal region of rat brains using RT-PCR assays. This detection was possible despite our inability to visualize β-galactosidase protein by X-gal staining of infected hippocampal brain slices (Chen, Fink, and Glorioso, manuscript in preparation). The low levels of transgene expression from LAP2 suggest that it is capable of driving long-term expression of a foreign transgene in both PNS and CNS, but will require further manipulation to boost the level of transgene expression in brain. Such modifications may include the introduction of additional enhancer elements

Figure 6 Production of a functional transactivator from a HSV vector. (A) The HCMV IE promoter was first modified by the placement of five tandem GAL4 binding sites at position −161 and then positioned upstream of both the GAL4 : VP16 transactivator protein and a reporter (or therapeutic) gene in independent constructs. The HCMV IE promoter drives the expression of the GAL4 : VP16 protein which will then bind to its own promoter, stimulating its own activity. Similarly, the GAL4 : VP16 protein will also activate the second HCMV IE promoter located upstream of the reporter gene in the alternate construct. The model proposes that the continuous autostimulation and production of the GAL4 : VP16 activator will sustain production of the reporter gene, setting up an autogene system. (B) The modified HCMV IE promoter containing five GAL4 binding sites was placed upstream of the GAL4 : VP16 transactivator sequences and recombined into the d120 viral backbone at the tk locus through the cell-free *cre–lox* recombination system (Sauer *et al.*, 1987; Gage *et al.*, 1992). (C) To test whether vector-expressed GAL4 : VP16 could transactivate a promoter located within a viral genome, a construct containing a GAL4 : VP16-sensitive promoter driving the tyrosine hydroxylase (TH) gene [d120, (ICP4$^-$), TK$^-$::G$_5$HCMV-TH] was used to infect TH-negative rat B103 neuroma cells with the TA virus. Infected cells were analyzed by HPLC for the synthesis of L-DOPA, the immediate product of TH activity. Coinfection with the TA virus resulted in significant stimulation of TH activity from the TH virus. (TH, infection with the TH virus alone; TH + TA, coinfection of the TA and TH viruses; TH + control, a coinfection of the TH and control virus which expresses *lacZ*).

into the LAP region or the fusion of a strong promoter sequence to LAP to produce a chimeric promoter capable of strong activity that is latency specific. For some applications, however, low levels of expression should be effective for gene therapy. This possibility awaits further study.

B. Inducible Foreign Gene Expression

An alternate approach toward the achievement of persistent long-term expression from HSV vectors is the application of a transcriptional activation system to amplify foreign gene expession. Such a system could overcome the problem of promoter "shut off" by employing a recombinant gene whose product is capable of transactivating its own promoter as well as that of other susceptible promoters which are juxtaposed to the therapeutic gene of interest. One such transcriptional activator is the fusion protein GAL4 : VP16, which is composed of the DNA-binding region of the yeast transactivator GAL4 and the potent acidic activation domain of the HSV VP16 protein (Sadowski *et al.*, 1988). GAL4 : VP16 acts in a two-step process. First, the GAL4 component binds to promoters which contain multiple copies of the GAL4 binding site. GAL4 can accomplish this binding even when the site is bound by nucleosomes and the promoter is transcriptionally silent due to its inaccessibility to transcription factors (Giniger *et al.*, 1985). GAL4 binding results in a displacement of the nucleosomes away from promoter regions (Croston *et al.*, 1991). Second, the VP16-activating domain acts to attract other transcription factors to the promoter in order to initiate and stimulate transcription (Lin *et al.*, 1991; Roberts *et al.*, 1993). Not only is GAL4 : VP16 able to stimulate sensitive promoters which contain GAL4 binding sites in *in vitro* transient assays (Carey *et al.*, 1990; Chasman *et al.*, 1989; Sadowski *et al.*, 1988), but it can also overcome the nucleosome repression of promoters during transcriptional activation in frog oocytes and yeast (Axelrod *et al.*, 1993; Xu *et al.*, 1993). These features of the fusion activator molecule could be effective in maintaining gene expression during latency where the viral genome is sequestered as chromatin.

We have adapted the GAL4 : VP16 system to HSV vectors as a means to effect continuous expression of the delivered foreign gene during viral latency (Bender *et al.*, manuscript in preparation). The experimental model was to design a promoter which would contain five GAL4 binding sites and would also serve to drive expression of the GAL4 : VP16 protein (see Fig. 6A). This would render the modified promoter sensitive to GAL4 : VP16 stimulation, enabling GAL4 : VP16 to maintain its own transcription. If a second susceptible promoter driving a foreign gene were present in the infected cell, either on a second viral vector or at a secondary locus of the same virus, it would also be sensitive to stimulation by GAL4 : VP16, potentially providing for persistent expression of both the transactivator and the foreign gene. To apply this autogene system to HSV gene expression, the strong HCMV IE promoter was modified to contain five GAL4 binding sites and placed upstream of the coding

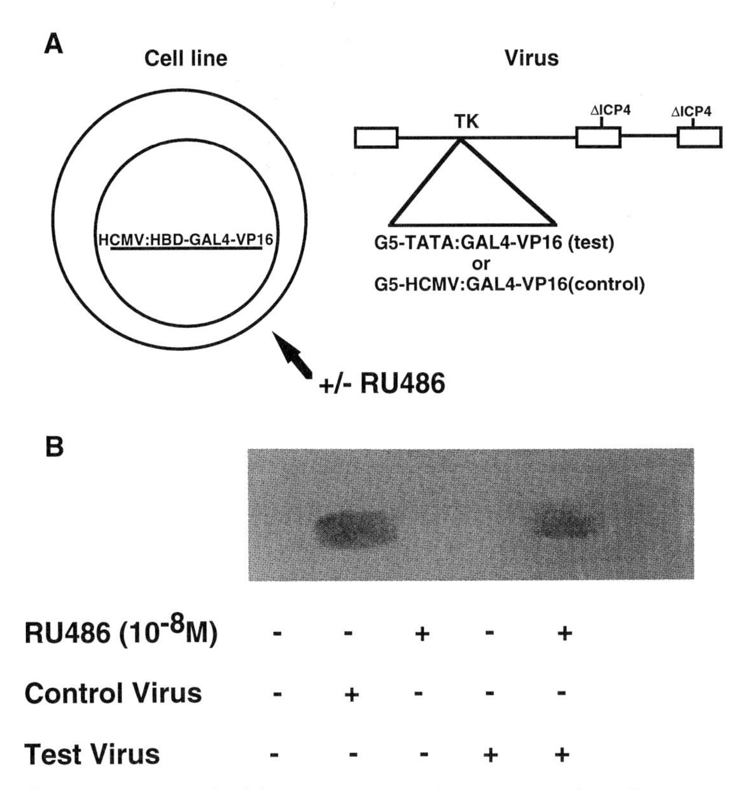

RU486 (10^{-8}M)	-	-	+	-	+
Control Virus	-	+	-	-	-
Test Virus	-	-	-	+	+

Figure 7 Drug-inducible transactivation of a HSV vector-based promoter. In collaboration with Dr. Bert O'Malley, we are applying drug-sensitive regulation to the GAL4 : VP16 transactivation system using a truncated form of the hormone-binding domain of the progesterone receptor to the GAL4 : VP16 transactivator. The shortened HBD no longer binds to progesterone, but can bind the steroid analog RU486 (Vegeto *et al.,* 1992) resulting in a transcriptionally active molecule. (A) The T1.21 cell line (Wang *et al.,* 1994) contains a stable integration of a GAL4 : VP16 expression cassette consisting of the strong HCMV IE promoter driving expression of the chimeric gene for HBD: GAL4 : VP16. The T1.21 cell line was infected with a test virus which contains the GAL4 : VP16 sequences driven by a minimal promoter containing a TATA box and five upstream GAL4 binding sites inserted at the tk locus of the d120 virus [d120, ($ICP4^-$), TK^-::G_5TATA-GAL4 : VP16]. This virus does not produce detectable GAL4 : VP16 because of its weak promoter, but is sensitive to GAL4 : VP16 stimulation by virtue of the five GAL4 binding sites. The positive control virus, d120 ($ICP4^-$), TK^-::G5HCMV-GAL4 : VP16, produces a functional transactivator. The abilities of these two viruses to produce the GAL4 : VP16 protein was analyzed after infection of the T1.21 cell line with or without RU486. (B) Drug-sensitive production of GAL4 : VP16 from the minimal promoter test virus or control virus was detected by Northern blot analysis. The control virus does not require exposure to RU486 to produce GAL4 : VP16; however, only in the presence of 10^{-8} *M* RU486 can the test virus support the synthesis of GAL4 : VP16 mRNA.

sequences for GAL4 : VP16. This cassette was then recombined into the nonreplicating d120 (ICP4$^-$) viral backbone (Fig. 6B) creating the GAL4 : VP16 HSV vector or transactivating (TA) virus. We then tested the GAL4 : VP16 virus for its ability to transactivate GAL4-sensitive promoters. In coinfection experiments with a d120 viral construct containing the GAL4-sensitive HCMV promoter driving the rat cDNA for tyrosine hydroxylase (TH), the GAL4 : VP16 virus (TA) was capable of stimulating virally encoded synthesis of TH (Fig. 6C). During *in vivo* infections of the brain, however, the synthesis of GAL4 : VP16 mRNA paralleled the transient kinetics of other genes driven by the HCMV and other promoters. Thus, the transactivation potential of the GAL4 : VP16 protein, though functional, is not able to overcome the silencing of viral promoters that occurs during the establishment of latency. An experimental system in which the latency promoter is exploited to express the TA molecule is now being tested as a means to enhance TA function.

A second approach to the autogene system involves the use of an inducible transactivator. Dr. Bert O'Malley and co-workers have fused the GAL4 : VP16 sequences to a truncated hormone-binding domain (HBD) of the progesterone receptor. This alteration does not allow it to bind to progesterone, but rather to the antiprogestin RU486 (Vegeto *et al.*, 1992), a small molecule that can pass the blood–brain barrier (Nieman *et al.*, 1987). Only when RU486 binds to the intracellular HBD receptor component of the fusion protein does the chimeric molecule become activated, thus creating a drug-inducible transactivation system. We obtained a cell line (T1.21) developed in the laboratory of Dr. O'Malley which has a stable insertion of the HBD : GAL4 : VP16 chimeric protein driven by the HCMV IE promoter (Wang *et al.*, 1994). This cell line was infected with a test virus similar to the GAL4 : VP16 transactivator virus, except that GAL4 : VP16 is driven by a minimum promoter consisting only of a TATA box and five GAL4 binding sites. This weak promoter is inactive unless stimulated by the GAL4 : VP16 TA product. As shown in Fig. 7, this activation occurs only when the cells are exposed to RU486, demonstrating that the transactivation function of the integrated HBD : GAL4 : VP16 gene is drug sensitive. Such an approach could be applied for foreign gene delivery to many tissues in the body, allowing for drug-inducible control of transgene expression. This approach also has the potential advantage that the length and level of therapeutic gene expression can be regulated by a receptor-activating drug regimen thereby increasing the versatility and effectiveness of gene therapy.

References

Akli, S., Cailland, C., Vigne, E., Stratford-Perricaudet, L. D., Poenaru, L., Perricaudent, M., and Peschanski, M. R. (1993). Transfer of a foreign gene into the brain using adenovirus vectors. *Nature Genet.* 3, 224–228.

Andersen, J. K., Garber, D. A., Meaney, C. A., and Breakefield, X. O. (1992). Gene transfer into mammalian central nervous system using herpesvirus vectors: Extended expression of bacterial *lacZ* in neurons using the neuron-specific enolase promoter. *Hum. Gene Ther.* **3,** 487–499.

Anderson, W. F. (1992). Human gene therapy. *Science* **256,** 808–813.

Axelrod, J. D., Reagan, M. S., and Majors, J. (1993). GAL4 disrupts a repressing nucleosome during activation of GAL 1 transcription *in vivo. Genes Dev.* **7,** 857–869.

Bak, I. J., Markhan, C. H., and Cook, M. L. (1977). Intra-axonal transport of herpes simplex virus in the rat central nervous system. *Brian Res.* **136,** 415–429.

Batchelor, A., and O'Hare, P. O. (1990). Regulation and cell-type-specific activity of a promoter located upstream of the latency-associated transcripts of herpes simplex virus type 1. *J. Virol.* **64,** 3269–3279.

Batchelor, A. H., and O'Hare, P. O. (1992). Localization of cis-acting sequence requirements in the promoter of the latency-associated transcripts of herpes simplex virus type 1 required for cell-type-specific activity. *J. Virol.* **66,** 3573–3582.

Batterson, W., Furlong, D., and Roizman, B. (1983). Molecular genetics of herpes simplex virus. VII. Further characterization of a *ts* mutant defective in release of viral DNA and in other stages of viral reproductive cycle. *J. Virol.* **45,** 397–407.

Block, T. M., Spivack, J. G., Steiner, I., Deshmane, S., McIntosh, M. T., Lirette, R. P., and Fraser, N. W. (1990). A herpes simplex virus type 1 latency-associated transcript mutant reactivates with normal kinetics from latent infection. *J. Virol.* **64,** 3417–3426.

Carey, M., Leatherwood, J., and Ptashne, M. (1990). A potent GAL4 derivative activates transcription at a distance *in vitro. Science* **247,** 710–712.

Chasman, D. I., Leatherwood, J., Carey, M., Ptashne, M., and Kornberg, R. D. (1989). Activation of yeast RNA polymerase II transcription by a herpesvirus VP16 and GAL4 derivative *in vitro. Mol. Cell. Biol.* **9,** 4746–4749.

Chiocca, A. E., Choi, B. B., Cai, W., DeLuca, N. A., Schaffer, P. A., DeFiglia, M., Breakefield, X. O., and Martuza, R. L. (1990). Transfer and expression of the *lacZ* gene in rat brain neurons by herpes simplex virus mutants. *New Biol.* **2,** 739–746.

Cook, M. L., and Stevens, J. G. (1973). Pathogenesis of herpetic neuritis and ganglionitis in mice: Evidence of intra-axonal transport of infection. *Infect. Immunol.* **7,** 272–288.

Cordingly, M. G., Campbell, M. E. M., and Preston, C. M. (1983). Functional analysis of a herpes simplex virus type 1 promoter: Identification of far-upstream regulatory sequences. *Nucleic Acids Res.* **11,** 2347–2365.

Croston, G. E., Laybourn, P. J., Paranjapa, S. M., and Kadonaga, J. T. (1991). Mechanism of transcriptional antirepression by GAL4/VP16. *Genes Dev.* **6,** 2270–2281.

Davidson, B. L., Allen, E. D., Kozarsky, K. F., Wilson, J. M., and Roessler, B. J. (1993). A model system for *in vivo* gene transfer into the central nervous system using an adenoviral vector. *Nature Genet.* **3,** 219–223.

DeLuca, N. A., McCarthy, A. M., and Schaffer, P. A. (1985). Isolation and characterization of deletion mutants of herpes simplex virus type 1 in the gene encoding immediate-early regulatory protein ICP4. *J. Virol.* **56,** 558–570.

DeLuca, N. A., and Schaffer, P. A. (1985). Activation of immediate-early, early, and late promoters by temperature-sensitive and wild-type forms of herpes simplex virus type 1 protein ICP4. *Mol. Cell. Biol.* **5,** 558–570.

DeLuca, N. A., and Schaffer, P. A. (1988). Physical and functional domains of the herpes simplex virus transcriptional regulatory protein ICP4. *J. Virol.* **62,** 732–743.

Deshmane, S. L., and Fraser, N. W. (1989). During latency herpes simplex virus type 1 DNA is associated with nucleosomes in a chromatin structure. *J. Virol.* **63,** 943–947.

Devi-Rao, G. B., Goddart, S. A., Hecht, L. M., Rochford, R., Rice, M. K., and Wagner, E. K. (1991). Relationship between polyadenylated and nonpolyadenylated herpes simplex virus type 1 latency-associated transcripts. *J. Virol.* **65,** 2179–2190.

Dixon, R. A. F., and Schaffer, P. A. (1980). Fine-structure mapping and functional analysis of temperature-sensitive mutants in the gene encoding the herpes simplex virus type 1 immediate early protein VP175. *J. Virol.* **36,** 189–203.

Dobson, A. T., Sedarati, F., Devi-Rao, G., Flanagan, W. M., Farrell, M. J., Stevens, J. G., Wagner, E. K., and Feldman, L. T. (1989). Identification of the latency-associated transcript promoter by expression of rabbit β-globin mRNA in mouse sensory nerve ganglia latently infected with a recombinant herpes simplex virus. *J. Virol.* **63,** 3844–3851.

Dobson, A. T., Margolis, T. P., Sedarati, F., Stevens, J. G., and Feldman, L. T. (1990). A latent nonpathogenic HSV-1 derived vector stably expresses β-galactosidase in mouse neurons. *Neuron* **5,** 353–360.

Doerig, C., Pizer, L. I., and Wilcox, C. L. (1991). An antigen encoded by the latency-associated transcripts in neuronal cell cultures latently infected with herpes simplex virus type 1. *J. Virol.* **65,** 2724–2727.

Dressler, G. R., Rock, D. L., and Fraser, N. W. (1987). Latent herpes simplex virus type 1 DNA is not extensively methylated *in vivo*. *J. Gen. Virol.* **68,** 1761–1765.

Farrell, M. J., Dobson, A. T., and Feldman, L. T. (1991). Herpes simplex virus latency-associated transcript is a stable intron. *Proc. Natl. Acad. Sci. USA* **88,** 790–794.

Fink, D. J., Sternberg, L. R., Weber, P. C., Mata, M., Goins, W. F., and Glorioso, J. C. (1992). In vivo expression of β-galactosidase in hippocampal neurons by HSV-mediated gene transfer. *Hum. Gene Ther.* **3,** 11–19.

Gage, P. J., Sauer, B., Levine, M., and Glorioso, J. C. (1992). A cell-free recombination system for site-specific integration of multigenic shuttle plasmids into the herpes simplex virus type 1 genome. *J. Virol.* **66,** 5509–5515.

Giniger, R., Varnum, S. M., and Ptashne, M. (1985). Specific DNA binding of GAL4, a positive regulatory protein of yeast. *Cell* **40,** 767–774.

Glorioso, J. C., Goins, W. F., and Fink, D. J. (1992). Herpes simplex virus-based vectors. *Semin. Virol.* **3,** 265–276.

Glorioso, J. C., Goins, W. F., Meaney, C. A., Fink, D. J., and DeLuca, N. A. (1994). Gene transfer to brain using herpes simplex virus vectors. *Ann. Neurol.* **35,** S28–S34.

Goins, W. F., Sternberg, L. R., Coren, K. D., Krause, P. R., Hendricks, R. L., Fink, D. J., Straus, S. E., Levine, M., and Glorioso, J. C. (1994). A novel latency active promoter is contained within the herpes simplex virus type 1 U_L flanking repeats. *J. Virol.* **68,** 2239–2252.

Grossman, M., Raper, S. E., Kozarsky, K., Stein, E. A., Engelhardt, J. F., Muller, D., Lupien, P. J., and Wilson, J. M. (1994). Successful *ex vivo* gene therapy directed to liver in a patient with familial hypercholesterolaemia. *Nature Genet.* **6,** 335–341.

Hill, J. M., Sedarati, F., Javier, R. T., Wagner, E. K., and Stevens, J. G. (1990). Herpes simplex virus latent phase transcription facilitiates *in vivo* reactivation. *Virology* **174,** 117–125.

Hill, T. J. (1985). Herpes simplex virus latency. In "The Herpesviruses" (B. Roizman, Ed.), pp. 175–240. Plenum Press, New York.

Ho, D. Y., and Mocarski, E. S. (1989). Herpes simplex virus latent RNA (LAT) is not required for latent infection in the mouse. *Proc. Natl. Acad. Sci. USA* **86,** 7596–7600.

Hoffman, E. K., Trusko, S. P., Freeman, N. A., and George, D. L. (1987). Structural and functional characterization of the promoter region of the mouse c-Ki-ras gene. *Mol. Cell. Biol.* **7,** 2592–2596.

Honess, R. W., and Roizman, B. (1974). Regulation of herpes virus macromolecular synthesis. I. Cascade regulation of the synthesis of three groups of viral proteins. *J. Virol.* **14,** 8–19.

Ishii, S., Xu, X.-H., and Stratton, R. H. (1985). Characterization and sequence of the promoter region of the human epidermal growth factor receptor gene. *Proc. Natl. Acad. Sci. USA* **82,** 4920–4924.

Javier, R. T., Stevens, J. G., Dissette, V. B., and Wagner, E. K. (1988). A herpes simplex virus transcript abundant in latently infected neurons is dispensible for establishment of the latent state. *Virology* **166,** 254–257.

Johnson, P., Miyanohara, A., Levind, F., Cahill, T., and Friedmann, T. (1992). Cytotoxicity of a replication defective mutant of herpes simplex virus type 1. *J. Virol.* **66**, 2952–2965.

Kolluri, R., Torrey, T. A., and Kinniburgh, A. J. (1992). A CT promoter element binding protein: Definition of a double-strand and a novel-strand DNA binding motif. *Nucleic Acids Res.* **20**, 111–116.

Kwong, A. D., and Frenkel, N. (1987). Herpes simplex virus-infected cells contain a function(s) that destablizes both host and viral mRNAs. *Proc. Natl. Acad. Sci. USA* **84**, 1926–1930.

Le Gal La Salle, G., Robert, J. J., Berrard, S., Ridoux, V., Stratford-Perricaudet, L. D., Perricaudet, M., and Mallet, J. (1993). An adenovirus vector for gene transfer into neurons and glia in the brain. *Science* **259**, 988–990.

Leib, D. A., Coen, D. M., Bogard, C. L., Hicks, K. A., Yager, D. R., Knipe, D. M., Tyler, K. L., and Schaffer, P. A. (1989). Immediate early regulatory mutants define stages in the establishment and reactivation of herpes simplex virus latency. *J. Virol.* **63**, 759–768.

Leiden, J. M., Frenkel, N., and Rapp, F. (1980). Identification of the herpes simplex virus DNA sequences present in six herpes simplex virus thymidine kinase-transformed mouse cell lines. *J. Virol.* **33**, 272–285.

Lin, Y.-S., Ha, I., Maldonado, E., Reinberg, D., and Green, M. R. (1991). Binding of a general transcription factor TFIIB to an acidic activating region. *Nature* **353**, 569–571.

Mackem, S., and Roizman, B. (1982). Structural features of the herpes simplex virus alpha gene 4, 0, and 27 promoter-regulatory sequences which confer alpha regulation on chimeric thymidine kinase genes. *J. Virol.* **44**, 939–949.

Margolis, T. P., Bloom, D. C., Dobson, A. T., Feldman, L. T., and Stevens, J. G. (1993). Decreased reporter gene expression during latent infection with HSV LAT promoter constructs. *Virology* **197**, 585–592.

McFarland, D. J., Sikora, E., and Hotchkin, J. (1986). The production of focal herpes encephalitis in mice by stereotaxic inoculation of virus. Anatomical and behavioral effects. *J. Neurol. Sci.* **72**, 307–318.

Meeker, T. C., Loeb, J., Ayres, M., and Sellers, W. (1990). The human PIM-1 gene is selectively transcribed in different hemato-lymphoid cell lines in spite of a G+C-rich housekeeping promoter. *Mol. Cell. Biol.* **10**, 1680–1688.

Natarajan, R., Deshmane, S., Valyi-Nagy, T., Everett, R., and Fraser, N. W. (1991). A herpes simplex virus type 1 mutant lacking the ICP0 introns reactivates with normal efficiency. *J. Virol.* **65**, 5569–5573.

Nicosia, M., Deshmane, S. L., Zabolotny, J. M., Valyi-Nagy, T., and Fraser, N. W. (1993). Herpes simplex virus type 1 latency-associated transcript (LAT) promoter deletion mutants can express a 2-kilobase transcript mapping to the LAT region. *J. Virol.* **67**, 7276–7283.

Nieman, L. K., Choate, T. M., Chrousos, G. P., Healy, D. L., Morin, M., Renquist, D., Merriam, G. R., Spitz, I. M., Bardin, C. W., Baulieu, E.-E., and Loriaux, D. L. (1987). The progesterone antagonist RU486. A potential new contraception agent. *N. Engl. J. Med.* **316**, 187–191.

O'Hare, P., and Hayward, G. S. (1985). Three trans-acting regulatory proteins of herpes simplex virus modulate immediate-early gene expression in a pathway involving positive and negative feedback regulation. *J. Virol.* **56**, 723–733.

Oroskar, A. A., and Read, G. S. (1989). Control of mRNA stability by the virion host shut-off function of herpes simplex virus. *J. Virol.* **63**, 1897–1906.

Palella, T. D., Silverman, L. J., Schroll, C. T., Homa, F. L., Levine, M., and Kelley, W. M. (1988). Herpes simplex virus-mediated human hypoxanthinae–guanine phosphoribosyl-transferase gene transfer into neuronal cells. *Mol. Cell. Biol.* **8**, 457–460.

Palella, T. D., Hidaka, Y., Silverman, L. J., Levine, M., Glorioso, J. C., and Kelly, W. M. (1989). Expression of human HPRT mRNA in brains of mice infected with a recombinant herpes simplex virus type 1 vector. *Gene* **80**, 137–144.

Panning, B., and Smiley, J. R. (1989). Regulation of cellular genes transduced by herpes simplex virus. *J. Virol.* **63**, 1929–1937.

Preston, C. M. (1979). Control of herpes simplex virus type 1 mRNA synthesis in cells infected with wild-type virus or the temperature-sensitive mutant tsK. *J. Virol.* **29**, 275–284.

Ramakrishnan, R., Fink, D. J., Guihua, J., Desai, P., Glorioso, J. C., and Levine, M. (1994). Competitive quantitative polymerase chain reaction (PCR) analysis of herpes simplex virus type 1 DNA and LAT RNA in latently infected cells of brain. *J. Virol.* **68**, 1864–1870.

Roberts, S. G., Ha, I., Maldonado, E., Reinberg, D., and Green, M. R. (1993). Interaction between an acidic activator and transcription factor TFIIB is required for transcriptional activation. *Nature* **363**, 741–744.

Roizman, B., and Furlong, D. (1974). The replication of herpesviruses. In "Comprehensive Virology" (H. Fraenkel-Conrat and R. R. Wagner, Eds.), Vol. 3, pp. 229–403. Plenum Press, New York.

Roizman, B., and Sears, A. E. (1990). Herpes simplex viruses and their replication. In "Virology, Second Edition" (B. N. Fields, D. M. Knipe, R. M. Chanock, M. S. Hirsch, J. L. Melnick, T. P. Monath, and B. Roizman, Eds.), pp. 1795–1841. Raven Press, New York.

Roizman, B., and Sears, A. E. (1993). Herpes simplex viruses and their replication. In "The Human Herpesviruses" (B. Roizman, R. J. Whitley, and C. Lopez, Eds.), pp. 11–68. Raven Press, New York.

Sacks, W. R., Greene, C. C., Aschman, D. A., and Schaffer, P. A. (1985). Herpes simplex virus type 1 ICP27 is an essential regulatory protein. *J. Virol.* **55**, 796–805.

Sadowski, I., Ma, J., Triezenberg, S., and Ptashne, M. (1988). GAL4-VP16 is an unusally potent transcriptional activator. *Nature* **335**, 563–564.

Sauer, B., Whealy, M., Robbins, A., and Enquist, L. (1987). Site-specific insertion of DNA into a pseudorabies virus vector. *Proc. Natl. Acad. Sci. USA* **84**, 9108–9112.

Sawtell, N. M., and Thompson, R. L. (1992). Herpes simplex virus type 1 latency-associated transcription unit promotes anatomical site-dependent establishment and reactivation from latency. *J. Virol.* **66**, 2157–2169.

Sedarati, F., Izumi, K. M., Wagner, E. K., and Stevens, J. G. (1989). Herpes simplex virus type 1 latency-associated transcript plays no role in establishment or maintenance of a latent infection in murine sensory neurons. *J. Virol.* **63**, 4455–4458.

Sehgal, A., Patil, N., and Chao, M. (1988). A constitutive promoter directs expression of the nerve growth factor receptor gene. *Mol. Cell. Biol.* **8**, 3160–3167.

Seiler, M., and Schwab, M. E. (1984). Specific retrograde transport of nerve growth factor (NGF) from neocortex to nucleus basalis in the rat. *Brain Res.* **300**, 33–39.

Shih, M.-F., Arsenakis, M., Tiollais, P., and Roizman, B. (1984). Expression of hepatitis B virus S gene by herpes simplex virus type 1 vectors carrying α- and β-regulated gene chimeras. *Proc. Natl. Acad. Sci. USA* **81**, 5867–5870.

Smiley, J. R., Smibert, C., and Everett, R. D. (1987). Expression of a cellular gene cloned in herpes simplex virus: Rabbit β-globin is regulated as an early viral gene in infected fibroblasts. *J. Virol.* **61**, 2368–2377.

Spivack, J. G., and Fraser, N. W. (1987). Detection of herpes simplex virus type 1 transcripts during latent infection in mice. *J. Virol.* **61**, 3841–3847.

Spivack, J. G., and Fraser, N. W. (1988). Expression of herpes simplex virus type 1 latency-associated transcripts in the trigeminal ganglia of mice during acute infection and reactivation of latent infection. *J. Virol.* **62**, 1479–1485.

Spivack, J. G., Woods, G. M., and Fraser, N. W. (1991). Identification of a novel latency-specific splice donor signal within HSV type 1 2.0-kilobase latency-associated transcript (LAT). Translation inhibition of LAT open reading frames by the intron within the 2.0-kilobase LAT. *J. Virol.* **65**, 6800–6810.

Steiner, I., Spivack, J. G., Deshmane, S. L., Ace, C. I., Preston, C. M., and Fraser, N. W. (1990). A herpes simplex virus type 1 mutant containing a nontransinducing Vmw 65 protein establishes latent infection *in vivo* in the absence of viral replication and reactivates efficiently from explanted trigeminal ganglia. *J. Virol.* **64**, 1630–1638.

Steiner, I., Spivack, J. G., Lirette, R. P., Brown, S. M., MacLean, A. R., Subak-Sharpe, J. H., and Fraser, N. W. (1989). Herpes simplex virus type 1 latency-associated transcripts are evidently not essential for latent infection. *EMBO J.* **8**, 505–511.

Stevens, J. G., Wagner, E. K., Devi-Rao, G. B., Cook, M. L., and Feldman, L. T. (1987). RNA complementary to a herpesvirus α gene mRNA is prominent in latently infected neurons. *Science* **235**, 1056–1059.

Stevens, J. G. (1989). Human herpesviruses: A consideration of the latent state. *Microbiol. Rev.* **53**, 318–332.

Stroop, W. G., and Schaefer, D. C. (1987). Herpes simplex virus type 1 invasion of the rabbit and mouse nervous systems revealed by *in situ* hybridization. *Acta Neuropathol.* **74**, 124–132.

Timmusk, T., Palm, K., Metsis, M., Reintam, T., Paalme, V., Saarma, M., and Persson, H. (1993). Multiple promoters direct tissue specific expression of the rat BDNF gene. *Neuron* **10**, 475–489.

Vegeto, E., Allan, G. F., Schrader, W. T., Tsai, M.-J., McDonnell, D. P., and O'Malley, B. W. (1992). The mechanism of RU486 antagonism is dependent on the conformation of the carboxy-terminal tail of the human progesterone receptor. *Cell* **69**, 703–713.

Wagner, E. K., Flanagan, W. M., Devi-Rao, G. B., Zhang, Y. F., Hill, J. M., Anderson, K. P., and Stevens, J. G. (1988). The herpes simplex virus latency-associated transcript is spliced during the latent phase of infection. *J. Virol.* **62**, 4577–4585.

Wang, Y., O'Malley, B. W., Jr., Tsai, S. Y., O'Malley, B. W. (1994). A novel regulatory system for gene transfer. *Proc. Natl. Acad. Sci. USA* **91**, 8180–8184.

Watson, R. J., and Clements, J. B. (1978). Characterization of transcription-deficient temperature sensitive mutants of herpes simplex virus type 1. *Virology* **91**, 364–379.

Wolfe, J. H., Deshmane, S. L., and Fraser, N. W. (1992). Herpes virus vector gene transfer and expression of β-glucuronidase in the central nervous system of MPS VII mice. *Nature Genet.* **1**, 379–384.

Xu, L., Schaffner, W., and Rungger, D. (1993). Transcription activation by recombinant GAL4/VP16 in the Xenopus oocyte. *Nucleic. Acids Res.* **21**, 2775.

Zwaagstra, J. C., Ghiasi, H., Nesburn, A. B., and Wechsler, S. L. (1989). In vitro promoter activity associated with the latency-associated transcript gene of herpes simplex virus type 1. *J. Gen. Virol.* **70**, 2163–2169.

Zwaagstra, J. C., Ghiasi, H., Nesburn, A. B., and Wechsler, S. L. (1991). Identification of a major regulatory sequence in the latency-associated transcript (LAT) promoter of herpex simplex virus type 1 (HSV-1). *Virology* **182**, 287–297.

Zwaagstra, J. C., Ghiasi, H., Slanina, S. M., Nesburn, A. B., Wheatley, S. C., Lillycrop, K., Wood, J., Latchman, D. S., Patel, K., and Wechsler, S. L. (1990). Activity of herpes simplex virus type 1 latency-associated transcript (LAT) promoter in neuron-derived cells: evidence for neuron specificity and for a large LAT transcript. *J. Virol.* **64**, 5019–5028.

CHAPTER 2

Biology of Herpes Simplex Virus (HSV) Defective Viruses and Development of the Amplicon System

Ann D. Kwong
Schering Plough Research Institute
Kenilworth, New Jersey

Niza Frenkel
Department of Cell Research and Immunology
Tel Aviv University
Tel Aviv, Israel

I. Introduction

The herpes simplex virus (HSV) defective virus vector amplicon system was designed approximately 15 years ago in our laboratory at the University of Chicago. At that time, we envisioned the amplicon as shuttle vector for moving DNA between prokaryotic cells and eukaryotic cells, tissues and organs. The HSV defective virus vector was termed an "amplicon" to delineate the fact that multiple copies of a DNA sequence of interest can be *amplified* in a head-to-tail arrangement in concatemeric defective virus genomes and packaged into HSV virions. This is in contrast to HSV whole virus vectors which were also being developed at the same time by Roizman and colleagues (Herz *et al.*,1983; Arsenakis *et al.*, 1987) in which a single copy of a gene of interest was inserted into a recombinant HSV genome. HSV amplicon vectors use the functions of their helper virus to enter, replicate, and express genes in a variety of cells both *in vitro*, and *in vivo*, depending on the helper virus and the promoter used in the construct. In this review, we will first discuss the origin of the amplicon system in studies of naturally occurring defective viral genomes where the delineation of many of the basic virological aspects of the system were first established. Second, we will discuss the essential *cis*-acting components which are required and the methods for generating and analyzing high-titer amplicon defective virus stocks. Last, we will discuss some of the first studies in which the HSV amplicon system was used to express a foreign gene and some of the parameters which affect expression. This review will not attempt to cover the more recent work of many colleagues in the field which has served to further develop HSV amplicon vectors for the transfer of genes into cells of the nervous system both in *in vitro* and *in vivo;* for recent reviews on this topic see (Geller *et al.*, 1991; Geller, 1993; Kaplitt *et al.*, 1993; Kennedy *et al.*, 1993; Frenkel *et al.*, 1994).

II. Basic Biology of HSV Defective Viruses

A. Structure of Standard Helper Virus Genomes

The 153-kb HSV genome has been fully sequenced (McGeoch *et al.*, 1988), and its structure has been studied in great detail (Hayward *et al.*, 1975; rev. in Roizman and Sears, 1992). Standard HSV genomic DNA contains two components, S and L, which invert relative to each other, giving rise to four isomers. The unique regions of the S and the L components are flanked by inverted repeats: ca and c′a′ bound the unique sequence of S (U_S) and ba and b′a′ bound the unique sequence of L (U_L). The terminal repeat *a* sequence contains reiterated as well as unique sequence elements. It is present in a variable number of copies and ranges in size from approximately 250 to 500 bp in different HSV strains (Locker and Frenkel, 1979; Roizman and Sears, 1992). Schematically displayed in Fig. 1 are the three lytic origins of DNA replication which include two copies of ori_S, which is located in the S inverted repeat sequences, and the ori_L origin, which is located in the U_L

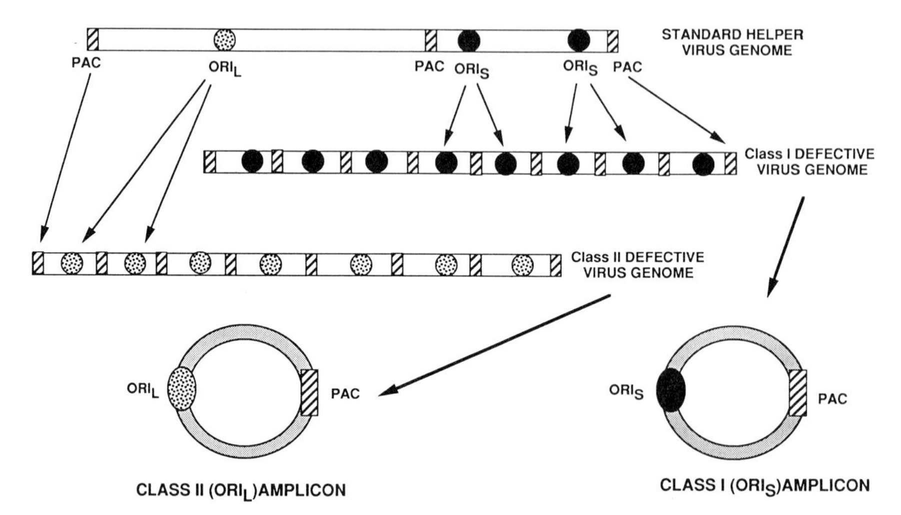

Figure 1 The structure of standard helper virus DNA and class I and class II defective virus DNA and amplicons.

component (Locker and Frenkel, 1979; Frenkel *et al.*, 1980; Locker *et al.*, 1982; Stow, 1982; Lockshon and Galloway, 1988).

B. Structure of Naturally Occurring Defective Virus Genomes

The propagation of standard virus stocks at a high multiplicity of infection (m.o.i.) results in the generation of defective virus genomes which probably arise by rolling circle replication of a single repeat unit (Frenkel *et al.*, 1976; Locker and Frenkel, 1979; Frenkel *et al.*, 1980; Locker *et al.*, 1982). Because defective virus genomes are replicated by machinery supplied by the helper virus and contain a simpler genomic structure, they have been instrumental in increasing our understanding of HSV DNA structure, replication, and gene expression. As summarized in Fig. 1, mapping of the DNA in the repeat units of the naturally occurring defective virus stocks of many different HSV-1 and HSV-2 strains revealed that there are of two types of structures. Class I defective virus genomes contain sequences from the S component of standard HSV DNA, and class II defective virus genomes contain sequences from the middle of U_L. The DNA of both classes of defective virus genomes contain the *a* sequence from the junction and ends of the helper virus genome, but differ with respect to the rest of their repeat unit sequence (Locker and Frenkel, 1979; Frenkel *et al.*, 1980, 1981; Locker *et al.*, 1982).

C. Two Separate Signals Are Required in *cis* (ori and pac)

The examination of the structure of naturally occurring defective virus genomes has facilitated the analysis of replication functions in standard virus DNA. The cloning of class I and class II defective virus genomes made it possible to develop assays to map the essential *cis* acting replication signals for defective virus propagation. In general, the assay involved the cotransfection of cells with helper virus DNA and a test plasmid containing the putative propagation signals. The resultant virus stocks were passed at a high m.o.i. for a limited number of passages and tested for the presence of concatemeric defective genomes derived from the test plasmid. Additional confirmation that one of the signals was indeed a replication origin was obtained using a *Dpn*I resistance assay which took advantage of the differential methylation pattern found in prokaryotic and eukaryotic cells. These analyses revealed that both an origin of DNA replication and a signal for the cleavage and packaging of viral DNA was required for the amplification of test amplicon plasmid DNA (Frenkel *et al.*, 1980; Vlazny *et al.*, 1981; Stow *et al.*, 1982; Spaete and Frenkel, 1985; Deiss *et al.*, 1986).

D. Origin of DNA Replication (ori_S and ori_L)

Detailed analyses of the structure of the origins of DNA replication have been pursued by a number of laboratories (rev. in Challberg, 1991). Ori_S and

ori_L are closely related and contain a palindromic inverted repeat sequence with a central AT-rich sequence. Ori_L contains a large (~100 bp) repeat sequence which is frequently deleted in bacteria. In contrast, the inverted repeat sequence in ori_S is ~70 bp in size and is relatively stable in bacteria. The core sequence of ori_S contains the binding sites for UL9, the HSV origin binding protein, as well as flanking sequences which increase the efficiency of replication (Stow, 1982; Stow and McMonagle, 1983; Stow, 1985; Elias and Lehman, 1988; Challberg, 1991; Olivio *et al.*, 1991). Viral DNA replication has been shown to be enhanced by the binding of one or more transcriptional factors to consensus binding sites in sequences flanking the core sequence (rev. in Challberg, 1991; Wong and Schaffer, 1991).

E. Cleavage and Packaging Signal (pac)

As shown in Fig. 1, the second *cis*-acting signal required for defective virus propagation is the signal for cleavage and packaging of viral DNA (pac). The minimal DNA sequence which can mediate this function was mapped to the *a* sequence within the pac signal at the ends and junctions of standard helper virus and defective virus genomes (Vlazny and Frenkel, 1981; Vlazny *et al.*, 1982; Spaete and Frenkel, 1985; Deiss *et al.*, 1986; Deiss and Frenkel, 1986). The replicated genomes were shown to be cleaved at the *a* sequence junctions between adjacent repeat units, concurrent with the packaging of viral DNA into virions. Two separate conserved signals (pac-1 and pac-2) in the *a* sequence serve as a measuring point from which cleavage takes place at a defined distance (Deiss *et al.*, 1986). Both pac-1 and pac-2 are well conserved in the terminal sequences of the genomes of several herpesviruses.

F. Class I and Class II Defective Virus Gene Expression

Standard virus replication in the HSV infected cell involves more than 70 genes, which have been grouped into a cascade regulation of α, β, and γ (immediate early, early, and late) infected cell polypeptides (ICP) (Honess and Roizman, 1975; Roizman and Sears, 1992). Mapping of the DNA sequences amplified in class I defective virus genomes derived from HSV-1 strain Justin revealed the presence of the ICP4 gene which is proximal to ori_S in the helper virus genome (Locker and Frenkel, 1979; Locker *et al.*, 1982). ICP4 was overexpressed in cells infected with virus stocks containing high proportions of the Justin defective genomes compared to cells which were infected with the helper virus alone (Frenkel *et al.*, 1980). When cells were infected with different Justin defective virus stocks containing varying ratios of helper to defective virus, ICP4 expression was linearly related to the abundance of the defective virus genomes. To determine whether the overproduced ICP4 was indeed expressed from the Justin defective virus genomes, the HSV-1 strain

Justin helper virus was replaced by HSV-2 strain G helper virus. Indeed, cells infected with these mixed virus stocks overproduced the HSV-1 form of ICP4 and not the HSV-2 form of ICP4 which runs slightly slower in electrophoretic mobility (Locker and Frenkel, 1979; Frenkel *et al.*, 1980). The expression of the overproduced ICP4 from the defective virus followed immediate early kinetics, indicating that it was regulated in a similar manner to ICP4 expressed from the helper virus.

Mapping of the DNA sequences amplified in class II defective virus genomes derived from serial passaging of HSV-1 strain tsLB2 revealed the presence of the sequence for ICP8, which is proximal to ori_L in the helper virus genome. Similar to the results obtained with the class I Justin defective virus, cells infected with tsLB2 stocks containing a high abundance of the defective virus genomes overproduced ICP8 compared to cells infected with helper virus alone (Frenkel *et al.*, 1980). The pattern of expression of the overproduced ICP8 was identical to that of the helper virus.

III. Development of the HSV Amplicon System

A. General Properties of the HSV Amplicon System

As shown in Fig. 2, full-length defective virus genomes, containing head-to-tail repeat units, can be generated from cloned seed amplicon plasmids. The amplicon must contain a HSV origin of replication (ori) and a cleavage/packaging signal (pac). To generate amplicon defective virus stocks, cells are transfected with the seed amplicon DNA. Helper virus is required for *trans*-acting DNA replication and packaging functions necessary for the replication of defective viruses. Helper virus can be supplied to the transfected amplicon either by superinfection or by cotransfection with helper viral DNA. The resultant virus stocks is passaged at a high m.o.i. for several rounds of infection in order to amplify the ratio of defective to helper virus. Finally, the state of amplification of the input amplicon repeat unit in the resultant defective virus stock is analyzed by restriction enzyme digestion of the total helper plus defective viral DNAs. As shown in Fig. 2, the resultant defective virus genomes consists of head to tail concatemers of the input seed HSV amplicon.

In the first set of studies designed to test whether cloned amplicons could be used as efficient vectors for expression of genes in eukaryotic cells, class II defective viral DNA amplified in HSV-1 strain Patton was cloned into the bacterial plasmid PKC7 to construct the amplicons pP2-102 and pP2-201 (Spaete and Frenkel, 1982). A restriction enzyme analysis of DNA from the resultant defective virus stocks is shown in Fig. 3. Lanes 1 and 20 shows the pattern of DNA from cells infected with HSV-1 strain Justin helper virus alone.

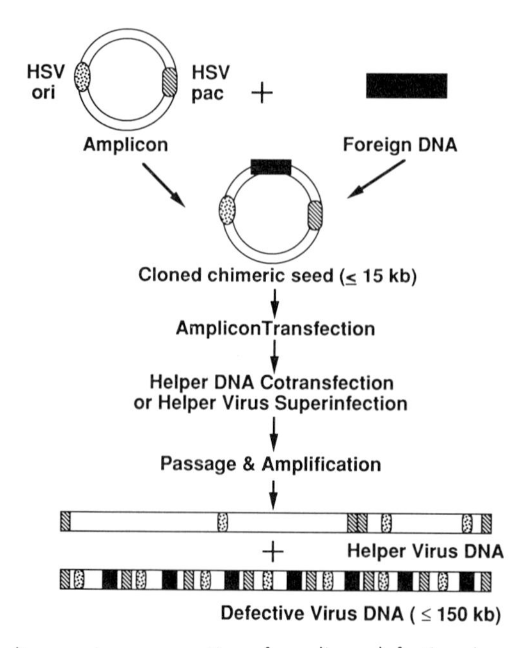

Figure 2 Introduction to the HSV amplicon system: generation of amplicon defective virus stocks. Defective virus stocks can be used to introduce genes into a wide range of eukaryotic cells.

In DNA from cells infected with defective virus stocks derived from the pP2-102 and pP2-201 amplicon transfections, there was a prominent new DNA band (Fig. 3, lanes 2–19, 21–38) which was not present in the helper DNA alone. The high titer of defective virus stocks which remained over many infected cell passages indicated that the amplicon derived genomes replicated stably and efficiently. Furthermore, the amplicon repeat unit containing PKC7 was excised from the defective viral DNA concatemers back into bacteria, verifying the structure of recombinant genomes present in the replicated amplicon-derived defective viruses.

B. Dynamics of Helper and Defective Virus Replication

When defective virus stocks are serially passaged, a constant number of virions is produced per cell (Frenkel *et al.*, 1980), most likely reflecting the total synthetic capacity of the cells. However, the population of helper and defective virus genome will fluctuate over the course of serial passaging of a defective virus stock. Because the cells are dually infected with varying ratios of the standard helper virus and defective virus, there is a corresponding fluctuation in the total infectious virus titer. The virions produced contain either standard helper virus genomes or the defective virus genomes, which interfere with the replication of the helper virus (Frenkel *et al.*, 1980; Frenkel, 1981).

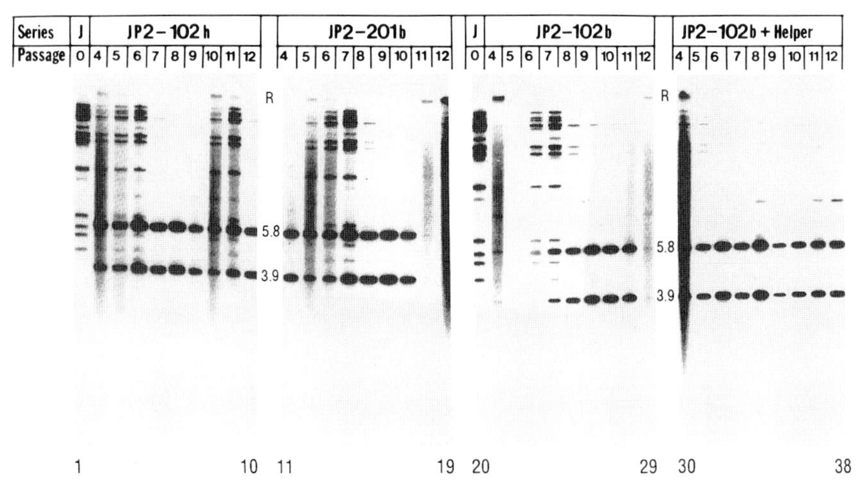

Figure 3 Effect of serial passaging on stability of defective virus genome titer. Autoradiograph of gels containing electrophoretically separated *BglII* fragments of ^{32}P-labeled DNA from cells infected with plaque-purified HSV-1 strain Justin (lanes 1 and 20) or with passages 4–12 of the transfection-derived series shown. Each of the passages of the JP2-102b + Helper series (lanes 30–38) received an additional 1 pfu/cell of helper virus in addition to the defective virus stock from the previous passage of the series. Data taken from Spaete and Frenkel, 1982.

Figure 3 shows representative restriction enzyme digest of ^{32}P-labeled DNA prepared from cells infected with passages 4–12 of four defective virus stocks. As seen in the figure, the size of the chimeric defective virus genomes was stable during the serial propagation. However, there was a cyclic fluctuation in the relative ratios of defective to helper virus DNA. Passages 7–9 of JP2-102h (Fig. 3, lanes 5–7) are examples of passages containing amplicon defective virus genomes in excess of 90% with a correspondingly low concentration of helper virus genomes. In contrast, passages 11–12 of JP2-201b (Fig. 3, lanes 18 and 19) illustrate the opposite end of the cycle with less than 1% defective virus genomes. The alternation in defective virus and helper virus titer observed in Fig. 3 is schematically illustrated in Fig. 4. Thus, amplicon-derived defective viral genomes exhibit the typical cycling characteristic of defective interfering particles described initially by von Magnus (von Magnus, 1954) and observed in other DNA and RNA viruses including HSV (Huang, 1973; Frenkel, 1981).

In order to develop the HSV amplicon as a useful viral vector system, we wanted to be able to stably maintain a defective virus stock with a high ratio of defective to helper virus over many passages without losing the defective virus titer. As seen in Fig. 3, lanes 30–38, the addition of 1 plaque forming unit (pfu) per cell of helper virus during the serial passaging of the "JP1-102b + Helper" series greatly reduced the fluctuation in the level of defective virus genomes. Thus, it is possible to stably maintain a high titer of defective virus

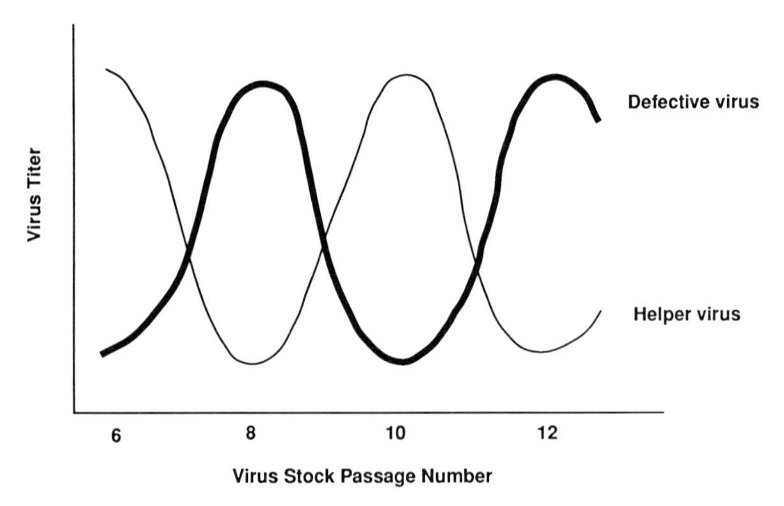

Figure 4 Schematic illustration of dynamic fluctuation in helper virus and defective virus titers over time during serial passaging.

over many passages if additional helper virus is added to the infected cells. This is consistent with von Magnus's hypothesis (von Magnus, 1954) that the cycling in the proportion of defective interfering particles through undiluted virus propagation results at least in part from the absence of sufficient helper virus to support defective virus replication.

In order to determine the temporal relationship between helper and defective viral DNA replication, [^{3}H]thymidine labeled DNA was harvested from cells at different times post infection with a high-titer class I defective virus stock derived from HSV-1 strain Justin (Vlazny and Frenkel, 1981). Because the defective viral DNA repeat unit contained a higher GC content than helper virus DNA, the different DNAs could be separated on CsCl density gradients where the defective viral DNA equilibrated at a higher buoyant density. As shown in Fig. 5, the initial rate of defective and helper viral DNA replication was similar up to 10 hr postinfection. However, at later times postinfection, the amount of defective viral DNA synthesized exceeded that of the helper virus. This suggests that defective virus genomes may be more efficient than their helper virus counterparts in utilizing the viral DNA replication machinery.

C. Generation and Analysis of High Titer Amplicon Virus Stocks

During the generation of all the initial amplicon defective virus stocks, the helper virus was supplied by cotransfection of CsCl density gradient purified viral DNA. Because it is much easier to generate a helper virus stock than to

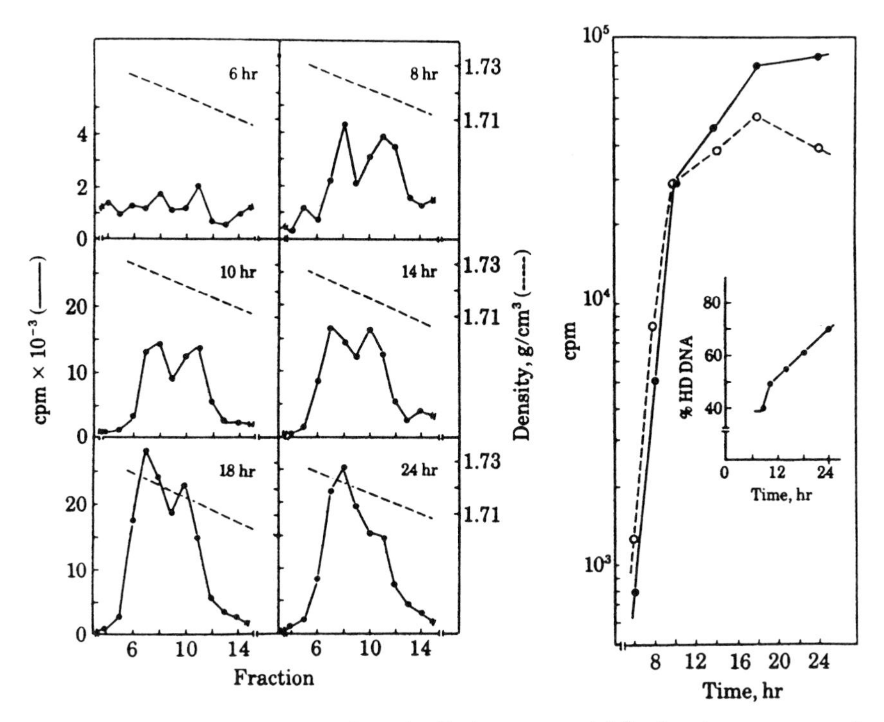

Figure 5 Time course analysis of standard helper virus and defective virus genome replication. Replicate cultures were infected with HSV-1 strain Justin standard virus and Justin passage 15 (P15) defective virus stocks. Beginning at 2 hr after infection, the DNA was labeled with [^{3}H]thymidine (1 μCi/ml) and harvested at the times shown. The DNA was extracted with SDS and proteinase K and centrifuged to equilibrium in CsCl. (Left) Regions of the gradients containing HSV DNA. The buoyant densities were estimated from the refractive indices and are in slight error due to the presence of SDS in the gradients. The heavy and light peaks in each gradient represent defective and helper viral DNA, respectively. (Right) Distribution of counts in heavy (●) and light (○) viral DNA as determined from the corresponding gradients shown on the left. (Inset) Percentage of viral DNA present as defective virus genomes at different times postinfection. HD, heavy density. Data taken from Vlazny and Frenkel, 1981.

purify helper virus DNA, a comparison of the efficacy of generating high-titer defective virus stocks using cotransfection with helper virus DNA versus superinfection with helper virus was performed. As shown in Fig. 6, the ratio of tsK helper virus DNA to amplicon DNA (pF1′-Pα, pF1′-Pα-CN1, pF1′-Pα-CN7) amplified in defective virus genomes was similar in virus stocks derived by cotransfection with helper DNA (lanes 3–9) compared to virus stocks derived by superinfection with helper virus (lanes 11–16). Therefore, superinfection with helper virus is just as efficacious as cotransfection with helper viral DNA in generating high titer defective virus stocks.

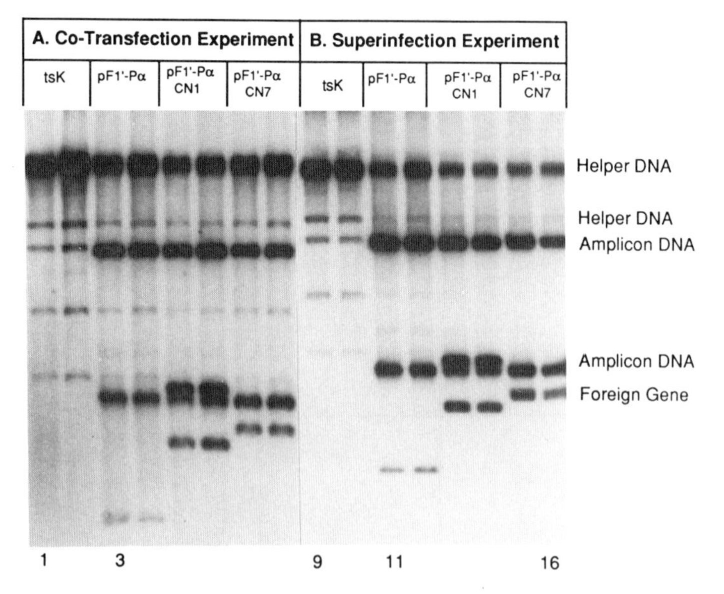

Figure 6 Effect of cotransfection versus superinfection of helper virus on generation of high-titer amplicon defective virus stocks. Autoradiograph of gels containing electrophoretically separated *Eco*RI fragments of ^{32}P-labeled DNA from cells infected with passage 4 of helper virus or amplicon defective virus stocks derived by Ca-PO_4-transfection of 25-cm^2 rabbit skin cell cultures with 5 μg of HSV-1 strain tsK viral DNA alone (lanes 1 and 2), or cotransfection of 0.5 μg of amplicon plasmid DNA(pF1′-Pα, pF1′-Pα-CN1, pF1′-Pα-CN7) with 5 μg of tsK viral DNA (lanes 3–9), superinfection with tsK alone (lanes 9–10), or transfection of the amplicons (pF1′-Pα, pF1′-Pα-CN1, pF1′-Pα-CN7) followed by superinfection with 0.1 pfu/cell of tsK helper virus. pF1′-Pα is from Kwong and Frenkel, 1985. pF1′-Pα-CN1 and pF1′-Pα-CN7 are from Danovich, Kwong, and Coffin, unpublished data.

There are at least three different methods which can be used to quantitatively measure the ratio of defective to helper virus in any given virus stock. In the first two methods, the ratio of defective to helper *virus* is inferred from the ratio of defective viral *DNA* to helper virus *DNA*. As illustrated in the experiment shown in Fig. 5, the first method involves the physical separation of [^{3}H]thymidine labeled DNA by equilibrium density gradient centrifugation. Both Fig. 3 and Fig. 6 illustrate the second method which involves the separation ^{32}P-labeled helper and defective viral DNA by agarose gel electrophoresis after restriction enzyme digest. The molar ratio of the helper to defective viral DNA in the dried gel can be quantitated with an Ambiss phosphoimager.

Figure 7 illustrates a third way to measure the defective virus titer when the gene which is amplified in the defective genome has a phenotype which can be enzymatically measured in infected cells. The pHCl and pHENK amplicon both contain the lac Z gene which expresses β-galactosidase under the control of the CMV IE promoter and the rat preproenkephalin promoter,

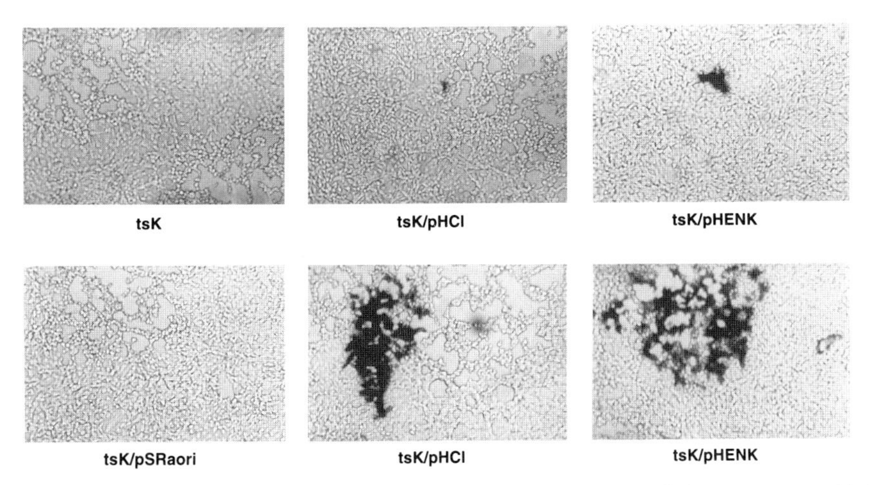

Figure 7 HSV amplicon X-gal plaque assay. Photographs are of plaque assays with the virus stocks shown at 33°C (permissive temperature for the tsK helper virus). Following fixation and staining with X-galactosidase, blue cells can be identified in cells infected with defective virus containing lac Z and expressing β-galactosidase, but not in cells infected with the helper virus alone (tsK) or the control amplicon defective virus stock (tsK/pSRaori). pSRaori is a control amplicon which does not express lac Z. pHCl and pHENK contain the lac Z gene for β-galactosidase under the control of the CMV IE promoter and the rat preproenkephalin promoter, respectively. Magnification, ×40.

respectively (Kaplitt *et al.*, 1991, 1994). The concentration of defective virus was quantitated by titration in a plaque assay followed by histochemical staining with X-gal substrate which stains cells blue after cleavage with β-galactosidase (Kaplitt *et al.*, 1991, 1993). Only cells which received defective virus stocks derived from amplicons containing the lac Z gene were stained blue, but not cells which received the tsK helper virus alone, or the pSRaori control amplicon vector defective virus stock.

All viral vectors have a limitation in the size of DNA which can be inserted into the vector. This size limitation is a function of the balance between the maximum packaging capacity of the viral capsid and the mimimum amount of DNA which must be retained in the vector for replication and packaging. In addition, for ease for manipulation, some additional DNA in the design of a viral vector is usually reserved for bacterial plasmid DNA sequences. Because HSV virions can package a 153-kb genome, the size limitation for efficiently packaging foreign DNA in HSV amplicon vectors is greater than the one for smaller viruses such as SV40, retroviral, adenoviral, or adeno-associated virus vectors. Figure 8 shows the results from an experiment which was performed to test the effect of overall amplicon size on the stability and efficient amplification of amplicon derived defective virus stocks (Kwong and Frenkel, 1984). Amplicons ranging in overall size from 7.8 kb (Fig. 8, lanes 2 and 3) to 15 kb

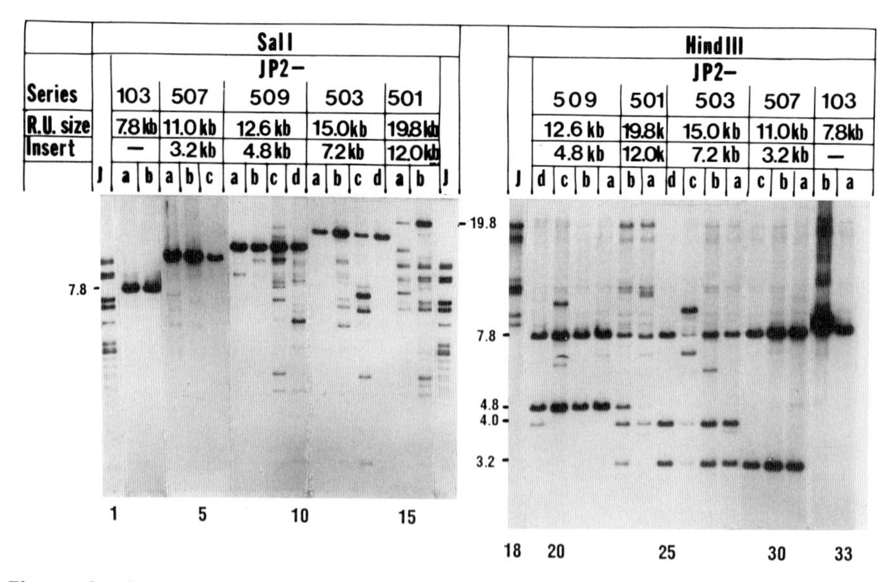

Figure 8 Generation of defective virus stocks from different sized amplicons. Autoradiograph of gels containing electrophoretically separated *Sal*I and *Hind*III fragments of ^{32}P-labeled DNA cells infected with HSV-1 strain Justin plaque purified virus (lane 15) or passage 3 of the defective virus shown. Each lane contains DNA from a separate series derived after transfection of a 25-cm^2 rabbit skin cell culture with 5 μg of HSV-1 (Justin) helper virus DNA and 1 μg of the test amplicons pP2-103, -507, -509, -503, and -501. *Sal*I linearizes each of the input amplicon plasmids and the repeat unit size (R.U.) is shown (7.8, 11, 12.6, 15, and 19.8 kb). Data taken from Kwong and Frenkel, 1984.

(lanes 11–14) were able to generate high-titer defective virus stocks. This can be assessed by comparing the ratio of helper virus DNA to defective viral DNA which has been digested with *Sal*I in each lane. The pattern for helper virus DNA alone is shown in lane 1. The ratio of defective viral DNA compared to helper viral DNA was lowest in cells infected with defective virus stocks generated from the 19.8-kb amplicon (lanes 15 and 16), suggesting that this amplicon is not as efficient as the smaller ones in replication. Further analyses also revealed the presence of multiple arrangements and deletions in the defective virus stocks which contained the 19.8-kb amplicon. Taken together, these observations suggest the existence of a size constraint of ~15 kb in the total size of a HSV amplicon. This is most likely due to the fact that each amplicon vector is packaged as a head-to-tail concatemer into HSV virions with a total genome size constraint of ~153 kb. An amplicon 15 kb in overall size is a simple multiple of ten times the total genome size. If there is not much extra room in the HSV capsid, 15 kb would be the practical size limitation because the next multiple of 30 kb is not easy to stably propagate in bacteria.

D. Factors Affecting Gene Expression in the HSV Amplicon System

To test whether a non-viral "foreign" gene could be expressed in the HSV amplicon system, we inserted a truncated chicken ovalbumin gene under the control of the ICP4 promoter in the HSV amplicon pF1′-pα (Kwong and Frenkel, 1985) to generate pF1′-pα-518. As shown in Fig. 9A, equivalent levels of high titer defective virus stocks were generated with both HSV-1 strain KOS

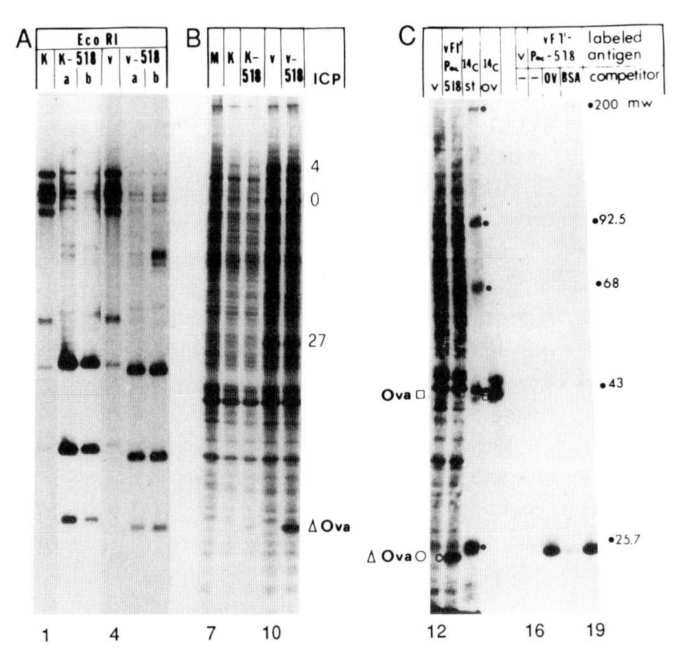

Figure 9 Expression of α-ovalbumin in the HSV amplicon system. (A) Autoradiograph of gels containing electrophoretically separated *Eco*RI fragments of ^{32}P-labeled DNA from cells infected with HSV-1 strain KOS virus (lane 1), *vhs*-1 virus (lane 4) helper virus alone or passage 5 of KOS-pF1′-pα-518 (lanes 2 and 3), or vhs-pF1′-pα-518 (lanes 5 and 6) defective virus stocks. The pF1′-pα-518 amplicon plasmid contains a truncated chicken ovalbumin cDNA gene under the control of the HSV IE ICP4 promoter. This promoter expresses genes with α kinetics. (B) Autoradiograph of [^{35}S]methionine-labeled proteins separated by SDS–PAGE. Total cell lysates of mock infected cells (M) or cells infected with KOS helper virus alone (K), *vhs*-1 helper virus alone (v), or the KOS-pF1′-pα-518 (K-518, lane 9) or vhs-pF1′-pα-518 (v-518, lane 10) defective virus stocks. ICPs 4, 0, and 27 are noted. (C) Autoradiograph of gels of [^{35}S]methionine-labeled protein immunoprecipitated with antibody to chicken ovalbumin (lanes 16–19). Total lysates are shown in lanes 12 and 13 and a ^{14}C MW standard is shown in lane 14. Lanes 16–19 contain immunoprecipitated samples from *vhs*-1-infected cells (lane 16) or vhs-pF1′-pα-518-infected cells (lanes 17–19). Unlabeled ovalbumin and BSA were added to the precipitations shown in lanes 18 and 19, respectively. Some of the data is taken from Kwong and Frenkel, 1985.

and strain *vhs*-1 helper virus. Virus strain *vhs*-1 is mutated in the virion host shut-off (vhs) function of UL41 which acts to increase the rate of degradation of both host and viral mRNAs in cells infected by wild type virus strains such as KOS (Read and Frenkel, 1983; Schek and Bachenheimer, 1985; Kwong and Frenkel, 1987; Oroskar and Read, 1987, 1989; Strom and Frenkel, 1987; Kwong *et al.*, 1988; Fenwick *et al.*, 1990a,b; Krikorian and Read, 1991; Sorenson *et al.*, 1991; Smibert *et al.*, 1992; Becker *et al.*, 1993). The *vhs*-1 virus is mutated in the UL41 gene, resulting in a G to A transition at codon 214 of the 56-kDa vhs gene product (A. D. Kwong and N. Frenkel, unpublished results), and exhibits a marked increase in the half life of host and viral mRNAs, as well as increased synthesis of host and viral polypeptides in the infected cells.

As a control for the specificity of expression and detection of the α-ovalbumin gene from amplicon defective virus DNA, lysates of cells infected with the *vhs*-1 α-ova defective virus stock were incubated with antisera to chicken ovalbumin. The truncated α-ovalbumin polypeptide was specifically immunoprecipitated (Fig. 9C, lane 17) as indicated by the ability of purified ovalbumin but not BSA to compete for binding (Fig. 9C, lanes 18 and 19).

To test whether the expression of the α-ovalbumin gene in the amplicon was regulated by the vhs function in different cell lines, we compared the level of truncated α-ovalbumin expression in HEp-2, Vero, and CHE cells infected with defective virus stocks generated with KOS or *vhs*-1 helper virus. As shown in Fig. 9B (HEp-2 cells) and Fig. 10 (Vero and CHE cells), the level of ovalbiumin gene expression was significantly higher in all the cells infected with α-ova defective virus stocks containing *vhs*-1 helper virus (Fig. 9B, lane 11; Fig. 10, lanes 5, 10, 15, and 20) compared to cells infected with α-ova defective virus stocks containing KOS helper virus (Fig. 9B, lane 9; Fig. 10, lanes 3, 8, 13, and 18). These results demonstrate that the expression of a foreign gene in the HSV amplicon system is affected by the vhs phenotype of the helper virus.

IV. Summary

The HSV amplicon is a defective virus vector which can be used to introduce foreign genes into eukaryotic cells. Of special relevance concerning the potential use of HSV amplicon vectors for gene transfer into neurons is the fact that the HSV is a neurotropic virus which resides latently in the nerve cell ganglia of humans. In the presence of helper virus, HSV amplicon vectors are packaged as head-to-tail concatemers up to 153 kb in length into defective virus particles. Because the amplicon defective virus genomes are packaged within HSV particles and HSV has a wide host range, the HSV amplicon can transfer genes into many types of cultured cells and animal tissues which

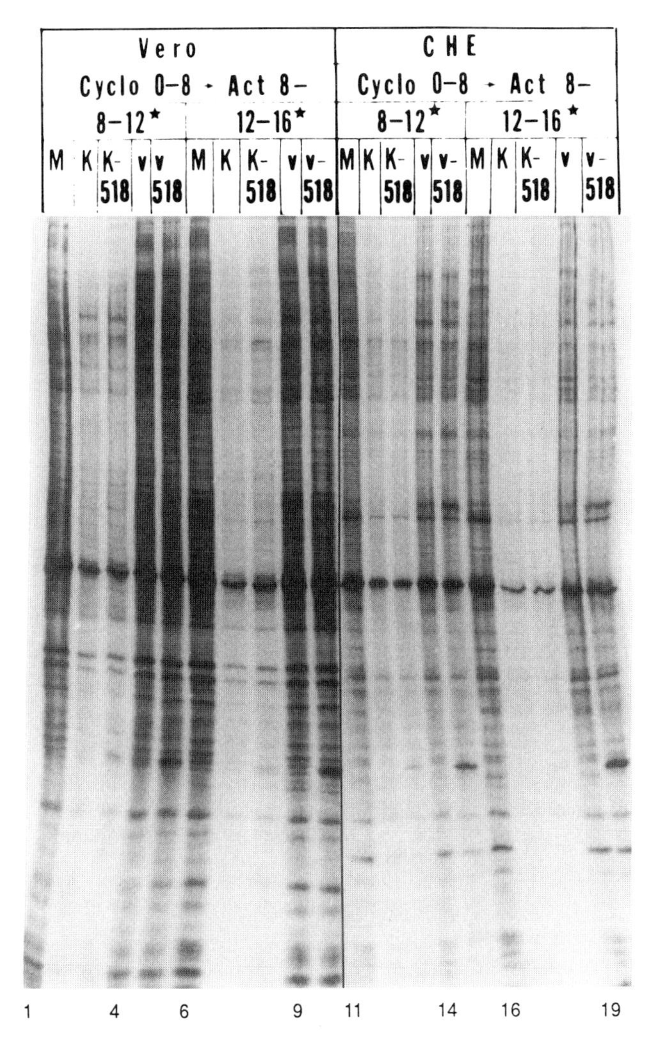

Figure 10 Effect of the virion host shut-off function on expression of α-ovalbumin in the HSV amplicon system. Authoradiograph of [^{35}S]methionine-labeled protein from African green monkey kidney Vero cells (lanes 1–10) or Chinese hamster embryo CHE cells (lanes 11–19) which were mock infected (M) or infected with KOS helper virus alone (K), *vhs*-1 helper virus alone (v), KOS-pF1′-pα-518 (K-518), or vhs-pF1′-pα-518 (v-518) defective virus stocks in the presence of cycloheximide. Proteins were labeled either early (8–12 hr) or late (12–16 hr) after reversal of the cycloheximide block into actinomycin D to obtain HSV α gene expression. Data taken from Frenkel *et al.* (1994).

may be useful for gene therapy or in the development of animal model systems.

References

Arsenakis, M., Poffenberger, K. L., and Roizman, B. (1987). Novel herpes simplex virus genomes: Construction and application. *In* "Herpesvirus: UCLA Symposia on Molecular and Cellular Biology, New Series," Vol. 43, p. 427. A. R. Liss, New York.

Becker, Y., Tavor, E., Asher, Y., Berkowitz, C., and Moyal, M. (1993). Effect of herpes simplex virus type-1 UL41 gene on the stability of mRNA from the cellular genes: β-Actin, fibronectin, glucose transporter-1, and docking protein, and on virus intraperitoneal pathogenicity to newborn mice. *Virus Genes* 7(2), 133–143,

Challberg, M. D. (1991). Herpes simplex virus DNA replication. *Semin. Virol.* **2,** 247–256.

Deiss, L. P., Chou, J., and Frenkel, N. (1986). Functional domains within the a sequence involved in the cleavage/packaging of herpes simplex virus DNA. *J. Virol.* **59,** 605–618.

Deiss, L. P., and Frenkel, N. (1986). The herpes simplex virus amplicon: Cleavage of concatemeric DNA is linked to packaging and involves amplification of the terminally reiterated a sequence. *J. Virol.* **57,** 933–941.

Elias, P., and Lehman, I. R. (1988). Interaction of origin binding protein with an origin of replication of herpes simplex virus type 1. *Proc. Natl. Acad. Sci. USA* **85,** 2959–2963.

Fenwick, M. L., and Everett, R. D. (1990a). Inactivation of the shutoff gene (UL41) of herpes simplex virus types 1 and 2. *J. Gen. Virol.* **71,** 2961–2967.

Fenwick, M. L., and Everett, R. D. (1990b). Transfer of UL41, the gene controlling virion-associated host cell shutoff, between different strains of herpes simplex virus. *J. Gen. Virol.* **71,** 411–418.

Frenkel, N. (Ed.) (1981). Defective interfering herpesviruses. *In* "The Human Herpesviruses—An Interdisciplinary Prospective." Elsevier–North Holland, Inc., New York.

Frenkel, N., Locker, H., Batterson, B., Hayward, G. S., and Roizman, B. (1976). Anatomy of herpes simplex virus. VI. Defective DNA originates from the S component. *J. Virol.* **20,** 527–531.

Frenkel, N., Locker, H., and Vlazny, D. A. (1980). Studies of defective herpes simplex viruses. *Ann. N.Y. Acad. Sci.* **354,** 347–370.

Frenkel, N., Locker, H., and Vlazny D. (Eds.) (1981). Structure and expression of class I and class II defective interfering HSV genomes. *In* "Herpesvirus DNA: Recent Studies of the Viral Genome." Nijihoff, The Hague.

Frenkel, N., Singer, O., and Kwong, A. D. (1994). Minireview: The herpes simplex virus amplicon—A versatile defective virus vector. *Gene Ther.* (February 1994 Supplement), S40–S46.

Geller, A. I. (1993). Herpesviruses: Expression of genes in postmitotic brain cells. *Curr. Opinion Genet. Dev.* **3,** 81–85.

Geller, A. I., During, M. J., and Neve, R. L. (1991). Molecular analysis of neuronal physiology by gene transfer into neurons with herpes simplex virus vectors. *Trends Neurosci.* **14**(10), 428–432.

Hayward, G. S., Jacob, R. J., Wadsworth, S. C., and Roizman, B. (1975). Anatomy of herpes simplex virus DNA: Evidence for four populations of molecules that differ in the relative orientation of their long and short segments. *Proc. Natl. Acad. Sci. USA* **72,** 4243–4247.

Herz, C., and Roizman, B. (1983). The alpha promoter regulator–ovalbumin chimeric gene resident in human cells is regulated like the authentic alpha 4 gene after infection with herpes simplex virus 1 mutants in alpha 4 gene. *Cell* **33,** 145–151.

Honess, R. W., Roizman, B. (1975). Regulation of herpesvirus macromolecular synthesis: Sequential transition of polypeptide synthesis requires functional viral polypeptides. *Proc. Natl. Acad. Sci. USA* **72,** 1276.

Huang, A. S. (1973). Defective interfering viruses. *Ann. Rev. Microbiol.* **27,** 101–117.

Kaplitt, M. G., Kwong, A. D., Kleopoulos, S. P., Mobbs, C. V., Rabkin, S. D., and Pfaff, D. W. (1994). Preproenkephalin promoter yields region-specific and long-term expression in adult brain after direct in vivo gene transfer via a defective herpes simplex viral vector. *Proc. Natl. Acad. Sci. USA* **91,** 8979–8983.

Kaplitt, M. G., Pfaus, J. G., Kleopoulos, S. P., Hanlon, B. A., Rabkin, S. D., Pfaff, D. W. (1991). Expression of a functional foreign gene in adult mammalian brain following in vivo transfer via a herpes simplex virus type 1 defective viral vector. *Mol. Cell. Neurosci.* **2,** 320–330.

Kaplitt, M. G., Rabkin, S. D., and Pfaff, D. W. (1993). Molecular alterations in nerve cells: Direct manipulation and physiological mediation. *Curr. Topics Neuroendocrinol.* **11,** 169–191.

Kennedy, P. G. E., and Steiner, I. (1993). The use of herpes simplex virus vectors for gene therapy in neurological disease. *Q. J. Med.* **86,** 697–702.

Krikorian, C. R., and Read, G. S. (1991). In vitro mRNA degradation system to study the virion host shutoff function of herpes simplex virus. *J. Virol.* **65**(1), 112–122.

Kwong, A. D., and Frenkel, N. (1984). Herpes simplex virus amplicon: Effect of size on replication of constructed defective genomes containing eucaryotic DNA sequences. *J. Virol.* **51**(3), 595–603.

Kwong, A. D., and Frenkel, N. (1985). The herpes simplex virus amplicon IV: Efficient expression of chimeric chicken ovalbumin gene amplified within defective virus genomes." *Virology* **142,** 421–425.

Kwong, A. D., and Frenkel, N. (1987). Herpes simplex virus-infected cells contain a function(s) that destabilizes both host and viral mRNAs. *Proc. Natl. Acad. Sci. USA* **84,** 1926–1930.

Kwong, A. D., Kruper, J. A., and Frenkel, N. (1988). Herpes simplex virus virion host shutoff function. *J. Virol.* **62,** 912–921.

Locker, H., and Frenkel, N. (1979). Structure and origin of defective genomes contained in serially passaged herpes simplex virus type 1 (Justin). *J. Virol.* **29,** 1065–1077.

Locker, H., Frenkel, N., and Halliburton, I. (1982). Structure and expression of class II defective herpes simplex virus genomes encoding infected cell polypeptide number 8. *J. Virol.* **43,** 574–593.

Lockshon, D., and Galloway, D. A. (1988). Cloning and characterization of OriL2, a large palindromic DNA replication origin of herpes simplex virus type 2. *J. Virol.* **58,** 513–521.

McGeoch, D. J., Dalyrymple, M. A., Davison, A. J., Dolan, A., Frame, M. C., McNab, D., Perry, L. J., Scott, J. E., and Taylor, P. J. (1988). The complete DNA sequence of the long unique region in the genome of herpes simplex virus type 1. *J. Gen. Virol.* **69,** 1531–1754.

Olivio, P. D., Nelson, N. J., and Challberg, M. D. (1991). Herpes simplex virus DNA replication: The UL9 gene encodes an origin-binding protein. *Proc. Natl. Acad. Sci. USA* **85,** 5414–5418.

Oroskar, A. A., and Read, G. S. (1987). A mutant of herpes simplex virus type 1 exhibits increased stability of immediate-early (alpha) mRNAs. *J. Virol.* **61,** 604–606.

Oroskar, A. A., and Read, G. S. (1989). Control of mRNA stability by the virion host shutoff function of herpes simplex virus. *J. Virol.* **63**(5), 1897–1906.

Read, G. S., and Frenkel, N. (1983). Herpes simplex virus mutants defective in the virion-associated shutoff of host polypeptide synthesis and exhibiting abnormal synthesis of α (immediate early) viral polypeptides. *J. Virol.* **61,** 498–512.

Roizman, B., and Sears, A. (1992). Herpes simplex viruses and their replication. *Virology* **2**, 1795–1887.

Schek, N., and Bachenheimer, S. L. (1985). Degradation of cellular mRNAs induced by a virion-associated factor during herpes simplex virus infection of Vero cells. *J. Virol.* **55**, 601–610.

Smibert, C., Johnson, D. C., and Smiley, J. R. (1992). Identification and characterization of the virion-induced host shutoff product of herpes simplex virus gene UL41. *J. Gen. Virol.* **73**, 467–470.

Sorenson, C. M., Hart, P. A., and Ross, J. (1991). Analysis of herpes simplex virus-induced mRNA destabilizing activity using an *in vitro* mRNA decay system. *Nucleic Acids Res.* **19**(16), 4459–4465.

Spaete, R. R., and Frenkel, N. (1982). The herpes simplex virus amplicon: A new eucaryotic defective-virus cloning-amplifying vector. *Cell* **30**, 295–304.

Spaete, R. R., and Frenkel, N. (1985). The herpes simplex virus amplicon: Analyses of cis-acting replication functions. *Proc. Natl. Acad. Sci. USA* **82**, 694–698.

Stow, N. D. (1982). Localization of an origin of DNA replication in the TRs/IRs repeated region of the herpes simplex virus type I genome. *EMBO J.* **1**, 863–867.

Stow, N. D. (1985). Mutagenesis of a herpes simplex virus origin of DNA replication and it effect of viral interference. *J. Gen. Virol.* **66**, 31–42.

Stow, N. D., and McMonagle, E. C. (1982). Propagation of foreign DNA sequences linded to a herpes simplex virus origin of replication. *In* "Eucaryotic Viral Vectors. (Gluzman, Y., Ed.), pp. 199–204. Cold Spring Harbor Laboratory, Cold Spring Harbor, NY.

Stow, N. D., and McMonagle, E. C. (1983). Characterization of the TRs/IRs origin of DNA replication of herpes simplex virus type 1. *Virology* **130**, 427–438.

Strom, T., and Frenkel, N. (1987). Effects of herpes simplex virus on mRNA stability. *J. Virol.* **61**, 2198–2207.

Vlazny, D. A., and Frenkel, N. (1981). Replication of herpes simplex virus DNA: Localization of replication recognition signals within defective virus genomes. *Proc. Natl. Acad. Sci. USA* **78**, 742–746.

Vlazny, D. A., Kwong, A. D., and Frenkel, N. (1982). Site-specific cleavage/packaging of herpes simplex virus DNA and the selective maturation of nucleocapsids containing full-length viral DNA. *Proc. Natl. Acad. Sci. USA* **79**, 1423–1427.

von Magnus, P. (1954). Incomplete forms of influenza virus. *Adv. Virus Res.* **2**, 59–78.

Wong, S. W., and Schaffer, P. A. (1991). Elements in the transcriptional regulatory region flanking herpes simplex virus type 1 oriS stimulate origin function. *J. Virol.* **65**(5), 2601–2611.

CHAPTER 3

Group C Adenoviruses as Vectors for Gene Therapy

Thomas Shenk
Howard Hughes Medical Institute
Department of Molecular Biology
Princeton University
Princeton, New Jersey

I. Introduction

Group C adenoviruses, which include the closely related adenovirus type 2 and type 5, are attractive candidates for use as vectors in gene therapy. The genetic makeup, gene functions, and interactions with the infected host cell are relatively well understood (reviewed in Shenk, in press). Procedures have been developed for the manipulation of the adenovirus genome and propagation of the resulting variants (reviewed in Shenk and Williams, 1984). This makes possible the insertion of therapeutic genes, deletion of undesirable viral genes, and propagation of viral stocks to high titers (reviewed in Berkner, 1988; Graham and Prevec, 1992; Goff and Shenk, 1993; Kozarsky and Wilson, 1993). Adenoviruses are able to express genes carried on the viral chromosome within nondividing cells such as airway epithelial cells and cells of the central nervous system (reviewed in Kozarsky and Wilson, 1993; Neve, 1993; Brody and Crystal, 1994). Most importantly, administration of adenovirus to humans appears to be safe. Most children exhibit immunological evidence of infection with group C adenoviruses at an early age with no associated serious disease, and adenovirus vaccines have been used extensively in the military without mishap (reviewed in Horwitz, in press).

Adenoviruses have been used to transfer and express a variety of genes in experimental animals, and first-generation adenovirus vectors, which have been rendered replication defective for normal human cells by deletion of their E1A and E1B genes, have been used to deliver the cystic fibrosis transmembrane

VIRAL VECTORS

conductance regulator (*CFTR*) gene to a defined area of the nasal airway epithelium in the human, alleviating the chloride transport defect in the infected cells of patients with cystic fibrosis (Zabner *et al.*, 1993), and several clinical trials are ongoing.

II. An Adenovirus Primer

Adenoviruses were first isolated in the 1950s by two groups of investigators who were searching for the etiologic agents of acute respiratory infections (Rowe *et al.*, 1953; Hilleman and Werner, 1954). Subsequently, over 100 members of the adenovirus group have been identified which infect a wide range of mammalian and avian hosts. All of these viruses contain a linear double-stranded DNA genome (Fig. 1) of approximately 36,000 base pairs encapsidated in an icosahedral protein shell that is about 70 nm in diameter. In 1962, human adenovirus type 12 was found to induce malignant tumors following inoculation into newborn hamsters (Trentin *et al.*, 1962). So far, epidemiological studies have not linked adenoviruses with malignant disease in the human, and extensive searches have failed to find adenovirus DNA or RNA in human tumors (Mackey *et al.*, 1976; Green *et al.*, 1980). As the interest in adenoviruses as tumor viruses intensified, their virtues as an experimental system became evident. The prototype human adenoviruses are easily propagated, and the viral genome is readily manipulated, facilitating the study of adenovirus gene functions by mutational analysis.

The adenovirus replication cycle is divided by convention into early and late phases which are separated by the onset of viral DNA replication. Early events begin as soon as the infecting virus interacts with the host cell, and

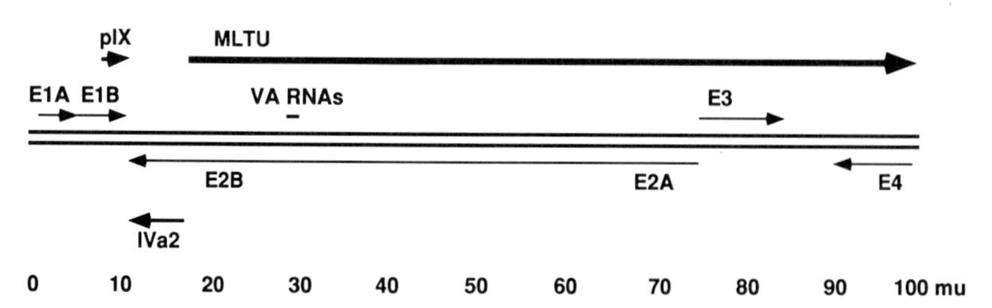

Figure 1 Transcription map of the group C adenovirus genome. The ~36,000-base pair adenovirus DNA molecule is shown in the center of the figure as two parallel lines, and transcription units are designated by arrows. Thin arrows represent the early units E1A, E1B, E2, E3, and E4, arrows of intermediate thickness indicate intermediate units pIX and IVa2, a thick arrow identifies the major late transcription unit (MLTU), and the region encoding two small polymerase III-transcribed VA RNAs is marked by a short line. Map units (mu) on the conventional adenovirus map are shown at the bottom of the figure.

include adsorption, penetration, transcription, and translation of an early set of genes and induction of host macromolecular synthesis. The late phase of the cycle begins with the onset of DNA replication and initiates expression of an additional set of viral genes leading to the assembly of progeny virions.

Adsorption of adenovirus to cells is a two step process. First, the carboxy-terminal knob of the fiber protein that projects from each of the 20 vertices of the icosahedral virion binds to a cellular receptor that has not been identified. Then, the penton base protein which is also present at each of the vertices binds to members of a family of heterodimeric cell surface receptors, termed integrins (Wickham *et al.*, 1993). The second interaction occurs through an arg–gly–asp sequence present in penton base, the same motif present in a variety of extracellular adhesion molecules that bind to integrins. After adsorption, viral particles are internalized by receptor-mediated endocytosis. This process is highly efficient: 80–85% of adsorbed virus is internalized, and about 90% of the virus within endosomes successfully moves to the cytosol (Greber *et al.*, 1993). About 40 min after penetration, virus particles can be visualized at nuclear pores, suggesting that DNA release occurs at the nuclear membrane. When the viral DNA enters the nucleus, it interacts with the nuclear matrix through its terminal protein, and nuclear matrix association is correlated with efficient transcriptional activation of the viral genome (Schaack *et al.*, 1990).

E1A is the first viral gene to be expressed after the viral chromosome reaches the nucleus. Its expression is controlled by a constitutively active promoter. The E1A gene encodes two mRNAs which contain identical 5′ and 3′ ends but differ internally as a result of splicing. As a result, the E1A mRNAs encode proteins that are identical except for an additional 46-amino acid segment that is present in the larger polypeptide. The two polypeptides are termed the 12S and 13S E1A proteins, names derived from the sedimentation coefficients of their mRNAs. The E1A proteins bind to cellular proteins and modulate their function. When viral mutants with defective E1A genes were examined, they were found to be deficient in the accumulation of early mRNAs (Berk *et al.*, 1979; Jones and Shenk, 1979); E1A proteins acted to dramatically increase the rate of transcription (Nevins, 1981). The E1A proteins activate transcription by binding to and modifying the function of a variety of transcription factors and transcriptional regulatory proteins. Since the E1A proteins can activate other viral genes in *trans*, they are often referred to as *trans*-activators. Several of the interactions of the E1A proteins with transcriptional regulatory proteins induce quiescent cells to enter the S phase of the cell cycle. The first cell cycle regulatory protein found in a complex with E1A proteins was the retinoblastoma tumor-suppressor protein, pRB (Whyte *et al.*, 1988). pRB can inhibit cell cycle progression, causing cells to arrest in mid to late G_1, and growth arrest correlates with the ability of pRB to bind to the E2F cellular transcription factor and block E2F-dependent transcriptional activation. When E1A proteins bind to pRB, E2F is released from its association with pRB, and free E2F can activate transcription of genes that contain E2F binding sites in their promoters, including genes important for S phase and cell growth (reviewed in Nevins, 1992).

The adenovirus E1B gene, which is one of the early viral genes whose expression is induced by the E1A protein, encodes a 55-kDa protein that, like E1A proteins, modulates cell cycle progression. The E1B protein targets the cellular p53 tumor-suppressor protein which regulates progression from G_1 to S. p53 is a DNA-binding protein that can activate transcription when it binds to p53 response elements. The current view of p53 is that it transcriptionally activates cellular genes that prevent entry into S phase (reviewed in Zambetti and Levine, 1993). The E1B–55-kDa protein binds to p53 within infected cells (Sarnow *et al.*, 1982), and it can block transcriptional activation by p53.

In addition to activating viral gene expression and inducing cellular proliferation, E1A proteins trigger apoptosis. E1A-induced apoptosis involves the induction of p53 levels, which not only can block cell cycle progression but also can induce apoptosis (Yonish-Rouach *et al.*, 1991). The E1B–55-kDa protein blocks apoptosis, probably as a result of its ability to bind to p53 and alter its function. A second adenovirus E1B gene product, the E1B–19-kDa protein also can block E1A-induced apoptosis (Rao *et al.*, 1992). Apoptosis is a cellular response to infection that has the potential to block viral spread within an infected organism. However, adenoviruses have evolved to encode gene products that effectively block the cellular defense.

As adenovirus E2 gene products accumulate and the infected cell enters the S phase of the cell cycle, the stage is set for viral DNA replication. The inverted terminal repeats of the adenovirus chromosome serve as replication origins. Adenovirus DNA replication takes place in two stages (Lechner and Kelly, 1977). First, synthesis is initiated at either end of the linear DNA and proceeds continuously to the other end of the genome. Only one of the two DNA strands serves as a template for the synthesis, so the products of the replication are a duplex DNA plus a displaced single strand of DNA. In the second stage of replication, a complement to the displaced single strand is synthesized. The single-stranded template circularizes through its self-complementary termini, and the resulting duplex "panhandle" has the same structure as the termini of the duplex viral genome.

The *cis*-acting replication origins are located within the inverted terminal repeats of the viral chromosome, and a complex of two viral proteins plus two cellular factors bind here: the E2-coded terminal protein and DNA polymerase plus cellular factors termed NFI and NFIII. The terminal protein serves as a primer for DNA replication (Rekosh *et al.*, 1977), preserving the integrity of the viral chromosome's terminal sequence during multiple rounds of DNA replication. NFI and NFIII appear to facilitate assembly of the primer–polymerase complex at the origin. Chain elongation requires two virus E2-coded proteins, the polymerase and single-stranded DNA-binding protein, plus the cellular factor NFII. The polymerase is highly processive in the presence of the DNA-binding protein, which coats single-stranded replication intermediates, and this is probably the basis for its requirement in chain elongation. The highly processive nature of the viral polymerase in the presence of the DNA-binding protein likely enables it to travel the entire length of the chromosome after an initiation event at the terminus. NFII copurifies with a cellular DNA topoisomerase activity.

Adenovirus late genes begin to be expressed efficiently at the onset of viral DNA replication. The majority of adenovirus late coding regions are organized into a single large transcription unit whose primary transcript is about 29,000 nucleotides in length (Evans *et al.*, 1977). This transcript is processed by differential polyadenylation and splicing to generate about 18 distinct mRNAs. Expression of this family of late mRNAs is controlled by the major late promoter, which exhibits a low level of activity early after infection and becomes substantially more active at late times.

When DNA replication begins and the full spectrum of late mRNAs are synthesized, the cytoplasmic accumulation of cellular mRNAs is blocked (Beltz and Flint, 1979). Synthesis and processing of cellular transcripts continue, but the mRNAs fail to accumulate in the cytoplasm, suggesting their transport is inhibited. The block to cellular mRNA accumulation is mediated by a complex of the the E1B–55-kDa polypeptide and the E4–34-kDa polypeptide (Babiss *et al.*, 1984; Halbert *et al.*, 1985; Pilder *et al.*, 1986). The same viral proteins are required for efficient cytoplasmic accumulation of viral mRNAs late after infection. Thus, a complex that includes two viral proteins both inhibits accumulation of cellular mRNAs and facilitates cytoplasmic accumulation of viral mRNAs. In addition to their facilitated transport from the nucleus, viral mRNAs are preferentially translated when they reach the cytoplasm late after infection (reviewed in Zhang and Schneider, 1993).

The replication of viral DNA, coupled with the accumulation of structural polypeptides, sets the stage for virus assembly. Studies of mutant viruses combined with analysis of the kinetics with which polypeptides are incorporated into virions have provided a rough outline of the adenovirus assembly process (reviewed in Philipson, 1984; Hasson et al., 1992). Assembly appears to begin with the formation of an empty capsid and, subsequently, a viral DNA molecule enters the capsid. The DNA–capsid recognition event is mediated by the packaging sequence, a *cis*-acting DNA element that is centered about 260 bp from the left end of the viral chromosome (Tibbetts, 1977; Hammarskjold *et al.*, 1980; Hearing and Shenk, 1987). Presumably one or more proteins bind at the packaging sequence and mediate the interaction between DNA and capsid, but they remain unidentified. Encapsidation of the chromosome is polar, beginning with the left end of the viral DNA. A virus-coded cysteine proteinase functions late in the assembly process. It cleaves at least four virion constituents, generating the mature viral particle.

Adenovirus can maintain a long-term association with its human host, persisting for years after an initial infection. The virus encodes three gene products that are known to antagonize antiviral responses of the host to facilitate long-term persistence: E1A proteins and VA RNAs which inhibit the cellular response to α- and β-interferons, and E3-coded products that protect infected cells from lysis by cytotoxic T lymphocytes and tumor necrosis factor.

The E1A proteins block activation of interferon response genes (Reich *et al.*, 1988). This appears to result from inhibition of the DNA-binding activity of ISGF3, a cellular transcription factor that is activated when α- or β-interferon binds to its receptor. The small, abundant, polymerase III-transcribed, adenovirus-coded VA RNAs also antagonize the antiviral effects of interferons (reviewed in Mathews and Shenk, 1991). VA RNA binds to the interferon-induced PKR protein kinase and blocks its activation, preventing phosphorylation of eIF-2α, a translational initiation factor. In the absence of VA RNA, the phosphorylation event leads to sequestration of eIF-2B in nonfunctional complexes with eIF-2a, and this causes the cessation of translation.

Adenovirus E3-coded proteins inhibit cytolysis of infected cells by CTLs or TNF-α (reviewed in Wold and Gooding, 1991; Gooding, 1992). Infected cells expressing the E3-coded 19-kDa glycoprotein (E3–gp19-kDa) are considerably less sensitive to CTL-mediated lysis than cells infected with mutant viruses unable to express the E3 protein. The E3–gp19-kDa protein resides in the endoplasmic reticulum and binds directly to the peptide-binding domain of MHC class I antigens, and this interaction is believed to cause retention of class I antigen in the endoplasmic reticulum. A reduced level of class I antigen on the cell surface should protect against premature lysis of the infected cell by CTLs. In fact, such protection can be inferred from experiments with cotton rats in which pulmonary infection with a virus unable to express the E3–gp19-kDa protein induced a markedly increased late-phase inflammatory response (Ginsberg *et al.*, 1989). Either the E3–14.7-kDa protein or the complex formed between the E3–14.5-kDa and E3–10.4-kDa proteins can prevent cytolysis by TNF-α, but, as yet, the mechanism underlying protection remains a mystery.

All human adenoviruses that have been tested can transform cultured rodent cells, and a subset of human adenovirus serotypes can directly induce the formation of sarcomatous tumors within rodents. Three lines of evidence have demonstrated that the E1A and E1B genes act in concert to mediate oncogenic transformation (reviewed in Tooze, 1981). First, transformed cells always express these genes; second, cultured cells can be transformed with cloned E1A and E1B genes; and, third, mutant viruses with defective E1A or E1B genes fail to transform rodent cells. As discussed above, the E1A and E1B proteins can manipulate growth regulation of human cells, inducing quiescent cells to enter the S phase of the cell cycle, blocking apoptosis, and creating an environment conducive to viral replication. In the context of a rodent cell, these same events contribute to oncogenic transformation.

III. Adenovirus Vectors

The first generation of adenovirus vectors was constructed to lack the E1A, E1B, and, in some cases, the E3 genes. The E1A and E1B genes were deleted in these vectors because of their oncogenic potential and to render the vectors replication defective. As described above, E1A proteins are required for efficient activation of adenovirus transcription, and E1B proteins are required for efficient accumulation of viral mRNA. The E3 gene was deleted in some cases primarily to make space for the inclusion of a large transgene without making the viral genome too large to be encapsidated.

The first generation of vectors has not performed well in animal studies. In one study, Ad*CFTR* (a recombinant adenovirus containing a *CFTR* transgene and lacking the viral E1A, E1B, and E3 genes) was instilled into limited regions of the lungs of baboons (Simon *et al.*, 1993). The animals exhibited minimal inflammation when receiving virus (7 ml) at a concentration of 10^7 or 10^8 pfu/ml, but a prominent perivascular lymphocytic and histiocytic infiltrate with diffuse alveolar wall damage and intraalveolar edema was seen in animals receiving the highest dose of Ad*CFTR* (10^{10} pfu/ml). This result is consistent with earlier work showing that high doses of adenovirus administered directly to the airway epithelium can induce pneumonia in mice even though the human virus does not replicate in mice (Ginsberg *et al.*, 1991). The inflammation produced by these vectors results from continued low-level expression of viral genes in the absence of the E1A and E1B genes, and this could be predicted because viral transcription occurs in the absence of E1A when cells are infected with $E1A^-$ mutants at a high-input multiplicity (Shenk *et al.*, 1979) or after an extended period when cells are infected at lower input multiplicities (Nevins, 1981).

However, inflammation has not been observed in all studies with first generation adenoviruses. Cotton rats that received 4×10^9 pfu Ad2/CFTR-1 (contains a *CFTR* transgene and lacks the viral E1A and E1B genes, Rich *et al.*, 1993) by nasal instillation to their lungs and rhesus monkeys that received 2×10^9 pfu by application to their nasal epithelium did not develop detectable inflammation at the sites of administration (Zabner *et al.*, 1994). Although it is difficult to compare between studies, it is likely that the difference in pathogenesis observed for the two recombinant adenoviruses is due to early region 3 (E3). Inflammation was observed for Ad*CTFR* (Simon *et al.*, 1993), which lacks the adenovirus E3 gene, but not for Ad2/CFTR-1 (Zabner *et al.*, 1994), which retains the E3 gene. As discussed above, the E3 gene encodes a series of gene products that protect the infected cell against the host immune response. However, it seems unlikely that retention of E3 will completely solve the problem. For example, in the experiment with Ad2/CFTR-treated rhesus monkeys which did not exhibit a detectable inflammatory response, CFTR mRNA was not detected within infected lung cells after 18 days using an extremely sensitive PCR assay (Zabner *et al.*, 1994). It seems likely that the infected cells were nevertheless killed because longer-term expression from adenovirus vectors can be achieved in athymic nude mice that are unable to mount a class 1-restricted cytotoxic T lymphocyte (CTL) response to infection (Yang *et al.*, 1994). This result is again consistent with earlier work demonstrating that the pathogenic response to wild-type adenovirus is significantly reduced and altered in nude compared to immunocompetent mice (Ginsberg *et al.*, 1991).

Recently, a second-generation adenovirus vector, Ad.CBlacZ (lacks E1A, E1B, a portion of E3, and contains a temperature-sensitive point mutation within the E2A gene) was described (Engelhardt *et al.*, 1994). This virus retains the ability to express the E3–19-kDa polypeptide, but expression of that protein and all other viral proteins was predicted to be lower than that in first-generation vectors since the E2A gene was also mutated. The E2A protein encodes the viral DNA-binding protein and, as described above, is required for adenovirus DNA replication. It also is a potent transcriptional activator (Chang and Shenk, 1990). This new vector appears promising: CTL infiltration was diminished and expression of the marker transgene continued for >70 days in a mouse liver model (Engelhardt *et al.*, 1994). These results support the view that adenovirus vectors (especially the first-generation vectors lacking E1A, E1B, and E3) continue to express viral genes at low levels, and this leads to an inflammatory response, death of the infected cells, and loss of transgene expression. It seems clear that the inflammatory response must be minimized in order for adenovirus vectors to be optimally useful in the treatment of human disease.

Is the second-generation $E1A^-/E1B^-/E2A^-/E3^+$ vector optimal? Even though it represents a significant advance over first generation vectors, the answer must be "no" for two reasons. First, even though Ad.ts125CBlacZ (Engelhardt *et al.*, 1994) mediated extended transgene expression with much less inflammation, it nevertheless induced detectable inflammation. Second, this vector retains the E4 region which should very likely be removed for safety purposes. In Ad9, a group D adenovirus, the E4orf1 protein is able to oncogenically transform cells (Javier *et al.*, 1992; Javier, 1994). Although the group C Ad2 and Ad5 E4orf1 proteins did not transform cells, they are highly related in sequence to the Ad9 protein, and they probably exhibit some functions in common. Thus, it would be prudent to remove the E4 gene from viral vectors for safety reasons, and its removal will also dramatically inhibit viral gene expression and DNA replication (e.g., Halbert *et al.*, 1985). An optimal vector might have the genotype $E1A^-/E1B^-/E2A^-/E4^-$.

Should the E3 gene be retained in future generations of adenovirus vectors? It would be possible to do so, placing it under control of the constitutively active E1A promoter, so that the E3 gene will be expressed in the absence of E1A and E2A proteins which normally serve to induce its expression. If the combined mutations in the E1A, E1B, E2A, and E4 genes completely block viral gene expression, then the E3 gene might best be deleted, in part because we do not yet understand the function of all of its coding regions or the full implications of E3–19-kDa activity. It is too soon to be certain whether E3 should be retained or excluded, so it will be important to construct, characterize, and compare third-generation vectors with two genotypes: $E1A^-/E1B^-/E2A^-/E3^-/E4^-$ and $E1A^-/E1B^-/E2A^-/E3^+/E4^-$.

Such third-generation vectors, lacking four different genes that are essential for viral growth (E1A, E1B, E2A, E4; E3 is not essential for viral growth in cultured cells), could prove difficult to propagate. Adenoviruses carrying E1A and E1B mutations are routinely grown in 293 cells, which are human embryonic kidney cells that have been transformed with a fragment of viral DNA containing these two genes (Graham *et al.*, 1977). Mutant viruses with lesions in the E2A gene have been propagated in HeLa cells that contain an inducible E2A gene (Klessig *et al.*, 1984; Brough *et al.*, 1992), and Vero cells have been generated that constitutively express the E4 gene and support the growth of viruses lacking that gene (Weinberg and Ketner, 1993). So, in theory, it should be possible to add properly regulated E2A and E4 genes to 293 cells, and use the resulting cell line to complement the growth of third-generation adenovirus vectors. These vectors should be extremely defective within normal human cells and, therefore, unlikely to express significant levels of viral products that could induce inflammation in gene therapy applications.

In the longer term, the challenge will be to develop the means for efficient propagation of an advanced adenovirus vector containing no *trans*-acting viral genes. Its genome would retain only the *cis*-acting sequences essential for viral growth (replication origins and packaging element) flanking one or more transgenes. If the viral proteins comprising the particles that deliver this genome to the infected cell do not themselves induce an inflammatory response, then this adenovirus vector should safely and efficiently deliver transgenes to human cells where they can be expressed for extended periods.

References

Babiss, L. E., and Ginsberg, H. S. (1984). Adenovirus type 5 early region 1b gene product is required for efficient shutoff of host protein synthesis. *J. Virol.* **50**, 202–212.

Beltz, G. A., and Flint, S. J. (1979). Inhibition of Hela cell protein synthesis during adenovirus infection. *J. Mol. Biol.* **131**, 353–373.

Berk, A. J., Lee, F., Harrison, T., Williams, J., and Sharp, P. A. (1979). Pre-early adenovirus 5 gene product regulates synthesis of early viral messenger RNAs. *Cell* **17**, 935–944.

Berkner, K. L. (1988). Development of adenovirus vectors for the expression of heterologous genes. *Biotechniques* **6**, 616–629.

Brody, S. L., and Crystal, R. G. (1994). Adenovirus-mediated *in vivo* gene transfer. *Ann. NY Acad. Sci.* **716**, 90–101; discussion, 101–103.

Brough, D. E., Cleghon, B., and Klessig, D. F. (1992). Construction, characterization, and utilization of cell lines which inducibly express the adenovirus DNA-binding protein. *Virology* **190**, 624–634.

Chang, L.-S., and Shenk, T. (1990). The adenovirus DNA binding protein can stimulate the rate of transcription directed by adenovirus and adeno-associated virus promoters. *J. Virol.* **64,** 2103–2109.

Engelhardt, J. F., Ye, X., Doranx, B., and Wilson, J. M. (1994). Ablation of E2A in recombinant adenoviruses improves transgene persistence and decreases inflammatory response in mouse liver. *Proc. Natl. Acad. Sci. USA* **91,** 6196–6200.

Evans, R. M., Fraser, N., Ziff, E., Weber, J., Wilson, M., and Darnell, J. E. (1977). The initiation sites for RNA transcription in Ad2 DNA. *Cell* **12,** 733–739.

Ginsberg, H. S., Moldawer, L. L., Sehgal, P. B., Redington, M., Kilian, P. L., Chanock, R. M., and Prince, G. A. (1991). A mouse model for investigating the molecular pathogenesis of adenovirus pneumonia. *Proc. Natl. Acad. Sci. USA* **88,** 1651–1655.

Ginsberg, H. S., Lundholm-Beauchamp, U., Horswood, R. L., Pernis, B., Wold, W. S. M., Chanock, R. M., and Prince, G. A. (1989). Role of early region 3 (E3) in pathogenesis of Adenovirus disease. *Proc. Natl. Acad. Sci. USA* **86,** 3823–3827.

Goff, S. P., and Shenk, T. (1993). Sleeping with the enemy: Viruses as gene transfer agents. *Curr. Opin. Genet. Dev.* **3,** 71–73.

Gooding, L. R. (1992). Virus proteins that counteract host immune defenses. *Cell* **71,** 5–7.

Graham, F. L., and Prevec, L. (1992). "Vaccines: New Approaches to Immunological Problems" (R. W. Ellis, Ed.), pp. 363–390. Butterworth–Heinemann, Boston.

Graham, F. L., Smiley, J., Russell, W. C., and Nairu, R. (1977). Characteristics of a human cell line transformed by DNA from human adenovirus type 5. *J. Gen. Virol.* **36,** 59–72.

Greber, U. F., Willetts, M., Webster, P., and Helenius, A. (1993). Stepwise dismantling of adenovirus 2 during entry into cells. *Cell* **75,** 477–486.

Green, M., Wold, W. S. M., Brackmann K. H., *et al.* (1980). Human adenovirus transforming genes: Group relationships, integration, expression in transformed cells and analysis of human cancers and tonsils. In: "7th Cold Spring Harbor Conference on Cell Proliferation Viruses in Naturally Occurring Tumors" (M. Essex, G. Todaro, and H. zurHausen, Eds.), pp. 373–397. Cold Spring Harbor Laboratory Press, Cold Spring Harbor, NY.

Halbert, D. N., Cutt, J. R., and Shenk, T. (1985). Adenovirus early region 4 encodes functions required for efficient DNA replication, late gene expression, and host cell shutoff. *J. Virol.* **56,** 250–257.

Hammarskjold, M. L., and Winberg, G. (1980). Encapsidation of adenovirus 16 DNA is directed by a small DNA sequence at the left end of the genome. *Cell* **20,** 787–795.

Hasson, T. B., Ornelles, D. A., and Shenk, T. (1992). Adenovirus L1 52- and 55-kilodalton proteins are present within assembling virions and colocalized with nuclear structures distinct from replication centers. *J. Virol.* **66,** 6133–6142.

Hearing, P., Samulski, R., Wishart, W., and Shenk, T. (1987). Identification of a repeated sequence element required for efficient encapsidation of the adenovirus type 5 chromosome. *J. Virol.* **61,** 2555–2558.

Hilleman, M. R., and Werner, J. H. (1954). Recovery of new agents from patients with acute respiratory illness. *Proc. Soc. Exp. Biol. Med.* **85,** 183–188.

Horwitz, M. (1995). Adenoviruses. "Fields Virology" (B. N. Fields, D. M. Knipe, and P. M. Howley, Eds.). Raven Press, New York.

Javier, R. (1994). Adenovirus type 9 E4 open reading frame 1 encodes a transforming protein required for the production of mammary tumors in rats. *J. Virol.* **68,** 3917–3924.

Javier, R., Raska, K., and Shenk, T. (1992). The adenovirus type 9 E4 region is required for production of mammary tumors. *Science* **257,** 1267–1271.

Jones, N., and Shenk, T. (1979). An adenovirus type 5 early gene function regulates expression of other early viral genes. *Proc. Natl. Acad. Sci. USA* **76,** 3665–3669.

Klessig, D. F., Brough, D. E., and Cleghorn, V. G. (1984). Introducing stable integration, and controlled expression of a chimeric adenovirus gene whose product is toxic to the recipient human cell. *Mol. Cell. Biol.* **4,** 1354–1362.

Kozarsky, K., and Wilson, J. M. (1993). Gene therapy: Adenovirus vectors. *Curr. Opin. Genet. Dev.* **3**, 499–503.

Lechner, R. L., and Kelly, T. J., Jr. (1977). The structure of replicating adenovirus 2 DNA molecules. *Cell* **12**, 1007–1020.

Mackey, J. K., Rigden, P. M., and Green, M. (1976). Do highly oncogenic group A human adenoviruses cause human cancer? Analysis of human tumors for adenovirus 12 transforming DNA sequences. (Proc. Natl. Acad. Sci. USA **73**, 4675–4661.

Mathews, M. B., and Shenk, T. (1991). Adenovirus virus-associated RNA and translation control. *J. Virol.* **65**, 5657–5662.

Neve, R. L. (1993). Adenovirus vectors enter the brain. *Trends Neurosci* **16**, 252–253.

Nevins, J. R. (1981). Mechanism of activation of early viral transcription by the adenovirus E1A gene product. *Cell* **26**, 213–220.

Nevins, J. R. (1992). E2F: A link between the Rb tumor suppressor protein and viral oncoproteins. *Science* **258**, 424–429.

Philipson, L. (1984). Adenovirus assembly. In "The Adenoviruses" (H. Ginsberg, Ed.). Plenum, New York.

Pilder, S., Moore, M., Logan, J., and Shenk, T. (1986). The adenovirus E1B-55K transforming polypeptide modulates transport or cytoplasmic stabilization of viral and host cell mRNAs. *Mol. Cell. Biol.* **6**, 470–476.

Rao, L., Debbas, M., Sabbatini, P., Hockenbery, D., Korsmeyer, S., and White, E. (1992). The adenovirus E1A proteins induce apoptosis, which is inhibited by the E1B 19-kDa and Bcl-2 proteins. *Proc. Natl. Acad. Sci. USA* **89**, 7742–7746. [Published erratum appears in *Proc. Natl. Acad. Sci. USA* 1992 Oct 15;**89**(20):9974].

Reich, N., Pine, R., Levy, D., and Darnell, J. E. (1988). Transcription of interferon-stimulated genes is induced by adenovirus particles but is suppressed by E1A gene products. *J. Virol.* **62**, 114–119.

Rekosh, D. M. K., Russell, W. C., Bellet, A. J. D., and Robinson, A. J. (1977). Identification of a protein linked to the ends of adenovirus DNA. *Cell* **11**, 283–295.

Rich, D. P., Couture, L. A., Cardoza, L. M., Guiggio, V. M., Armentano, D., Espino, P. C., Hehir, K., Welsh, M. J., Smith, A. E., and Gregory, R. J. (1993). Development and analysis of recombinant adenoviruses for gene therapy of cystic fibrosis. *Hum. Gene Ther.* **4**, 461–476.

Rowe, W. P., Huebner, R. J., Gilmore, L. K., Parrott, R. H., and Ward, T. G. (1953). Isolation of a cytopathogenic agent from human adenoids undergoing spontaneous degeneration in tissue culture. *Proc. Soc. Exp. Biol. Med.* **84**, 570–573.

Sarnow, P., Ho, Y.-S., Williams, J., and Levine, A. J. (1982a). Adenovirus D1b-58kd tumor antigen and SV40 large tumor antigen are physically associated with the same 54 kd cellular protein in transformed cells. *Cell* **28**, 387–394.

Schaack, J., Ho, W. Y.-W., Freimuth, P., and Shenk, T. (1990). Adenovirus terminal protein mediates both nuclear matrix association and efficient transcription of adenovirus DNA. *Genes Devel.* **4**, 1197–1208.

Shenk, T. (1995). Adenoviridae and their replication. In "Fields Virology" (B. N. Fields, D. M. Knipe, and P. M. Howley, Eds.). Raven Press, New York.

Shenk, T., Jones, N., Colby, W., and Fowlkes, D. (1979). Functional analysis of adenovirus type 5 host-range deletion mutants defective for transformation of rat embryo cells. *Cold Spring Harbor Symp. Quant. Biol.* **44**, 367–375.

Shenk, T., and Williams, J. (1984). Genetic analysis of adenoviruses. *Curr. Topics. Microbiol. Immunol.* **111**, 1–39.

Simon, R. H., Engelhardt, J. F., Yang, Y., Zepeda, M., Weber-Pendleton, S., Grossman, M., and Wilson, J. M. (1993). Adenovirus-mediated transfer of the CFTR gene to lung of nonhuman primates: toxicity study. *Hum. Gene Ther.* **4**, 771–780.

Tibbetts, C. (1977). Viral DNA sequences from incomplete particles of human adenovirus type 7. *Cell* **12**, 243–249.

Tooze, J. (1981). "DNA Tumor Viruses," 2nd ed., pp. 943–1054. Cold Spring Harbor, Laboratory Press, Cold Spring Harbor, NY.

Trentin, J. J., Yabe, Y., and Taylor, G. (1962). The quest for human cancer viruses. *Science* **137**, 835–849.

Weinberg, D., and Ketner, G. (1983). A cell line that supports the growth of a defective early region 4 deletion mutant of human adenovirus type 2. *Proc. Natl. Acad. Sci. USA* **80**, 5383–5386.

Whyte, P., Buchkovich, K. J., Horowitz, J. M., Friend, S. H., Raybuck, M., Weinberg, R. A., and Harlow, E. (1988). Association between an oncogene and an anti-oncogene: The adenovirus E1A proteins bind to the retinoblastoma gene product. *Nature* **334**, 124–129.

Wickham, T. J., Mathias, P., Cheresh, D. A., and Nemerow, G. R. (1993). Integrins alpha v beta 3 and alpha v beta 5 promote adenovirus internalization but not virus attachment. *Cell* **73**, 309–319.

Wold, W. S., and Gooding, L. R. (1991). Region E3 of adenovirus: A cassette of genes involved in host immunosurveillance and virus–cell interactions. *Virology* **184**, 1–8.

Yang, Y., Nunes, F., Berencsi, K., Furth, E., and Gonczol, E. (1994). Cellular immunity to viral antigens limits E1-deleted adenoviruses for gene therapy. *Proc. Natl. Acad. Sci. USA* **91**, 4407–4411.

Yonish-Rouach, E., Resnitzky, D., Lotem, J., Sachs, L., Kimchi, A., and Oren, M. (1991). Wild-type p53 induces apoptosis of myeloid leukaemic cells that is inhibited by interleukin-6. *Nature* **352**, 345–347.

Zabner, J., Couture, L. A., Gregory, R. J., Graham, S. M., Smith, A. E., and Welsh, M. J. (1993). Adenovirus-mediated gene transfer transiently corrects the chloride transport defect in nasal epithelia of patients with cystic fibrosis. *Cell* **75**, 207–216.

Zabner, J., Petersen, D., Puga, A. P., Graham, S. M., Couture, L. A., Keyes, L. D., Lukason, M. J., St. George, J. A., Gregory, R. J., Smith, A. E., and Welsh, M. J. (1994). Safety and efficacy of repetitive adenovirus-mediated transfer of CFTR cDNA to airway epithelia of primates and cotton rats. *Nature Genet.* **6**, 75–83.

Zambetti, G. P., and Levine, A. J. (1993). A comparison of the biological activities of wild-type and mutant p53. *FASEB J.* **7**, 855–865.

Zhang, Y., and Schneider, R. (1993). Adenovirus inhibition of cellular protein synthesis and the specific translation of late viral mRNAs. *Semin. Virol.* **4**, 229–236.

CHAPTER

Genetics and Biology of Adeno-Associated Virus

Jeffrey S. Bartlett
Richard J. Samulski
Gene Therapy Center
University of North Carolina at Chapel Hill
Chapel Hill, North Carolina

Adeno-associated virus (AAV) is a defective member of the parvovirus family. AAV can be propagated as a lytic virus or maintained as a provirus integrated into the host cell genome (Atchison *et al.*, 1965; Hoggan *et al.*, 1966, 1972) (Fig. 1). In a lytic infection, replication requires coinfection with either adenovirus (Atchison *et al.*, 1965; Hoggan *et al.*, 1966; Melnick *et al.*, 1965) or herpes simplex virus (Buller *et al.*, 1981; McPherson *et al.*, 1985), hence, the classification of AAV as a "defective" virus. Vaccinia virus has also been shown to provide some helper function (Schlehofer *et al.*, 1986). In the absence of a helper virus AAV is unable to replicate and rather establishes latency by persisting in the host cell genome as an integrated provirus (Berns *et al.*, 1975; Cheung *et al.*, 1980; Handa *et al.*, 1977; Hoggan *et al.*, 1972). This is one of the most interesting aspects of the AAV life cycle. The latent AAV proviruses are quite stable, although AAV can be rescued from the chromosome and re-enter the lytic cycle if these cells are super infected with wild-type helper virus (Fig. 1). This unique capacity to establish a latent infection which can later be rescued appears to be a means for ensuring the survival of AAV in the absence of helper virus.

The AAV genome is encapsidated as a single-stranded DNA molecule of plus or minus polarity. Strands of both polarities are packaged, but in separate virus particles (Berns and Adler, 1972; Berns and Rose, 1970; Mayor *et al.*, 1969; Rose *et al.*, 1969), and both strands are infectious (Samulski *et al.*, 1987). Five serotypes of AAV have been identified, but the most extensively characterized is AAV-2. The nonenveloped virion is icosohedral in shape and one of the smallest that has been described, about 20–24 nm in diameter (Hoggan, 1970; Tsao *et al.*, 1991).

VIRAL VECTORS

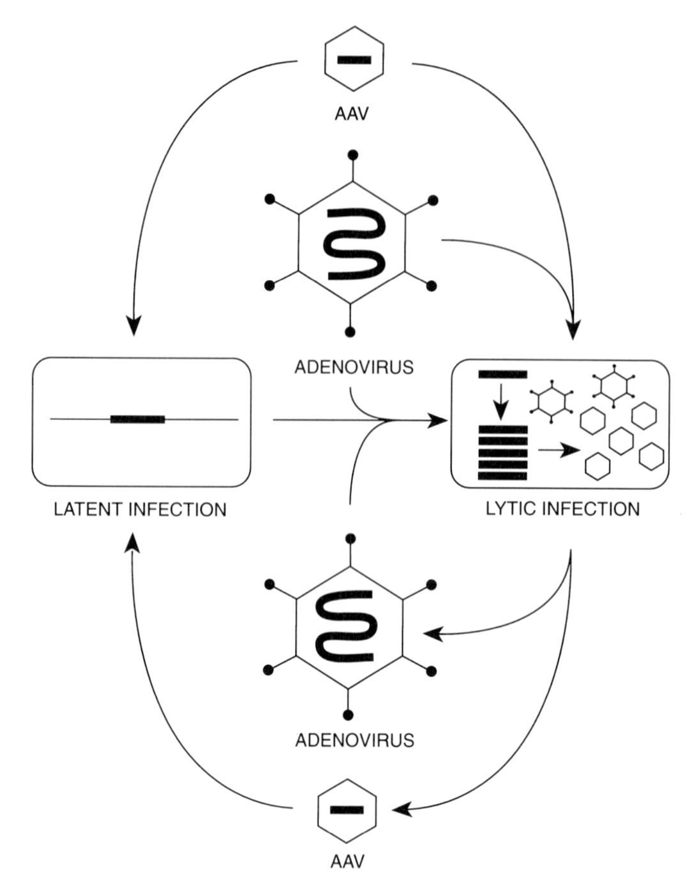

Figure 1 Adeno-associated virus (AAV) life cycle. The virus life cycle is biphasic. In the presence of a helper virus, such as adenovirus, AAV proceeds through a lytic infection and produces progeny virus. In the absence of helper virus, the AAV genome integrates into the host genome establishing latency. The latent virus genome is stable for many generations and can be "rescued" to enter the lytic phase upon subsequent helper virus superinfection.

I. Genetics of Adeno-Associated Virus

A. Genome Structure

The complete nucleotide sequence of AAV-2 consists of 4680 nucleotides (Srivastava *et al.*, 1983). The genome is single-stranded, linear DNA and contains two inverted terminal repeats that are 145 bp long (Gerry *et al.*, 1973; Koczot *et al.*, 1973; Lusby *et al.*, 1980). The first 125 bp of each repeat can form a T-shaped hairpin structure which is composed of two small internal palindromes (B and C) (Fig. 2) flanked by a larger palindrome (A). The terminal sequence can exist in either of two orientations termed *flip* (in which the B palindrome is adjacent to the 3′ end of the viral DNA) or *flop* (in which the C palindrome is closer to the 3′ end) (Fig. 2). Based on the viral replication model, both ends of the AAV genome have equal probability of being in either the flip or flop orientation (Lusby *et al.*, 1980). The internal portion of the AAV genome is divided genetically into two open reading frames (Fig. 3) that

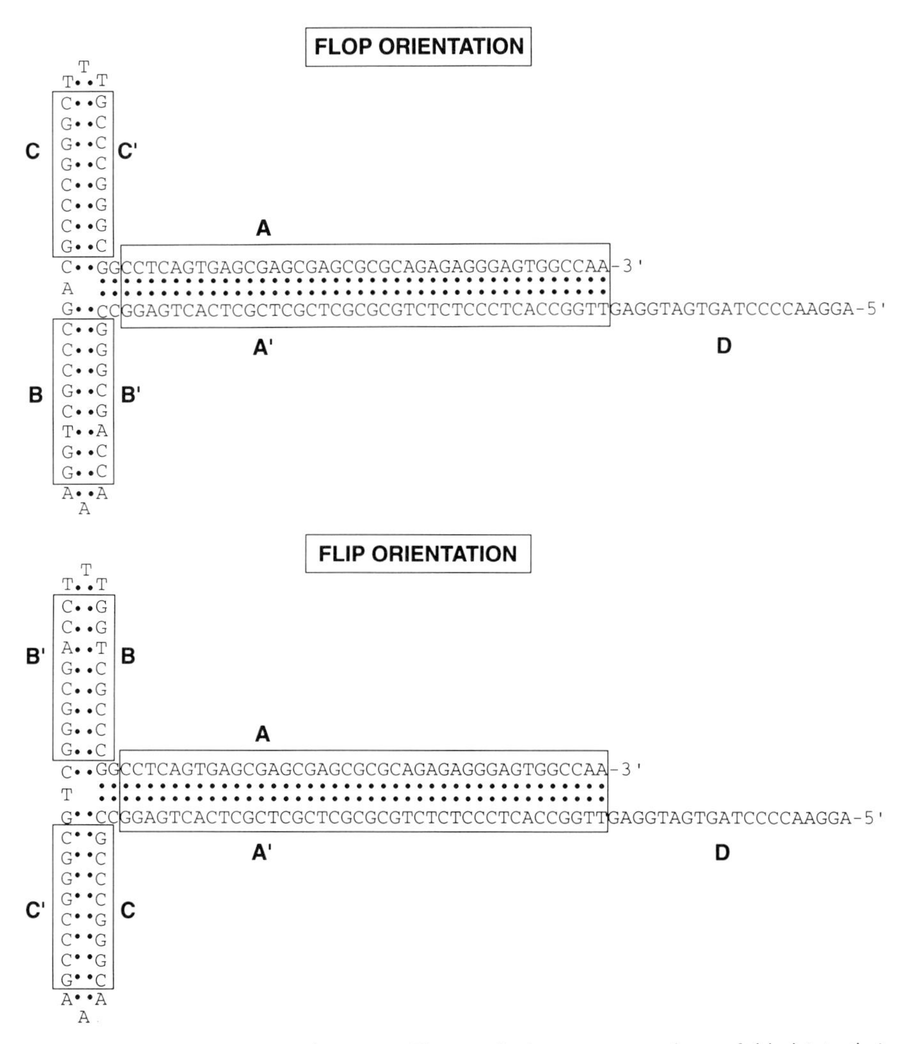

Figure 2 The AAV terminal repeats. The terminal repeats are shown folded into their proposed secondary structure in each orientations, *flip* and *flop*. Two small palindromes, B–B′ and C–C′ are embedded in a larger palindrome, A–A′.

encode the structural and nonstructural viral proteins. Three promoters have been identified and named according to their approximate map positions: p5, p19, and p40 (Green and Roeder, 1980a,b; Green *et al.*, 1980; Laughlin *et al.*, 1979; Lusby and Berns, 1982). Transcripts initiating at each of these promoters share a common intron and all terminate at the same polyadenylation site at map position 95 (Green and Roeder, 1980b; Laughlin *et al.*, 1979; Srivastava *et al.*, 1983). Both spliced and unspliced transcripts can be detected in infected cells (Green and Roeder, 1980a,b; Laughlin *et al.*, 1979), and at least one alternately spliced message is also present (Trempe and Carter, 1988a). This message initiates from the p40 promoter and uses the same splice donor site as the major intron but a different acceptor site. The resulting message is only slightly larger than the major 2.3-kb species, but allows for the expression of the entire right-hand open reading frame (Fig. 3).

The messages encoding the nonstructural replication (*Rep*) proteins are transcribed from the p5 and p19 promoters; whereas the structural, capsid (*Cap*) mRNAs are transcribed from the p40 promoter. The amino acid sequence of the major capsid protein VP3 is encoded within the same sequence as the two larger and less abundant capsid proteins VP1 and VP2 (Janik *et al.*, 1984). VP2 is synthesized from the same mRNA as VP3 using an upstream ACG start codon. Mutants with alterations within the VP3 region make very little capsid

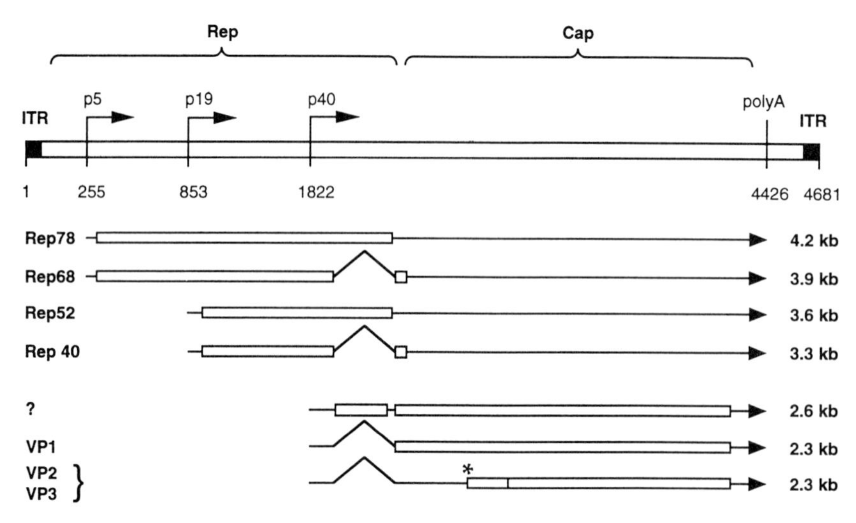

Figure 3 The AAV genome. (*Top*) Genetic map of AAV showing the location of the *Rep* and *Cap* regions. Also shown are the three AAV promoters (*arrows*) and the polyadenylation site. (*Bottom*) The known AAV RNAs. The size of each mRNA and the proteins synthesized from each mRNA are shown. The reading frames used in each mRNA are also shown. The question mark indicates that no known function exists for this message. VP2 is synthesized from an extended VP3 open reading frame by using an ACG start codon, indicated by the asterisk.

protein (Hermonat *et al.*, 1984) and produce only small amounts of progeny single-stranded viral DNA (Tratschin *et al.*, 1984). Therefore, it seems likely that in the absence of capsid protein, or procapsid assembly, nascently synthesized viral DNA is rapidly degraded (Hermonat *et al.*, 1984; Tratschin *et al.*, 1984). Mutations within the VP1 N-terminal region suggest that VP1 is important at some later stage of capsid processing (Hermonat *et al.*, 1984; Tratschin *et al.*, 1984). VP1 is synthesized from an alternatively spliced p40 transcript. Mutants lacking VP1 produce very low yields of infectious viral particles but are viable for viral DNA synthesis, suggesting that these mutants are making procapsid and packaging single-stranded viral DNA but are defective in a later stage of capsid assembly that is reliant on VP3 expression. It is not clear whether noninfectious or unstable viral particles can be made in the absence of VP1.

B. Capsid Structure

The AAV virion is composed of the three structural proteins: VP1, VP2, and VP3 (Johnson *et al.*, 1971, Johnson 1977, 1975; Rose *et al.*, 1971). These have molecular weights of 87, 73, and 61 kDa, respectively. VP3 represents about 90% of the total virion protein; VP2 and VP3 account for about 5% each. All three capsid proteins are N-acetylated (Becerra *et al.*, 1985). At present little is known about the structure of these proteins in the capsid. However, the crystal structure of a related canine parvovirus has recently been determined (Tsao *et al.*, 1991) and presumably the structure of AAV particles is similar.

There are at least four nonstructural AAV proteins collectively termed the *Rep* proteins. These are named according to their molecular weights: Rep 78, Rep 68, Rep 52, and Rep 40 (Mendelson *et al.*, 1986; Srivastava *et al.*, 1983; Trempe *et al.*, 1987). In addition, two smaller *Rep* proteins have also been observed in AAV-infected cells (Trempe *et al.*, 1987), presumably resulting from translational initiation at internal ATG codons within the mRNA that produces the Rep 52 and Rep 40 proteins. Heterogeneity has been observed in the apparent molecular weights of the Rep 68, Rep 78, and Rep 40 proteins in denaturing polyacrylamide gels (Im and Muzyczka, 1990; Redemann *et al.*, 1989). The reason for this heterogeneity is not known.

C. Replication

Rep 78 and Rep 68 control DNA replication (Hermonat *et al.*, 1984; Labow *et al.*, 1986, 1987; Tratschin *et al.*, 1986). As yet, the exact role the two Rep proteins expressed from the p19 promoter play during a productive viral infection has not been determined (Owens *et al.*, 1993). Both Rep 68 and Rep 78 have been expressed in nonmammalian systems and purified to homogeneity. Their activities include binding to the terminal hairpin structures

(Ashktorab and Srivastava, 1989; Chiorini *et al.*, 1994; Im and Muzyczka, 1989), ATP-dependent DNA helicase activity (Im and Muzyczka, 1990), strand- and sequence-specific DNA endonuclease activity (Im and Muzyczka, 1990), covalently binding to the 5′ end of their DNA substrate (Im and Muzyczka, 1990), and the ability to replicate viral DNA in a cell-free system (Chiorini *et al.*, 1994; Ni *et al.*, 1994). Interestingly, in the absence of helper virus coinfection, Rep 78 and Rep 68 are able to negatively regulate replication (Berns *et al.*, 1988) and can repress both viral and heterologous gene expression (Antoni *et al.*, 1991; Labow *et al.*, 1987; Mendelson *et al.*, 1988; Tratschin *et al.*, 1986; Trempe and Carter, 1988b; West *et al.*, 1987). It has been suggested that this effect is mediated at the translational level for repression of the *Cap* proteins from the p40 promoter (Trempe and Carter, 1988b). The mechanism for the decreased expression of the *Rep* proteins from the p5 and p19 promoters is not as clear. The idea that the *Rep* gene can autoregulate its own expression is attractive because it may help explain how *Rep* protein synthesis and DNA replication are turned off when the virus has integrated into the host genome.

II. Adeno-Associated Virus Gene Expression

A. *Rep* Protein Function

As discussed above, the *Rep* proteins appear to be capable of repressing both AAV and heterologous gene expression. Although this repression is considered to be dependent on expression from one of the viral promoters, the *rep* gene is also capable of repressing heterologous promoters. In cases where plasmid DNA containing a selectable marker under the control of a specific promoter was cotransfected with a plasmid containing the wild-type *rep* gene, a number of promoters were shown to be repressed. These included the neomycin resistance gene under control of the SV40 or mouse metallothionein promoters (Labow *et al.*, 1987), the neomycin resistance gene under AAV p40 control (Mendelson *et al.*, 1988), the human immunodeficiency virus type 1 (HIV-1) genes driven by the HIV long terminal repeat (LTR) (Antoni *et al.*, 1991), and the herpes thymidine kinase gene (Labow *et al.*, 1987). With the HIV LTR and SV40 gene promoters, repression appeared to be in part at the level of transcription (Antoni *et al.*, 1991; Labow *et al.*, 1987).

B. Helper Virus Function

Helper virus function has been best studied with adenovirus. Under normal circumstances the lytic phase, either upon primary infection or upon rescue, requires the expression of the adenovirus early gene products E1A (Chang *et al.*, 1989; Richardson and Westphal, 1981, 1984), E1B (Richardson and Westphal, 1984; Samulski and Shenk, 1988), E2A (Jay *et al.*, 1979), E4 (Carter *et al.*, 1983; Laughlin *et al.*, 1982; Richardson and Westphal, 1981, 1984), and VA RNA (Janik *et al.*, 1989; West *et al.*, 1987). The E1A region has been shown to *trans*-activate the AAV p5 and p19 promoters (Chang *et al.*, 1989; Flint and Shenk, 1989; Tratschin *et al.*, 1984), although this region also activates a large number of cellular genes. Therefore, its precise role in helper function has been difficult to determine. The E1B 55-kDa protein and the E4 34-kDa ORF-6 gene product influence the efficient accumulation of AAV mRNAs (Carter *et al.*, 1983; Huang and Hearing, 1989; Laughlin *et al.*, 1982; Ostrove and Berns, 1980; Richardson and Westphal, 1981; Samulski and Shenk, 1988). Apparently they either stabilize the AAV RNA or they facilitate its transport to the cytoplasm. The adenovirus E2A DNA binding protein and the adenovirus VA RNAs are primarily involved in the efficient translation of the AAV capsid mRNAs initiated from the p40 promoter (Janik *et al.*, 1989; Jay *et al.*, 1981; McPherson *et al.*, 1982; Myers *et al.*, 1980; Richardson and Westphal, 1981; Straus *et al.*, 1976; West *et al.*, 1987). It should be noted that all of these adenovirus proteins perform a similar role during the expression of adenovirus genes in the absence of AAV. In addition to facilitating their translation, the presence of the VA RNA genes also increases the steady-state level of AAV transcripts (Janik *et al.*, 1989; West *et al.*, 1987).

C. Cellular Function Required for Replication and Gene Expression

Although a true lytic AAV infection is usually thought to require the presence of a helper virus, no helper virus genes appear to be directly involved in AAV DNA replication, nor do they appear to be absolutely essential for a permissive AAV infection. The role of the helper virus is clouded by the fact that certain cells can become permissive for AAV DNA replication in the absence of helper virus if they are appropriately perturbed. This effect has been observed by transforming cells with viral or cellular oncogenes, as well as treating cells with reagents that transiently arrest cellular DNA synthesis (e.g., hydroxyurea or uv irradiation) (Bantel-Schall and zur Hausen, 1988a,b; Yacobson *et al.*, 1987, 1989; Yalkinoglu *et al.*, 1988). Additional agents that induce a generalized cellular stress response have also been implicated in providing helper function (e.g., heat shock or cell synchronization). Helper virus may be affecting cellular gene expression in such a way that the genes needed for AAV DNA replication are activated. A direct effect of adenoviral gene expression on AAV DNA replication is further brought into question by the observations that adenovirus containing mutations in the adenoviral polymerase gene are normal for AAV DNA replication, and that plasmid constructs containing the adenoviral polymerase and adenovirus DNA binding protein coding regions are not required for AAV helper function in mixed transfection assays (Janik *et al.*, 1981; Myers *et al.*, 1980; Straus *et al.*, 1976). Studies have pointed to an aphidocolin-sensitive cellular DNA polymerase, presumably polymerase α or δ, that is required for AAV DNA synthesis (Ikegami *et al.*, 1979). Additionally, polymerase δ seems to be required for the terminal resolution step of AAV DNA replication (Snyder *et al.*, 1990). However, other than these few examples the cellular gene products that are required for a productive AAV infection are unknown.

III. Adeno-Associated Virus Integration

A. Site-Specific Integration

One of the most interesting aspects of the AAV life cycle is the ability of the virus DNA to integrate into the host genome in the absence of a helper virus. AAV integration appears to have no effect on cell growth, and despite its propensity to integrate into the cellular genome as a rescueable provirus, there is no evidence of AAV functioning as a tumor virus (Handa *et al.*, 1977). The mechanism of viral integration is not known. It appears that AAV integration is novel, since there are several aspects that distinguish it from other well-characterized viral integration events. First, no viral gene expression is required for integration to occur (McLaughlin *et al.*, 1988; Samulski *et al.*, 1989). Only the AAV terminal repeats appear to be essential for integration (Hermonat and Muzyczka, 1984; McLaughlin *et al.*, 1988; Samulski *et al.*, 1989; Srivastava *et al.*, 1989; Tratschin *et al.*, 1984). Second, latently infected cells are very stable and capable of maintaining the integrated viral DNA for thousands of passages (Berns *et al.*, 1982; Samulski *et al.*, 1991). Third, concatemers consisting of two to four tandem copies often exist at the integration locus regardless of the initial multiplicity of infection, suggesting that at least a limited amount of viral replication may precede the integration event (Laughlin *et al.*, 1986; McLaughlin *et al.*, 1988). Finally, integration of the wild-type AAV genome seems to prefer a target sequence located on human chromosome 19q13.3-qter (Kotin *et al.*, 1990, 1991; Samulski *et al.*, 1991). This site-specific integration has been documented in a number of cell types including human T cells, colon, lung, bone marrow stem cells, and monkey kidney cells (Goodman *et al.*, 1994; Zhu *et al.*, unpublished observations).

B. Proviral Structure

Determination of the AAV proviral structure has come from genomic hybridization experiments and sequence analysis of cloned AAV–cellular DNA junctions. Since an intact AAV provirus has not been isolated, we still do not have a fine structure map of the integrated virus. From the genomic hybridization experiments it appears that the main proviral structure consists of two to four tandem copies of the AAV genome integrated in a single chromosome location (Laughlin *et al.,* 1986; McLaughlin *et al.,* 1988). Interestingly, the same copy number of AAV genomes is seen regardless of the multiplicity of infection, and at very high multiplicities of infection, only a very limited number of separate chromosome integration events occurs. It has been suggested that the tandem arrays are the result of viral replication rather than the concatenation of multiple AAV molecules (Laughlin *et al.,* 1986; McLaughlin *et al.,* 1988). Other kinds of proviral structures have also been identified (Laughlin *et al.,* 1986; McLaughlin *et al.,* 1988). Sometimes only a single copy of the AAV genome appears to be present, and other times the tandem concatemers contain amplified cellular as well as viral DNA (McLaughlin *et al.,* 1988). In addition, several other cells lines have been isolated which contain only partial AAV DNA sequences or AAV DNA sequences interrupted by rearranged cellular DNA sequences (Kotin and Berns, 1989). Recently, several cellular DNA–AAV and AAV–AAV junctions that are present in the tandem proviral arrays have been cloned and sequenced. It has been shown that recombination at the terminal repeats between viral and cellular sequences is imprecise and regions of the terminal repeats are almost always deleted (Grossman *et al.,* 1985; Grossman *et al.,* 1984). However, the sequence between the two AAV copies in a tandem array usually contains an intact terminal repeat sequence (Muzyczka, 1992), leading to the suggestion that the formation of a concatemer is a mechanism for insuring the integrity of at least one copy of the AAV genome during integration (Muzyczka, 1992).

Introduction of AAV DNA into cells in a viral form also appears necessary for some aspects of AAV integration involving the AAV terminal repeats (Grossman *et al.,* 1984). There may be a requirement for the single-stranded AAV DNA molecule that is present in the viral particles or some component of the AAV capsid as suggested by Muzyczka (1992).

C. Rescue of Integrated Virus

The exact mechanism of AAV rescue remains unknown. Speculation has centered on the assumption that *Rep* protein can make the same cuts that it makes during terminal resolution to excise the provirus from the genome (Samulski *et al.,* 1983; Senepathy *et al.,* 1984). Interestingly, a cellular enzyme, endo R, has been shown to excise AAV sequences from plasmid DNA at a low frequency by cutting the DNA within the AAV terminal repeats (Gottlieb and Muzyczka, 1988, 1990a,b). However, endo R is a relatively nonspecific nuclease and will produce double-stranded cuts within any sequence that contains a high G : C content. Therefore, there is no direct evidence that either

AAV *Rep* protein or endo R is directly involved in the rescue of AAV DNA from mammalian chromosomes.

IV. Recombinant Adeno-Associated Vectors

A. Recombinant AAV Plasmid Vectors

As a prerequisite for vector construction it was necessary to clone a double-stranded version of the virus into a plasmid backbone (Laughlin *et al.*, 1983; Samulski *et al.*, 1982). Since wild-type AAV could be rescued from an integrated chromosome and enter the lytic cycle following adenovirus infection, it was important to determine if such a recombinant AAV plasmid could be used for generating wild-type AAV virus. Once it was shown that the viral genome could be rescued from a plasmid backbone when adenovirus was provided as helper virus, the doors were opened for the use of AAV as a gene transduction vector. Infectious plasmid clones have been the template for all subsequent vector constructions. The ability to generate large quantities of plasmid DNA which is basically inert until introduced into adenovirus infected human cells also provides a safe and efficient way of manipulating this system.

The first use of AAV as a vector for the transduction of a foreign gene was demonstrated by Hermonat and Muzyczka in 1984 (Hermonat and Muzyczka, 1984). A recombinant AAV (rAAV) viral stock was produced using an infectious plasmid vector similar to that described above in which the neomycin resistance gene (*neo*^{r}) was substituted for the AAV capsid genes. To supply the missing capsid proteins a second plasmid containing the wild-type *Cap* region was cotransfected into the same cells. The rAAV that was produced was able to transduce neomycin resistance to both murine and human cell lines (Hermonat and Muzyczka, 1984). Others subsequently constructed and tested similar vectors (Handa *et al.*, 1977; Laughlin *et al.*, 1986; Lebkowski *et al.*, 1988; Tratschin *et al.*, 1985). Typically, the recombinant genome was found integrated into the host cell DNA and was stable for hundreds of passages. However, if the cells containing the integrated provirus were subsequently superinfected with adenovirus, the recombinant AAV could be rescued and amplified within the cell since these early constructs still contained the AAV *Rep* region (Hermonat and Muzyczka, 1984; McLaughlin *et al.*, 1988; Samulski *et al.*, 1989; Tratschin *et al.*, 1985).

B. Size Constraints

Although the studies described above demonstrated the potential use of AAV as a vector, several technical problems remained, such as the need for efficient packaging systems, methods for producing recombinant virus stock free of wild-type AAV, and the identification of minimum AAV sequences required for transduction. This last hurdle would have direct impact on the size of foreign DNA inserts. In attempts to solve these problems, constructs which retained only a limited number of nucleotides from the viral terminal sequences were tested (McLaughlin *et al.*, 1988; Samulski *et al.*, 1987, 1989). One such vector, *psub201* (Samulski *et al.*, 1987), which was a derivative of the original recombinant viral vector described above (Samulski *et al.*, 1982), contained only the left and right AAV terminal repeats and 45 bp of nonrepeated sequences adjacent to the right terminal repeat. Other vectors contained slightly more nonrepeated sequences. However, all vectors proved capable of transducing foreign DNA into mammalian cells (McLaughlin *et al.*, 1988; Samulski *et al.*, 1989). These constructs demonstrated that the terminal repeats (and possibly a few nucleotides adjacent to the right terminal repeat) were the only sequences required in *cis* for AAV DNA replication, packaging, integration, and rescue.

C. Host Range and Transduction of Nondividing Cells

Little is known about the relationship between viral infection and the cell cycle. In fact, the mode of viral uptake has not been established and no cellular receptor has been identified. In tissue culture AAV has been shown to infect all established human cell lines thus far examined (Laughlin *et al.*, 1986; Lebkowski *et al.*, 1988; McLaughlin *et al.*, 1988; Samulski *et al.*, 1989; Tratschin *et al.*, 1985). Recombinant AAV has also shown a similarly broad cellular host range *in vivo*, although this has not been characterized as extensively (Samulski *et al.*, unpublished observations). In addition, the ability of AAV to infect terminally differentiated or nondividing cells has not been resolved satisfactorily. Retroviral vectors based on the murine leukemia virus require cell division for efficient transduction (Miller *et al.*, 1990). For this reason there has been intense interest in the ability of recombinant AAV vectors to transduce nondividing cells. Recently, several reports have begun to shed light on this possibility. Wong *et al.* (1993) reported that recombinant AAV was able to transduce growth arrested human fibroblasts and 293 cells. However, since the cells were allowed to resume growing prior to being scored for transduction, there remains the possibility that the virus simply remained episomal and integrated after the cells entered S phase. A second report seems to confirm this possibility: it was determined that the vector genomes can persist in stationary phase cells, but transduction preferentially occurs in cells after they have entered S phase (Russell, *et al.*, 1994). In this case transduction was assayed without subsequent stimulation and cell division. Although proliferating cells may be preferentially transduced by AAV vectors, proliferation may not be absolutely required for transduction. In fact, *in vivo* experiments have shown that recombinant AAV vectors are able to transduce cell populations that are thought to be largely quiescent at remarkably high efficiencies. These include human bone marrow progenitors (Goodman *et al.*, 1994; Miller *et al.*, 1994; Zhou *et al.*, 1993) in culture and rat brain *in vivo* (Kaplitt *et al.*, 1994; Xiao and Samulski, unpublished observation). Although, the human bone marrow progenitor ($CD34^+$) cells were maintained in media containing growth factors, interleukin (IL)-3, IL-6, and stem-cell factor, analysis of colonies derived from these progenitors has allowed the determination of the integration frequency in the primary hematopoeitic cells that gave rise to each colony. Histological analysis of rat brains injected with recombinant virus encoding β-galactosidase has demonstrated long-term expression in many regions of the rat brain. This finding represents the most conclusive demonstration that recombinant AAV vectors can transduce nondividing cells. It should be noted that AAV transduction measured by gene expression, as described in the majority of the published reports, does not reflect viral integration, since gene expression can take place off episomal or integrated templates.

D. Generation of Recombinant Virus

The present method for producing stocks of recombinant AAV utilizes a two component plasmid system divided in terms of the *cis* and *trans* components necessary for replication, expression, and encapsidation of the recombinant virus. As mentioned above, the viral terminal repeats are the only elements required in cis and in the current packaging system flank the transgene on one of two plasmids. The second plasmid supplies the necessary *Rep* and *Cap* gene products in *trans*. An important consideration is that the two plasmid DNAs must be sufficiently nonhomologous so as to preclude homologous recombination events which could generate wild-type AAV (Samulski *et al.*, 1989). Although there are many variations on this theme, a widely used method for recombinant AAV production involves cotransfection of the human cell line 293 with a plasmid containing the internal region of the AAV genome, with adenovirus type 5 terminal repeats in place of the normal AAV termini, and a second plasmid consisting of a heterologous gene (containing appropriate regulatory elements) flanked by the AAV inverted terminal repeats (Samulski *et al.*, 1989). When infected with adenovirus, rescue, replication, and packaging of the foreign gene into AAV particles occurs. The adenovirus genome activates the adenovirus terminal repeats on the helper plasmid which enhances expression of the AAV genes. The *Rep* gene products recognize the AAV *cis*-acting terminal repeats on the recombinant vector containing the foreign gene, rescue the recombinant segments out of the plasmid, and begin to replicate them. The AAV capsids begin to accumulate, recognize the AAV *cis*-acting packaging signals located in the AAV terminal repeats and encapsidate the recombinant viral DNA into an AAV virion. The result of such a packaging scheme is an adenovirus helper virus and AAV particles carrying the recombinant heterologous genes. Adenovirus can then be removed by any of a number of physical separation strategies. In this packaging system one can generate helper-free stocks of recombinant AAV at titers of 10^4 to 10^5 transducing viruses per milliliter.

The general feeling has been that the relatively low amount of recombinant virus produced by this system is due to the low levels of *Rep* and *Cap* gene products produced from the transfected plasmid templates (Kotin, 1994). In a wild-type lytic infection, the viral genome is amplified several-fold, which provides a much larger number of templates for *Rep* and *Cap* transcripts than are available following plasmid transfection. Consequently, several attempts have been made to increase the number of AAV genome equivalents available for *Rep* and *Cap* production. These have included the incorporation of the *Rep* and *Cap* genes into replication competent plasmid or viral vectors. As well as attempts to establish cell lines with high copy numbers of the *Rep* and/or *Cap* genes (Trempe and Yang, 1993; Vincent *et al.,* 1990). The construction of such cell lines, however, has been hampered by the apparent toxicity of *Rep* (Winocour *et al.,* 1992). One of these cell lines (HA25a) (Vincent *et al.,* 1990), which contains integrated copies of both the AAV *Rep* and *Cap* coding regions, was capable of generating recombinant stocks with titers of 10^3–10^4. The low virus titers produced were apparently due to the low copy number of the AAV gene sequences. In another case, a 293 cell line was made in which the *Rep* gene was placed under the inducible control of the metallothionein promoter (Trempe and Yang, 1993). Although the level of inducible *Rep* expression from this line was shown to complement rep$^-$ AAV replication, it is unclear whether any increase in recombinant viral titers will be possible. Recently another *Rep* cell line was established in which the *Rep* gene was placed under control of a minimal promoter sequence derived from the human cytomegalovirus promoter IE complexed with control elements of the tetracycline resistance operon (Ferrari and Samulski, unpublished). Although this cell line has also been shown to complement AAV replication, its ability to drive recombinant AAV production has yet to be determined.

Although no breakthroughs leading to an AAV packaging cell line have been realized from this work, several important phenomenon relating to AAV biology have been observed. The first has been the ability to uncouple expression of the *Rep* and *Cap* proteins and still maintain processing of the transcripts from each region and production of the different *Rep* and *Cap* products in ratios the same as those expressed during a wild-type lytic infection. The second is that the absolute levels of the *Rep, Cap,* and terminal repeat-containing DNA substrates present in an adenovirus infected cell do not appear to effect the efficiency of recombinant AAV packaging to the extent initially thought. If the relatively low amount of recombinant virus produced by plasmid transfection were due to the low levels of *Rep* and *Cap* gene products produced from these templates, then any increase in these levels should have resulted in a coordinate increase in recombinant virus titer. However, this has not been observed. It appears that additional unidentified elements may be required for the efficient packaging of the recombinant DNAs into AAV virions. Identification of these elements and the development of a new procedure for generating recombinant viral stocks of higher titer will be imperative for subsequent animal and human studies using AAV as a vector for gene therapy.

E. Gene Transduction and Gene Therapy

The development of gene transfer vectors from AAV has provided an efficient and effective way of delivering genes into mammalian cells. AAV possesses several unique properties which distinguish it from other vectors. Its advantages include stable and efficient integration of viral DNA into the host genome (Berns *et al.*, 1975; Cheung *et al.*, 1980; Hoggan *et al.*, 1972; Laughlin *et al.*, 1986; McLaughlin *et al.*, 1988), lack of any associated human disease (Berns *et al.*, 1982), a broad host range (Buller *et al.*, 1979; Casto *et al.*, 1967), the ability to infect growth arrested cells (Wong *et al.*, 1993), and the ability to carry nonviral regulatory sequences without interference from the viral genome (Miller *et al.*, 1993; Walsh *et al.*, 1992). In addition, there has not been any superinfection immunity associated with AAV vectors (Lebkowski *et al.*, 1988; McLaughlin *et al.*, 1988).

References

Antoni, B. A., Rabson, A. B., *et al.* (1991). Adeno-associated virus Rep protein inhibits human immunodeficiency virus type I production in human cells. *J. Virol.* **65**, 396–404.

Ashktorab, H., and Srivastava, A. (1989). Identification of nuclear proteins that specifically interact with the adeno-associated virus 2 inverted terminal repeat hairpin DNA. *J. Virol.* **63**: 3034–3039.

Atchison, R. W., Casto, B. C., *et al.* (1965). Adenovirus-associated defective virus particles. *Science* **149**, 754–756.

Bantel-Schall, U., and zur Hausen, H. (1988a). Adeno-associated viruses inhibit SV40 DNA amplification and replication of Herpes simplex virus in SV40-transformed hamster cells. *Virology* **164**, 64–74.

Bantel-Schall, U., and zur Hausen, H. (1988b). Dissociation of carcinogen-induced amplification and amplification of AAV DNA in a Chinese hamster cell line. *Virology* **166**, 113–122.

Becerra, S. P., Rose, J. A., *et al.* (1985). Direct mapping of adeno-associated virus proteins B and C: a possible ACG initiation codon. *Proc. Natl. Acad. Sci. USA* **82**, 7919–7923.

Berns, K. I., and Adler, S. (1972). Separation of two types of adeno-associated virus particles containing complementary polynucleotide chains. *J. Virol.* **5**, 693–699.

Berns, K. I., Cheung, A., *et al.* (1982). Adeno-associated Virus Latent Infection. In "Virus Persistence" (Mahy, B. W. J., Minson, A. C., *et al.*, Eds.), p. 249. Cambridge University Press, Cambridge.

Berns, K. I., Kotin, R. M., *et al.* (1988). Regulation of adeno-associated virus DNA replication. *Biochem. Biophys. Acta.* **951**, 425–429.

Berns, K. I., Pinkerton, T. C., *et al.* (1975). Detection of adeno-associated virus (AAV)-specific nucleotide sequences in DNA isolated from latently infected Detroit 6 cells. *Virology* **68**, 556–560.

Berns, K. I., and Rose, J. A. (1970). Evidence for a single-stranded adeno-associated virus genome: Isolation and separation of complementary single strands. *J. Virol.* **5**, 693–699.

Buller, R. M., Janik, J. E., *et al.* (1981). Herpes simplex virus types I and 2 completely help adenovirus-associated virus replication. *J. Virol.* **40**, 241–247.

Buller, R. M., Straus, S. E., *et al.* (1979). Mechanism of host restriction of adenovirus-associated virus replication in African green monkey kidney cells. *J. Gen. Virol.* **43**, 663–672.

Carter, B. J., Marcus-Sekura, C. J., *et al.* (1983). Properties of an adenovirus type 2 mutant, Addl807, having a deletion near the right-hand genome terminus: Failure to help AAV replication. *Virology* **126**, 505–516.

Casto, B. C., Armstrong, J. A., *et al.* (1967). Studies on the relationship between adeno-associated virus type I (AAV-I) and adenoviruses. II. Inhibition of adenovirus plaques by AAV: Its nature and specificity. *Virology* **33**, 452–458.

Chang, L.-S., Shi, Y., *et al.* (1989). Adeno-associated virus p5 promoter contains an adenovirus EIA inducible element and a binding site for the major late transcription factor. *J. Virol.* **63**, 3479–3488.

Cheung, A. K., Hoggan, M. D., *et al.* (1980). Integration of the adeno-associated virus genome into cellular DNA in latently infected human Detroit 6 cells. *J. Virol.* **33**, 739–748.

Chiorini, J. A., Weitzman, M. D., *et al.* (1994). Biologically active Rep proteins of adeno-associated virus type 2 produced as fusion proteins in *Escherichia coli*. *J. Virol.* **68**, 797–804.

Flint, J., and Shenk, T. (1989). Adenovirus EIA protein paradigm viral transactivator. *Annu. Rev. Genet.* **23**, 141–161.

Gerry, H. W., Kelly, T. J. J., *et al.* (1973). Arrangement of nucleotide sequences in adeno-associated virus DNA. *J. Mol. Biol.* **79**, 207–225.

Goodman, S., Xiao, X., *et al.* (1994). Recombinant adeno-associated virus-mediated gene transfer into hematopoietic progenitor cells. *Blood* **84**(5), 1492–1500.

Gottlieb, J., and Muzyczka, N. (1988). In vitro excision of adeno-associated virus DNA from recombinant plasmids: Isolation of an enzyme fraction from HeLa cells that cleaves DNA at poly(G) sequences. *Mol. Cell. Biol.* **6**, 2513–2522.

Gottlieb, J., and Muzyczka, N. (1990a). Purification and characterization of HeLa endonuclease R: A G-specific mammalian endonuclease. *J. Biol. Chem.* **265**, 10836–10841.

Gottlieb, J., and Muzyczka, N. (1990b). Substrate specificity of HeLa endonuclease R: A G-specific mammalian endonuclease. *J. Biol. Chem.* **265**, 10842–10850.

Green, M. R., and Roeder, R. G. (1980a). Definition of a novel promoter for the major adenovirus-associated virus mRNA. *Cell* **1**, 231–242.

Green, M. R., and Roeder, R. G. (1980b). Transcripts of the adeno-associated virus genome: Mapping of the major RNAs. *J. Virol.* **36**, 79–92.

Green, M. R., Straus, S. E., *et al.* (1980). Transcripts of the adenovirus-associated virus genome: Multiple polyadenylated RNAs including a potential primary transcript. *J. Virol.* **35**, 560–565.

Grossman, Z., Berns, K. I., *et al.* (1985). Structure of simian virus 40-adeno-associated virus recombinant genomes. *J. Virol.* **56**, 457–465.

Grossman, Z., Winocour, E., *et al.* (1984). Recombination between simian virus 40 and adeno-associated virus: Virion coinfection compared to DNA cotransfection. *Virology* **134**, 125–137.

Handa, H., Shiroki, K., *et al.* (1977). Establishment and characterization of KB cell lines latently infected with adeno-associated virus type 1. *Virology* **82**, 84–92.

Hermonat, P. L., Labow, M. A., *et al.* (1984). Genetics of adeno-associated virus: Isolation and preliminary characterization of adeno-associated virus type 2 mutants. *J. Virol.* **51**, 329–333.

Hermonat, P. L., and Muzyczka, N. (1984). Use of adeno-associated virus as a mammalian DNA cloning vector: Transduction of neomycin resistance into mammalian tissue culture cells. *Proc. Natl. Acad. Sci. USA* **81**, 6466–6470.

Hoggan, M. D. (1970). Adeno-associated viruses. *Prog. Med. Virol.* **12**, 211–239.

Hoggan, M. D., Blacklow, N. R., *et al.* (1966). Studies of small DNA viruses found in various adenovirus preparations: Physical, biological, and immunological characteristics. *Proc. Natl. Acad. Sci. USA* **55**, 1457–1471.

Hoggan, M. D., Thomas, G. F., *et al.* (1972). Continuous carriage of adenovirus associated virus genome in cell culture in the absence of helper adenovirus. In "Proceedings of the Fourth Lepetite Colloquium," Cocoyac, Mexico.

Huang, M., and Hearing, P. (1989). Adenovirus early region 4 encodes two gene products with redundant effects in lytic infection. *J. Virol.* **63**, 2605–2615.

Ikegami, S., Taguchi, T., *et al.* (1979). Aphidocolin prevents mitotic cell division by interfering with the activity of DNA polymerase-α. *Nature* **275**, 458–459.

Im, D.-S., and Muzyczka, N. (1989). Factors that bind to the AAV terminal repeats. *J. Virol.* **63**, 3095–3104.

Im, D.-S., and Muzyczka, N. (1990). The AAV origin binding protein Rep68 is an ATP-dependent site-specific endonuclease with DNA helicase activity. *Cell* **61**, 447–457.

Janik, I. E., Huston, M. M., *et al.* (1984). Adeno-associated virus proteins: Origin of the capsid components. *J. Virol.* **52**, 591–597.

Janik, J. E., Huston, M. M., *et al.* (1989). Efficient synthesis of adeno-associated virus structural proteins requires both adenovirus DNA binding protein and VA I RNA. *Virology* **168**, 320–329.

Janik, J. E., Huston, M. M., *et al.* (1981). Locations of adenovirus genes required for the replication of adenovirus-associated virus. *Proc. Natl. Acad. Sci. USA* **78**, 1925–1929.

Jay, F. T., De La Maza, L. M., *et al.* (1979). Parvovirus RNA transcripts containing sequences not present in mature mRNA: A method for isolation of putative mRNA precursor sequences. *Proc. Natl. Acad. Sci. USA* **76**, 625–629.

Jay, F. T., Laughlin, C. A., *et al.* (1981). Eukaryotic translational control: Adeno-associated virus protein synthesis is arrested by a mutation in the adenovirus DNA-binding protein. *Proc. Natl. Acad. Sci. USA* **78**, 2927–2931.

Johnson, F. B., Ozer, H. L., *et al.* (1971). Structural proteins of adenovirus-associated virus type 3. *J. Virol.* **8**, 860–863.

Johnson, F. B., Thomson, T. A., *et al.* (1977). Molecular similarities among the adenovirus-associated virus polypeptides and evidence for a precursor protein. *Virology* **82**, 1–13.

Johnson, F. B., Whitaker, C. W., *et al.* (1975). Structural polypeptides of adenovirus-associated virus top component. *Virology* **65**, 196–203.

Kaplitt, M. G., Leone, P., *et al.* (1994). Adeno-associated virus vectors yield safe delivery and long term expression of potentially therapeutic genes into the mammalian brain. *Nature Genet.* **8**, 148–154.

Koczot, F. J., Carter, B. J., *et al.* (1973). Self-complementarity of terminal sequences within plus or minus strands of adenovirus-associated virus DNA. *Proc. Natl. Acad. Sci. USA* **70**, 215–219.

Kotin, R. M. (1994). Prospects for the use of adeno-associated virus as a vector for human gene therapy. *Hum. Gene Ther.* **5**, 793–801.

Kotin, R. M., and Berns, K. I. (1989). Organization of adeno-associated virus DNA in latently infected Detroit 6 cells. *Virology* **170**, 460–467.

Kotin, R. M., Menninger, J. C., *et al.* (1991). Mapping and direct visualization of a region-specific viral DNA integration site on chromosome 19q 13-qter. *Genomics* **10**, 831–834.

Kotin, R. M., Siniscalco, M., *et al.* (1990). Site-specific integration by adeno-associated virus. *Proc. Natl. Acad. Sci. USA* **87,** 2211–2215.

Labow, M. A., Graf, L. H., *et al.* (1987). Adeno-associated virus gene expression inhibits cellular transformation by heterologous genes. *Mol. Cell. Biol.* **7,** 1320–1325.

Labow, M. A., Hermonat, P. L., *et al.* (1986). Positive and negative autoregulation or the adeno-associated virus type 2 genome. *J. Virol.* **60,** 251–258.

Laughlin, C. A., Cardellichio, C. B., *et al.* (1986). Latent infection of KB cells with adeno-associated virus type 2. *J. Virol.* **60,** 515–524.

Laughlin, C. A., Jones, N., *et al.* (1982). Effects of deletions in adenovirus region I genes upon replication of adeno-associated virus. *J. Virol.* **41,** 868–876.

Laughlin, C. A., Tratschin, J.-D., *et al.* (1983). Cloning of infectious adeno-associated virus genomes in bacterial plasmids. *Gene* **23,** 65–73.

Laughlin, C. A., Westphal, H., *et al.* (1979). Spliced adenovirus-associated virus RNA. *Proc. Natl. Acad. Sci. USA* **76,** 5567–5571.

Lebkowski, J. S., McNally, M. M., *et al.* (1988). Adeno-associated virus: A vector system for efficient introduction of DNA into a variety of mammalian cell types. *Mol. Cell. Biol.* **8,** 3988–3996.

Lusby, E., and Berns, K. I. (1982). Mapping of the 5′ termini of two adeno-associated virus 2 RNAs in the left half of the genome. *J. Virol.* **41,** 518–526.

Lusby, E., Fife, K. H., *et al.* (1980). Nucleotide sequence of the inverted terminal repetition in adeno-associated virus DNA. *J. Virol.* **34,** 402–409.

Mayor, H. D., Torikai, K., *et al.* (1969). Plus and minus single-stranded DNA separately encapsidated in adeno-associated satellite virions. *Science* **166,** 1280–1282.

McLaughlin, S. K., Collis, P., *et al.* (1988). Adeno-associated virus general transduction vectors: Analysis of proviral structures. *J. Virol.* **62,** 1963–1973.

McPherson, R. A., Ginsburg, H. S., *et al.* (1982). Adeno-associated virus helper activity of adenovirus DNA binding protein. *J. Virol.* **44,** 666–673.

McPherson, R. A., Rosenthal, L. J., *et al.* (1985). Human cytomegalovirus completely helps adeno-associated virus replication. *Virology* **147,** 217–222.

Melnick, J. L., Mayor, H. D., *et al.* (1965). Association of 20 millimicron particles with adenoviruses. *J. Bacteriol.* **90,** 271–274.

Mendelson, E., Smith, M. G., *et al.* (1988). Effect of a viral rep gene on transformation of cells by an adeno-associated virus vector. *Virology* **166,** 612–615.

Mendelson, E., Trempe, J. P., *et al.* (1986). Identification of the trans-active rep proteins of adeno-associated virus by antibodies to a synthetic oligopeptide. *J. Virol.* **60,** 823–832.

Miller, D. G., Adam, M. A., *et al.* (1990). *Mol. Cell. Biol.* **10,** 4239–4242.

Miller, J. L., Donahue, R. E., *et al.* (1994). Recombinant adeno-associated virus (rAAV) mediated expression of a human γ-globin gene in human progenitor derived erythroid cells. *Proc. Natl. Acad. Sci. USA,* **91,** 10,183–10,187.

Miller, J. L., Walsh, C. E., *et al.* (1993). Single-copy transduction and expression of human gamma-globin in K562 erythroleukemia cells using recombinant adeno-associated virus vectors: The effect of mutations in NF-E2 and GATA-1 binding motifs within the hypersensitivity site 2 enhancer. *Blood* **82,** 1900–1906.

Muzyczka, M. (1992). Use of adeno-associated virus as a general transduction vector for mammalian cells. *Curr. Top. Microbiol. Immunol.* **158,** 97–129.

Myers, M. W., Laughlin, C. A., *et al.* (1980). Adenovirus helper function for growth of adeno-associated virus: Effect of temperature-sensitive mutations in adenovirus early gene region 2. *J. Virol.* **35,** 65–75.

Ni, T.-H., Zhou, X., *et al.* (1994). In vitro replication of adeno-associated virus DNA. *J. Virol.* **68,** 1128–1138.

Ostrove, J., and Berns, K. I. (1980). Adenovirus early region IB gene function required for rescue of latent adeno-associated virus. *Virology* **104,** 502–506.

Owens, R. A., Weitzman, M. D., *et al.* (1993). Identification of a DNA-binding domain in the amino terminus of adeno-associated virus Rep protein. *J. Virol.* **67**(2), 997–1005.
Richardson, W. D., and Westphal, W. D. (1981). A cascade of adenovirus early functions is required for expression of adeno-associated virus. *Cell* **27**, 133–141.
Richardson, W. D., and Westphal, W. D. (1984). Requirement for either early region 1A or early region 1B adenovirus gene products in the helper effect for adeno-associated virus. *J. Virol.* **51**, 404–410.
Rose, J. A., Berns, K. I., *et al.* (1969). Evidence for a single-stranded adenovirus-associated virus genome: Formation of a DNA density hybrid on release of viral DNA. *Proc. Natl. Acad. Sci. USA* **64**, 863–869.
Rose, J. A., Maizel, J. K., *et al.* (1971). Structural proteins of adenovirus-associated viruses. *J. Virol.* **8**, 766–770.
Russell, D. W., Miller, A. D., *et al.* (1994). Adeno-associated virus vectors preferentially transduce cells in S phase. *Proc. Natl. Acad. Sci. USA* **91**, 8915–8919.
Samulski, R. J., Berns, K. I., *et al.* (1982). Cloning of adeno-associated virus into pBR322: Rescue of intact virus from the recombinant plasmid in human cells. *Proc. Natl. Acad. Sci. USA* **79**, 2077–2081.
Samulski, R. J., Chang, L.-S., *et al.* (1987). A recombinant plasmid from which an infectious adeno-associated virus genome can be excised in vitro and its use to study viral replication. *J. Virol.* **61**, 3096–3101.
Samulski, R. J., Chang, L.-S., *et al.* (1989). Helper-free stocks of recombinant adeno-associated viruses: Normal integration does not require viral gene expression. *J. Virol.* **63**, 3822–3828.
Samulski, R. J., and Shenk, T. (1988). Adenovirus EIB 55-M, polypeptide facilitates timely cytoplasmic accumulation of adeno-associated virus mRNAs. *J. Virol.* **62**, 206–210.
Samulski, R. J., Srivastava, A., *et al.* (1983). Rescue of adeno-associated virus from recombinant plasmids: Gene correction within the terminal repeats of AAV. *Cell* **33**, 135–143.
Samulski, R. J., Zhu, X., *et al.* (1991). Targeted integration of adeno-associated virus (AAV) into human chromosome 19. *EMBO J.* **10**, 3941–3950.
Schlehofer, J. R., Ehrbar, M., *et al.* (1986). Vaccinia virus, herpes simplex virus, and carcinogens induce DNA amplification in a human cell line and support replication of a helper virus dependent parvovirus. *Virology* **152**, 110–117.
Senepathy, P., Tratschin, J.-D., *et al.* (1984). Replication of adeno-associated virus DNA: Complementation of naturally occurring rep$^-$ mutants by a wild-type genome or an ori$^-$ mutant and correction of terminal deletions. *J. Mol. Biol.* **179**, 1–20.
Snyder, R. O., Im, D.-S., *et al.* (1990). Evidence for covalent attachment of the adeno-associated virus Rep protein to the ends of the AAV genome. *J. Virol.* **64**, 6204–6213.
Srivastava, A., Lusby, E. W., *et al.* (1983). Nucleotide sequence and organization of the adeno-associated virus 2 genome. *J. Virol.* **45**, 555–564.
Srivastava, C. H., Samulski, R. J., *et al.* (1989). Construction of a recombinant human parvovirus B19: Adeno-associated virus 2 (AAV) DNA inverted terminal repeats are functional in an AAV–B19 hybrid virus. *Proc. Natl. Acad. Sci. USA* **86**, 8078–8082.
Straus, S. E., Ginsburg, H. S., *et al.* (1976). DNA-minus temperature-sensitive mutants of adenovirus type 5 help adenovirus-associated virus replication. *J. Virol.* **17**, 140–148.
Tratschin, J.-D., Miller, I. L., *et al.* (1984). Genetic analysis of adeno-associated virus: Properties of deletion mutants constructed in vitro and evidence for an adeno-associated virus replication function. *J. Virol.* **51**, 611–619.
Tratschin, J.-D., Miller, I. L., *et al.* (1985). Adeno-associated virus vector for high-frequency integration, expression, and rescue of genes in mammalian cells. *Mol. Cell. Biol.* **5**, 3251–3260.
Tratschin, J.-D., Tal, J., *et al.* (1986). Negative and positive regulation in trans of gene expression from adeno-associated virus vectors in mammalian cells by a viral Rep gene product. *Mol. Cell. Biol.* **6**, 2884–2894.

Tratschin, J.-D., West, M. H. P., *et al.* (1984). A human parvovirus, adeno-associated virus, as a eukaryotic vector: Transient expression and encapsidation of the prokaryotic gene for chloramphenicol acetyltransferase. *Mol. Cell. Biol.* **4**, 2072–2081.

Trempe, J. P., and Carter, B. J. (1988a). Alternate mRNA splicing is required for synthesis of adeno-associated virus VPI capsid protein. *J. Virol.* **62**, 3356–3363.

Trempe, J. P., and Carter, B. J. (1988b). Regulation of adeno-associated virus gene expression in 293 cells: Control of mRNA abundance and translation. *J. Virol.* **62**, 68–74.

Trempe, J. P., Mendelson, E., *et al.* (1987). Characterization of adeno-associated virus rep proteins in human cells by antibodies raised against rep expressed in *Escherichia coli. Virology* **161**, 18–28.

Trempe, J. P., and Yang, Q. (1993). Characterization of a cell line that expresses the AAV replication proteins. In "Proceedings, Fifth Parvovirus Workshop," Crystal River, FL. [Abstract]

Tsao, J., Chapman, M. S., *et al.* (1991). The three-dimensional structure of canine parvovirus and its functional implications. *Science* **25**, 1456–1464.

Vincent, K. A., Moore, G. K., *et al.* (1990). Replication and packaging of HIV envelope genes in a novel adeno-associated virus vector system. *Vaccine* **90**, 353–359.

Walsh, C. E., Liu, J. M., *et al.* (1992). Regulated high level expression of a human γ-globin gene introduced into erythroid cells by an adeno-associated virus vector. *Proc. Natl. Acad. Sci. USA* **89**, 7257–7261.

West, M. H. P., Trempe, J. P., *et al.* (1987). Gene expression in adeno-associated virus vectors: The effects of chimeric mRNA structure, helper virus, and adenovirus VAI RNA. *Virology* **160**, 38–47.

Winocour, E., Puzis, L., *et al.* (1992). Modulation of the cellular phenotype by integrated adeno-associated virus. *Virology* **190**(1), 316–329.

Wong, K. K., Podsakoff, G., *et al.* (1993). High efficiency gene transfer into growth arrested cells utilizing an adeno-associated virus (AAV)-based vector. In "Proceedings, American Society of Hematology," St. Louis, MO. [Abstract]

Yacobson, B., Hrynko, T. A., *et al.* (1989). Replication of adeno-associated virus in cells irradiated with UV light at 254 nm. *J. Virol.* **63**, 1023–1030.

Yacobson, B., Koch, T., *et al.* (1987). Replication of adeno-associated virus in synchronized cells without the addition of helper virus. *J. Virol.* **61**, 972–981.

Yalkinoglu, A. O., Heilbronn, R., *et al.* (1988). DNA Amplification of adeno-associated virus as a response to cellular genotoxic stress. *Cancer Res.* **48**, 3123–3125.

Zhou, S. Z., Broxmeyer, H. E., *et al.* (1993). Adeno-associated virus 2-mediated gene transfer in murine hematopoietic progenitor cells. *Exp. Hematol.* **21**(7), 928–933.

C H A P T E R

Genetics and Biology of Retroviral Vectors

Rajat Bannerji
Department of Molecular Biology
Memorial–Sloan Kettering Cancer Center
New York, New York

In this chapter, retroviral vectors, based on the Moloney murine leukemia virus, are discussed as a method for delivering genes into target cells. This vector system offers the advantages of highly efficient gene transfer and stable gene expression in the target cell by virtue of the integration of the vector genome into the host chromosomal DNA. Other vector systems including adenovirus vectors, vaccinia virus vectors, and nonviral methods, such as liposomes, may allow only transient gene expression (Mulligan, 1993). Adeno-associated virus vectors do have the ability to stabily integrate into the host genome and will be discussed elsewhere (see Samulski, this volume). Helper-free retroviral vectors infect only once and do not spread in vivo, in contrast to other viral vector methods which employ attenuated viruses that may retain the ability to infect other cells. Retroviral vectors have a wide host range, both in terms of species and cell types (Hartley and Rowe, 1976; Rasheed *et al.*, 1976; Luciw and Leung, 1992), but are limited to infecting only dividing cells (Varmus *et al.*, 1977; Fritsch and Temin, 1977; Humphries *et al.*, 1981). Overall, retroviral vectors offer an excellent gene delivery system available for stable gene transfer and expression.

Retroviruses are single-stranded RNA viruses encoding a characteristic RNA-dependent DNA polymerase (reverse transcriptase, see Baltimore, 1970; Temin and Mizutani, 1970). They initially were identified as RNA tumor viruses by Peyton Rous, who was working at the Rockefeller Institute on a spontaneous chicken sarcoma (Rous, 1911). Since the discovery of the Rous sarcoma virus (RSV), many tumor causing retroviruses have been found in a variety of vertebrates including reptiles, birds, and mammals. These viruses have been shown to transform cells either by expression of viral oncogenic sequences originally transduced from host-cell genomes or by integration near

VIRAL VECTORS

a cellular oncogene and its subsequent activation (Bishop and Varmus, 1984). Retroviruses may also be pathogenic by nononcogenic mechanisms as in the case of acquired immunodeficiency syndrome in humans caused by the human immunodeficiency virus (Fauci, 1988).

Retroviruses which contain oncogenic sequences lead to a rapid induction of tumor formation and are termed acutely transforming retroviruses. For example, the Harvey sarcoma virus containing the Ha-*ras* oncogene can induce a sarcoma within 2 weeks after infection of a newborn mouse (Kozak and Ruscetti, 1992). These viruses are the result of nonhomologous recombination events between replication-competent retroviruses and cellular proto-oncogenes. The Harvey sarcoma virus is a recombinant of the Moloney murine leukemia virus and the Ha-*ras* proto-oncogene (Scolnick and Parks, 1974). The rate of nonhomologous recombination in retroviruses was recently calculated to be 5×10^{-5} per replication cycle (Zhang and Temin, 1993). Recombination usually results in a replication-defective acutely transforming virus which has lost some or all of its viral genes and gained the transforming oncogene. Throughout the 1970s and early 1980s many oncogenes were discovered in the context of acutely transforming retroviruses including *mos* (Scolnick *et al.*, 1975), Ki-*ras* (Scolnick *et al.*, 1976), *abl* (Goff *et al.*, 1980), *fos* (Curran *et al.*, 1982), *raf* (Rapp *et al.*, 1983), and others.

Integration of a retrovirus near a cellular proto-oncogene may drive its transcription under the influence of viral promoter or enhancer elements. In this case, disease is the result of a relatively rare integration event and is reflected in a relatively long lag time between infection and tumor formation. As an example, thymic lymphomas are seen only after 3 months following infection of newborn mice with the nonacute transforming Moloney murine leukemia virus (Moloney, 1960a,b). Study of the integration sites of these viruses has resulted in the discovery of additional oncogenes such as *myc* (Hayward *et al.*, 1981), *wnt-1* (Nusse and Varmus, 1982), *myb* (Shen-Ong *et al.*, 1984), and *lck* (Voronova and Sefton, 1986) among others. Viral integration may also lead to inactivation of tumor-suppressor genes as in the case of *p53* (Mowat *et al.*, 1985). The nonacute transforming viruses are replication-competent retroviruses with an intact viral genome.

A third mechanism of transformation may be occuring in certain human retroviruses. Human retroviruses were first discovered by Gallo and co-workers in cutaneous T-cell lymphomas/leukemias and named human T-cell leukemia virus-1 (HTLV-I; Poiesz *et al.*, 1980) followed by the discovery of HTLV-II in a T-cell type hairy cell leukemia (Kalyanaraman *et al.*, 1982). HTLV-I and HTLV-II do not contain oncogenes nor do they integrate near any cellular proto-oncogenes in the host cell genome. Transcriptional activation of host genes by virally encoded proteins have been postulated as the mechanism of transformation (Yoshida and Seiki, 1990; Cann *et al.*, 1990).

I. Genomic Organization and Life Cycle of Retroviruses

This discussion will use the C-type Molony murine leukemia virus (MoMLV) (Moloney, 1960a,b) as the prototypic retrovirus, since it is the source of the retroviral vectors described in detail later in this chapter. The structure of the proviral DNA is shown in Fig. 1A. The genome is 9 kb long (Gilboa, 1979a) and contains two long terminal repeats (LTR) made up of U3, R, and U5 segments. The unique 3′ region or U3 contains *cis*-acting signals for integration, viral promoter, and viral enhancer signals. U3 is 449 bases long in MoMLV. R is a short sequence, of 68 bases, repeated at both ends of the viral genome. R is crucial to the process of reverse transcription and, in addition, R contains the polyadenylation signal. U5 is the unique 5′ sequence (76 bases long) and contains the binding site for the tRNA primer required for initiation of viral DNA synthesis by the enzyme reverse transcriptase (Coffin, 1984; Luciw and Leung, 1992). The 5′ LTR is followed by the untranslated packaging signal sequence essential for packaging of the genomic RNA into virions (Mann *et al.,* 1983). The viral *gag* gene encodes the internal structural proteins of the virion, the *pol* gene encodes reverse transcriptase and integrase (IN), and the *env* gene encodes the virion envelope proteins (Coffin, 1984; Varmus and Swanstrom, 1984, and Dickson et al., 1984). The promoter in the U3 region of the 5′ LTR initiates transcription by cellular RNA *Pol*II of full-length viral RNAs and singly spliced viral RNAs containing 5′-terminal m^7Gppp caps and 3′ poly(A) tails. The full-length mRNA encodes the *gag* polypeptide and the *gag–pol* fusion protein. The latter is formed by an occasional translational frameshift that regulates the ratio of *gag* to *pol* products (Yoshinaka *et al.,* 1985; Jacks and Varmus, 1985; Hatfield *et al.,* 1992). The full-length transcript is also the source of the viral genomic RNA packaged into virions. The spliced transcript is translated to produce the *env* polypeptide.

Proteolytic cleavage of the three precursor polypeptides by a protease encoded in the *pol* gene generates the 8 to 10 mature viral proteins including the viral protease, reverse transcriptase, and integrase enzymes as well as the matrix, capsid, nucleocapsid, surface, and transmembrane structural proteins (Miller, 1992; Dickson *et al.,* 1984). The viral proteins assemble to form an icosahedral virion core containing two copies of the single-stranded RNA genome (Kung *et al.,* 1976; Bender and Davidson, 1976: Bender *et al.,* 1978). The virion buds from the cell surface surrounded by a lipid membrane studded with the envelope proteins (Varmus and Swanstrom, 1984). The life cycle of the virus is shown in Fig. 1B. The virion's host range is determined by the presence or absence of cell surface receptors on the target cell compatible with the envelope glycoproteins present on the surface of the virus (Albritton *et al.,* 1989). The tropism of MLV

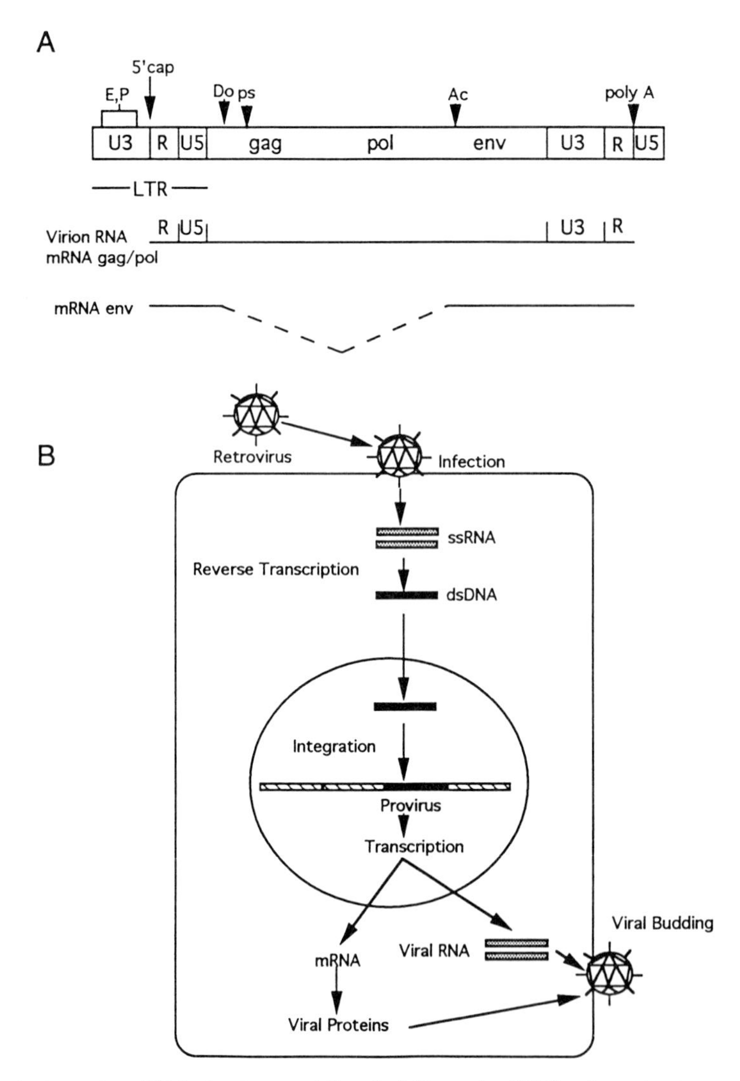

Figure 1 Retroviral provirus DNA structure and the viral life cycle. (A) Provirus structure (see text for a discussion of the elements of the retroviral genome). (B) Viral life cycle. The virion contains two copies of the ssRNA viral genome. After infection of a susceptible cell, the RNA is reverse transcribed into dsDNA which then migrates into the cell nucleus and integrates into a host chromosome to form the provirus. The integrated provirus provides the template for transcription of viral mRNAs and viral RNA. Viral proteins translated from the viral mRNAs form new virions containing two copies of the viral RNA. These virions then bud from the host cell to infect a new target cell.

strains can be xenotropic, in which the virus is unable to infect mouse strains but can infect a wide range of other species; ecotropic, in which viral infection is restricted to mouse cells; or amphotropic, in which viruses can infect both murine and nonmurine cells (Weiss, 1984). Productively infected cells also can be resistent to superinfection by the same viral strain by a mechanism in which envelope glycoproteins bind to the cellular virus receptor in the endoplasmic reticulum and thus abrogate superinfection (Delwart and Panganiban, 1989).

Upon infection, the single-stranded viral RNA is copied into a double-stranded DNA by reverse transcriptase carried within the virion particle (for a detailed understanding of this mechanism see Gilboa, 1979b). This linear, double-stranded DNA is transported into the nucleus as the precursor to the integrated proviral form of the viral DNA (Brown *et al.*, 1987). The virion-supplied IN protein catalyzes removal of two nucleotides from the 3′ ends of the linear viral DNA, generation of staggered nicks in the host DNA, and integrative recombination resulting in proviral insertion with a short, characteristic duplication of host target sequences at the site of integration (Grandgenett and Mumm, 1990). The site of integration is not entirely random because DNaseI-hypersensitive sites are clearly favored, suggesting a preference for integration into open chromatin domains (Rohdewohld *et al.*, 1987). The integrated provirus then expresses the viral genes and new virions are produced constitutively and released from the infected cell without any deleterious effects to the cell itself.

A large body of evidence indicates that viral replication, integration, and production of progeny virus depend on host cell progression through the cell cycle. Replication and integration are particularly dependent on cells being in S phase at the time of infection (Varmus and Swanstrom, 1984). The importance to viral integration of the host cell progressing through S phase was seen in early experiments using BrdU to density-label genomic DNA at the time of retroviral infection. In an unsychronized population of target cells retroviral DNA was only associated with BrdU-labeled cellular DNA. Since this label is only incorporated during DNA replication these results suggest that retroviral integration occured only in cells which had passed through S phase (Varmus *et al.*, 1977, 1979). More recent experiments have exploited a *lacZ* reporter gene in a retroviral vector to study viral integration and the host cell cycle. Hajihosseini *et al.* (1993) find only half the daughter cells of an infected target cell express the *lacZ* gene, indicating that viral integration occured after completion of S phase, since integration before S phase would have resulted in all the daughter cells carrying the *lacZ* gene.

II. Retroviruses as Gene Transfer Vectors

Although retroviruses were discovered as pathogenic agents, the vast majority of retroviruses do not cause disease nor do they kill the infected cell.

Their wide host range, ability to carry exogenous genes, and ability to integrate stably into an infected cell's genome make them ideal vectors for the transfer of "foreign" genes into target cells (Temin, 1986).

Retroviral vectors may be either replication-competent or replication-defective. RSV is an example of a replication-competent vector that carries the *src* oncogene in addition to viral genes. However, most applications require the vector not be spread after the initial infection of the target cell. Replication-defective vectors are limited to one cycle of infection (Miller, 1992). Wild-type virus encodes both *trans* and *cis* functions, whereby *trans* functions can be complemented by information on a separate DNA molecule while cis functions cannot. The viral *gag, pol,* and *env* genes encode diffusable proteins which act *in trans* and therefore can be deleted from the vector and replaced with exogenous genes. The missing functions can be supplied by another virus, which acts as a *helper*. Successful viral production and integration cannot occur without the *cis* functions which are present in the viral RNA. Examples of *cis* functions include the LTRs, packaging signal, primer binding site, and, if a spliced product is required, the splice donor and splice acceptor sites (Gilboa, 1986).

Virions containing replication-defective vectors are generated in *packaging* cell lines carrying helper provirus, shown in Fig. 2. The host range of the virions produced can be increased by replacing the *env* gene of ecotropic helper MoMLV with the *env* gene of the amphotropic virus 4070A (Miller *et al.*, 1985). To ensure that the progeny virions do not contain helper virus, the helper provirus must be modified to prevent its RNA from being packaged into virions. This was initially accomplished by deleting the packaging signal region of the helper virus (Mann *et al.*, 1983; Watanabe and Temin, 1983). Unfortunately, these packaging lines often release helper virus, apparently after a recombination event between the vector and the helper virus restores the packaging signal sequence to the helper virus (Miller *et al.*, 1986). Packaging lines in which the helper contains deletions of both packaging signal and the 3′ LTR (e.g., PA317; Miller and Buttimore, 1986) also occasionally lead to helper virus production, presumably after two recombination events. More recently, helper-free amphotropic and ecotropic MoMLV-derived packaging lines have been developed. These constructs employ a strategy in which the *gag* and *pol* genes are expressed on a single plasmid and the *env* gene is expressed on a second plasmid, essentially eliminating the production of helper virus via recombination (Markowitz *et al.*, 1988a,b; Danos and Mulligan, 1988).

The usefulness of retroviral vectors is enhanced by the inclusion of a dominant selectable marker, such as the gene for neomycin resistance, *neo* (Hwang and Gilboa, 1984; Joyner and Bernstein, 1983). Selection of cells with the drug G418 allows for the isolation of *neo*-resistant colonies which contain the retroviral vector. The efficiency of gene transfer is dependent upon the *titer* of the vector. The titer of a virus producing cell line is assayed by infecting

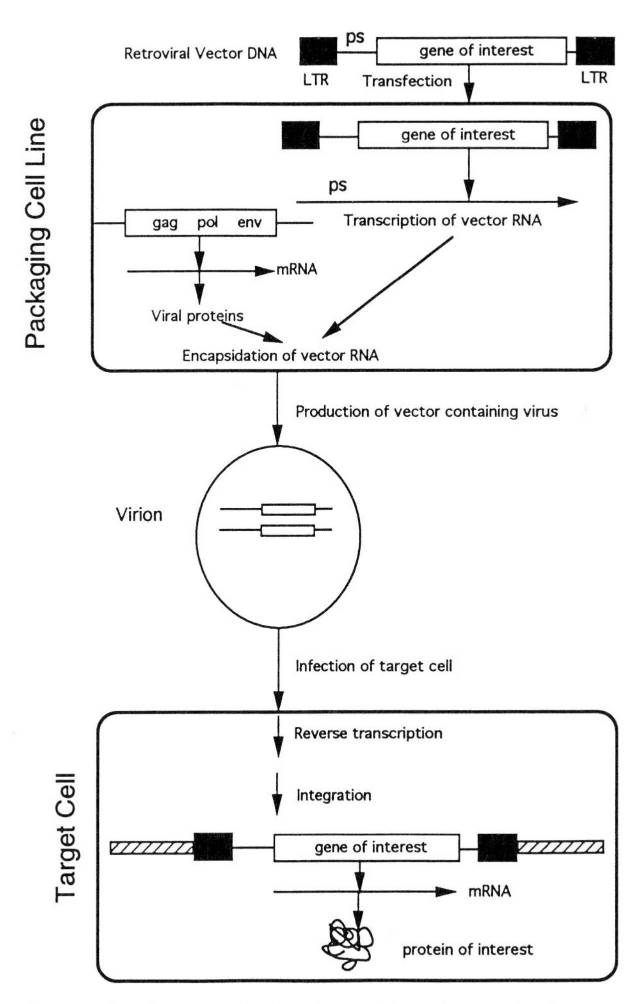

Figure 2 The use of packaging cell lines in retroviral-vector mediated gene transfer. The vector DNA contains the gene(s) of interest as well as the *cis* elements required for replication and packaging of the vector RNA into virions. The vector DNA lacks the viral genes encoding the viral proteins needed to form infectious virus. Therefore, the vector DNA is transfected into a packaging cell line which provides the viral proteins in trans. The packaging cell line normally does not produce infectious virus because it does not contain any packaging signals. When the packaging cell is transfected with a retroviral vector containing the packaging signal (ps), the vector RNA is packaged into virions and is secreted as vector-containing virus to be used to infect target cells. See text for details.

target cells with vector containing culture supernatant from the producer cell line and selecting for *neo*^{r} colonies. Viral titer is calculated as colony-forming units per milliliter of culture supernatant used for infection (cfu/ml). Virus

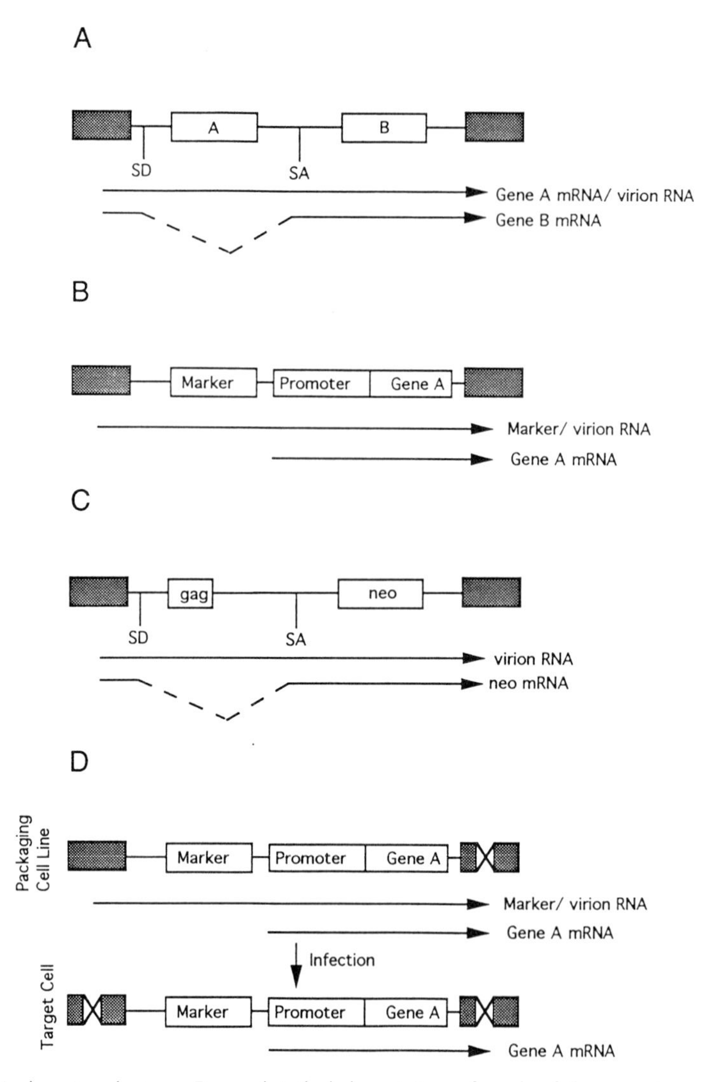

Figure 3 Retroviral vector designs. For a detailed description of each of these vectors see the text. (A) Double expression (DE) vector. (B) Vectors with internal promoters (VIP). (C) N2 vector. (D) Self-inactivating (SIN) vector showing the deletion of LTR promoter elements in the integrated provirus. (E) Double copy (DC) vector showing the duplication, in the integrated provirus 5′LTR, of the gene cloned into the vector's 3′ LTR. (F) DC-reverse vector.

titer can be influenced by vector size in which vectors too small or too large are not packaged efficiently into virions. Titer also can be influenced by *cis* elements. For example, MoMLV vectors containing packaging signal plus a 5′ portion of *gag* sequences (packaging signal^{+}) produce up to 200-fold higher titers than corresponding vectors with only the packaging signal (Adam and

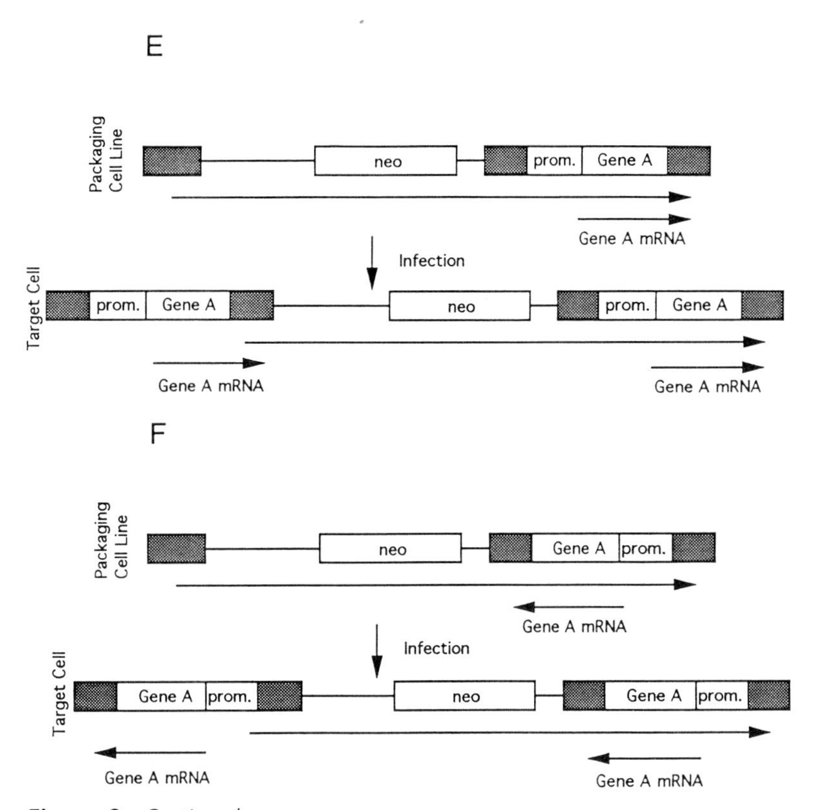

Figure 3 *Continued*

Miller, 1988; Armentano *et al.*, 1987; Bender *et al.*, 1987). Titers of MoMLV vectors have been reported to be as high as 1–2 × 10^7 cfu/ml, but are more often in the range of 0.5–1 × 10^6 cfu/ml when assayed on NIH-3T3 cells (Miller, 1992; Markowitz *et al.*, 1988a,b; Danos and Mulligan, 1988).

The utility of a vector depends not only on the number of cells infected, but also on the level of expression of the transduced gene. A variety of vector designs have been developed, as seen in Fig. 3. Double expression (DE) vectors (Fig. 3A) carry two genes, a selectable marker gene and the cloned gene of interest, both of which are driven by the 5′ LTR promoter (Cepko *et al.*, 1984). This design depends on proper splicing to occur because one of the genes can only be expressed as a spliced mRNA. LTR shut-off also limits the usefulness of this design (Gilboa, 1986). Vectors with internal promoters (VIP) express the selectable marker using the 5′ LTR promoter and the cDNA of interest from a promoter cloned 3′ to the marker gene (Fig. 3B). This design allows for the use of a variety of promoters to drive the cDNA; however, the choice of internal promoter may have a negative effect on viral titer, which only can be determined empirically (Emerman and Temin, 1984a,b). The VIP-type vector

N2 (Fig. 3C) contains the packaging signal sequence plus 418 bp of *gag* sequence fused to *neo* and produces 10- to 50-fold higher titers than similar vectors without the *gag* sequences (Armentano *et al.*, 1987).

Gene expression of the downstream cDNA may be inhibited in VIP vector designs because the internal promoter is within the LTR-driven transcription unit. Strong upstream promoters have been shown to interfere with transcription from downstream promoters (Cullen *et al.*, 1984; Kadesh and Berg, 1986; Proudfoot, 1986; Emerman and Temin, 1984a, 1986). Self-inactivating (SIN) vectors (Fig. 3D) eliminate LTR-driven transcription in the provirus allowing both increased expression by the internal promoter and a reduced risk of activating cellular oncogenes by the LTR-encoded enhancers (Yu *et al.*, 1986). This strategy takes advantage of the use, by reverse transcriptase, of the U3 region of the viral RNA as the template for U3 of both the 5′ and 3′ LTRs in the provirus. U3 contains both the promoter and enhancer functions of the LTR. A 299-bp deletion eliminates these functions from the U3 of the vector 3′ LTR, resulting in a provirus which contains the deletion in both the 5′ and 3′ LTRs (Yu *et al.*, 1986). The titer of SIN vectors, 10^3–10^4 cfu/ml, is too low for this design to be useful in gene transfer (Gilboa, 1986).

Double copy (DC) vectors (Fig. 3E) also attempt to bypass transcriptional interference from the LTR. In this design, the transduced cDNA and a promoter are cloned into U3 of the vector 3′ LTR (Hantzopoulos *et al.*, 1989). As in SIN vectors, the modification is copied into both the 5′ and 3′ LTRs of the provirus. The copy in the 5′ LTR is upstream of the retroviral transcriptional unit and thus avoids any transcriptional interference from the LTR (Hantzopoulos *et al.*, 1989). In addition, the 5′ gene transcript runs through the promoter/enhancer sequences of the 5′LTR U3 before reaching the poly(A) signal in the 5′LTR R segment. This may cause interference of the 5′ LTR driven viral transcript and may, therefore, indirectly relieve transcriptional interference of the downstream gene, potentially allowing for even higher gene expression. The DC-reverse design (Fig. 3F) prevents any interference between the 5′ gene transcript and the LTR-driven transcript. Highly successful modifications of the DC design include the use of RNA *Pol*III-driven tRNA promoters for the efficient expression of small RNA transcripts (Sullenger *et al.*, 1990a,b, 1991; Lee *et al.*, 1992) and the combination of two genes between the LTRs and one gene in the double copy position to express three different cDNAs (including the marker) in a single vector (Rosenthal *et al.*, 1993).

Retroviral vectors provide a highly efficient method to transfer and stabily express genes in target cells. Their limitations include difficulties in obtaining high titer producer lines, the time involved in establishing a stable clonal producer cell line, and gene inactivation due to a high rate of mutation per replication cycle. Improvements in vectorology continue to decrease the importance of these limitations. New producer lines allow transient transfection protocols requiring only 48–72 hr, resulting in high viral titers (Pear *et al.*,

1993). A recent study on the rate of gene inactivation in MoMLV calculated a rate of 7.7% which the authors suggest "is likely to be acceptable for many gene transfer protocols" (Varela-Echavarria *et al.*, 1993). The ubiquitous use of retroviral vectors and their continual improvement makes them a fundamental component of the repertoire of gene transfer technologies.

References

Adam, M. A., and Miller, A. D. (1988). Identification of a signal in a murine retrovirus that is sufficient for packaging of nonretroviral RNA into virions. *J. Virol.* **62,** 3802–3806.

Albritton, L. M., Tseng, L., Scadden, D., and Cunningham, J. M. (1989). A putative murine ecotropic retrovirus receptor gene encodes a multiple membrane-spanning protein and confers susceptibility to virus infection. *Cell* **57,** 659–666.

Armentano, D., Yu, S.-F., Kantoff, P. W., von Ruden, T., Anderson, W. F., and Gilboa, E. (1987). Effects of internal viral sequences on the utility of retroviral vectors. *J. Virol.* **61,** 1647–1650.

Baltimore, D. (1970). RNA-dependent DNA polymerase in virions of RNA tumor viruses. *Nature* **226,** 1209–1211.

Bender, W., and Davidson, N. (1976). Mapping of the poly(A) sequences in the electron microscope reveals unusual structure of type C oncornavirus RNA molecules. *Cell* **7,** 595–607.

Bender, W., Chien, Y. H., Chattopadhyay, S., Vogt, P. K., Gardner, M. B., and Davidson, N. (1978). High-molecular-weight RNAs of AKR, NZB, and wild mouse viruses and avian reticuloendotheliosis virus all have similar dimer structures. *J. Virol.* **25,** 888–896.

Bender, M. A., Palmer, T. D., Gelinas, R. E., and Miller, A. D. (1987). Evidence that the packaging signal of moloney murine leukemia virus extends into the gag region. *J. Virol.* **61,** 1639–1646.

Bishop, J. M., and Varmus, H. (1984). "Functions and Origins of Retroviral Transforming Genes: RNA Tumor Viruses." Cold Spring Harbor Laboratory, Cold Spring Harbor, NY.

Brown, P. O., Bowerman, B., Varmus, H. E., and Bishop, J. M. (1987). Correct integration of retroviral DNA in vitro. *Cell* **49,** 347–356.

Cann, A. J., Rosenblatt, J. D., Wachsman, W., and Chen, I. S. Y. (1990). Molecular biology and pathogenesis of HTLV-II. In "Retrovirus Biology and Human Disease" (Gallo and Wong, Eds.). Dekker, New York.

Cepko, C. L., Roberts, B. E., and Mulligan, R. C. (1984). Construction and applications of a highly transmissible murine retrovirus shuttle vector. *Cell* **37,** 1053–1062.

Coffin, J. (1984). Structure of the retroviral genome. In "RNA Tumor Viruses" (Furmanski, P., *et al.*, Eds.) Cold Spring Harbor Laboratory, Cold Spring Harbor, NY.

Cullen, B. R., Lomedico, P. T., and Ju, G. (1984). Transcriptional interference in avian retroviruses—Implications for the promoter insertion model of leukaemogenesis. *Nature* **307,** 241–245.

Curran, T., Peters, G., van Beveren, C., Teich, N. M., and Verma, I. M. (1982). FBJ murine osteosarcoma virus: Identification and molecular cloning of biologically active proviral DNA. *J. Virol.* **44,** 674.

Danos, O., and Mulligan, R. C. (1988). Safe and efficient generation of recombinant retroviruses with amphotropic and ecotropic host ranges. *Proc. Natl. Acad. Sci. USA* **85,** 6460–6464.

Delwart, E. L., and Panganiban, A. T. (1989). Role of reticuloendotheliosis virus envelope glycoprotein in superinfection interference. *J. Virol.* **63,** 273–280.

Dickson, C., Eisenman, R., Fan, H., Hunter, E., and Teich, N. (1984). Protein biosynthesis and assembly. In "RNA Tumor Viruses" (Furmanski, P., *et al.*, Eds.). Cold Spring Harbor Laboratory, Cold Spring Harbor, NY.

Emerman, M., and Temin, H. M. (1984a). Genes with promoters in retrovirus vectors can be independently suppressed by an epigenetic mechanism. *Cell* **39**, 459–467.

Emerman, M., and Temin, H. M. (1984b). High-frequency deletion in recovered retrovirus vectors containing exogenous DNA with promoters. *J. Virol.* **50**, 42–49.

Emerman, M., and Temin, H. M. (1986). Quantitative analysis of gene suppression in integrated retrovirus vectors. *Mol. Cell. Biol.* **6**, 792–800.

Fauci, A. S. (1988). The human immunodeficiency virus: Infectivity and mechanisms of pathogenesis. *Science* **239**, 617–622.

Fritsch, E., and Temin, H. M. (1977). Inhibition of viral DNA synthesis in stationary chicken embryo fibroblasts infected with avian retroviruses. *J. Virol.* **24**, 461–469.

Gilboa, E., Goff, S., Shields, A., Yoshimura, F., Mitra, S., and Baltimore, D. (1979a). In vitro synthesis of a 9kb terminally redundant DNA carrying the infectivity of Moloney murine leukemia virus. *Cell* **16**, 863–874.

Gilboa, E., Mitra, S. W., Goff, S., and Baltimore, D. (1979b). A detailed model of reverse transcription and tests of crucial aspects. *Cell* **18**, 93–100.

Gilboa, E. (1986). Retrovirus vectors and their uses in molecular biology. *BioEssays* **5**, 252–258.

Goff, S. P., Gilboa, E., Witte, O. N., and Baltimore, D. (1980). Structure of the Abelson murine leukemia virus genome and the homologous cellular gene: Studies with cloned viral DNA. *Cell* **22**, 777–785.

Grandgenett, D. P., and Mumm, S. R. (1990). Unraveling retrovirus integration. *Cell* **60**, 3–4.

Hajihosseini, M., Iavachev, L., and Price, J. (1993). Evidence that retroviruses integrate into post-replication host DNA. *EMBO J.* **12**, 4969–4974.

Hantzopoulos, P. A., Sullenger, B. A., Ungers, G., and Gilboa, E. (1989). Improved gene expression upon transfer of the adenosine deaminase minigene outside the transcriptional unit of a retroviral vector. *Proc. Natl. Acad. Sci. USA* **86**, 3519–3523.

Hartley, J. W., and Rowe, W. P. (1976). Naturally occurring murine leukemia viruses in wild mice: Characterization of a new "amphotropic" class. *J. Virol.* **19**, 19–25.

Hatfield, D. L., Levin, J. G., Rein, A., and Oroszlan, S. (1992). Translational suppression in retroviral gene expression. *Adv. Virus Res.* **41**, 193.

Hayward, W. S., Neel, B. G., and Astrin, S. M. (1981). Activation of a cellular onc gene by promoter insertion in ALV-induced lymphoid leukosis. *Nature* **290**, 475–480.

Humphries, E. H., Glover, C., and Reichmann, M. E. (1981). Rous sarcoma virus infection of synchronized cells establishes provirus integration during S phase DNA synthesis prior to cell division. *Proc. Natl. Acad. Sci. USA* **78**, 2601–2605.

Hwang, L. S., and Gilboa, E. (1984). Expression of genes introduced into cells by retroviral infection is more efficient than that of genes introduced into cells by DNA transfection. *J. Virol.* **50**, 417–424.

Jacks, R., and Varmus, H. E. (1985). Expression of the Rous sarcoma virus pol gene by ribosomal frameshifting. *Science* **230**, 1237.

Joyner, A. L., and Bernstein, A. (1983). Retrovirus transduction: Generation of infectious retroviruses expressing dominant and selectable genes is associated with in vivo recombination and deletion events. *Mol. Cell. Biol.* **3**, 2180–2190.

Kadesh, T., and Berg, P. (1986). Effects of the position of the simian virus 40 enhancer on expression of multiple transcription units in a single plasmid. *Mol. Cell. Biol.* **6**, 2593–2601.

Kalyanaraman, V. S., Sarngadharan, M. G., Robert-Guroff, M., Miyoshi, I., Blayney, D., Golde, D., and Gallo, R. C. (1982). A new subtype of human T-cell leukemia virus (HTLV-II) associated with a T-cell variant of hairy cell leukemia. *Science* **218**, 571–575.

Kozak, C. A., and Ruscetti, S. (1992). Retroviruses in rodents. In "The Retroviridae," Vol I. Plenum Press, New York.

Kung, H. J., Hu, S., Bender, W., Bailey, J. M., Davidson, N., Nicolson, M. O., and McAllister,

R. M. (1976). RD-114, baboon, and wooly monkey viral RNAs compared in size and structure. *Cell* 7, 609–620.

Lee, T. C., Sullenger, B. A., Gallardo, H. F., Ungers, G. E., and Gilboa, E. (1992). Overexpression of RRE-derived sequences inhibits HIV-1 replication in CEM cells. *New Biol.* **4**, 66–74.

Luciw, P. A., and Leung, N. J. (1992). Mechanisms of retrovirus replication. In "The Retroviridae," Vol I. Plenum Press, New York.

Mann, R., Mulligan, R. C., and Baltimore, D. (1983). Construction of a retrovirus packaging mutant and its use to produce helper-free defective retrovirus. *Cell* **33**, 153–159.

Markowitz, D., Goff, S., and Bank, A. (1988a). Construction and use of a safe and efficient amphotropic packaging cell line. *Virology* **167**, 400–406.

Markowitz, D., Goff, S., and Bank, A. (1988b). A safe packaging line for gene transfer: Separating viral genes on two different plasmids. *J. Virol.* **62**, 1120–1124.

Miller, A. D., Law, M.-F., and Verma, I. M. (1985). Generation of helper-free amphotropic retroviruses that transduce a dominant-acting, methotrexate-resistant dihydrofolate reductase gene. *Mol. Cell. Biol.* **5**, 431–437.

Miller, A. D., and Buttimore, C. (1986). Redesign of retrovirus packaging cell lines to avoid recombination leading to helper virus production. *Mol. Cell. Biol.* **6**, 2895–2902.

Miller, A. D., Trauber, D. R., and Buttimore, C. (1986). Factors involved in production of helper virus-free retrovirus vectors. *Som. Cell Mol. Genet.* **12**, 175–183.

Miller, A. D. (1992). Retroviral vectors. *Curr. Topics Microbiol. Immunol.* **158**, 1–24.

Moloney, J. B. (1960a). Biological studies on a lymphoid-leukemia virus extracted from Sarcoma-37. I. Origin and introductory investigations. *J. Natl. Cancer Inst.* **24**, 933–951.

Moloney, J. B. (1960b). Properties of a leukemic virus. *Natl. Cancer Inst. Monogr.* **4**, 7–38.

Mowat, M., Cheng, A., Kimura, N., Bernstein, A., and Benchimol, S. (1985). Rearrangements of the cellular p53 gene in erythroleukemic cells transformed by Friend virus. *Nature* **214**, 633.

Mulligan, R. C. (1993). The basic science of gene therapy. *Science* **260**, 926–932.

Nusse, R., and Varmus, H. E. (1982). Many tumors induced by the mouse mammary tumor virus contain a provirus integrated in the same region of the host genome. *Cell* **31**, 99.

Pear, W. S., Nolan, G. P., Scott, M. L., and Baltimore, D. (1993). Production of high-titer helper-free retroviruses by transient transfection. *Proc. Natl. Acad. Sci. USA* **90**, 8392–8396.

Poiesz, B. J., Ruscetti, F. W., Gazdar, A. F., Bunn, P. A., Minna, P. A., and Gallo, R. C. (1980). Detection and isolation of type C retrovirus particles from fresh and cultured lymphocytes of a patient with cutaneous T-cell lymphoma. *Proc. Natl. Acad. Sci. USA* **77**, 7415–7419.

Proudfoot, N. J. (1986). Transcriptional interference and termination between duplicated alpha-globin gene constructs suggests a novel mechanism for gene regulation. *Nature* **322**, 562–565.

Rapp, U. R., Goldsborough, M. D., Mark, G. E., Bonner, T. I., Groffen, J., Reynolds, F. H., and Stephenson, J. R. (1983). Structure and biological activity of v-*raf*, a unique oncogene transduced by a retrovirus. *Proc. Natl. Acad. Sci. USA* **80**, 4218.

Rasheed, S., Gardner, M. B., and Chan, E. (1976). Amphotropic host range of naturally occurring wild mouse leukemia virus. *J. Virol.* **19**, 13–18.

Rohdewohld, H., Weiher, H., Reik, W., Jaenisch, R., and Breindl, M. (1987). Retrovirus integration and chromatin structure: Moloney murine leukemia proviral integration sites map near DNase I-hypersensitive sites. *J. Virol.* **61**, 336–343.

Rosenthal, F. M., Cronin, K., Bannerji, R., Salvadori, S., Zier, K., and Gansbacher, B. (1993). Synergistic induction of cytotoxic effector cells by tumor cells transduced with a retroviral vector carrying both the interleukin-2 and interferon-gamma cDNAs. Submitted for publication.

Rous, P. (1911). A sarcoma of the fowl transmissible by an agent separable from the tumor cells. *J. Exp. Med.* **13**. 397–411.

Scolnick, E. M., and Parks, W. P. (1974). Harvey sarcoma virus: A second murine type-C sarcoma virus with rat genetic information. *J. Virol.* **13**, 1211–1219.

Scolnick, E. M., Howk, R. S., Anisowicz, A., Peebles, P. T., Scher, C. D., and Parks, W. P. (1975). Separation of sarcoma virus-specific and leukemia virus-specific genetic sequences of Moloney sarcoma virus. *Proc. Natl. Acad. Sci. USA* **72**, 4650–4654.

Scolnick, E. M., Goldberg, R. J., and Williams, D. (1976). Characterization of rat genetic sequences of Kirsten sarcoma virus: Distinct class of endogenous rat type C viral sequences. *J. Virol.* **18**, 559–566.

Shen-Ong, G. L. C., Potter, M., Mushinski, J. F., Lavu, S., and Reddy, E. P. (1984). Activation of the c-*myb* locus by viral insertional mutagenesis in plasmacytoid lymphosarcomas. *Science* **226**, 1077.

Sullenger, B. A., Gallardo, H. F., Ungers, G. E., and Gilboa, E. (1990a). Overexpression of TAR sequences renders cells resistant to human immunodeficiency virus replication. *Cell* **63**, 601–608.

Sullenger, B. A., Lee, T. C., Smith, C. A., Ungers, G. E., and Gilboa, E. (1990b). Expression of chimeric tRNA-driven antisense transcripts renders NIH 3T3 cells highly resistant to moloney murine leukemia virus replication. *Mol. Cell. Biol.* **10**, 6512–6523.

Sullenger, B. A., Gallardo, H. F., Ungers, G. E., and Gilboa, E. (1991). Analysis of trans-acting response decoy RNA-mediated inhibition of human immunodeficiency virus type 1 transactivation. *J. Virol.* **65**, 6811–6816.

Temin, H. M., and Mizutani, S. (1970). RNA-directed DNA polymerase in virions of Rous sarcoma virus. *Nature* **226**, 1211–1213.

Temin, H. M. (1986). Retrovirus vectors for gene transfer: Efficient integration into and expression of exogenous DNA in vertebrate cell genomes. In "Gene Transfer" (Kucherlapati, R., Ed.). Plenum, New York.

Varela-Echavarria, A., Prorock, C. M., Ron, Y., and Dougherty, J. P. (1993). High rate of genetic rearrangement during replication of a Moloney murine leukemia virus-based vector. *J. Virol.* **67**, 6357–6364.

Varmus, H. E., Padgett, T., Heasley, S., Simon, G., and Bishop, J. M. (1977). Cellular functions are required for the synthesis and integration of avian sarcoma virus-specific DNA. *Cell* **11**, 307–319.

Varmus, H. E., Shank, P. R., Hughes, S. E., Kung, H.-J., Heasley, S., Majors, J., Vogt, P. K., and Bishop, J. M. (1979). Synthesis, structure, and integration of the DNA of RNA tumor viruses. *Cold Spring Harbor Symp. Quant. Biol.* **43**, 851–864.

Varmus, H., and Swanstrom, R. (1984). Replication of retroviruses. In "RNA Tumor Viruses" (Furmanski, P., *et al.*, Eds.). Cold Spring Harbor Laboratory, Cold Spring Harbor, NY.

Voronova, A. F., and Sefton, B. M. (1986). Expression of a new tyrosine protein kinase is stimulated by retrovirus promoter insertion. *Nature* **319**, 682.

Watanabe, S., and Temin, H. M. (1983). Construction of a helper cell line for avian reticuloendotheliosis virus cloning vectors. *Mol. Cell. Biol.* **3**, 2241–2249.

Weiss, R. (1984). Experimental biology and assay of RNA tumor viruses. In "RNA Tumor Viruses" (Furmanski, P., *et al.*, Eds.). Cold Spring Harbor Laboratory, Cold Spring Harbor, NY.

Yoshida, M., and Seiki, M. (1990). Molecular biology of HTLV-I. In "Retrovirus Biology and Human Disease" (Gallo and Wong-Staal, Eds.). Dekker, New York.

Yoshinaka, Y., Katoh, I., Copeland, T. D., and Oroszlan, S. (1985). Murine leukemia protease is encoded by the gag-pol gene and is synthesized through suppression of an amber termination codon. *Proc. Natl. Acad. Sci. USA* **82**, 1618–1622.

Yu, S.-F., von Ruden, T., Kantoff, P. W., Garber, C., Seiberg, M., Ruther, U., Anderson, W. F., Wagner, E. F., and Gilboa, E. (1986). Self-inactivating retroviral vectors designed for transfer of whole genes into mammalian cells. *Proc. Natl. Acad. Sci. USA* **83**, 3194–3198.

Zhang, J., and Temin, H. M. (1993). Rate and mechanism of nonhomologous recombination during a single cycle of retroviral replication. *Science* **259**, 234–238.

CHAPTER 6

HSV Vectors in the Mammalian Brain: Molecular Analysis of Neuronal Physiology, Generation of Genetic Models, and Gene Therapy of Neurological Disorders

Matthew J. During
Molecular Pharmacology and Neurogenetics Laboratory
Yale University School of Medicine
New Haven, Connecticut

The ability to manipulate gene expression in the mammalian nervous system at defined times during development and aging, in specific brain regions and in selected cell-types, represents a significant technological advance which is likely to revolutionize neurobiology and clinical neurology. DNA viral vectors, with their capacity to efficiently transduce terminally differentiated, post-mitotic cells, may provide such a technique resulting in major breakthroughs in basic and clinical neuroscience.

Over the past decade numerous proteins which have specific functions in neuronal physiology and/or human neurological disease have been identified. Such molecules include ion channels, neurotransmitter receptors and transporters, signal transduction enzymes, transcription factors, proteins involved in neurotransmitter synthesis and release, and neurotrophic factors, to name a few. Although the cloning of these genes has advanced the field enormously, there is a relative paucity of studies on manipulating these genes *in vivo*. The majority of such studies have used transgenic or homologous recombination approaches. These techniques, with refinements including the use of regulatable and cell-type-specific promoters and recombinase systems to obtain tissue and cell specificity and even temporal control over expression have moved the field to a point where the roles of specific genes in regulating neuronal function have become increasingly established. Viral vectors are a complementary approach whereby the gene of interest can be expressed in a defined brain region and specific cells in both developing and adult animals. However, in comparison to the rapidly growing literature on transgenic and knockout animals, the number of neurobiological studies using a viral vector approach remains relatively limited. In this chapter I shall review our own experience with HSV-1 vectors in the brain. This work has been done in collaboration with several investigators including Drs. Alfred Geller and Rachael Neve of Harvard Medical School as well as Dr. Howard Federoff of Albert Einstein College of Medicine and Dr. Karen O'Malley at Washington University School of Medicine.

The small number of published studies using a vector approach reflects the difficulties that the brain imposes on somatic cell gene transfer. The greatest hurdle is the terminally differentiated, postmitotic nature of neurons in the adult brain. Although establishment of retroviral vectors revolutionized the field of gene transfer, retroviral vectors have a critical limitation for gene transfer in the CNS: they need at least one mitotic division for integration and stable expression. Retroviral vectors have however been used in several situations in the CNS, specifically for studies involving *ex vivo* gene transfer. For example, nerve cell progenitors have been genetically labeled and their progeny followed after transplantation into the brain (Sanes *et al.*, 1986; Price *et al.*,1987). Retroviruses have also been used to immortalize various neuronal-type cell lines by oncogene transfer (Cepko, 1988; Lendahl and McKay, 1990). Moreover, retroviral vectors are clearly established as the predominant system for gene therapy, and clinical trials using retroviruses expressing the HSV-1 *tk* gene are already ongoing for malignant brain tumors (Oldfield *et al.*, 1993). Furthermore, retroviruses are also being used as the primary vector for *ex vivo* approaches to neurological gene therapy in disease models ranging from Parkinson's and Alzheimer's Disease to hereditary neurometabolic disorders (see Wolff, 1993).

The inherent limitation of retroviral vectors in obtaining efficient and stable gene transfer in neurons has imposed major constraints on *in vivo* genetic intervention studies in the adult brain. The characterization of a defective herpes simplex type 1 (HSV-1) vector for efficient and stable gene transfer into neurons was therefore a significant advance (Geller and Breakefield, 1988). Over the past 6 years since this original report, numerous studies on viral vector gene transfer have been conducted. Such studies have ranged in scope from gene therapy and the development of disease models to promoter analysis and studies investigating the molecular basis of neuronal physiology (Federoff *et al.*, 1992; Wolfe *et al.*, 1992; Andersen *et al.*, 1993; Geller *et al.*, 1993, 1995; Ho *et al.*, 1993; Kaplitt *et al.*, 1994; During *et al.*, 1994).

Several basic principles apply for gene transfer studies of neuronal function. The introduced gene must have a dominant phenotype for an effect on cell physiology to be apparent. The recombinant gene must result in either a gain or loss of function. The virally encoded gene may result in a dominant positive effect where the endogenous protein is regulated whereas the vector-encoded mutant protein is constitutively expressed. For example, a vector encoding the catalytic domain of adenylate cyclase when used to infect neurons will result in unregulated, continuous activation of the enzyme; i.e., it will have a dominant effect, whereas the wild-type enzyme is very tightly regulated and is only activated by specific stimuli. Therefore, overexpressing a full-length wild-type enzyme has no effect on neuronal function (Geller *et al.*, 1993). In contrast, dominant negative mutants result in a protein with altered function which also interferes with function of the wild-type protein. An example of this strategy would be expression of an altered receptor subunit which, if it formed homomeric receptors, would have altered properties; in addition, the vector-encoded subunit would interfere with the function of heteromeric receptors, including both normal and vector-derived (mutant) subunits. Such a strategy involving glutamate receptors is described below.

In addition to the ability to study dominant positive or dominant negative mutants, the vector-encoded gene can be engineered to direct the protein to specific regions of polarized cells. A nuclear targeting signal can be used to restrict expression to the nucleus, whereas other sequences may be used to target dendrites (for example, the adult-specific form of microtubule-associated protein, MAP2; Papandrikopoulon *et al.*, 1989) or to axons (for example, a 10-amino-acid domain at the N terminal of the neuronal growth associated protein, GAP43; Zuber *et al.*, 1989). Therefore, the vector system may be used both to explore the effect of regional expression as well as to help define those sequences critical in the targeting process.

I. Study of Mammalian Neuronal Physiology Using HSV Vectors

Several points need to be considered when using viral vectors *in vivo:* the type and location of the transduced cells, the promoter used to drive expression of the recombinant gene, the level and stability of expression (which are largely dependent on the promoter), and the recombinant gene itself.

The specificity and location of the cells which are transduced is largely dependent on both the inherent tropism of the virus and the method of delivery. HSV-1 is a neurotrophic virus and although HSV-1 vectors can infect both neurons and glia, they clearly have greater efficiency in transducing neurons (During *et al.*, 1994; Geller *et al.*, 1995). In contrast, adenoviral vectors may have some degree of glial tropism (Horrelou *et al.*, 1994; Freese *et al.*, submitted for publication). However, by far the greatest determinant of which cells are transduced is the method of the virus delivery. HSV-1 vectors do not cross the blood–brain barrier; therefore, systemic, intravenous delivery does not result in CNS expression. However, approaches to penetrate the blood–brain barrier which include hyperosmolar infusions may result in some degree of brain expression using adenoviral vectors (Doran *et al.*, in press). HSV-1 is significantly larger than adenovirus and it is therefore not clear whether HSV-1 vectors would similarly be able to cross the blood–brain barrier under these conditions.

Herpes encephalitis is due to infection of the brain with wild-type HSV-1. It is generally believed that entry of the virus is via an ascending intraaxonal, transsynaptic route from periphery to brain. Certainly, recombinant HSV vectors have been applied to mucosal surfaces and shown to result in expression in the relevant sensory ganglion cells (Davar *et al.*, 1994). However, the majority of studies involving the use of HSV-1 vectors in the mammalian nervous system have used direct intraparenchymal injection. The region of injection is therefore the primary determinant of which cells are transduced. The methods of injection, including the volume of the virus, the gauge of the needle, and the rate of infusion, are critical both in terms of minimizing trauma and determine the area of viral diffusion and/or convection and the number and location of transduced cells. As a general rule, the finer the injection needle and the slower the infusion rate of virus (0.5 μl/min or less), the smaller the toxicity, the greater the viral diffusion, and the higher the level of expression (Horrelou *et al.*, 1994). Small volumes (ca. 1 μl) of an HSV-1 vector injected directly into specific brainstem nuclei using a stereotactic frame can restrict expression to a specific neuronal population, for example, the substantia nigra pars compacta (Song *et al.*, 1993). Alternatively, as HSV vectors are transported retrogradely, injections can be made into the projection area, resulting in uptake in the terminals and expression in the afferent cell bodies (Leone *et al.*, 1993). Although real progress has been made in cell targeting of retroviral vectors (Kasahara *et al.*, 1994) and efforts are currently underway with both adenoviral and adeno-associated virus (AAV) vectors, it is unlikely that HSV vectors will be engineered in the near future to bind to specific cell surface receptors to restrict or augment expression in a specific cell type. Current strategies to limit expression to a specific class of neuron depend largely on the use of cell-type-specific promoters. Specific examples that we have used include the use of the neuron specific promoter of tyrosine hydroxylase (TH). Following the stereotactic injection of an HSV vector using the TH promoter, relative expression in TH immunoreactive neurons is greatly enhanced (Song *et al.*, 1994). Several other groups have used cell-type-specific promoters including the enkephalin promoter (Kaplitt *et al.*, 1994) and the neuron-specific enolase promoter (Andersen *et al.*, 1993). The vast majority of studies, however, have used strong viral promoters which are consititutively expressed in all cell types.

The stability and level of expression is critical for viral vector studies, in particular for gene therapy but also for experiments investigating the role of specific genes in neuronal physiology. The level of expression is largely dependent upon upstream regulatory sequences including the promoter, enhancer and silencer elements, and sites for transcriptional activation and methylation. Typically, early viral promoters are stronger than cellular promoters and constitutively active, hence, their popularity in vector studies. However, viral promoters tend to downregulate over time, perhaps due in part to methylation (Challita and Kohn, 1994). In contrast, although cellular promoters are weaker in studies of transient expression, in long-term studies, cellular promoters may have an advantage as although they also lose strength over time, the loss of expression is much less (Kaplitt *et al.,* 1994). Cellular promoters also have potential advantages in obtaining both temporal and developmental control over expression. As promoters are analyzed and sequences which suppress or enhance expression in specific cell types are further established, synthetic promoters may be tailored for the specific indication. For example, the use of promoters which have small molecule response elements for transcriptonal activation (Wang *et al.,* 1994; Lu and Federoff, in press).

The key variable in viral vector studies is the nature of the recombinant gene itself. To date, HSV vectors have been used to introduce enzymes of inherited metabolic disorders (Wolfe *et al.,* 1992), glucose transporters (Ho *et al.,* 1993), transmitter synthesizing enzymes (During *et al.,* 1994; Geller *et al.,* 1995), neuropeptides (Andersen *et al.,* 1993), signal transduction enzymes (Geller *et al.,* 1993), glutamate receptors (Bergold *et al.,* 1993), neurotrophic factors (Federoff *et al.,* 1992), and the growth-associated protein, GAP43 (Verhaagen *et al.,* 1994), to name a few.

HSV-1 vectors provide an attractive technique for the investigation of the molecular basis of neuronal physiology (Geller *et al.,* 1991). In collaboration with Drs. Alfred Geller and Rachael Neve, we have used HSV vectors which express enzymes involved in different signal transduction pathways to look at the effect on neuronal function. As discussed above, for a vector-encoded gene to influence cell function it has to result in either a gain or loss of function. We selected to express enzymes which have a domain structure—both regulatory and catalytic regions whereby deletion of the regulatory domain results in an active, unregulated catalytic protein, i.e., a dominant positive, gain of function mutant. The rationale behind such an intervention was to test the hypothesis that stable activation of specific signal transduction pathways could result in a long-term change in neuronal physiology, suggesting that such a mechanism might underlie stable changes in neuronal function *in vivo,* e.g., the possible molecular basis for synaptic strengthening underlying long-term potentiation and learning and memory. Such a hypothesis is supported by studies in invertebrates where it has been shown that the cAMP pathway is involved in both short-term, i.e., the presynaptic facilitation of the gill withdrawal reflex in *Aplysia* (Kandel & Schwartz, 1982), and also long-term changes in neuronal physiology as in *Drosophila,* where a mutation in a cAMP phosphodiesterase results in the *dunce* phenotype, where associative learning is markedly impaired (Dudai, 1989), and *rutabaga,* which has a learning deficiency associated with a defect in calcium/calmodulin-sensitive adenylate cyclase (Livingstone *et al.*, 1984).

As a first step to elucidate the potential of an HSV vector-based approach to determine the role of the cAMP pathway, we constructed a vector which encoded the catalytic domain of yeast adenylate cyclase (HSVcyr). This vector was used to transduce a variety of cells *in vitro* including PC-12 cells and sympathetic neurons. Infection of these cells with HSVcyr resulted in an increase in cAMP levels, protein phosphorylation, activation of protein kinase A as determined by specific phosphorylation of the PKA site (Ser) on tyrosine hydroxylase, and catecholamine release. Moreover, the increased release of neurotransmitter was both calcium- and activity-dependent and persisted for at least 1 week following transduction, suggesting that activation of the cAMP may mediate long-term changes in neuronal physiology (Geller *et al.*, 1993). More recently, we have investigated the protein kinase C (PKC) pathway and have expressed an unregulated, active catalytic domain of PKC (the $\beta 2$ isoform) under control of a 6-kb rat tyrosine hydroxylase promoter. This PKC mutant results in activity-dependent enhancement of neurotransmitter release from primary cells *in vitro* (During, 1992). We subsequently used this vector in the rat and directly injected the vector into the hippocampus. Preliminary results suggest that the vector is retrogradely transported with expression obtained in the TH-positive afferents (predominantly the locus coeruleus noradrenergic projection to the hippocampus). Moreover, norepinephrine release is enhanced and GAP-43 and MARCKS, two major substrates of PKC, have increased phosphorylation. Of most significance is that the rats exhibit significant improvement in learning and memory in a hippocampal-dependent spatial navigation task (Leone *et al.*, 1993). These data therefore demonstrate that an HSV vector strategy may be complementary to the homologous recombination approach where calcium–calmodulin kinase II, PKC gamma, and CREB mutant mice have hippocampal-dependent behavioral deficits (Silva *et al.*,1992; Abeliovich *et al.*, 1993; Bourchuladze *et al.*, 1994).

II. Generation of a Model of Human Neurological Disease Using HSV Vectors

Simple mendelian, single gene diseases are perhaps best modeled using a homologous recombination approach whereby the gene of interest is "knocked out." Such an approach has already been used to generate models of Lesch Nyan and Gaucher disease. Although in general these knockout models parallel the phenotype of the human disease this is not always the case. For example, in the hypoxanthine–guanine phosphoribosyltransferase (HPRT) knockout mouse there is no phenotype typical of Lesch Nyan's despite HPRT being the defective enzyme in the disease (Kolberg, 1992).

Complex or acquired genetic diseases are not readily modeled using a homologous recombination strategy. However, a viral vector approach shows some potential. In collaboration with Drs. Howard Federoff at Albert Einstein College of Medicine, Peter Bergold at State University of New York at Brooklyn, and Anne Williamson at Yale we have developed a genetic model of human temporal lobe epilepsy (TLE) using an HSV vector-based approach. In the most simplistic analysis, TLE is a disease of excitation/inhibition imbalance favoring excessive excitation due to increased glutamatergic neurotransmission (During and Spencer, 1993). One of the most commonly used animal models of TLE is that of systemic and local application of the glutamate analog, kainate. Although there is no evidence for kainate toxicity or an endogenous kainate-like toxin in the pathogenesis of human TLE (apart from the rare case of domoic acid poisoning), an alternative possibility is that excessive glutamate release and/or an increased expression of specific glutamate receptors may be involved in the pathogenesis of TLE. We have previously shown that epileptogenic tissue has excessive release of glutamate, particularly in the preictal period and during seizures (During and Spencer, 1993). Moreover, in TLE there is some evidence of increased expression of specific glutamate receptors, with perhaps the most consistent result that of increase in kainic acid binding (de Lanerolle *et al.*, 1993).

Kainic acid is a potent excitotoxin and convulsant (Olney *et al.*, 1974). Both parental and intracerebral administration of kainate produce limbic seizures and neuronal injury, and both approaches have been used to generate a potentially useful model of human TLE (Nadler, 1981; Ben-Ari, 1985). Moreover, the hippocampus is strongly implicated in kainate-induced seizures. Furthermore, spontaneous limbic motor seizures persist for weeks following the initial seizure.

We tested the hypothesis that an alteration in the expression of a specific glutamate receptor results in hyperexcitability characteristic of TLE. We chose to use the GluR6 subunit of the kainate receptor as kainate receptors are clearly implicated in epilepsy (as described above). Furthermore, the GluR6 subunit forms homo-oligomeric channels. Such channels are formed following enhanced expression of GluR6 cDNA in a variety of cells (Egebjerg *et al.*, 1991; Bettler *et al.*, 1992; Burnashev *et al.*, 1992), and the expression of these channels results in increased excitability of the cells. GluR5 and GluR7 are also members of the same family as defined by sequence homology and pharmacology. They all bind glutamate agonists with a similar order of affinity (kainate > quisquilate > glutamate; Meyer, 1992). Moreover, expression of GluR6 alone in either oocytes or human fibroblasts produces a rapidly desensitizing ion channel gated by kainate, quisqualate, or glutamate but not AMPA or NMDA (Bettler *et al.*, 1992; Egebjerg *et al.*, 1991). Three ionotropic glutamate receptors, GluR2, 5, and 6, undergo RNA editing in the second transmembrane domain (TM2), a proposed channel-forming segment. The editing generates an arginine residue in a critical position (the Q/R site) with the edited TM2 diminishing calcium permeability. Kohler *et al.* (1993) described two further sites of RNA editing in the transmembrane segment TM1 of GluR6. Moreover, in contrast to GluR2 and the editing of TM2, the fully TM1 edited form of GluR6 encodes a channel with high calcium permeability regardless of editing of TM2. This degree of plasticity with multiple site RNA editing is another reason why GluR6 is a reasonable first choice to investigate in terms of determining the potential role of altered glutamate receptor expression in the pathophysiology of epilepsy. Altered glutamatergic transmission and hyperexcitability might result from not only an overall increase in the expression of GluR6 but also a change in editing favoring the fully edited, calcium-permeable receptor.

To increase expression of GluR6 within a defined region of the brain, we used a defective HSV-1 vector. HSVGluR6 directs the expression of the fully edited GluR6 cDNA (GluR6 V,C,R) driven from the HSV IE 4/5 promoter. Hippocampal slice culture experiments indicate that increased GluR6 expression causes epileptiform discharge that persists after loss of transduced cells (Bergold *et al.*, 1993), suggesting that enhanced glutamate receptor expression generates long-term hyperexcitability in nontransduced neurons.

To examine whether increased GluR6 expression would induce epilepsy, HSVGluR6 was injected into the rat hippocampus. GluR6 expression was determined by RT-PCR analysis of mRNA extracted from brain regions. Vector-derived transcripts were amplified from HSVGluR6-injected brain regions and not from uninjected regions from the same animal.

Animals were monitored by EEG for 8–12 hr postinjection. Neither HSVlac, which expresses *Escherichia coli* β-galactosidase, nor phosphate buffered saline-treated control animals exhibited any seizure activity. However, all animals in the GluR6 group had limbic seizures.

Hippocampal glutamate and GABA release was determined *in vivo* (During and Spencer, 1993). In HSVGluR6-injected animals, glutamate levels began to rise approximately 1 hr prior to seizure onset and 3 hr after HSVGluR6 injection. Levels of GABA remain stable during this preictal period. Both glutamate and GABA rose acutely during the initial seizure. In HSVlac-injected animals no neurochemical changes were observed.

Anne Williamson of Yale examined the electrophysiological characteristics of CA1 and CA3 pyramidal cells as well as of dentate granule cells in animals 6 hr, 24 hr, 1 week, 1 month, and 2 months following the injection of the vector.

In all animals studied, CA1 and CA3 cells exhibited robust spontaneous burst firing at rest. All cells studied had membrane properties analogous to endogenous burst-firing cells such as those found in CA3 and deep layer 5 neocortical cells (Connors and Gutnick, 1990). Depolarizing stimuli from membrane potentials hyperpolarized below −65 mV produced a burst followed by single action potential firing. The bursts produced by current injection were similar in shape and duration to the spontaneous events. However, spontaneous bursting was not initially associated with evoked synaptic hyperexcitability. In most cells, there was no burst firing with orthodromic stimulation in any area, and these responses were followed by robust biphasic inhibitory potentials. In the CA1 cells from animals studied 1 and 2 months postinjection, the IPSP amplitudes were smaller and bursts could be evoked with synaptic stimulation. None of these changes were noted in the electrophysiological properties of the HSVlac-injected and noninjected rats. Moreover, dentate granule cells were normal in all animals.

In HSVGluR6-injected animals, progressive mossy fiber sprouting into the inner molecular layer of both blades of the dentate gyrus was noted by using a Timm stain (Sloviter, 1982). In addition, cell loss in both CA1 and CA3 was readily apparent after 1 month.

These results suggest that hippocampal injection of HSVGluR6 can induce local expression of vector encoded GluR6 resulting in increased glutamate release, spontaneous seizures, cell loss, and mossy fiber sprouting as well as cellular hyperexcitability.

The HSVGluR6 vector expresses fully edited GluR6, which in nonneuronal cells forms homomeric channels with high calcium permeability. This form of GluR6 is the most abundant type in the adult rat hippocampus (Lowe and Smith, 1993). Coexpression of fully edited and unedited mRNA forms in nonneuronal cells leads to receptors with lower calcium permeability than when either form is expressed alone (Kohler *et al.*, 1993). In our experiments, the stoichiometry of edited GluR6 relative to endogenous receptor forms is unknown. The strong promoter used in HSVGluR6 is likely to lead to high levels of expression. Thus, it is probable that in most transduced cells homomeric channels form, and this may underlie the initial increase in excitability that produces glutamate release and seizures. The rise in extracellular glutamate commencing approximately 3 hr following HSVGluR6 injection probably reflects increased receptor expression in glutamatergic neurons, which results in enhanced synaptic transmission. This progressive increase in glutamate release may also underlie the synchronization resulting in an ictal event as has been proposed in human TLE (During and Spencer, 1993).

The electrophysiological analysis of HSVGluR6-injected animals indicates that both CA1 and CA3 cells develop spontaneous bursting. This characteristic was seen in the presence of normal synaptic inhibition and persisted for periods of up to 2 months following injection. These data indicate that overexpression of GluR6 can alter the physiological phenotype of hippocampal pyramidal cells. This type of bursting in both CA1 and CA3 will serve to entrain hippocampal activity to a set rhythm which may be critical in the generation of limbic seizures. In hippocampal slices studied at later time points (1 and 2 months), there was a progressive reduction in inhibition with concomitant synaptically evoked bursting, a pattern which more closely resembles the changes observed in acute TLE slices. It remains to be determined if changes similar to those seen in the HSVGluR6-injected animals occur early in the development of TLE.

Recorded cells were generally several hundred micrometers to 2 mm from the injection site. X-gal staining of HSVlac-treated animals show that virus spread is limited to within approximately 200–300 μm around the injection site, suggesting that the recorded cells were unlikely to be transduced. Thus, it is likely that the physiologically altered CA1 cells were not vector-transduced but had acquired hyperexcitability. This notion is sponsored by the observation that animals injected only in CA3 also developed bursting in CA1 cells. HSVGluR6 transduction into CA3 of organotypic slice cultures also produces acquired hyperexcitability (Bergold *et al.*, 1993).

These data indicate that the localized overexpression of a single glutamate receptor subunit gene can be used to generate a novel model of epilepsy. While the early physiological changes seen in HSVGluR6-injected animals are unique, the development of synaptic excitability we observed may provide a useful model for the developmental changes which are presumed to occur in TLE.

Somatic transfer of the GluR6 gene induces seizures and histopathologic changes similar to TLE. The expression of a receptor channel highly permeable to calcium suggests a general mechanism for initiating increased excitability. Furthermore, our experiments suggest that the long-term increase in hyperexcitability resides, in part, within cells that do not express the vector. An advantage of this specific approach is that it will be possible to test whether expression of other GluR6 forms and other glutamate receptors have the ability to induce or inhibit hyperexcitability.

This HSVGluR6 model of TLE demonstrates the potential of viral vectors to model human disease and strongly supports the hypothesis of altered gene expression in the pathogenesis of TLE. However, although viral vectors clearly may be used to generate disease models as illustrated above, a major advantage of a viral vector approach over transgenic and homologous recombination techniques is the potential for gene therapy.

III. Gene Therapy Using Defective HSV-1 Vectors

Neurological disorders present significant hurdles for gene therapy. First, as discussed above, the vast majority of cells in the brain are terminally differentiated and postmitotic. The more established retroviral vectors are therefore unable to transduce these cells. However, there does appear to be a small population of pluripotent stem cells in the CNS leading to both glial and neuronal progenitors, but this stem cell population in the adult brain is small and not readily harvested. However, isolation and *ex vivo* transfer of therapeutic genes in these immortalized progenitor cells holds promise as an alternative gene therapy strategy (Snyder *et al.*, 1994). Second, the relative inaccessibility of the brain presents a major delivery problem with systemic administration of a vector not reaching the CNS without pertubation of the blood–brain barrier. Currently the only efficient method for direct *in vivo* gene therapy of neurological disorders is intraparenchymal injection of the viral vector. Since such delivery will result in somatic cell transfer restricted to the site of injection and some limited diffusion and axonal transport, such an approach is unlikely to be successful for neurological genetic diseases where global brain delivery is likely to be needed. The neurological disorder which has served as the prototype disease for many gene therapy strategies in the brain is Parkinson's Disease (PD). PD is perhaps the simplest neurodegenerative disease and is characterized by an accelerated loss of substantia nigral dopamine cells, resulting in a movement disorder of tremors, rigidity, and akinesia. Although pharmacological treatments exist (based on facilitating dopaminergic transmission in the caudate–putamen), the mainstay of current therapy, L-dopa, loses efficacy over time and patients have significant morbidity. Perhaps the simplest genetic intervention strategy would be to attempt to permanently restore the dopamine deficit in nigrastriatal target region, the caudate–putamen. Intrinsic cells of the caudate–putamen do not normally contain dopamine, nor do they express dopamine biosynthetic enzymes. As genetically engineered noneuronal cells which have been stably transduced with the rate-limiting dopamine synthesizing enzyme, TH (which catalyzes the conversion of tyrosine to L-dopa, the immediate precursor of dopamine) has been shown to be effective in rodent models of PD, we chose TH as our first candidate gene for direct *in vivo* gene therapy of PD.

Together with Drs. Alfred Geller, Jan Naegele, and Karen O'Malley, we generated a defective HSV-1 vector which expressed a full-length human cDNA for TH, HSVth. In our initial analysis we characterized HSVth *in vitro*. HSVth was used to infect both cultured fibroblasts and striatal cells. In fibroblasts, HSVth resulted in both tyrosine hydroxylase immunoreactivity (TH-IR) and TH enzyme activity. Moreover, both the expression and enzyme activity persisted for the duration of the culture (greater than 1 week). Primary cultures of dissociated striatal cells were also infected with HSVth. In HSVth-treated cells, but not in mock-infected or pHSVlac-treated cultures, there was detectable TH mRNA, TH-IR, and L-dopa release. Moreover, there were also very low levels of dopamine, suggesting some endogenous aromatic amino acid decarboxylase activity in the striatal cells. Furthermore, the rate of L-dopa release from transduced striatal cells was comparable to that of genetically engineered cells which restore function following transplantation in the 6-hydroxydopamine (6-OHDA) rodent model of PD (Geller *et al.*, 1995). Therefore, it appeared likely that a direct *in vivo* approach whereby striatal cells were transduced *in situ* by HSVth might also restore neurological deficits in the rat model of PD. In order to directly test this hypothesis we followed up our *in vitro* characterization study with a long-term *in vivo* study. We used the well-established rodent model of PD whereby rats undergo a unilateral lesion of the nigrostriatal system resulting in striatal dopamine depletion, postsynaptic dopamine receptor asymmetry, and neurological deficits, most notably rotational behavior following systemic dopaminergic agonist administration. We selected rats with near complete unilateral lesions which exhibited stable rotational behavior following injection of the dopamine agonist, apomorphine. Rats were randomized into three surgical groups—a PBS control group, a HSVlac group, and the HSVth-treated group. Injections were made using a stereotactic frame directly into multiple sites in the denervated striatum. Animals were followed by repeated behavioral analysis and in addition a group of animals underwent freely moving, *in vivo* microdialysis to estimate both basal and stimulated dopamine release as well as L-dopa accumulation following decarboxylase inhibition, an index of *in vivo* TH activity. Groups of animals were sacrificed at varying times following vector administration with several animals maintained for up to 16 months. All HSVth animals exhibited recovery with a mean of 64% reduction in rotational behavior, furthermore this recovery persisted for the entire duration of the study (12 months; Fig. 1A). The two HSVth-treated rats that were maintained to 16 months had persistent recovery of 56 and 62% from baseline. In contrast, neither HSVlac or PBS-treated animals had any behavioral recovery. The microdialysis data demonstrated a 60% increase in the accumulation of L-dopa following inhibition of endogenous AADC activity with the drug NSD 1015 in HSVth-treated rats vs controls (Fig. 1B). This is an *in vivo* index of TH activity. Furthermore, both basal and potassium-stimulated release of dopamine was increased by 120% (Fig. 1C). PCR and RT-PCR analysis demonstrated persistence of vector DNA and mRNA expression, respectively (Fig. 2). Immunocytochemical analysis demonstrated TH-IR striatal cells surrounding the injection site (Fig. 3). This study therefore suggested

Figure 1 Rotation rates and striatal L-dopa or DA concentrations after stereotactic injection of HSVth, HSVlac, or PBS into the denervated striatum in 6-OHDA-lesioned rats (During *et al.*, 1994). (A) Rats were tested for apomorphine-induced asymmetrical rotations at repeated intervals, and the values represent the percentage of the baseline rotation rate for each group. (B) Striatal L-dopa concentrations were measured by microdialysis after perfusion with the aromatic amino acid decarboxylase inhibitor, NSD 1015. The accumulation of L-dopa is an index of *in vivo* TH activity. (C) Striatal dopamine levels were measured by microdialysis in low potassium (3 m*M*) and during perfusion with depolarizing concentrations of potassium (56 m*M*).

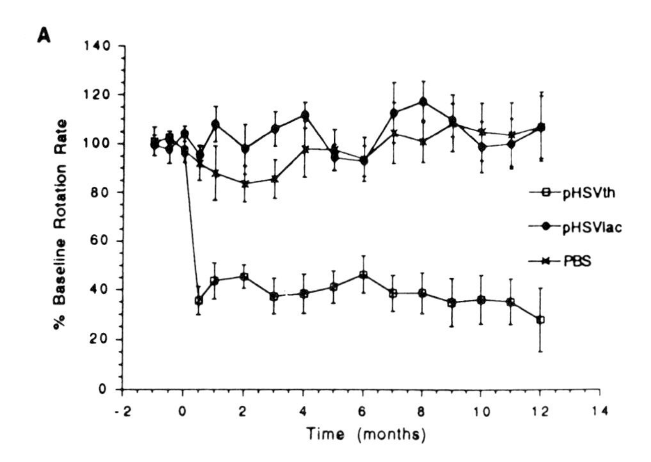
A
% Baseline Rotation Rate
140
120
100
80
60
40
20
0
-2
0
2
4
6
8
10
12
14
Time (months)
pHSVth
pHSVlac
PBS

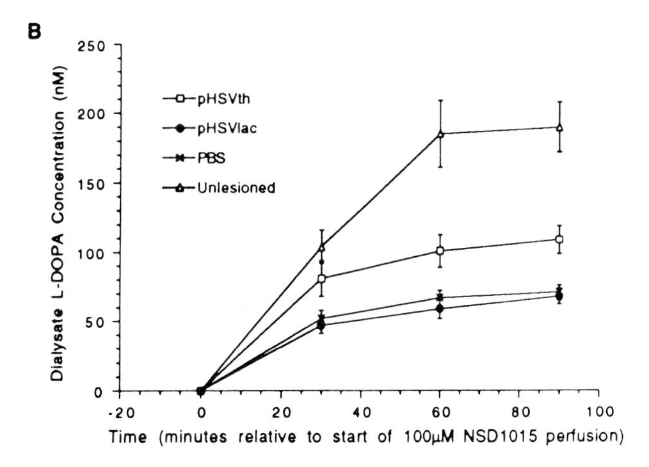
B
Dialysate L-DOPA Concentration (nM)
250
200
150
100
50
0
-20
0
20
40
60
80
100
Time (minutes relative to start of 100μM NSD1015 perfusion)
pHSVth
pHSVlac
PBS
Unlesioned

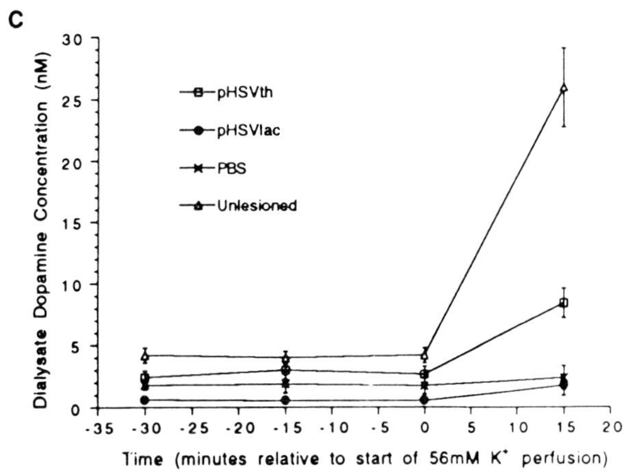
C
Dialysate Dopamine Concentration (nM)
30
25
20
15
10
5
0
-35
-30
-25
-20
-15
-10
-5
0
5
10
15
20
Time (minutes relative to start of 56mM K⁺ perfusion)
pHSVth
pHSVlac
PBS
Unlesioned

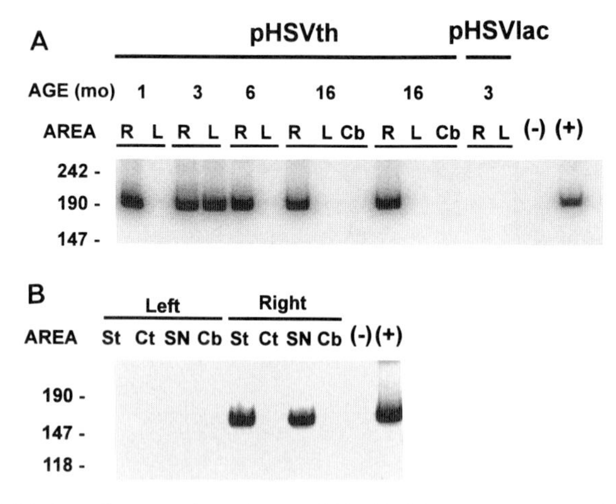

Figure 2 Persistence of HSVth DNA and expression of TH mRNA. (A) DNA was extracted from sections and subjected to PCR with the use of primers specific to the human TH gene, and the products were electrophoresed. Age is the time following gene transfer when the rats were analyzed (1, 3, 6, and 2 animals at 16 months for HSVth and 3 months for HSVlac). Brain areas: R, right injected striatum; L, left striatum; Cb, cerebellum. Minus sign represents no DNA; plus sign indicates HSVth DNA isolated from *E. coli,* which should direct expression of a 186-bp fragment (number of base pairs is shown at left). (B) RT-PCR analysis of RNA isolated from specific brain areas 1 month after injection of HSVth into the right striatum. Brain areas: St, striatum; Ct, cortex; SN, midbrain, substantia nigra; Cb, cerebellum. Minus sign represents no DNA; plus sign indicates HSVth DNA isolated from *E. coli.* The methods used should generate a 160-bp fragment (number of base pairs is shown on left). Courtesy of Dr. Karen O'Malley.

that HSV vectors can direct persistent expression of potentially therapeutic genes and may be considered for gene therapy. The major caveat with the current HSV amplicon system is the inherent toxicity of the system. In our study up to 10% of aminals died within the first 2 weeks following vector administration. Subsequent analysis showed that this toxicity was due to wild-type HSV-1 infection. Current generation HSV-1 amplicon vectors require a helper virus for packaging. These helper viruses have either mutations or deletions in essential genes and are therefore replication incompetent. However, there is a significant reversion frequency to wild-type and as both the helper virus and the amplicon vector are identical physically; the helper virus cannot be separated. If techniques existed to eliminate helper virus or sufficient deletion mutants existed to eliminate recombination to wild-type, then HSV-1 amplicons might be considered for human gene therapy applications in the CNS. The safety and efficacy of AAV vectors would suggest that AAV may be the current first choice for CNS gene therapy (Kaplitt *et al.*, 1994a; Kaplitt and During, Chapter 12, this volume). HSV vectors do however have some advantages which make them an attractive choice for studies in experimental animals.

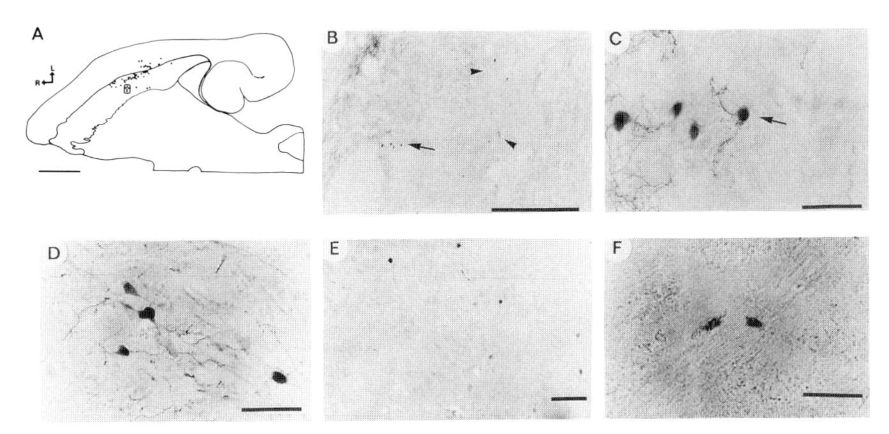

Figure 3 TH immunoreactivity was detected with an antibody to TH and β-gal was detected with X-gal. (A–C) A HSVth-injected rat which was analyzed at 6 months following gene transfer. (A) Composite drawing of charted sections, showing the positions of 48 cells containing TH immunoreactivity in the striatum and neocortex. Every third section was analyzed. L, lateral; R, rostral; scale bar, 2 mm. (B) Low-magnification photomicrograph of clusters of cells containing TH-IR. Arrowheads point to two clusters, and the arrow indicates a third cluster (boxed in A); scale bar, 500 μm. (C) High-magnification view of a cluster of striatal cells containing TH-IR with neuronal morphology (boxed in A); scale bar, 50 μm. (D and E) Another rat sacrificed at 6 months. (D) A cluster of pallidal neurons containing TH-IR; scale bar, 50 μm. (E) A cluster of cortical neurons (agranular frontal cortex, layers 3 and 5) containing TH-IR; scale bar, 100 μm. (F) High magnification view of X-gal-positive striatal cells from an HSVlac-treated rat (analyzed at 12 months); scale bar, 50 μm. Courtesy of Dr. Janio Naegele.

First, they have far greater packaging capability (15 kb or more) than either AAV (5 kb) or adenoviral and retroviral vectors (7–8 kb). Therefore, for diseases such as Duchenne's muscular dystrophy where the cDNA of the defective dystropin gene is 14 kb, HSV may be the only suitable vector, although attempts are being made to use a Becker dystrophin minigene in adenoviral vectors (Hauser *et al.*, 1994). An additional advantage of HSV vectors is the efficiency of neuronal infection and rapidity of expression. For example, Bergold *et al.* (1993), found that following HSVGluR6 infection of hippocampal organotypic cultures the protein was expressed and the receptor transported into the cell membrane within 2 hr. We similarly found mRNA and protein expression within 3 hr following *in vivo* administration. HSV vectors are the only viral vectors used in the brain to date which have been shown to result in DNA persistence and gene expression essentially for the life of the animal (out to 16 months following somatic cell gene transfer in adult rats; During *et al.*, 1994), although AAV may also share similar DNA stability (with expression persisting for at least 6 months; Kaplitt *et al.*, 1994a). Furthermore, HSV are neurotropic viruses and have evolved to efficiently infect neurons in contrast to the predominantly respiratory tract viruses, AAV and adenovirus, the latter virus which appears to have glial tropism (Horrelou *et al.*, 1994; Freese *et al.*, submitted for publication).

In summary, HSV amplicon vectors are proving valuable in establishing that a viral vector approach can complement more established methods to investigate neuronal physiology, characterize regulation of gene expression in the brain, generate novel genetic models of neurological diseases, and show promise for gene therapy in the nervous system.

References

Abeliovich, A., Chen, C., Goda, Y., Silva, A. J., Stevens, C. F., and Tonegawa, S. (1993). PKC gamma mutant mice exhibit mild deficits in spatial and contextual learning. *Cell* **75**, 1253–1262.

Andersen, J. K., Frim, D. M., Isacson, O., and Breakefield, X. O. (1993). Herpesvirus-mediated gene delivery into the rat brain: Specificity and efficiency of the neuron-specific enolase promoter. *Cell Mol. Neurobiol.* **13**, 503–515.

Ben-Ari, Y. (1985). Limbic seizure and brain damage produced by kainic acid: Mechanisms and relevance to human temporal lobe epilepsy. *Neuroscience* **14**, 375–403.

Bergold, P. J., Casaccia-Bonnefil, P., Zeng, X. L., and Federoff, H. J. (1993). Transsynaptic neuronal loss induced in hippocampal slice cultures by a herpes simplex virus vector expressing the GluR6 subunit of the kainate receptor. *Proc. Natl. Acad. Sci. USA* **90**, 6165–6156.

Bettler, B., Boulter, J., Hermans-Borgmeyer, I., O'Shea-Greenfield, A., Deneris, E. S., Moll, C., Borgmeyer, U., Hollman, M., and Heineman, S. (1992). Cloning of a novel glutamate receptor subunit, GluR5: Expression in the nervous system during development. *Neuron* **8**, 257–265.

Bourtchuladze, R., Frenguelli, B., Blendy, J., Cioffi, D., Schutz, G., and Silva, A. J. (1994). Deficient long-term memory in mice with a targeted mutation of the cAMP-responsive element binding protein. *Cell* **79**, 59–68.

Burnashev, N., Monyer, H., Seeburg, P. H., and Sakmann, B. (1992). Divalent permeability of AMPA receptors is dominated by the edited form of a single subunit. *Neuron* **8**, 189–198.

Cepko, C. (1988). Immortilization of neural cells via oncogene transformation. *Trends Neurosci.* **11**, 6–8.

Challita, P-M., and Kohn, D. B. (1994). Lack of expression from a retroviral vector after transduction of murine hematopoietic stem cells is associated with methylation *in vivo*. *Proc. Natl. Acad. Sci. USA* **91**, 2567–2571.

Connors, B. W., and Gutnick, M. H. (1990). Intrinsic firing patterns of diverse neocortical neurons. *TINS* **13**, 99–104.

Davar, G., Kramer, M. F., Garber, D., *et al.* (1994). Comparative efficacy of expression of genes delivered to mouse sensory neurons with herpes virus vectors. *J. Comp. Neurol.* 339, 3–11.

de Lanerolle, N. C., Brines, M. L., Kim, J. H., Williamson, A., Philips, M. F., and Spencer, D. D. (1992). Neurochemical remodelling of the hippocampus in human temporal lobe epilepsy. *Epilepsy Res.* **9**(Suppl.), 205–219.

Doran, S. E., Ren, X. D., Betz, A. L., Pagel, M. A., Neuwelt, E. A., Roessler, B. J., and Davidson, B. L. Gene expression from recombinant viral vectors in the CNS following blood-brain-barrier disruption. *Neurosurgery,* in press.

Dudai, Y. "The Neurobiology of Memory: Concepts, Findings, Trends." Oxford Univ. Press, Oxford.

During, M. J. (1992). Genetic manipulation of signal transduction pathways in the adult mammalian brain. *Soc. Neurosci. Abstr.* **19,** 453.3.

During, M. J., and Spencer, D. D. (1993). Extracellular hippocampal glutamate and spontaneous seizure in the conscious human brain. *Lancet* **341,** 1607–1610.

During, M. J., Naegele, J. R., O'Malley, K. L., and Geller, A. I. (1994). Long-term behavioral recovery in Parkinsonian rats by an HSV vector expressing tyrosine hydroxylase. *Science* **266,** 1399–1403.

Egebjerg, J., Bettler, B., Hermans-Borgmeyer, I., and Heineman, S. (1991). Cloning of a cDNA for a glutamate receptor subunit activated by kainate but not AMPA. *Nature* **351,** 745–748.

Federoff, H. J., Geschwind, M. D., Geller, A. I., and Kessler, J. A. (1992). Expression of NGF *in vivo,* from a defective HSV-1 vector, prevents effects of axotomy on sympathetic neurons. *Proc. Natl. Acad. Sci. USA* **89,** 1636–1640.

Freese, A., Davidson, B., O'Connor, W. M., During, M. J., Kaplitt, M. G., and O'Connor, M. J. Preferential transfection of dividing glial cells by adenoviral vectors in mixed neuronal–glial cultures. Submitted for publication.

Geller, A. I., and Breakfield, X. O. (1988). A defective HSV-1 vector expresses *Escherichia coli* β-galactosidase in cultured peripheral neurons. *Science* **241,** 1667–1669.

Geller, A. I., During, M. J., and Neve, R. L. (1991). Molecular analysis of neuronal physiology by gene transfer into neurons with herpes simplex virus vectors. *Trends Neurosci.* **14,** 428–432.

Geller, A. I., During, M. J., Haycock, J. W., Freese, A., and Neve, R. (1993). Long-term increases in neurotransmitter release from neuronal cells expressing a constitutively active adenylate cyclase from a herpes simplex virus type 1 vector. *Proc. Natl. Acad. Sci. USA* **90,** 7603–7607.

Geller, A. I., During, M. J., Oh, Y. J., Freese, A., and O'Malley, K. L. (1995). An HSV-1 vector expressing tyrosine hydroxylase causes production and release of L-dopa from cultured rat striatal cells. *J. Neurochem.* **64,** 487–496.

Hauser, M. A., Phelps, F., Hauschka, S. D., and Chamberlain, J. S. (1994). Towards gene therapy for Duchenne muscular dystrophy. In "Cold Spring Harbor Lab. Meeting on Gene Therapy," abstract 67.

Ho, D. Y., Mocarski, E. S., and Sapolsky, R. M. (1993). Altering central nervous system physiology with a defective simplex virus vector expressing the glucose transporter gene. *Proc. Natl. Acad. Sci. USA* **90,** 3655–3659.

Horellou, P., Vigne, E., Castel, M-N., Barneoud, P., Colin, P., Perricaudet, M., Delaere, P., and Mallet, J. (1994). Direct intracerebral gene transfer of an adenoviral vector expressing tyrosine hydroxylase in a rat model of Parkinson's disease. *Neuroreport* **6.**

Kandel, E. R., and Schwartz, J. H. (1982). Molecular biology of learning: Modulation of transmitter release. *Science* **218,** 433–434.

Kaplitt, M. G., Leone, P., Xiao, X., Samulski, R. J., Pfaff, D. W., O'Malley, K. L., and During, M. J. (1994a). Long-term gene expression and phenotypic correction using adeno-associated virus vectors in the mammalian brain. *Nature Genet.* **8,** 148–154.

Kaplitt, M. G., Kwong, A. D., Kleopoulos, S. P., Mobbs, C. V., Rabkin, S. D., and Pfaff, D. W. (1994b). Preproenkephalin promoter yields region specific and long-term expression in adult brain after direct *in vivo* gene transfer via a defective herpes simplex viral vector. *Proc. Natl. Acad. Sci. USA* **91,** 8979–8983.

Kasahara, N., Dozy, A. M., and Kan, Y. W. Tissue-specific targeting of retroviral vectors through ligand–receptor interactions. *Science* **266,** 1373–1376.

Kohler, M., Burnashev, N., Sakmann, B., and Seeburg, P. H. (1993). Determinants of Ca permeability in both TM1 and TM2 of high affinity kainate receptor channels: Diversity by RNA editing. *Neuron* **10,** 491–500.

Kolberg, R. (1992). Animal models point the way to human clinical trials. *Science* **256,** 772–773.

Lendahl, U., and McKay, R. D. G. (1990). The use of cell lines in neurobiology. *Trends Neurosci.* **13,** 132–137.

Leone, P., Dragunow, M., Davis, K. E., Ullrey, D., O'Malley, K. L., Neve, R., Geller, A. I., and During, M. J. (1993). Hippocampal injection of a HSV-1 vector expressing an unregulated PKC from the TH promotor increases NE release, induces KROX 24 expression in dentate granule cells and improves spatial navigation performance in rats. *Soc. Neurosci. Abstr.* **19,** 328.1.

Livingstone, D. S., Sziber, P. P., and Quin, W. G. (1984). Loss of calcium/calmodulin responsiveness in adenylate cyclase of rutabaga. *Cell* **37,** 205–215.

Lowe, D., and Smith, D. O. (1993). Developmental changes in mRNA editing of glutamate receptors. *Soc. Neurosci. Abstr.* **19,** 381.12.

Lu, B., and Federoff, H. J. Herpes simplex virus type 1 amplicon vectors with glucocorticoid inducible gene expression. *Hum. Gene Ther.,* in press.

Meyer, M. L. (1992). Fine focus on glutamate receptors. *Curr. Biol.* **2,** 23–25.

Nadler, J. V. (1981). Kainic acid as a tool for the study of temporal lobe epilepsy. *Life Sci.* **29,** 2031–2042.

Oldfield, E. H., Ram, Z., Culver, K. W., Blaese, R. M., DeVroom, H. L., and Anderson, W. F. (1993). Gene therapy for the treatment of brain tumors using intra-tumoral transduction with the thymidine kinase gene and intravenous ganciclovir. *Hum. Gene Ther.* **4,** 39–69.

Olney, J. W., Rhee, V., and Ho, O. L. (1974). Kainic acid: A powerful neurotoxic analogue of glutamate. *Brain Res.* **77,** 507–512.

Papandrikopoulon, A., Doll, T., Tucker, R. P., Garner, C. C., and Matus, M. (1989). Embryonic MAP2 lacks the cross-linking sidearm sequences and dendritic targeting signal of adult MAP2. *Nature* **340,** 650–652.

Price, J., Turner, D., and Cepko, C. (1987). Lineage analysis in the vertebrate nervous system by retrovirally-mediated gene transfer. *Proc. Natl. Acad. Sci. USA* **84,** 156–160.

Sanes, J. R., Rubenstein, J. L. R., and Nicolas, J. (1986). Use of a recombinant retrovirus to study postimplantation cell lineage in mouse embryos. *EMBO J.* **5,** 3133–3142.

Silva, A. J., Paylor, R., Wehner, J. M., and Tonegawa, S. (1992). Impaired spatial learning in alpha-calcium-calmodulin kinase II mutant mice. *Science* **257,** 206–211.

Sloviter, R. S. (1982). A simplified Timm stain procedure compatible with formaldehyde fixation and routine paraffin embedding of rat brain. *Brain Res. Bull.* **8,** 771–774.

Snyder, E. Y., Macklis, J. D., Wolfe, J. H., *et al.* (1994). CNS precursor cells as gene delivery vehicles and mediators of repair. "Proceedings, Cold Spring Harbor Lab. Meeting on Gene Therapy," pp. 186.

Song, S., Leone, P., Wang, Y., Hartley, D., Bryan, J., Ullrey, D., Bak, S., Davis, K., Haycock, J., O'Malley, K., Neve, R., Geller, A., and During, M. (1993). Rotational behavior in adult rats is induced by selective activation of the protein kinase C pathway in substantia nigra pars compacta neurons with a HSV-1 vector. *Soc. Neurosci. Abstr.* **19,** 328.2.

Song, S., Wang, Y., Hartley, D., Ullrey, D., Bak, S., Ashe, O., Bryan, J., Haycock, J., During, M., Neve, R., O'Malley, K., and Geller, A. (1994). Cell type specific expression from the tyrosine hydroxylase promoter may account for the rotational behavior induced by activation of the protein kinase C pathway in substantia nigra pars compacta neurons with a HSV-1 vector. *Soc. Neurosci. Abstr.* **20,** 586.10.

Verhaagen, J., Hermens, W., Oestricher, B., Rabin, S. D., Pfaff, D., Gispen, W. H., and Kaplitt, M. G. (1994). Expression of B-50 (GAP-43) via a defective herpes simplex virus vector yields morphological changes in cultured non-neuronal cells. *Mol. Brain Res.*

Wang, Y., O'Malley, B. W., Jr., Tsai, S. Y., and O'Malley, B. W. (1994). A regulatory system for use in gene transfer. *Proc. Natl. Acad. Sci. USA* **91,** 8180–8184.

Wolfe, J. H., Deshmane, S. L, and Fraser, N. W. (1992). Herpesvirus vector gene transger and expression of β-glucoronidase in the central nervous system of MPS VII mice. *Nature Genet.* **1**, 379–384.

Wolff, J. A. (1993). Postnatal gene transfer into the central nervous system. *Curr. Biol.* **3**, 743–748.

Zuber, M. X., Strittmatter, S. M., and Fishman, M. C. (1989). A membrane targeting signal in the amino terminus of the neuroprotein, GAP-43. *Nature* **341**, 345–348.

CHAPTER 7

The Use of Defective Herpes Simplex Virus Amplicon Vectors to Modify Neural Cells and the Nervous System

Howard J. Federoff
Departments of Medicine and Neuroscience
Albert Einstein College of Medicine
Bronx, New York

I. The Origin of Herpes Simplex Virus (HSV) Amplicon Vectors

Herpes simplex virus is a 150-kb DNA virus that contains approximately 70 genes (Roizman, 1990; Roizman and Sears, 1990). The virus has a broad host range and appears to be capable of infecting most mammalian differentiated cell types. A number of genes within the wild-type HSV genome are not essential for its growth in tissue culture. This feature was initially exploited by Roizman and his colleagues to develop recombinant HSV viruses as vehicles for the transfer of heterologous genes (Roizman and Jenkins, 1985; Shih *et al.*, 1984). This recombinant HSV vector approach has subsequently been used by a number of investigators and remains a viable gene-transfer method that complements the amplicon vector approach to be described (Andersen *et al.*, 1992; Dobson *et al.*, 1990; Fink *et al.*, 1992).

The observations that led to the amplicon vector arose from studies on the genomes of defective interfering HSV particles (Spaete and Frenkel, 1982; Stow and McMonagle, 1982) that accumulated in HSV stocks that were passaged at high multiplicities of infection (m.o.i.). Analysis of these genomes revealed that they were composed of relatively simple reiterations of DNA sequences from the wild-type HSV genome. Principally, the genomes contained an origin of DNA replication and a cleavage/packaging site (Spaete and Frenkel, 1985; Spaete *et al.*, 1982; Stow *et al.*, 1982). The critical observation was that a plasmid containing these two DNA sequences could be replicated and packaged into virions if it was transfected into a cell and supplied with HSV replication and virion assembly functions by a superinfecting wild-type virus. Analysis of the genomes in these amplicon stocks indicated that they remained simple and were predominantly 150-kb molecules composed of concatenated units of the original plasmid (Spaete *et al.*, 1985, 1982; Stow *et al.*, 1982; Vlazny *et al.*, 1982).

II. Amplicon Vectors as Gene-Transfer Vehicles

The production of amplicon vectors requires a copropagated HSV helper virus. The use of wild-type helper virus greatly limits the utility of the amplicon as gene-transfer vehicle since it produces cell lysis. Two related approaches have been developed to produce amplicon vector stocks in which the helper virus is replication incompetent or impaired. Both approaches rely on the use of helper viruses that carry mutations within an essential immediate-early HSV gene, typically IE3 (Geller *et al.*, 1990; Paterson and Everett, 1990). Helper virus with a missense mutation within the IE3 gene has a temperature-sensitive phenotype: replication occurs at 34°C but not at 39°C (Preston, 1979a,b). Deletion of all or part of the IE3 gene yields viruses that are incapable of growing on normal tissue cells but can be grown on cell lines that stably express integrated copies of the IE3 gene. These complementing cell lines have been generated in different cell types and with different transfected gene segments (DeLuca *et al.*, 1985; DeLuca and Schaffer, 1987; Paterson *et al.*, 1990). When a helper virus carrying a partial deletion of the IE3 gene is grown on a complementing cell line recombinant wild-type particles arise. The frequency that they arise (10^{-5} to 10^{-6}) appears related to the extent of overlapping homology between the integrated gene and the residual IE3 sequences flanking the deletion (DeLuca *et al.*, 1985, 1987; Paterson *et al.*, 1990). Typical amplicon stocks will have titers between 1 and 10 $\times$ 10^6 infectious particles (i.p.) of amplicon and between 0.1 and 5 $\times$ 10^7 i.p. of helper virus.

The minimal vector contains four genetic elements: (1) a plasmid backbone with a Col E1 origin and drug-resistance gene (typically β-lactamase) for growth in *Escherichia coli*, (2) an HSV origin of replication, (3) a cleavage/packaging sequence, and (4) a transcription unit. Initial amplicon vectors were constructed so that a viral promoter, usually an HSV immediate-early (IE) promoter (Geller and Breakefield, 1988) or human CMV IE promoter (Ho *et al.*, 1993; Kaplitt *et al.*, 1991), was driving the expression of the gene of interest. Recently, some investigators have begun to construct amplicon vectors with cellular promoters (Kaplitt *et al.*, 1993). Most amplicon vectors are designed to contain an intron although the requirement of this feature for efficient gene expression has not been systematically studied.

Amplicon vectors can transfer genes into postmitotic cells, such as neurons and myocytes, and also into dividing primary cells and established lines (Boothman *et al.*, 1989; Federoff *et al.*, 1992; Geller, *et al.*, 1988; Geller and Freese, 1990; Geschwind *et al.*, 1994; Federoff, unpublished observations). Gene transfer into neurons *in vivo* appears to result in stable maintenance of the episomal amplicon vector in some cells. However, a quantitative analysis of persistent amplicon genomes over time postinfection has not been reported. Using PCR-based assays amplicon genomes have been detected up to 12 months after infection (M. During, personal communication). In cultured sensory neurons, amplicon vectors persist for many weeks (Geller *et al.*, 1993; Federoff, unpublished observations). In dividing cells amplicons are mitotically segregated leading to a progressive dilution of the cells harboring the replication-incompetent vector.

Most data suggest that *in vivo* expression of transferred genes from amplicon vectors declines within a month (Federoff, unpublished data) although some studies report expressing cells after a year (During *et al.*, 1992; M. During, personal communication). In these studies gene expression was driven by HSV promoters. At this time it is unclear whether decreasing expression reflects loss of amplicon genomes, downregulation of viral promoter transcription, or both. Studies of viral promoter-driven gene expression in different types of vectors suggest that downregulation is a common problem, thus leading to the speculation that a similar mechanism operates in the amplicon vector.

The nature of the episomal amplicon genome in the transduced cell has not been examined. In sensory neurons latently infected with wild-type HSV the genome is associated with nucleosomes (Deshmane and Fraser, 1989). By analogy we presume the amplicon associates with nucleosomes. More importantly, we must determine whether the organization of a cellular transcription unit in the episomal amplicon context is similar to the native gene in its chromosomal context.

III. Use of Amplicon Vectors to Modify Neuron Function

A. Altering Neuron Survival and Phenotype by Transduction of NGF

We constructed an amplicon vector that placed a NGF minigene under the transcription control of the HSV IE 4/5 promoter. The vector, HSVngf, was first characterized in cultures of fibroblasts. In these inital studies we demonstrated that transduction led to transcription of the vector-encoded NGF gene, processing of the primary transcript, and synthesis and secretion of NGF (Federoff *et al.*, 1992; Geschwind *et al.*, 1994). In subsequent studies we determined whether primary sympathetic neurons that require exogenous NGF for survival could be transduced with the HSVngf vector and whether paracrine neurotrophic factor secretion would obviate the need for exogenous NGF. We observed that neuron survival was related to the amplicon m.o.i. and that when optimized survival was comparable to nontransduced cultures treated with saturating amounts of exogenous NGF (Geschwind *et al.*, 1994).

Our experiments in cultured sympathetic neurons led us to evaluate the HSVngf vector in sympathetic ganglia *in vivo*. We chose to use an axotomy model of the superior cervical ganglion (SCG). The cell bodies of sympathetic neurons in the SCG send postganglionic axons to the structures in the head and neck. Axotomy of these axons disconnects the neurons from their target cell-derived NGF and within days produces a decline in the catecholamine biosynthetic enzyme tyrosine hydroxylase (TH). In our experiment we questioned whether transduction of SCG with HSVngf virus before axotomy would lead to NGF expression and secretion and protect against the effects of subsequent axotomy. In the study, HSVngf or control virus was injected and allowed to express for 4 days prior to performing postganglionic axotomy. After an additional 10 days ganglia were harvested and TH levels measured. In ganglia that were transduced with HSVngf axotomy produced no decline in TH content. Whereas, in control virus-transduced or saline-injected ganglia, axotomy produced a significant decline in TH (Federoff *et al.*, 1992). These experiments demonstrated the *in vivo* efficacy of HSV vectors and provided a basis for other studies on neuronal modification.

B. Converting Neurons to a NGF-Responsive State: Expression of *trkA*

NGF action is transduced through a receptor tyrosine kinase, *trkA*. The expression of the *trkA* gene product in nonneuronal cells is sufficient to produce a high-affinity receptor with complete signaling capability (Ip *et al.*, 1993). However, in neuronal cells and neurons an accessory receptor, the low-affinity NGF receptor (LNGFR), a 75-kDa protein modulates *trkA* receptor function (Hempstead *et al.*, 1991; Ip *et al.*, 1993).

In our experiments we wanted to test whether expression of *trkA* in neurons that are unresponsive to NGF would render them NGF responsive. We constructed HSVtrk, an amplicon vector that expresses the rat *trkA* gene (Xu *et al.*, 1994). Transduction of cultured sensory neurons from the nodose ganglion and spinal motor neurons with HSVtrk established an activated NGF signal transduction pathway. Treatment of transduced cultures with NGF produced neuron survival in both neuron types and also upregulated the expression of the acetylcholine biosynthetic enzyme cholineacetyl transferase in spinal motor neurons (Xu *et al.,* 1994). Although these experiments demonstrated that the expression of *trkA* within nodose and spinal motor neurons was sufficient to create NGF responsiveness it is uncertain whether this observation will hold for all neuron types. Since both the nodose and the spinal motor neurons were responsive to other members of the neurotrophin family of growth factors it is likely that the vector encoded *trkA* gene product utilized the existing downstream signal transduction components.

C. Modification of Neuronal Excitability: Expression of the Kainate Receptor GluR6

Amplicon vectors can be used to express genes in dissociated neurons *in vivo* and in organotypic cultures prepared from brain. The potential advantages of the latter are that these preparations are stable for weeks in culture and maintain many of the physiological properties of acute slices (Gahwiler, 1981, 1984; Gahwiler and Brown, 1987; Stoppini *et al.,* 1991; Xu *et al.,* 1994).

For amplicon gene transfer into hippocampal slice cultures to be a useful adjunctive method for physiological experiments required that the virus be capable of producing regional infections. Bath application of virus resulted in gene transfer and expression predominantly in surface glia and little expression within neurons (Casaccia-Bonnefil *et al.,* 1993). Taking a different approach, we used a micropipette to deliver nanoliter quantities of virus directly to regions of the slice culture. Staining of cultures showed that expression was limited to the microapplication site. With this method there is a linear relationship between the number of virions applied and the number of transduced cells (Casaccia-Bonnefil *et al.,* 1993).

In our initial studies to modify hippocampal slice physiology we chose to express the edited form of the glutamate receptor GluR6 (Egebjerg *et al.,* 1991). In this vector, HSVGluR6, the GluR6 gene was under the trancriptional control of the strong and rapidly activated HSV IE 4/5 promoter. GluR6 is kainate-activated ionotropic receptor that is capable of forming homomeric channels that are highly permeable to calcium. Activation of the GluR6 channel by kainate or glutamate causes an inward current carried by sodium or calcium ions. Calcium influx through GluR6 and other related kainate-activated channels is presumed to underlie their consvulsant and excitotoxic actions (reviewed in Seeburg, 1993).

Transduction of fibroblasts with HSVGluR6 led to accumulation of the ionotropic receptor and rendered them sensitive to kainate toxicity (Bergold *et al.,* 1993b). When introduced into CA3 pyramidal neurons of hippocampal slice cultures, HSVGluR6 produced biphasic neuronal cell loss. Early cell loss occurred within 6 hr and involved cells at the virus microapplication site (Bergold *et al.,* 1993a). Delayed cell loss occurred in the remaining CA3 region and at times in part of the hilar region (Bergold *et al.,* 1993a,b). Electrophysiological recording and pharmacological experiments provided insight into this pattern of cell death. Within 2.5 hr after virus delivery, slices developed spontaneous epileptiform discharge that was recorded away from the microapplication site. The discharge persisted after transduced cells were lost at the microapplication site. Therefore, nontransduced cells had acquired increased excitability. To determine which glutamate receptor subtypes were involved in each phase of the process we used glutamate antagonists. Treatment of HSVGluR6-transduced slice cultures with CNQX, a non-NMDA receptor antagonist that blocks kainate receptors including GluR6, blocked epileptiform discharge and both early and delayed cell loss. Washout of CNQX after 6 hr resulted in the rapid appearance of epileptiform activity. However, treatment of HSVGluR6-transduced slice cultures with APV, an NMDA antagonist, blocked delayed but not early cell loss. These data suggested that transduced cells initiate epileptiform discharge, produce acquired excitability in nontransduced cells via a process that requires NMDA receptor activation, and then die.

Introduction of HSVGluR6 virions into the hippocampal formation of the rat produced limbic seizures (During *et al.*, 1993). The mean latency before induction of epilepsy was similar to that observed in slice cultures. Acute hippocampal slices were prepared from rats 1 day after virus injection and studied electrophysiologically. Spontaneous bursting was observbed in CA1 neurons recorded at a distance from the virus injection site. The bursting response was not accompanied by increased excitatory and diminished inhibitory postsynaptic potentials. These data suggested that, like the slice culture system, nontransduced neurons had acquired increased excitability. The pharmacologic and physiologic basis for the the bursting behavior of nontransduced neurons is currently under investigation. A morphological study of the hippocampus was performed on rats after several months of a persistent seizure disorder. Staining for zinc containing mossy fibers by the Timm method revealed sprouting, a histopathological finding in other models of epilepsy as well as in human temporal lobe epilepsy. The relevance of the HSVGluR6 model of epilepsy to the human condition remains to be clarifed. As a model, however, it allows for the molecular genetic dissection of the glutamate receptor function in triggering epileptiform discharge.

IV. Summary and Future Directions

HSV amplicon vectors are readily engineered and are an efficient mammalian expression system. They appear to infect most, if not all, mammalian cell types. When constructed with a strong HSV IE promoter, amplicon vectors turn on gene expression within hours and lead to the accumulation of large amounts of gene product. As illustrated in the previous sections, the HSV amplicons can be used to transfer genes into dissociated cell culture, organotypic preparations, and in the intact animal.

As a gene-transfer vehicle, amplicon vectors have some limitations. The requirement for helper virus, although itself replication defective, results in stocks containing two types of virions: vector and helper. The expression of IE genes from helper virus may produce some toxicity in certain cell types, particularly at higher m.o.i. (Johnson *et al.*, 1992). As mentioned, the relatively rare recombination event between helper and the packaging cell line results in wild type revertants.

Future development of amplicon vectors is necessary and certain areas need attention. First, amplicon vectors should be developed to study promoter function and identify *cis* elements that can target gene expression to specific subpopulations of cells. Such vectors could complement the transgenic animal approach for the study of promoter function. Second, amplicon vectors with inducible promoters would be important for studying the function of gene products over a defined concentration and time frame. For some *in vivo* work it is desirable to temporally separate amplicon vector delivery from the expression of the inducible gene product. Third, the practical limits of amplicon DNA carrying capacity should be evaluated. For example, will it be possible to use the 150 kb capacity to mobilize a large genomic fragment of DNA? Fourth, complementing packaging cell lines and helper virus combinations should be developed to reduce wild-type reversion and cytotoxicity. Although the ideal system would be one in which helper virus could be eliminated or removed, this is an unlikely possibility given our current understanding of HSV biology.

Although a number of other viral delivery systems have been developed to transfer genes into neurons, such as those based on adenovirus (Davidson *et al.*, 1993; Le Gal La Salle *et al.*, 1993) and adeno-associated virus (M. Kaplitt and M. During, personal communication), it is currently unclear whether these will supplant the amplicon vector approach. Given the relative simplicity of amplicon vector construction and manipulation it is likely that they will continue to be used as a gene transfer method.

References

Andersen, J., Garber, D., Meaney, C., and Breakefield, X. (1992). Gene transfer into mammalian central nervous system using herpes virus vectors: Extended expression of bacterial lacz in neurons using the neuron-specific enolase promoter. *Hum. Gene Ther.* **3**, 487–499.

Bergold, P., Casaccia-Bonnefil, P., Federoff, H., and Stelzer, A. (1993a). Transduction of CA3 neurons by a herpes virus vector containing a GluR6 subunit of the kainate receptor induces epileptiform discharge. *Soc. Neurosci. Abstr.* **19**, 21.

Bergold, P., Cassaccia-Bonnefil, P., Xiu-Liu, Z., and Federoff, H. (1993b). Transynaptic neuronal loss induced in hippocampal slice cultures by a herpes simplex virus vector expressing the GluR6 subunit of the kainate receptor. *Proc. Natl. Acad. Sci. USA* **90**, 6165–6169.

Boothman, D. A., Geller, A. I., and Pardee, A. B. (1989). Expression of the E. coli Lac Z gene from a defective HSV-1 vector in various human normal, cancer-prone and tumor cells. *FEBS Lett.* **258,** 159–162.

Casaccia-Bonnefil, P., Benedikz, E., Shen, H., Stelzer, A., Edelstein, D., Geschwind, M., Brownlee, M., Federoff, H., and Bergold, P. (1993). Localized gene transfer into hippocampal slice cultures and acute hippocampal slices. *J. Neurosci. Methods* **50,** 341–351.

Davidson, B., Allen, E., Kozarsky, K., Wilson, K., and Roessler, B. (1993). A model system for in vivo gene transfer into the central nervous system using an adenoviral vector. *Nature Genet.* **3,** 219–223.

DeLuca, N., McCarthy, A., and Schaffer, P. (1985). Isolation and characterization of deletion mutants of herpes simplex virus type 1 in the gene encoding immediate-early regulatory protein ICP4. *J. Virol* **56,** 558–570.

DeLuca, N., and Schaffer, P. (1987). Activities of herpes simplex virus type 1 (HSV-1) ICP4 genes specifying nonsense peptides. *Nucleic Acids Res.* **15,** 4491–4511.

Deshmane, S., and Fraser, N. (1989). During latency, herpes simplex virus type 1 DNA is associated with nucleosomes in a chromatin structure. *J Virol.* **63,** 943–947.

Dobson, A., Margolis, T. P., Sedarati, F., Stevens, J., and Feldman, L. T. (1990). A latent, nonpathogenic HSV-1 derived vector stably expresses beta-galactosidase in mouse neurons. *Neuron* **5,** 353–360.

During, M., Mirchandani, G., Leone, P., Williamson, A., de Lanerolle, N., Geschwind, M., Bergold, P., and Federoff, H. (1993). Direct hippocampal injection of a HSV-1 vector expressing GluR6 results in spontaneous seizures, hyperexcitability in CA1 cells, and loss of CA1, hilar and CA3 neurons. *Soc. Neurosci. Abstr.* **19,** 21.

During, M. J., Geller, A. I., Deutch, A. Y., and O'Malley, K.L. (1992). Recovery in the rate 6-hydroxydopamine model of Parkinson's Disease by direct intrastriatal injection of HSV-1 vectors which express the human tyrosine hydroxylase gene. *Soc. Neurosci. Abstr.* **18,** 782.

Egebjerg, J., Bettler, B., and Hermans-Borgmeyer, I. S. H. (1991). Cloning of a cDNA for a glutamate receptor subunit activated by kainate but not AMPA. *Nature* **351,** 745–748.

Federoff, H. J., Geschwind, M. D., Geller, A. I., and Kessler, J. A. (1992). Expression of nerve growth factor *in vivo*, from a defective HSV-1 vector, prevents effects of axotomy on sympathetic ganglia. *Proc. Natl. Acad. Sci. USA* **89,** 1636–1640.

Fink, D., Sternberg, L., Weber, P., Mata, M., Goins, W., and Glorioso, J. (1992). In vivo expression of beta-galactosidase in hippocampal neurons by HSV-1 mediated transfer. *Hum. Gene Ther.* **3,** 11–19.

Gahwiler, B. (1981). Organotypic monolayer cultures of nervous tissue. *J. Neurosci. Methods.* **4,** 329–342.

Gahwiler, B. (1984). Development of the hippocampus in vitro: Synapses and receptors. *Neuroscience* **11,** 751–760.

Gahwiler, B., and Brown, D. (1987). Muscarine affects calcium-currents in rat hippocampal pyramidal cells in vitro. *Neurosci. Lett.* 301–306.

Geller, A., During, M., Haycock, J., Freese, A., and Neve, R. (1993). Long-term increases in neurotransmitter release for neuronal cells expressing a constitutively acting adenylate cyclase from a HSV-1 vector. *Proc. Natl. Acad. Sci. USA* **90,** 7603–7607.

Geller, A. I., and Breakefield, X. O. (1988). A defective HSV-1 vector expresses *Escherichia coli* beta-galactosidase in cultured peripheral neurons. *Science* **241,** 1667–1669.

Geller, A. I., and Freese, A. (1990). Infection of cultured central nervous system neurons with a defective herpes simplex virus 1 vector results in stable expression of *Escherichia coli* beta-galactosidase. *Proc. Natl. Acad. Sci. USA* 87, 1149–1153.

Geller, A. I., Keyomarsi, K., Bryan, J., and Pardee, A. B. (1990). An efficient deletion mutant packaging system for defective herpes simplex vectors: Potential applications to human gene therapy and neuronal physiology. *Proc. Natl. Acad. Sci. USA* **87,** 8950–8954.

Geschwind, M., Kessler, J., Geller, A., and Federoff, H. (1994). Transfer of the nerve growth factor gene into cell lines and cultured neurons using a defective Herpes Simplex virus vector. *Mol. Brain Res* **24**, 327–335.

Hempstead, B. L., Martin-Zanca, D., Kaplan, D. R., Parada, L. F., and Chao, M. V. (1991). High affinity NGF binding requires co-expression of the trk proto-oncogene and the low affinity NGF receptor. *Nature* **350**, 678–683.

Ho, D., Mocarski, E., and Sapoloski, R. (1993). Altering central nervous system physiology with a defective herpes simplex virus vector expressing the glucose transporter gene. *Proc. Natl. Acad. Sci. USA* **90**, 3655–3659.

Ip, N., Stitt, T., Tapley, P., Klein, R., Glass, D., Fandl, J., Greene, L., Barbacid, M., and Yancopoulos, G. (1993). Similarities and differences in the way neurotrophins interact with the Trk receptors in neuronal and nonneuronal cells. *Neuron* **10**, 137–149.

Johnson, P., Miyanohara, A., Levine, F., Cahill, T., and Friedman, T. (1992). Cytotoxicity of replication-defective mutant of herpes simplex virus type 1. *J. Virol.* **66**, 2952–2965.

Kaplitt, M., Yin, J., Kwong, A., Kleopoulos, S., Mobbs, C., and Pfaff, D. (1993). In vivo deletion analysis of the rat pre-proenkephalin promoter using a defective herpes simplex viral vector. *Soc. Neurosci. Abstr.* **19**, 1259.

Kaplitt, M. G., Pfauss, J. G., Kleopoulous, S. P., Hanlon, B. A., Rabkin, S. D., and Pfaff, D. W. (1991). Expression of a functional foreign gene in adult mammalian brain following in vivo transfer via a herpes simplex virus defective viral vector. *Soc. Neurosci. Abstr.* **17**, 1285.

Le Gal La Salle, G., Robert, J., Berrard, S., Ridous, V., Stratford-Perricaudet, L., Perricaudet, M., and Mallet, J. (1993). An adenovirus vector for gene transfer into neurons and glia in the brain. *Science* **259**, 988–990.

Paterson, T., and Everett, R. (1990). A prominent serine-Rich region in Vmw175, the major transcriptional regulator portein of herpes simplex virus type 1, is not essential for virus growth in tissue culture. *J. Gen. Virol.* **71**, 1775–1783.

Preston, C. (1979a). Abnormal properties of an immediate early polypeptide in cells infected with the herpes simplex virus type 1 mutant tsK. *J. Virol.* **32**, 357–369.

Preston, C. (1979b). Control of herpes simplex virus type 1 mRNA synthesis in cells infected with wild type or the temperature sensitive mutant TsK. *J. Virol.* **29**, 275–284.

Roizman, B. (1990). Herpes viridae: A brief introduction. In "Virology" (B. Fields, and D. Knipe, Eds.), 2nd ed., pp. 1787–1794. Raven Press, New York.

Roizman, B., and Jenkins, F. (1985). Genetic engineering of novel genomes of large DNA viruses. *Science* **229**, 1208–1214.

Roizman, B., and Sears, A. (1990). Herpes simplex viruses and their replication. In "Virology" (B. Fields, and D. Knipe, Eds.), pp. 1795–1842. Raven Press, New York.

Seeburg, P. (1993). The TINS/TiPS lecture: The molecular biology of mammalian glutamate receptor channels. *Trends Neurosci.* **16**, 359–365.

Shih, M., Arsenakis, M., Tiollais, P., and Roizman, B. (1984). Expression of hepatitis B S gene by a herpes simplex type 1 vectors carrying α- and β-regulated gene chimeras. *Proc. Natl. Acad. Sci. USA* **81**, 5867–5870.

Spaete, R., and Frenkel, N. (1985). The herpes simplex virus amplicon: Analyses of cis-acting replication functions. *Proc. Nat. Acad. Sci. USA* **82**, 694–698.

Spaete, R., and Frenkel, N. (1982). The herpes simplex virus amplicon: A new eucaryotic defective-virus cloning-amplifying vector. *Cell* **30**, 305–310.

Stoppini, L., Buchs, P-A., and Muller, D. (1991). A simple method for organotypic culture of nervous tissue. *J. Neurosci. Methods* **37**, 173–182.

Stow, N., and McMonagle, E. (1982). Propagation of foreign DNA sequences linked to a herpes simplex virus origin of replication. In "Eucaryotic Viral Vectors" (Y. Gluzman, Eds.), pp. 199–204. Cold Spring Harbor Laboratory Press, Cold Spring Harbor, NY.

Vlazny, D., Kwong, A., and Frenkel, N. (1982). Site-specific cleavage/packaging of herpes simplex virus DNA and the selective maturation of nucleocapsids containing full-length viral DNA. *Proc. Natl. Acad. Sci. USA* 79, 1423–1427.

Xu, H., Federoff, H., Maragos, J., Parada, L., and Kessler, J. (1994). Viral transduction of *trkA* into cultured nodose and spinal motor neurons conveys NGF responsiveness. *Dev. Biol.* 163, 152–161.

C H A P T E R

Viral Vector-Mediated Gene Transfer in the Nervous System: Application to the Reconstruction of Neural Circuits in the Injured Mammalian Brain

J. Verhaagen
W. T. J. M. C. Hermens
A. J. G. D. Holtmaat
Netherlands Institute for Brain Research
Amsterdam, The Netherlands
Rudolf Magnus Institute for Neuroscience
Department of Pharmacology
Stratenum
Utrecht, The Netherlands

A. B. Oestreicher
W. H. Gispen
Rudolf Magnus Institute for Neuroscience
Department of Pharmacology
Stratenum
Utrecht, The Netherlands

M. G. Kaplitt
Laboratory of Neurobiology and Behavior
The Rockefeller University
New York, New York

I. Introduction

The mammalian nervous system consists of cellular networks formed by neurons and their projections and glial cells. The integrity of this complex cellular structure is essential for the proper execution of all body functions. Injury inflicted upon central nervous system (CNS) neurons usually poses major problems to patients and their neurologists, since damaged neurons do not normally regenerate and there are no effective treatments available to enhance neuroregeneration. Previous approaches to improve neuroregeneration were predominantly of a neurosurgical or pharmacological nature (Kline, 1990; Strand *et al.*, 1991). Although important progress has been made with these strategies, in particular in the peripheral nervous system (PNS), repair of the injured CNS remains an insurmountable problem.

The overall objective of our research is to develop experimental gene therapy strategies, based on gene transfer with defective viral vectors, to promote neuroregeneration. Direct viral vector-mediated transduction of genes to neural cells following injury could either augment existing or provide new functions to these cells. Thus, this approach aims at the direct delivery and expression of putative therapeutic genes to injured neurons and to glial cells in an injured nerve tract.

Our initial studies focus on viral vector-mediated gene transfer of B-50/GAP43, a neural growth-associated protein implicated in the formation of nerve fibers. The generation of herpes simplex and adenovirus vectors for this growth-associated protein together with their application *in vitro* and *in vivo* will be the main issue of this chapter.

Two crucial aspects of the repair process are considered as important future targets for genetic intervention. First, the (over)expression of neuron-survival promoting (e.g., nerve growth factor) and axonal guidance molecules (e.g., N-CAM, axonin-1/TAG1) in post-traumatic neurons could result in neuron rescue or could promote regenerative sprouting and nerve fiber fasciculation. Second, gene-transfer experiments will be devised to disrupt the cellular and molecular events that result in the formation of neural scar tissue. A neural scar acts as a barrier to newly formed sprouts, thus contributing significantly to the failure of neuroregeneration in the CNS. The simultaneous viral vector-directed stimulation of nerve fiber outgrowth and the modification of scar formation could provide a powerful novel way to promote the regenerative response of damaged neural tissues.

II. Overview: Regeneration in the Peripheral and Central Nervous System

A fundamental issue in modern neurobiology entails the study of the formation of neuronal fiber tracts and their potential for regeneration following nerve damage. After lesioning of a peripheral nerve, four essential stages in the repair process can be distinguished: (1) initiation of nerve fiber growth, (2) elongation of nerve fibers over long distances, (3) arrival of the newly formed sprouts in the target tissue and recognition of a specific target cell, and (4) functional maturation of the newly formed connections.

In contrast to the relatively favorable situation in the PNS disruption of the integrity of a CNS connection is usually not accompanied by regeneration, but results in nerve fiber degeneration and nerve cell death. Although some nerve fiber formation occurs in the CNS, the growth of newly formed sprouts is often aborted shortly after nerve damage. Thus, regenerative sprouts in the CNS do not grow over long distances, as is often seen in the PNS, and reconstruction of the original fiber tract does not occur.

There is no evidence to suggest that lost mammalian neurons can be replaced through cell division. Great progress, however, has been made in understanding the factors that contribute to the survival of neurons and the formation and regeneration of nerve tracts. In particular, the biochemical purification and molecular cloning of neurotrophins (NGF, BDNF, NT-3/5; Korsching, 1993), intraneuronal growth-associated proteins (B-50/GAP43, Tuα1; Skene, 1989; Miller *et al.*, 1989), and cell adhesion molecules (N-CAM, Ng-CAM, axonin-1/TAG1; Rutishauer and Jessell, 1988; Sonderegger and Rahtjen, 1992) has advanced our insight into the forces that stimulate the growth potential of nerve fibers. These molecules act in a partly overlapping fashion on distinct aspects of the regeneration process. The neurotrophins and their receptors are required for neuronal survival and differentiation. The growth-associated protein B-50/GAP-43 acts intraneuronally and *in vitro* experiments (see Sections III and IV) suggest that this protein may facilitate nerve fiber extension over long distances, while a cell adhesion molecule (e.g., N-CAM or axonin-1) promotes nerve fiber fasciculation and guides growing fibers to the target area.

The dichotomy between the relatively successful regeneration in the PNS and the poor regeneration in the CNS is due to differences in intrinsic properties (Skene, 1989; Fawcet, 1992) and to differences in the environment in which regenerating nerve fibers try to penetrate (Schwab *et al.*, 1993). PNS neurons have retained their capacity to upregulate the expression of a number of intrinsic proteins, including a juvenile form of tubulin (Tuα1; Miller *et al.*, 1989), and a growth-associated membrane protein (B-50/GAP43; Skene and Willard, 1981; Verhaagen *et al.*, 1986, 1988). These proteins are probably involved in the regulation of the stabilization of the cytoskeleton (Tuα1) or in signal transduction in the growing axon (B-50/GAP-43, see also Section III). Although growth-associated proteins are expressed during development in the CNS their expression is not or only transiently and incompletely upregulated in CNS neurons following trauma (Tetzlaff *et al.*, 1990, 1991; Doster *et al.*, 1991; Verhaagen *et al.*, 1993). In contrast, injury to PNS neurons does result in an induction and prolonged increase in expression of these proteins (Skene, 1989; Verhaagen *et al.*, 1986, 1988; Tetzlaff *et al.*, 1989; Van de Zee *et al.*, 1989).

In addition to these intrinsic differences between PNS and CNS neurons, many CNS neurons encounter an environment which does not permit regeneration. Insight into the molecular basis for this phenomenon is emerging, mainly from work by the groups of Schwab, Raper, and Goodman (Schwab *et al.*, 1993; Luo *et al.*, 1993; Kolodkin *et al.*, 1993). Two purified glycoproteins, neural inhibitory proteins with a relative molecular weight of 35 and 250 kDa (designated NI-35 and NI250), that apparently form a molecular complex and are expressed by oligodendrocytes inhibit the outgrowth of nerve fibers. Application of monoclonal antibodies directed against NI-35 allowed nerve fiber outgrowth over white matter *in vitro* (Caroni and Schwab, 1988) and resulted in a moderate degree of regeneration in the injured spinal cord *in vivo* (Schnell and Schwab, 1990). The recent cloning of a growth-cone collapse protein from chick (designated collapsin; Luo *et al.*, 1993) and the identification of the homologous proteins from drosophila and man (designated semaphorin; Kolodkin *et al.*, 1993) provide a solid molecular basis for the concept that both repulsive and stimulating forces are involved in nerve tract formation.

III. The Growth-Associated Protein B-50/GAP43

B-50/GAP43 is a nervous tissue-specific phosphoprotein (Zwiers *et al.*, 1976; Kristjansson *et al.*, 1982) associated with the cytoplasmic side of the neuronal plasma membrane through thioester bonds between fatty acids and two N-terminally located cysteine residues (Skene and Viràg, 1989). Protein kinase-C phosphorylates B-50/GAP43 at serine residue 41 (Coggins and Zwiers, 1989). This phospho site in B-50/GAP43 is part of a calmodulin binding site (Alexander *et al.*, 1987; De Graan *et al.*, 1990) and the association of B-50/GAP43 with calmodulin is phosphorylation dependent: phospho-B50 does not interact with calmodulin, while dephospho-B50 forms a complex with calmodulin. The aminoterminal 10–20 residues of B-50/GAP43 have been proposed as a domain that interacts and activates G_0, a G-protein abundantly expressed in neuronal growth cones (Strittmatter *et al.*, 1990, 1994). The phosphorylation/calmodulin binding site and the combined membrane anchor/G-protein activation site are probably crucial to the function of B-50/GAP43 since these regions of the molecule have been highly conserved during evolution (LaBate and Skene, 1989).

During neuroembryogenesis B-50/GAP43 is expressed throughout the developing peripheral and central nervous system and occurs according to a precise temporal and spatial expression pattern reflecting stages of nerve fiber outgrowth (Biffo *et al.*, 1990). The tight correlation between the expression of B-50/GAP43 and nerve fiber outgrowth is perhaps most clearly demonstrated in the neuroepithelium surrounding the primitive embryonic brain vesicles. This neuroepithelium consists of three layers: an ependymal layer where mitotic divisions occur, a mantle layer formed from migrating neuroblasts, and a marginal layer where neuroblasts differentiate and nerve fibers are formed. B-50/GAP43 mRNA is exclusively expressed in cells in the marginal layer from where the fine B-50/GAP43-positive pioneer fibers arise. This demonstrates that B-50/GAP43 is not expressed in neuronal stem cells or in neuroblasts, but appears when neuronal cells begin to elaborate neurites. B-50/GAP43 levels in injured peripheral neurons are normally upregulated 5- to 10-fold following a lesion and increased levels of the protein are associated with periods of nerve regeneration (Skene and Willard, 1981; Verhaagen *et al.*, 1986). The decline in B-50/GAP43 levels closely correlates with the completion of synapse formation and with the maturation of axon–glia interactions. This suggests that inhibitory signals associated with these events may play a role in downregulating B-50/GAP43 expression and that injury may interrupt this inhibitory influence resulting in the reexpression of the B-50/GAP43 gene (Skene, 1992).

IV. Viral Vector-Mediated Gene Transfer of B-50/GAP43

The striking growth-associated expression of B-50/GAP43 fuelled intense interest in determining what role, if any, B-50/GAP43 plays in the development and regeneration of nerve tracts. Some of the most direct experimental evidence supporting a role for B-50/GAP43 in mechanisms that govern neuronal process outgrowth was obtained in genetic intervention experiments *in vitro*. Transfection studies in COS cell with a B-50/GAP43 plasmid expression vector resulted in the formation of filopodial extensions very similar to those seen on growth cones (Zuber *et al.*, 1989; Widmer and Caroni, 1992). In addition, stable transfections of PC-12 cells (Yankner *et al.*, 1991) and retrovirus-mediated expression of B-50/GAP43 in neuro-2A cells (Morton and Buss, 1992) imparted these cells with a more rapid neurite outgrowth response following NGF (PC-12 cells) or retinoid acid treatment (neuro-2A cells). Selective inhibition of B-50 protein synthesis with antisense B-50/GAP43 oligonucleotides (Jap Tjoen San *et al.*, 1993) and blockade of the protein with specific anti-B-50/GAP43 antibodies (Shea *et al.*, 1991) caused a significant reduction of neurite outgrowth in these cells in culture.

A. Defective Herpes Viral Vector-Directed Expression of B-50/GAP43

An approach to demonstrate a direct effect of B-50/GAP43 on cell shape, nerve fiber extension, and regeneration would be to create a gene-transfer system that allows long-term high-level expression of this protein in cells in culture and *in vivo* in postmitotic injured neurons (Verhaagen *et al.*, 1992, 1994). Thus, as a first necessary step to analyze the effect of B-50/GAP43 (over)expression *in vivo,* defective viral vectors, including herpes and adeno viral vectors, were constructed for B-50/GAP43. A defective herpes simplex viral vector was generated using a previously described prototype amplicon pSRa-ori as starting material (Kaplitt *et al.*, 1991). Amplicon pHCB-50 was created by introducing a CMV-B-50 transcription unit into pSRa-ori, a plasmid containing the HSV-1 cleavage/packaging signal, and the HSV-2 origin of replication. Defective viral particles were generated with a temperature-sensitive helper virus using a procedure described previously (Kaplitt *et al.*, 1991) and explained in more detail in other chapters of this volume. Titers of the stocks were determined by B-50/GAP43 immunohistochemistry. Western blots of proteins extracted from dvHCB-50-infected vero cells revealed that B-50/GAP43 is readily detectable in infected vero cells and is absent from noninfected cells. B-50/GAP43 expressed in vero cells via dvHCB-50 migrates to the same position as B-50/GAP43 from mouse brain protein extracts suggesting that viral vector-mediated expression results in the synthesis of an intact B-50/GAP43 protein in vero cells.

Visual inspection of dvHCB-50-infected vero and rabbits skin cells revealed the occurrence of striking changes in the morphology of the B-50/GAP43 immunoreactive cells in a time-dependent fashion (Fig. 1). Such changes were not observed in the β-gal-positive cells. At 6, 10, 24, 48, and 72 hr postinfection, the surface area of the infected cells, the number of cells with a process, and the length of individual processes were measured using a computerized image analysis system. A process was defined as an extension longer than 15 μm. The surface areas of the B-50-positive cells was 3.0 times larger at 72 hr postinfection compared to the surface area of β-gal-expressing cells. B-50/GAP43-positive cells with processes were only rarely seen at 6 and 10 hr postinfection, while 25 (24 hr), 34 (48 hr), and 30% (72 hr) of the B-50/GAP43-positive cells expressed extensions on these later time points. The mean length of the processes increased from 136 $\pm$ 15.2 μm at 24 hr to 165 $\pm$ 14.5 μm at 72 hr postinfection.

These results demonstrate that long-term high-level expression of B-50/GAP43 can be achieved *in vitro* in nonneuronal cells via a defective herpes simplex virus vector. B-50/GAP43 induced profound and progressive morphological changes that persisted for several days in culture. The quantitative data reveal the existence of two distinct phenotypes in B-50/GAP43-positive cells: (1) virtually all B-50/GAP43-expressing cells were larger than the cells expression β-gal and exhibited ruffled membranes, and (2) a second class of cells exhibited additional long processes often terminating in club-shaped growth cone-like structures. First-order processes emanating directly from the cell surface occasionally branched to form second-order extensions. Examples of the cell shape changes induced by B-50/GAP43 are shown in Fig. 1.

Effects of B-50/GAP43 on cell shape have been shown before (Zuber *et al.*, 1989; Widmer and Caroni, 1992), but our results differ from these past results in two important aspects. First, the previously reported changes in cell shape were quite modest compared to the effects reported here. Second, the previous effects were of a transient nature, whereas in contrast, our data demonstrate changes in cell shape that persisted up to 72 hr after infection.

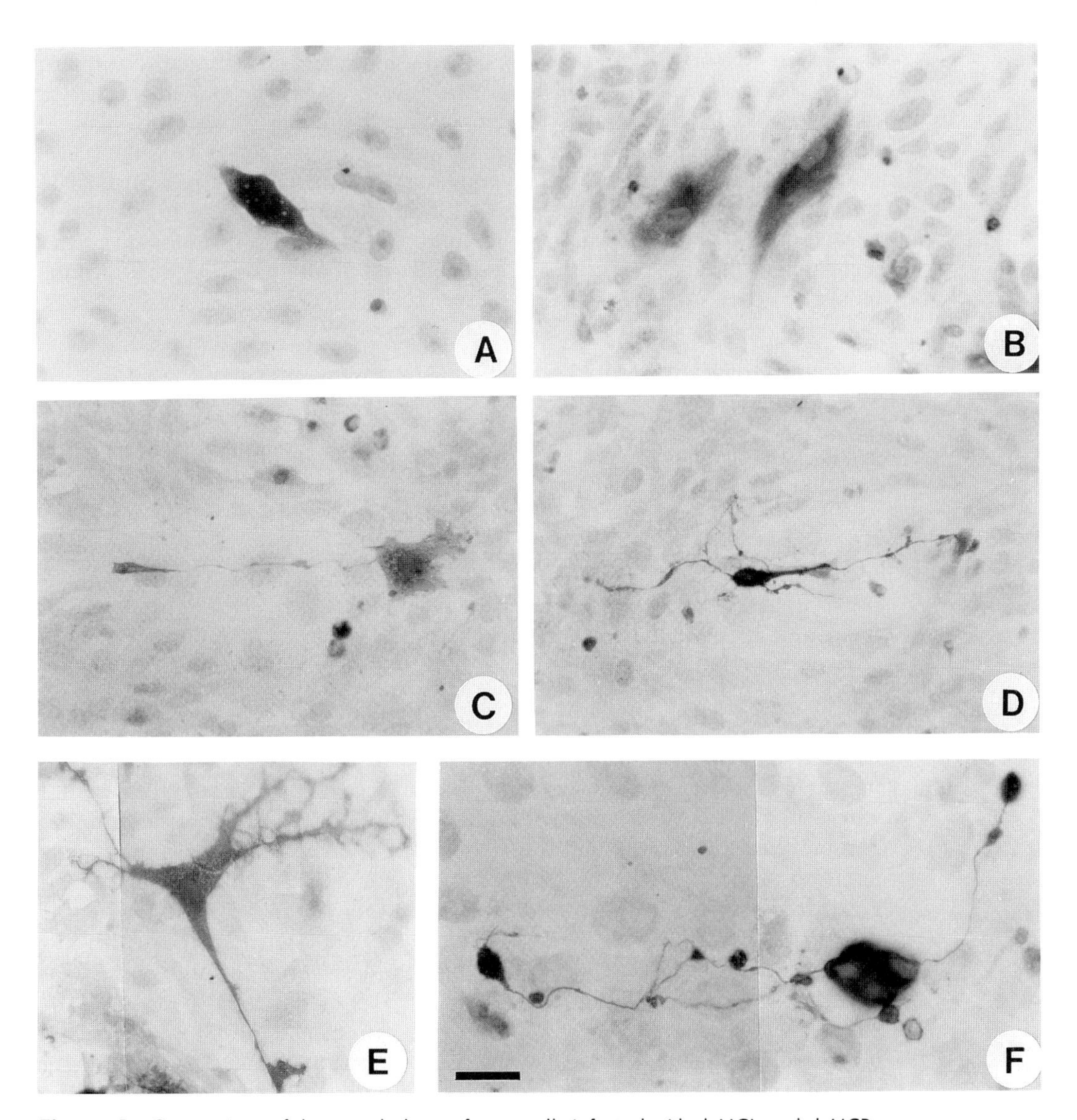

Figure 1 Comparison of the morphology of vero cells infected with dvHCL and dvHCB-50. (A,B) Vero cells expressing β-gal 48 hr after infection with dvHCL. (C–F) Vero cells expressing B-50/GAP43 48 hr (C,D) and 72 hr (E,F) following infection with dvHCB-50. Note the dramatic changes in cell shape in the B-50/GAP43-expressing cells. Changes as shown here were observed in nearly all cells expressing B-50/GAP43 between 24 and 72 hr following infection in four independent experiments. Bar is 50 μm for A–E and 19 μm for F.

The more profound effects of B-50/GAP43 are probably due to the high level of expression of B-50/GAP43 that is obtained using the HSV defective vector. The use of a strong viral promoter (CMV) to drive gene expression and transfer of the gene via a defective viral vector consisting of numerous copies of the plasmid up to 150 kb per defective viral genome apparently assures a very high level of expression. Each infected cell contains numerous copies of the B-50/GAP43 transcription unit. An additional reason for an effect of B-50/GAP43 on all cells may lie in the fact that gene transfer through viral infection results in greater uniformity of gene delivery than with transfection of plasmid DNA, in which the number of copies of a gene entering a cell can be quite variable.

B. Adenovirus Vector-Mediated B-50/GAP43 Gene Transfer

Although defective HSV vectors have been very effective gene-transfer agents to study the effect of B-50/GAP43 *in vitro* in individual cells, their usefulness *in vivo* has been limited due to their poor efficiency of infection and pathogenicity (Leib and Oliva, 1993). In order to enable the study of B-50/GAP43's role in nerve fiber growth *in vivo,* we created a recombinant adenovirus vector for B-50/GAP43 (Ad-B-50/GAP43; Hermens *et al.*, 1994). Adenovirus vectors have recently been shown to be highly efficient gene-transfer vectors for neural cells *in vivo* (Akli *et al.*, 1993). To generate Ad-B-50/GAP43, an adenovirus-targeting plasmid was created, containing the adenovirus type 5 (Ad-5) ITR, a stretch of Ad5 DNA (map units 9.2–15.5) needed for homologous recombination with Ad-5 genomic DNA and a transcription unit for B-50/GAP43 (Fig. 2). Recombinant Ad-B-50/GAP43 was generated by cotransfection of the Ad-B-50/GAP43-targeting plasmid and Ad-5 genomic DNA from which the ITR was removed by restriction with *Cla*I. Homologous recombination between the plasmid and Ad-5 genomic DNA results in the simultaneous insertion of the B-50/GAP43 expression cassette and detection of the Ad E1a gene. Stocks of this replication-defective recombinant Ad-B-50/GAP43 virus were prepared in 293 cells that express the E1a function. Titers of Ad-B-50/GAP43 viral stocks, as determined by infection of vero cells and subsequent counting of B-50/GAP43-immunostained cells, ranged from 10^9 to 4×10^9 viral particles per milliliter, B-50/GAP43-positive cells in culture again displayed ruffled membranes and neurite-like extensions (Fig. 3A), confirming the phenotypic changes seen in cells infected with dvHCB-50.

Recently, studies were started on the performance of Ad-B-50/GAP43 *in vivo*. Four adult rats received infusions of 25 μl of Ad-B-50/GAP43 (4×10^9 viral particles per milliliter) in the right nostril in an attempt to test the ability of the virus to express B-50/GAP43 in mature olfactory neurons in the olfactory neuroepithelium. B-50/GAP43 is normally exclusively expressed in the lower region of this neuroepithelium, in the differentiating neurons that are continuously formed in this neural tissue to replace lost mature olfactory neurons in the upper compartment of the epithelium (Fig. 3B; Verhaagen *et al.*, 1989).

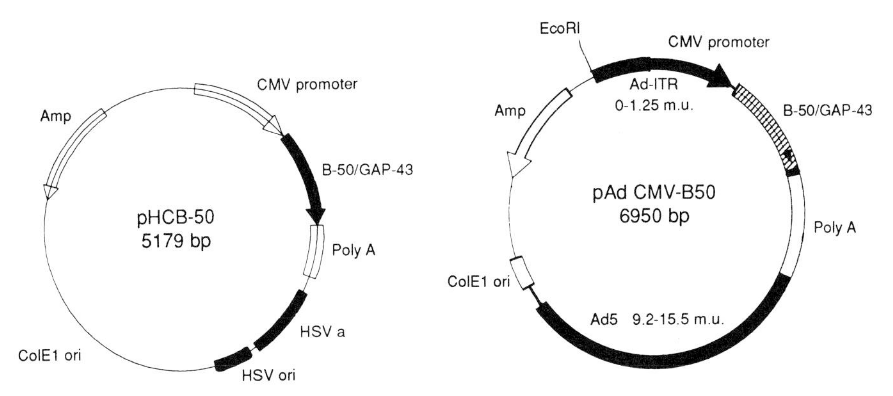

Figure 2 Diagram of B-50/GAP43 amplicon pHCB-50 and of B-50/GAP43 adenovirus-targeting plasmic pAdCMV-B-50. pHCB-50 is based on amplicon pSRa-ori (Kaplitt *et al.*, 1991) and contains the B-50/GAP43 coding sequence under the control of the CMV promoter. An SV-40 polyadenylation signal is located downstream of the transcription unit. The positions of the HSV cleavage/packaging signal (HSV"a") and the HSV origin of replication (HSVori) are indicated. In the presence of HSV proteins provided by the temperature-sensitive helper virus, the HSV ori allows replication of the amplicon and the HSV"a" sequence permits cleavage of 150 kb units of DNA and packaging into HSV viral particles. pAdCMV-B50 was created by inserting the CMV-B-50/GAP43 from pHCB-50 in a vector containing the Ad-5 ITR and a portion of the Ad-5 genome (map units 9.2–15.5). This vector was linearized with *Eco*RI, a unique restriction site located adjacent to the ITR, and the linearized fragment was cotransfected with Ad-5 genomic DNA restricted with *Cla*I to remove the ITR. Homologous recombination between pAdCMVB-50 and Ad5 DNA resulted in the recombinant Ad-B-50/GAP43 viral vector used for initial *in vitro* and *in vivo* experiments.

Mature neurons and the supporting or sustentacular cells do not express B-50/GAP43. Following unilateral infusion of Ad-B-50/GAP43, however, numerous mature olfactory neurons in the upper region of the epithelium acquired a B-50/GAP43-positive phenotype. Furthermore, B-50/GAP43-positive sustentacular cell profiles were observed in the experimental epithelium, while such cells were always devoid of B-50/GAP43 at the noninfected control side (Fig. 3C). These results clearly demonstrate that Ad-B-50/GAP43 is able to transduce and express the B-50/GAP43 gene to neural cells that do not normally synthesize this protein. Although these experiments are promising, inspection of tissue sections from this first series of animals reveals that the virus-directed expression of B-50/GAP43 in olfactory neuroepithelium following Ad-B-50/GAP43 vector infusion results in a patchy, incomplete pattern of infection. Roughly estimated, only about 10 to 20% of the cells in the upper region of the epithelium express B-50/GAP43. Studies on the effects of expression of B-50/GAP43 on olfactory neuron turnover, survival, and regeneration in the primary olfactory nerve are only feasible when a higher percentage of cells can be rendered transgenic. To achieve this we are currently performing experiments to improve the efficiency

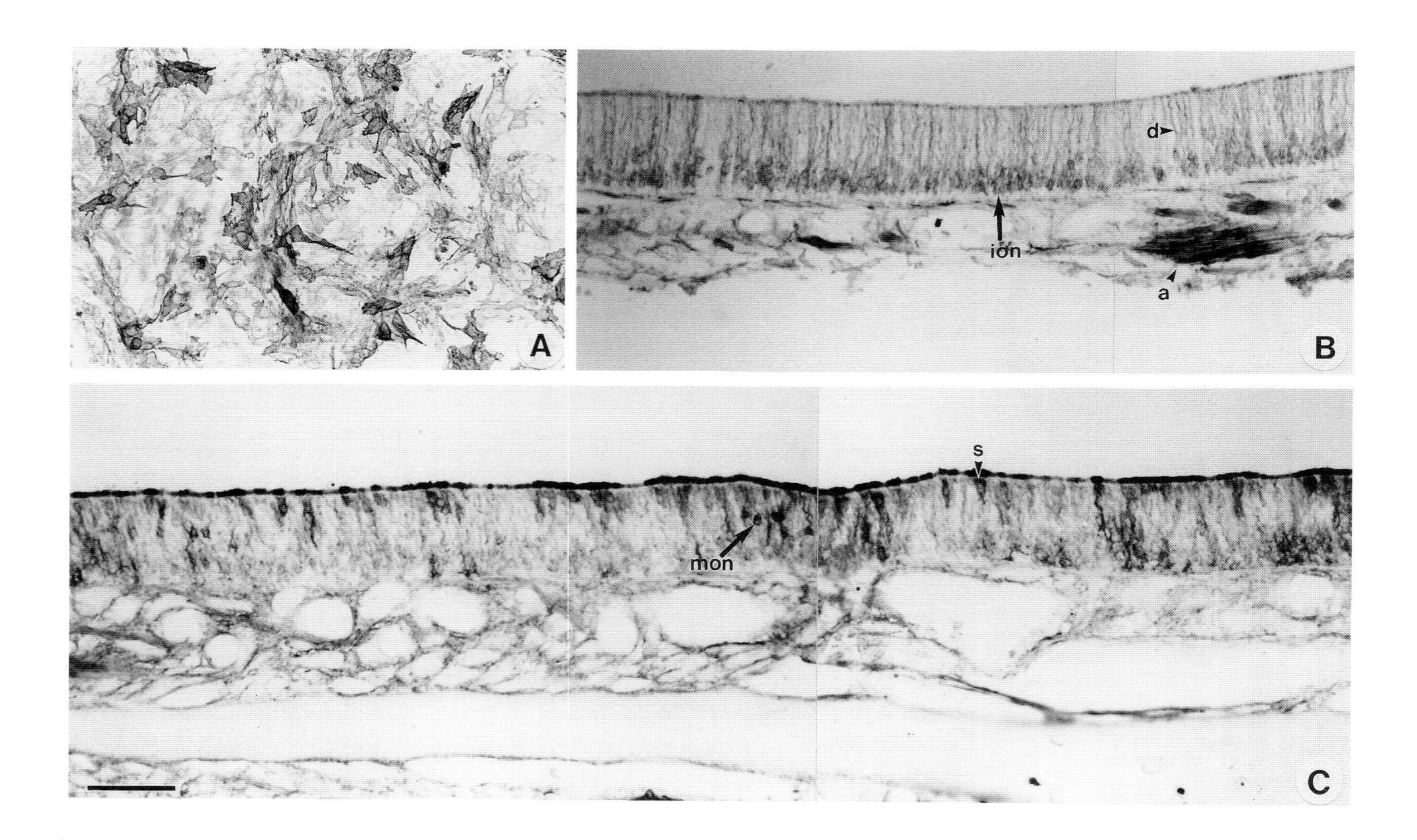
A
d
ion
a
B
s
mon
C

of virus delivery, e.g., by infusing multiple aliquots of virus into the nasal cavity and by increasing the titer of Ad-B-50/GAP43 viral stocks.

V. Future Research Directions

Successful genetic intervention in neurogeneration can probably only be achieved by simultaneous attempts to stimulate neuron survival and sprouting following trauma and by reversing and/or counteracting the nonpermissive environment in the area of the lesion. Studies with peripheral nerve grafts in the CNS provokingly demonstrated that such grafts provide an environment that permits growth of CNS neurons over long distances (Richardson *et al.*, 1980). Although the molecules in these grafts responsible for the favorable effects on nerve fiber regrowth have not been identified with certainty, these experiments highlight the importance of the availability of guidance cues and the apparent absence of repulsive neural growth-inhibitory molecules.

In view of this, in future studies gene-transfer vectors will be required to improve survival and sprouting and to attack the formation of the scar. Federoff and colleagues constructed a vector for NGF and demonstrated that viral vector-directed expression of NGF following denervation of peripheral neurons *in vivo* results in the maintenance of a neurotransmitter phenotype in these neurons that is otherwise lost following denervation (Federoff *et al.*, 1992, and Chapter 7). A viral vector directing the expression of this neurotrophin would also be an extremely valuable tool in the study of the effectiveness of NGF gene transfer in nerve regeneration. DvHCB-50 and Ad-B-50/GAP43 vectors, as described in this chapter, are another step on the road to exploring the possibility to promote the sprouting response of a traumatized nerve tract using a genetic strategy (Verhaagen *et al.*, 1994).

Figure 3 Performance of AdCMV-B50 *in vitro* and *in vivo.* (A) Infection of a monolayer of vero cells with the recombinant adenovirus vector AdCMV-B50 results in numerous B-50/GAP43-positive cells. The cells shown here were fixed 24 hr after infection. A significant proportion of the cells exhibit cell surface reactions as was seen in vero cells infected with dvHCB-50 in Fig. 2. (B,C) *In vivo* expression of B-50/GAP43 via AdCMV-B50 in the olfactory epithelium. In control olfactory epithelium (B) the previously documented distribution of B-50/GAP43 is observed: B-50/GAP43 is exclusively expressed in the differentiating immature olfactory neurons (ion) in the lower compartment of the epithelium (Verhaagen *et al.,* 1990). B-50/GAP43 is present throughout the immature neuron, including their cell bodies, dendrites (d), and axons (a). At the experimental side (C) B-50/GAP43 is expressed in numerous mature olfactory neurons (mon) and sustentacular cells (s) in the upper region of the epithelium. These preliminary data clearly demonstrated the feasibility of adenovirus vector-directed expression of B-50/GAP43 in neural cells. In order to investigate the effects of B-50/GAP43 on neuronal turnover and neuroregeneration we are currently working to improve the infection efficiency in this neuroepithelium. Bar is 25 μm for A and 100 μm for B and C.

The modulation of the gliotic response could be approached by the use of vectors that locally knock out oligodendryocytes, e.g., a vector containing an MBP promoter–diphtheriotoxin (DT) expression cassette. The infusion of an MBP-promoter–DT vector in a degenerating nerve tract would result in the selective local ablation of these myelin-forming cells that express neural growth-inhibiting molecules like NI-35 and NI-250 on their surface (Schnell and Schwab, 1990). As an alternative to this rather crude ablation of oligodendrocytes, a more selective downregulation of neural-inhibitory molecules could be envisioned using antisense viral vectors or vectors that express specific "hammerhead" ribozymes. Hence, future genetic strategies will hopefully offer ample opportunities for "molecular surgery" of traumatized neural tissue.

The application of gene transfer as a way to ameliorate neuroregeneration is still in its infancy. Genetic intervention aiming at the stimulation of the extremely complex process of nervous system regeneration is a difficult and ambitious endeavour. However, it is our belief that recent scientific developments, including the creation of efficient, nonpathogenic vectors (see other chapters of this volume), and recent insights in the role of specific cell types and molecules that either stimulate or impair nerve fiber formation, open up new avenues for neuroregeneration research.

Acknowledgments

This work was supported by grants from NWO/GMW (900-552-212), the Royal Netherlands Academy of Arts and Sciences, and the Utrecht Pharmacologic Studiefonds. We acknowledge Gerben van der Meulen for the preparation of the figures and Wilma Verweij and Tini Eikelboom for their expert secretarial assistance.

References

Akli, S., Caillaud, C., Vigne, E., Stratford-Perricaudet, L. D., Poenaru, L., Perricaudet, M., Kahn, A., and Peschanski, M. R. (1993). Transfer of a foreign gene into the brain using adenovirus vectors. *Nature Genet.* **3**, 224–228.

Alexander, K. A., Cimler, B. M., Meier, K. E., and Storus, D. R. (1987). Regulation of clamodulin binding to P-57. *J. Biol. Chem.* **262**, 6108–6113.

Biffo, S., Verhaagen, J., Schrama, L. H., Schotman, P., Danho, W., and Mangolis, F. L. (1990). B-50/GAP43 expression correlates with process outgrowth in the embryonic mouse nervous system. *Eur. J. Neurosci.* **2**, 487–499.

Caroni, P., and Schwab, M. E. (1988). Antibody against a myelin-associated inhibitor of neurite growth neutralizes nonpermissive substrate properties of CNS-white matter. *Neuron* **1**, 85–96.

Coggins, P. T., and Zwiers, H. (1989). Evidence for a single kinase-C mediated phosphorylation site in rat brain protein B-50. *J. Neurochem.* **53**, 1895–1901.

De Graan, P. N. E., Oestreicher, A. B., De Wit, M., Kroef, M., Schrama, L. H., and Gispen, W. H. (1990). Evidence for the binding of calmodulin to endogenous B-50 (GAP43) in native synapto somal plasma membranes. *J. Neurochem.* **55**, 2139–2141.

Doster, S. K., Lozano, A. M., Aguano, A. J., and Willard, W. B. (1991). Expression of the growth-associated protein GAP-43 in adult rat retinal ganglion cells following axon injury. *Neuron* **6**, 635–647.

Fawcett, J. W. (1992). Intrinsic neuronal determinants of regeneration. *TINS* **15**, 5–8.

Federoff, H. J., Geschwind, M. G., Geller, A. I., and Kessler, J. A. (1992). Expression of nerve growth factor in vivo from a defective herpes simplex virus vector prevents effects of axotomy on sympathetic ganglia. *Proc. Natl. Acad. Sci. USA* **89**, 1636–1640.

Hermens, W. T. J. M. C., Kaplitt, M. G., Pfaff, D. W., Gispen, W. H., and Verhaagen, J. (1994). Expression of the growth-associated protein B-50/GAP43 via defective herpes simplex virus (HSV) and adenovirus vectors results in a persistent neuron-like morophology in non-neuronal cells. *Neurology* **44**(2), A284.

Jap Tjoen San, E. R. A., Schmidt-Michels, M., Oestreicher, A. B., Gispen, W. H., and Schotman, P. (1993). Inhibition of B-50/GAP43 expression by antisense oligomers interferes with neurite outgrowth in PC-12 cells. *Biochem. Biophys. Res. Commun.* **187**, 839–846.

Kaplitt, M. G., Pfaus, J. G., Kleopoulos, S. P., Hanlon, B. A., Rabkin, S. D., and Pfaff, D. W. (1991). Expression of a functional foreign gene in adult mammalian brain following in vitro transfer via a Herpes Simplex virus Type 1 defective vector. *Mol. Cell. Neurosci.* **2**, 320–330.

Kline, D. G. (1990). Surgical repair of peripheral nerve injury. *Muscle Nerve* **13**, 843–853.

Kolodkin, A. L., Matthes, D. J., and Goodman, C. S. (1993). The semaphorin genes encode a family of transmembrane and secreted growth cone guidance molecules. *Cell* **75**, 1385–1399.

Korsching, S. (1993). The neurotrophic factor concept: A reexamination. *J. Neurosci.* **13**, 2739–2748.

Kristjansson, G. I., Zwiers, H., Oestreicher, A. B., and Gispen, W. H. (1982). Evidence that the synaptic phosphoprotein B-50 is localized exclusively in nerve tissue. *J. Neurochem.* **39**, 371–378.

LaBate, M. E., and Skene, J. P. H. (1989). Selective conservation of GAP-43 structure in vertebrate evolution. *Neuron* **3**, 299–310.

Leib, D. A., and Oliva, P. D. (1993). Gene delivery to neurons: Is herpes simplex virus the right tool for the job? *BioEssays* **15**, 547–554.

Luo, Y., Raible, D., and Raper, J. A. (1993). Collapsin: A protein in brain that induces the collapse and paralysis of neuronal growth cones. *Cell* **75**, 217–227.

Miller, F. D., Tetzlaff, W., Bisby, M. A., Fawcett, J. W., and Milner, R. J. (1989). Rapid induction of the major embryonic α-tubulin mRNA in adults following neuronal injury. *J. Neurosci.* **9**, 1452–1463.

Morton, A. J., and Buss, T. N. (1992). Accelerated differentiation in response to retinoic acid after retrovirally mediated gene transfer of GAP-43 into mouse neuroblastoma cells. *Eur. J. Neurosci.* **4**, 910–916.

Richardson, P. M., McGuinness, U. M., and Aguayo, A. J. (1980). Axons from CNS neurons regenerate into PNS grafts. *Nature* **284**, 264–265.

Rutishauer, U., and Jessell, T. M. (1988). Cell adhesion molecules in vertebrate neural development. *Physiol. Rev.* **68**, 819–857.

Schnell, L., and Schwab, M. E. (1990). Axonal regeneration in the rat spinal cord produced by an antibody against myelin-associated neurite growth inhibitions. *Nature* **343**, 269–272.

Schwab, M. E., Kampfhammer, J. P., and Bandtlow, C. E. (1993). Inhibitors of neurite growth. *Annu. Rev. Neurosci.* **16**, 565–595.

Shea, T. B., Perrone-Bizzozero, N. I., Beerman, M. L., and Benowitz, L. I. (1991). Phospholipid-mediated delivery of anti-GAP-43 antibodies into neuroblastoma cells prevents neuritogenesis. *J. Neurosci.* **11**, 1685–1690.

Skene, J. H. P., and Willard, M. (1981). Axonally transported proteins associated with axon growth in rabbit central and peripheral nervous system. *J. Cell Biol.* **89**, 96–103.

Skene, J. H. P. (1989). Axonal growth associated protein. *Annu. Rev. Neurosci.* **12**, 127–156.

Skene, J. H. P., and Virag, I. (1989). Post-translational membrane attachment and dynamic fatty acylation of a neuronal growth cone protein, GAP43. *J. Cell Biol.* **108**, 613–624.

Skene, J. H. P. (1992). Retrograde pathways controlling expression of a major growth cone component in the adult CNS. In "The Nerve Growth Cone" (P. C. Letourneau, S. B. Kater, and E. R. Macagno, Eds.). Raven Press, New York.

Sonderegger, P., and Rathjen, F. G. (1992). Regulation of axonal growth in the vertebrate nervous system by interaction between glycoproteins belonging to two subgroups of the immunoglobulin superfamily. *J. Cell Biol.* **119**, 1387–1394.

Strand, F. L., Rose, K. L., Zuccarelli, L. A., Kume, J., Alves, S. E., Antonawich, F. J., and Garrett, L. Y. (1991). Neuropeptide hormones as neurotrophic factors. *Physiol. Rev.* **71**, 1017–1046.

Strittmatter, S. M., Valenzuela, D., Kennedy, T. E., Neer, E. J., and Fishman, M. C. (1990). Go is a major growth cone protein subject to regulation by GAP43. *Nature* **344**, 836–841.

Strittmatter, S. M., Valenzuela, D., and Fishman, M. C. (1994). An amino terminal domain of the growth-associated protein GAP43 mediates its effect on filopodial formation and cell spreading. *J. Cell Sci.* **107**, 195–204.

Tetzlaff, W., Zwiers, H., Lederis, K., Cassar, L., and Bisby, M. A. (1989). Axonal transport and localization of B-50/GAP-43-like immunoreactivity in regenerating sciatic and facial nerves. *J. Neurosci.* **9**, 1303–1313.

Tetzlaff, W., Tsui, B. J., and Balfeur, J. K. (1990). Rubrospinal neurons increase GAP-43 and tubulin mRNA after cervical but not after thoracic axotomy. *Soc. Neurosci. Abstr.* **16**, 338.

Tetzlaff, W., Alexander, B. W., Miller, F. D., and Bisby, M. A. (1991). Response of facial and rubrospinal neurons to axotomy: Changes in mRNA expression for cytoskeletal proteins and GAP-43. *J. Neurosci.* **11**, 2528–2544.

Van der Zee, C. E. E. M., Vos, J. P., Nielander, H. B., Lopes da Silva, S., Verhaagen, J., Oestreicher, A. B., Schrama, L. H., Schotman, P., and Gispen, W. H. (1989). Expression of the growth-associated protein B-50/GAP43 in dorsal root ganglia and sciatic nerve during regenerative sprouting. *J. Neurosci.* **9**, 3505–3512.

Verhaagen, J., Van Hooff, C. O. M., Edwards, P. M., De Graan, P. N. E., Oestreicher, A. B., Jennekens, F. G. L., and Gispen, W. H. (1986). The kinase C substrate protein B-50 and axonal regeneration. *Brain Res. Bull.* **17**, 737–741.

Verhaagen, J., Oestreicher, A. B., Edwards, P. M., Veldman, H., Jennekens, F. G. I., and Gispen, W. H. (1988). Light and electron microscopical study of phosphoprotein B-50 following denervation and reinnervation of the rat soleus muscle. *J. Neurosci.* **8**, 1759–1766.

Verhaagen, J., Oestreicher, A. B., Grillo, M., Khew-Goodall, Y. S., Gispen, W. H., and Margolis, F. L. (1990). Neuroplasticity in the olfactory system: Differential effects of central and peripheral lesions of the primary olfactory pathway on the expression of B-50/GAP43 and the olfactory marker protein. *J. Neurosci. Res.* **26**, 31–44.

Verhaagen, J., Gispen, W. H., Hermens, W. T. J. M. C., Rabkin, S. D., Pfaff, D. W., and Kaplitt, M. G. (1992). Expression of B-50/GAP43 via a defective herpes simplex virus vector in cultured non-neuronal cells. *Soc. Neurosci. Abstr.* **18**, 604.

Verhaagen, J., Zhang, Y., Hamers, F. P. T., and Gispen, W. H. (1993). Elevated expression of B-50 (GAP-43) mRNA in a subpopulation of olfacotry bulb mitral cells following axotomy. *J. Neurosci. Res.* **35**, 162–169.

Verhaagen, J., Hermens, W. T. J. M. C., Oestreicher, A. B., Gispen, W. H., Rabkin, S. D., Pfaff, D. W., and Kapitt, M. G. (1994). Expression of the growth-associated protein B-50/GAP-43 via a defective herpes-simplex virus vector results in profound morphological changes in non-neuronal cells. *Mol. Brain Res.* **26**, 26–36.

Widmer, F., and Caroni, P. (1993). Phosphorylation-site mutagenesis of the growth-associated protein GAP-43 modulates its effect on cell spreading and morphology. *J. Cell Biol.* **120**, 503–512.

Yankner, G. A., Benowitz, L I., Villa Kamoroff, L., and Neve, R. L. (1990). Transfection of PC-12 cells with the human GAP-43 gene: Effects on neurite outgrowth and regeneration. *Mol. Brain Res.* **7**, 39–44.

Zuber, M. X., Goodman, D. W., Karns, L .R., and Fishman, M. C. (1989). The neuronal growth-associated protein GAP-43 induces filopodia in non-neuronal cells. *Science* **244**, 1193–1195.

Zwiers, H., Veldhuis, H. D., Schotman, P., and Gispen, W. H. (1976). Cyclic nucleotides and brain protein phosphorylation in vitro. *Neurochem. Res.* **1**, 669–677.

CHAPTER 9

Use of Herpes Simplex Virus Vectors for Protection from Necrotic Neuron Death

Dora Y. Ho
Matthew S. Lawrence
Timothy J. Meier
Sheri L. Fink
Rajesh Dash
Tippi C. Saydam
Robert M. Sapolsky
Department of Biological Sciences
Stanford University
Stanford, California

I. Introduction: The Glutamatergic Cascade of Necrotic Neuron Death

If we are crippled by an accident, if we lose our sight or hearing, if we are so weakened by heart disease as to be bed bound, we cease having many of the things that make our lives worth living. But when it is the brain which is damaged, we can often cease to exist as sentient individuals. This tragic nature of neurological damage is made even more severe by the postmitotic status of neurons and, thus, by the inability of the brain to replace neurons once they are lost. Approaches, such as transplantation of fetal neurons into the injured adult brain, or the stimulation of neurogenesis, may ultimately prove of broad clinical value. Nonetheless, at present, the most plausible routes for protecting the brain from neurological injury consist of interventions at the time of the insult.

Carrying out such effective interventions requires knowledge of the biology of neuron death following a neurological insult. Fortunately, a tremendous amount of information has been obtained in recent years concerning the patho-

VIRAL VECTORS

physiology of necrotic neuron death. Central to such neuron death are the excitatory amino acids (EAA) such as glutamate and aspartate. These are the most abundant excitatory neurotransmitters in the brain, and they play vital roles in phenomena such as long-term potentiation in the hippocampus. Normally, EAAs bind to a number of different receptor types, including the ionotropic NMDA and non-NMDA receptors, as well as inositol phosphate-coupled metabotropic receptors. Among the most important consequences of such receptor interactions is the mobilization of free cytosolic calcium concentrations from both extracellular and organelle sources. Calcium, in term, triggers a complex array of events that contribute to the plasticity of LTP.

Despite these salutary physiologic effects, the pathologic potential of EAAs is tremendous. Necrotic insults, such as global ischemia and infarct, seizure, or hypoglycemia all result in excessive synaptic concentrations of EAAs, leading to excessive mobilization of free cytosolic calcium (Choi, 1990). The deleterious consequences of such excess include promiscuous overactivation of calcium-dependent proteases, lipases, and nucleases (McBurney and Neering, 1987). Of probably the greatest pathologic significance is that the calcium excess leads to the generation of damaging radical oxygen species. The most important routes by which this occurs probably include calcium-dependent activation of xanthine oxidase (McCord, 1985), of nitric oxide synthase (Lipton *et al.*, 1993), and of phospholipase (Dumuis *et al.*, 1988), each of which can directly or indirectly generate oxygen radicals.

Energy plays a critical role in determining whether the EAAs and calcium subserve their normal physiologic roles or pathophysiologic ones (Novelli *et al.*, 1988; Cox *et al.*, 1989; Lysko *et al.*, 1989; Zeevalk and Nichlas, 1991, 1992). Necrotic neurological insults are ultimately energetic in nature (involving either disruption of substrate delivery, as in ischemia or hypoglycemia, or pathologic demand for energy, as in seizure), and they dramatically compromise the capacity of neurons to contain EAAs and calcium (Beal, 1992). During a necrotic insult, energy failure leads to depolarization and enhanced release of EAAs. Moreover, the reuptake and uptake of EAAs by neurons and glia, respectively, are extremely costly, depending upon steep ionic gradiants. As a result, during such necrotic insults, removal of EAAs from the synapse fails; at the extremes of energy failure, reuptake pumps will even reverse. Thus, the energy failure intrinsic to necrotic insults produces larger and more persistent waves of extracellular EAA concentrations (reviewed in Sapolsky, 1992).

Similarly, the energetic crisis of a necrotic insult will compromise the capacity of the postsynaptic neuron to contain the calcium consequences of EAA exposure (such that energy failure can even lead to postsynaptic neuron death in the face of normal EAA concentrations). Energy depletion will lead to neuronal depolarization, enhancing calcium mobilization via a number of complex routes. Moreover, the removal of cytosolic calcium, either through sequestration into organelles or extracellular efflux, is costly, depending upon an ATPase or steep ionic gradiants. Such removal fails during necrotic insults, resulting in larger and more persistent calcium waves (reviewed in Sapolsky, 1992).

The workings of the EAA/calcium/oxygen radical cascade and its dependence upon energy status suggest a number of plausible points of intervention that might prove neuroprotective. Broadly, they might involve decreasing EAA release or calcium mobilization, enhancing the removal of each from the synapse or cytoplasm, respectively, blockade of EAA receptors, enhancing the levels or activity of oxygen radical scavengers, or enhancing neuronal metabolism. A number of pharmacological interventions targeting some of these steps have proven protective when carried out immediately following experimental models of necrotic insults. These include use of phenytoin (an antiseizure drug which decreases EAA release), the NMDA receptor antagonist MK801, the calcium channel blocker nimodipine, calcium chelators, antioxidants such as tacopherol, and glucose supplementation. In some cases, some of these pharmacological interventions are now being tested clinically (Choi, 1990).

The broad array of interventions outlined at the start of the preceding paragraph might not only be accomplished through pharmacological means, but could potentially be brought about through gene-transfer strategies as well. For example, one might wish to overexpress the gene for a EAA transporter at the time of a necrotic insult, in order to facilitate glutamate reuptake, or for an oxygen radical scavenger such as superoxide dismutase. In the remainder of this chapter, we discuss our work with herpes simplex virus (HSV)-1 vectors that were designed for such gene-transfer studies. We review the progress we have made in bolstering neuronal energetics during necrotic insults, both *in vitro* and *in vivo,* by the transfer of the gene for the glucose transporter; in addition, we note some preliminary success we have had in using an HSV vector to overexpress the gene for a key calcium-binding protein and for an inhibitor of apoptotic neuron death.

II. Neuroprotection Using Monocistronic Vectors

Two different HSV vector systems have been developed; namely, the recombinant virus vector system and the amplicon-based vector system. The development and application of these systems are reviewed in Chapter 1 by J. Glorioso and in Chapter 2 by A. Kwong and N. Frenkel, respectively. The methodology of constructing these vectors has also been recently discussed in detail (Ho, 1994; Johnson and Friedmann, 1994). Briefly, the recombinant virus vector carries the gene of interest inserted into the viral genome, while an "amplicon" (for cloning–amplifying vector) is a plasmid which carries the gene of interest and the HSV origin of replication and packaging signals (Spaete and Frenkel, 1982). The HSV sequences allow it to replicate and be packaged into virions when complemented by a replication competent "helper" virus (reviewed in Frenkel, 1981; Ho, 1994). The resulting stock of viral progenies is thus inevitably a mixture containing the vectors of interest and the helper virus. We have chosen to use the amplicon system primarily because its relative ease of construction allows a more rapid testing of various promoters and different gene products. Furthermore, since immediate-early gene expression from the viral genome is cytotoxic (Johnson *et al.*, 1992a), every cell infected by the recombinant vector would inevitably be affected. In contrast, in a cell population infected by amplicon-based vectors, only those cells that received the helper virus would suffer from the cytotoxic effects of the viral gene expression.

A prototype of our amplicons is shown in Fig. 1. Given the energetic nature of the neurological insults that we study, we have been interested in testing the neuroprotective capability of enhancing neuronal glucose uptake. A glucose transporter gene (*gt*, GLUT-1 isoform) derived from the rat blood–brain barrier epithelium (Birnbaum *et al.*, 1986; Pardridge *et al.*, 1990) was cloned into the amplicon under the control of the human cytomegalovirus immediate-early gene 1 (HCMV *ie1*) promoter (Ho *et al.*, 1993). The resulting amplicon pIE1GT*ori* was propagated in vero cells using HSV-1 mutant ts756 (Hughes and Munyon, 1975) as the helper virus and the resulting vector was termed vIE1GT. ts756 has a temperature-sensitive mutation in the major immediate-early gene $\alpha 4$, whose gene product is a transcriptional regulator

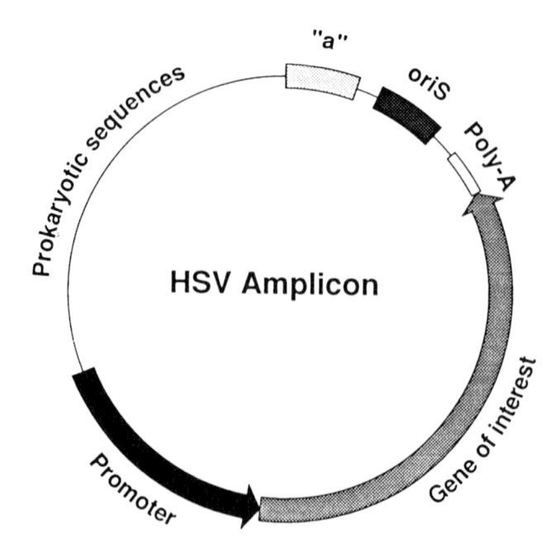

Figure 1 Schematic diagram of a prototype amplicon. The transcriptional unit in the amplicon is represented by the promoter, the gene of interest, and a polyadenylation signal. The two *cis*-acting sequences from HSV-1, the ori_s and the *a* sequences provide necessary signals for replication and packaging into virions. The prokaryotic sequences usually contain a bacterial origin of replication and a drug-selection marker that allow propagation and amplification in *Escherichia coli*.

essential for the induction of the early and late viral genes (see Roizman and Sears, 1990 for review). Without $\alpha4$ expression, infection with the virus would become abortive. A similar vector, vIE1β-gal, carrying the *Escherichia coli lacZ* gene in place of *gt* was constructed as a control.

vIE1GT and vIE1β-gal can efficiently infect many cell lines (including fibroblasts and neuroblastoma of primate or rodent origins) as well as primary neuronal or glial cultures from rat cortex, hippocampus, striatum, and spinal cord (data not shown). Infection with vIE1GT results in the expression of GLUT-1 concomitant with enhanced glucose uptake (Ho *et al.,* 1993). Depending on the multiplicity of infection (m.o.i.) and the cell type, an increase in uptake of up to 400% has been observed *in vitro* (data not shown). These vectors can also efficiently express their gene products in the adult rat brain. Using *in situ* hybridization techniques, we have previously demonstrated that microinfusion of vIE1GT into the adult rat hippocampus results in the expression of the exogenous GLUT-1. Furthermore, a significant 10% increase of glucose transport is observed at the vIE1GT infusion site compared to the contralateral side microinfused with vIE1β-gal (Ho *et al.,* 1993).

Infection with vIE1β-gal and subsequent reaction with X-gal (a chromogenic substrate of β-gal) have allowed us to identify the infected cell types based on their size and morphology. In both cell cultures and intact animals, the vectors clearly have the capability to infect many cell types. However, when injected into the adult rat brain, they seem to preferentially infect neurons. Furthermore, in the hippocampus, the virus predominantly infects the neurons of the dentate gyrus, although β-gal-expressing neurons in the CA1, CA2/3, and CA4 cell fields are also observed (Fig. 2; Lawrence *et al.,* 1994). This phenomenon is observed even when the injection is targeted away from the dentate gyrus. The mechanism responsible for such a selective infection of target cells is not clearly understood at this point. It is known that initial attachment of HSV-1 virions onto cells is mediated by binding to ubiquitous cell surface heparan sulfate proteoglycans (reviewed in Spear *et al.,* 1992); however, penetration into the cell involves further interaction of viral glycoproteins and other cell surface molecules. The presence or abundance of these cell surface molecules, which have not been clearly identified, may further determine the preferential infection of a certain cell type among a heterogeneous population of cells.

Given that infection with vIE1GT enhances glucose uptake, we then tested whether this enhancement could protect neurons from hypoglycemia. We measured the metabolic rate of mixed neuronal and glial cultures from the rat hippocampus using a Cytosensor microphysiometer (Molecular Devices, Menlo Park, CA), a device that allows real-time, sensitive measurement of cellular metabolism (McConnell *et al.,* 1992; Raley-Susman *et al.,* 1992). Cells grown on specialized coverslips are placed in a low-volume flow chamber, with constant perfusion of a nonbuffered Dulbecco's modified Eagle's medium

(DMEM). The extrusion of protons (generated in part by ATP hydrolysis and lactic acid production) from the cells is measured as an indicator of their metabolic rate. When uninfected mixed neuronal and glial cultures are subjected to hypoglycemic challenge by decreasing the glucose concentration of the circulating medium from 20 to 0.2 m*M*, a decline in metabolism is observed. Prior infection with vIE1GT, but not with vIE1β-gal or ts756, significantly attenuates this decline in metabolism (Fig. 3; Lawrence *et al.*, 1994). Subsequent staining of the vIE1β-gal-infected cultures indicates that the majority of β-gal-expressing cells are neuronal; thus, the attenuation of metabolic decline observed in vIE1GT-infected cultures most likely reflects increased glucose uptake in neurons.

We further determined whether increasing glucose uptake could actually prevent neurons from dying under hypoglycemic conditions. We maintained mixed hippocampal cultures in medium containing 30 m*M* glucose and chal-

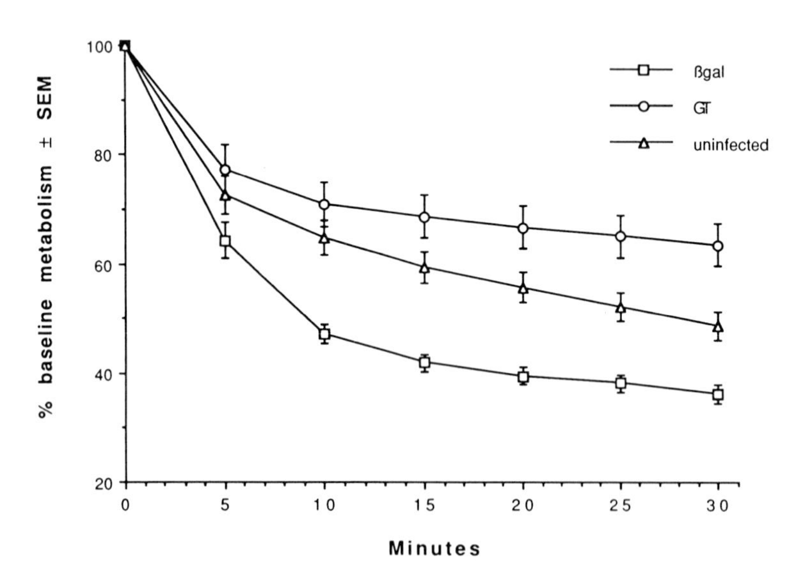

Figure 3 vIE1GT infection attenuates the metabolic decline of hippocampal cultures during hypoglycemic challenge. Five-day-old hippocampal cultures were mock infected or infected with vIE1GT or vIE1β-gal (m.o.i. = 0.2). Twenty hours later, the cultures were placed in a Cytosensor microphysiometer. Measurements of metabolic rates were made in medium containing 20 m*M* glucose for 1 hr to obtain a baseline metabolic rate before exposure to hypoglycemic conditions (0.2 m*M* glucose). Ten-second measurements were taken at 1-min intervals and averaged to arrive at 5-min data points, which were further averaged within treatment groups. The mean metabolic rates of vIE1GT- ($n = 10$), vIE1β-gal- ($n = 12$), or mock-infected cultures ($n = 11$) during the 30-min hypoglycemic period are reported as a percentage of baseline measures ± SEM. All groups were significantly different from each other ($P < 0.0001$; Fisher PLSD).

lenged them by dropping the glucose concentration of the medium to 5, 1, or 0.5 m*M*. Under such conditions, the neurons usually die within 8 to 16 hr, with the amount of cell death inversely related to the availability of glucose. Infection of the cultures with vIE1GT significantly enhances survivorship compared to infection with vIE1β-gal at a similar m.o.i. (Fig. 4; Lawrence *et al.*, 1994). Thus, vIE1GT infection increases glucose uptake of neurons, which in turn allows the cells to maintain metabolism and survive better during hypoglycemia.

The experiments that we have described above employed vIE1GT and vIE1β-gal, which were propagated with ts756 as the helper virus. The use of this mutant limits the experimental temperature to 39°C, which is nonpermissive for viral replication. However, using a higher-than-physiological temperature may not be desirable for some experimental designs. Furthermore, a rat's body temperature is not high enough to be completely nonpermissive, and the potential replication of the helper virus is a concern. We have indeed observed significant cytopathic effects induced by ts756 in adult rat hippocampus (see below). Therefore, in subsequent experiments, we employed d120 as the helper virus. d120 contains a 4.1-kb deletion in the $\alpha 4$ gene and can propagate only

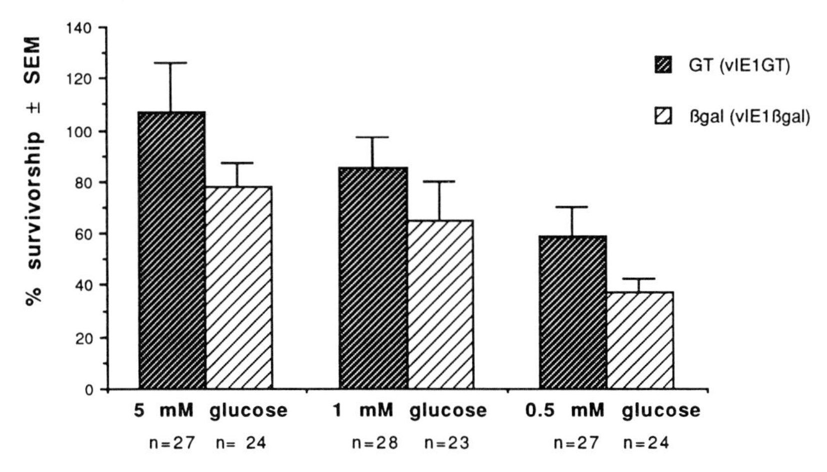

Figure 4 vIE1GT infection protects hippocampal cultures against hypoglycemia. Mixed hippocampal cultures were maintained in 30 m*M* glucose and treated with the antimitotic agent cytosine arabinoside at Day 4 to retard astrocyte growth. At Day 10 or 11, cells were infected with vIE1GT or vIE1β-gal at a m.o.i. of 0.2. Sixteen hours later, media were removed and replaced with DMEM containing 5, 1, or 0.5 m*M* glucose. Eight hours later, cells were fixed and stained for microtubule-associated protein 2 immunoreactivity. Cell survivorship was assessed by counting postive-staining neurons with intact processes. The number of surviving neurons was expressed as the percentage of mean values ± SEM of uninfected cultures that had not been subjected to any medium change. Significant differences were observed between vIE1GT- and vIE1β-gal-infected cultures ($P < 0.01$ by ANOVA).

in a complementing cell line, E5, which is stably transformed with $\alpha 4$ (DeLuca *et al.,* 1985). Cytopathic effects induced by d120 were markedly reduced compared to those by ts756 (Ho *et al.,* 1994, and see below).

We have further modified our vector system by using the HSV $\alpha 4$ promoter to express the gene of interest. Although the HCMV *ie1* promoter used in vIE1GT can confer strong expression in many cell types, this promoter may not express rapidly enough to counteract acute neurological insults; thus, a faster-acting promoter is desirable. Upon infection of a cell, the immediate-early genes (one of which is $\alpha 4$) of HSV are turned on maximally within the first few hours of infection. Such rapid induction of the immediate-early promoters is the result of transactivation by a virion component, VP16, also known as the α-*trans*-inducing factor (see Roizman and Sears, 1990 for review). Thus, we have used the $\alpha 4$ promoter in our subsequent amplicons to guarantee more rapid expression of the gene of interest. The resulting vectors that expressed GT and β-gal are termed vα4GT and v$\alpha 4\beta$-gal, respectively.

These two new vectors have been used to test potential neuroprotective effects in animals (Lawrence *et al.,* 1994). We have examined the protective effects of enhanced glucose transport (due to vα4GT infection) against kainic acid-induced toxicity, the severity of which has been shown to depend on the availability of energy substrates (Sapolsky and Stein, 1988). Microinfusion of kainic acid into the adult rat hippocampus induces seizures resembling human status epilepticus and damages neurons selectively in the CA3 cell field (Ben-Ari, 1985). In the first set of experiments, rats were made hypoglycemic, normoglycemic, or hyperglycemic. vα4GT or v$\alpha 4\beta$-gal was microinfused unilaterally into the hippocampus and the contralateral side received an equal volume of virus-free DMEM. Eight hours after infection, both sides of the hippocampus were microinfused with the same dosage of kainic acid. The animals were sacrificed 3 days later, and damage induced by the kainic acid at the CA3 area was quantified. We found that vα4GT, but not v$\alpha 4\beta$-gal, significantly protects the neurons against kainic acid toxicity (Fig. 5). The level of protection is most pronounced for the hypoglycemic rats, indicating that the increased glucose uptake capability can counteract the adverse effects of decreased circulating glucose. Furthermore, we subsequently showed that vα4GT confers significant protection against kainic acid-induced seizure, even when the vector is used up to 1 hr postseizure, rather than in anticipation of it.

III. Neuroprotection Using Bicistronic Vectors

One of the hurdles we have encountered using the monocistronic vectors described above is that cells infected by the vectors cannot be readily identified. To circumvent this problem, we have developed two versions of bicistronic

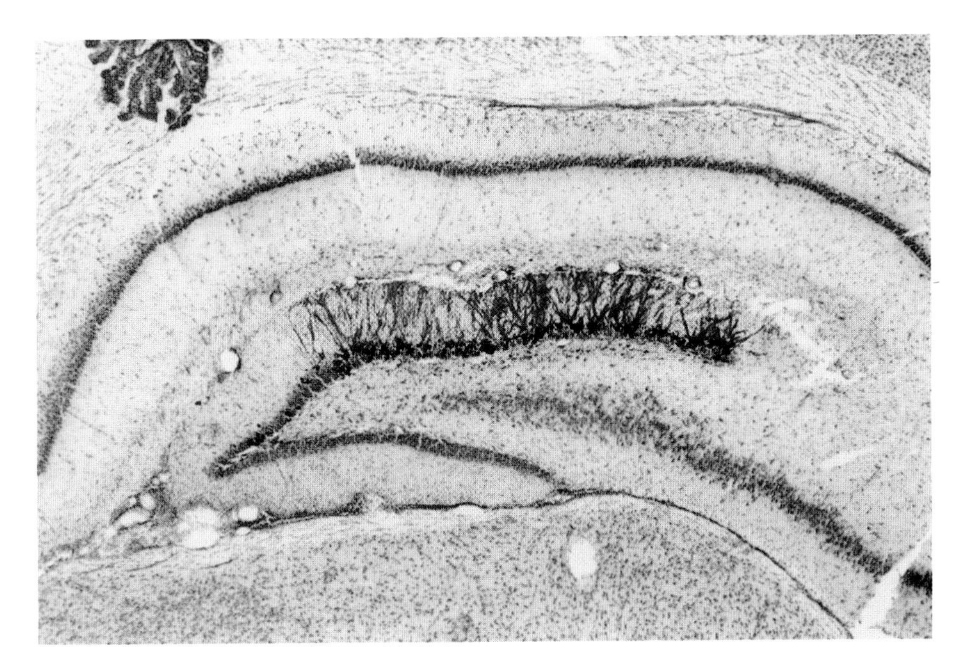

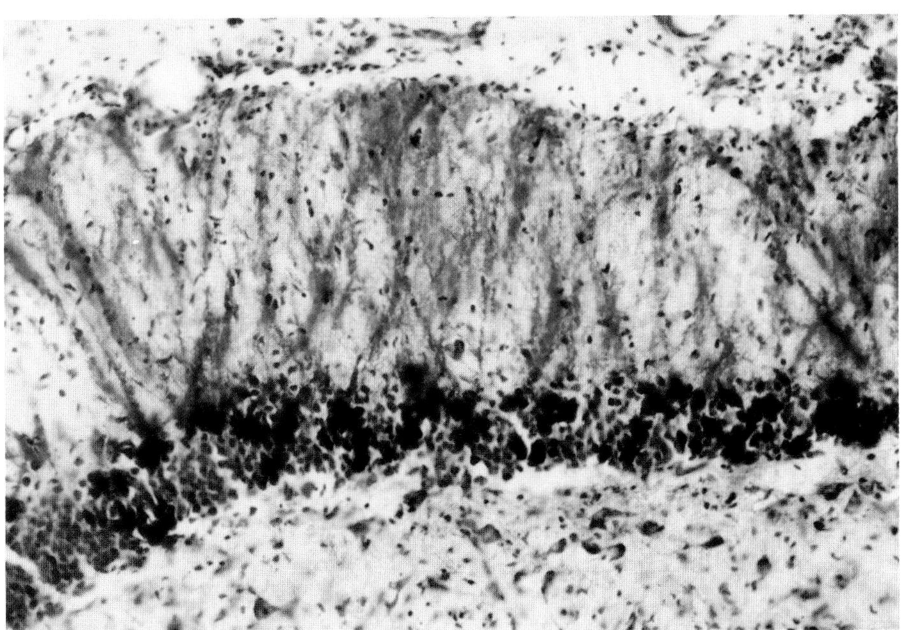

Figure 2 Photomicrographs of hippocampus demonstrating βgal expression from a HSV vector. A bicistronic vector vα22β-galα4GT (≈1 × 10^4 particles; see text for discussion on bicistronic vectors) was injected into the hippocampus at coordinates AP 4.1, ML 3.0, DV 3.0, bregma = lambda. Brain sections were prepared from the animal 24 hr postinfection and incubated with X-gal. Most of the β-gal-expressing cells are concentrated in the dentate gyrus (*upper*) and a higher magnification shows their neuronal morphology (*lower*).

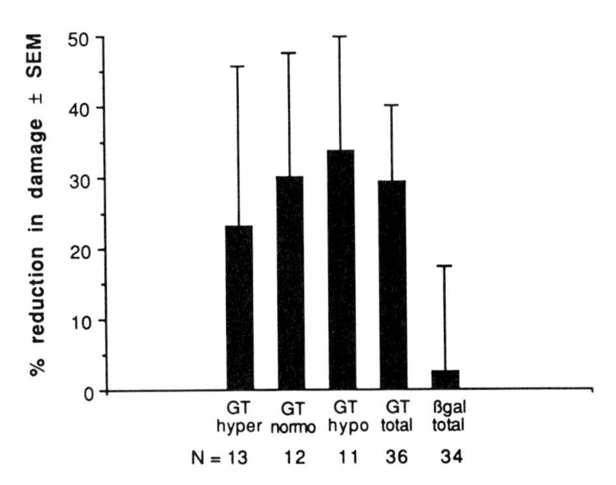

Figure 5 vα4GT infection protects against kainic acid toxicity in adult rat hippocampus. Male Sprague–Dawley rats were unilaterally microinfused with either 2 μl of vα4GT or vα4β-gal (vector titer: 1×10^7 particles/ml; helper virus titer: $1–5 \times 10^7$ PFU/ml), with the contralateral cell field receiving 2 μl DMEM (bregma=lamda: AP 4.1, ML 3.0, DV 3.0). Twelve hours later, kainic acid (0.035 μg in 1 μl phosphate-buffered saline) was microinfused bilaterally into the hippocampus. All microinfusions were made through stereotaxically implanted cannulae which allowed the delivery of the vectors and kainic acid to the same site. Two days after kainic acid adminstration, the animals were sacrificed and coronal sections of the brains from within 800–1000 μm anterior and posterior to the infusion site were prepared. Damage was quantified by measuring the length of the deletions in the CA3 cell fields as described previously (Stein and Sapolsky, 1988). Measurements in successive sections were integrated to arrive at a total volume of damage and were expressed as the percentage reduction in damage relative to the contralateral control cell field for each brain. Data are presented as the mean within each group ± SEM. Sample sizes are given beneath each bar. Microinfusion of vIE1GT resulted in a significant reduction in the damage of CA3 in hypoglycemic and normoglycemic rats ($P < 0.05$), but not in hyperglycemic rats.

vectors, both of which carry the gene of interest and *lacZ* as a marker gene. In the first approach, both genes are placed under a single transcriptional unit. The translation of the second cistron is made possible by employing an internal ribosomal entry site (IRES) from the encephalomyocarditis virus (Duke *et al.*, 1992; Jang *et al.*, 1989, 1988). The IRES provides sites for internal binding of ribosomes and thus directs cap-independent translation of the second cistron. The prototype vector that we constructed is vα4GTβ-gal, which contains the glucose transporter gene as the first cistron and the *lacZ* gene as the second cistron (Fig. 6). Infection with vα4GTβ-gal consistently provides coexpression of both genes in the same cells as observed by double immunofluorescence. However, by comparing the expression of *lacZ* as a first versus a second cistron, we have noticed that IRES-directed translation is less efficient than cap-dependent translation (data not shown). Thus, when *lacZ* constitutes the

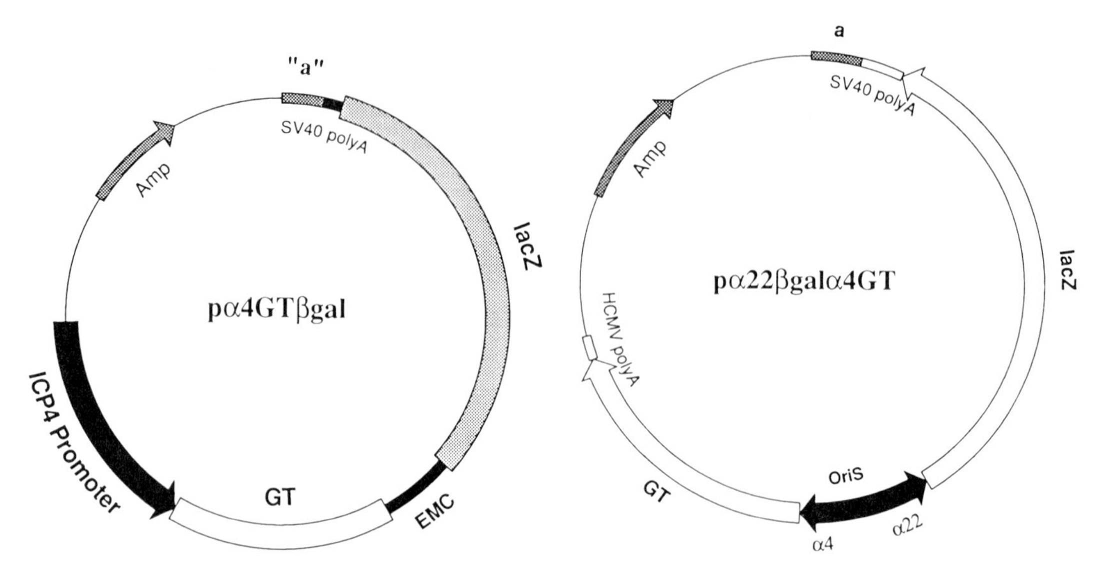

Figure 6 Schematic diagrams showing the structures of pα4GTβ-gal and pα22β-galα4GT. (*Left*) pα4GTβ-gal: both the *gt* and *lacZ* genes are in the same transcriptional unit under the control of the α4 promoter. The IRES sequence from the encephalomyocarditis virus (EMC) confers efficient translation of the second cistron. (*Right*) pα22β-galα4GT: the *gt* and *lacZ* genes are controlled by the α4 and α22/α47 promoters, respectively. The two promoters are separated by the ori_S, as in the HSV genome.

second cistron, β-gal expression is more conservative than that of GT, making detection of β-gal expression a guarantee of GT expression.

In the second approach, we employ two promoters to express the gene of interest and *lacZ* independently. The α4 and α22/α47 promoters from the HSV genome conveniently provide us with an ideal set of promoters. These two promoters are in opposite directions, separated by the ori_S. Furthermore, both of them belong to the immediate-early class and induction by VP16 ensures their strong expression (see Roizman and Sears, 1990 for review). We have studied the activities of these two promoters using the *lacZ* gene as an indicator gene. Our preliminary data suggested that the α4 promoter has higher activity than the α22/α47 promoter. However, the α4 promoter seems to have a more stringent requirement for induction by VP16, without which the α22/α47 promoter is more active. The prototype vector that we have constructed is vα22β-galα4GT (Fig. 6), which expresses the glucose transporter gene and the *lacZ* gene under the control of the α4 and the α22/α47 promoters, respectively. Compared to vα4GTβ-gal (the IRES approach), the GT and β-gal expresssion from vα22β-galα4GT may not have identical kinetics; however, using double immunoflorescence, we have demonstrated that this vector also faithfully coexpresses GT and β-gal in every cell observed (data not shown). Comparatively

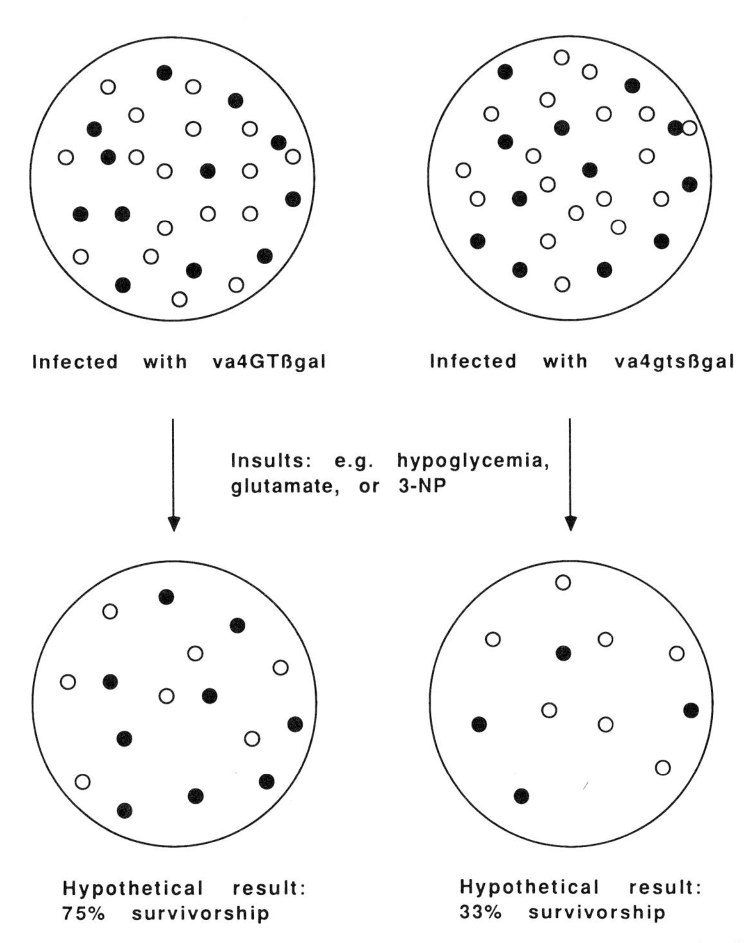

Figure 7 Experimental scheme using bicistronic vectors. Neuronal cultures (in 48-well plates) are infected with bicistronic vectors expressing the gene of interest or *lacZ*, (e.g., vα4GTβ-gal) or with control vectors expressing *lacZ* alone (e.g., vα4gtsβ-gal). Following an insult, the number of β-gal-expressing neurons is quantified directly as blue-staining cells after X-gal staining. A low m.o.i. ($\leq$0.1) is chosen such that each well will have a few hundred β-gal-expressing neurons. By comparing the number of β-gal-expressing neurons from the vα4GTβ-gal-infected cultures to that of the vα4gtsβ-gal-infected cultures, the effects of enhanced glucose transport can be monitored directly.

speaking, we believe that the IRES system may better guarantee that the expression of β-gal reflects that of the gene of interest. Unfortunately, because of the relatively weak β-gal expression under IRES-directed translation, this system is less sensitive and we have had difficulties detecting its expression in adult rat brain.

We have applied both bicistronic systems to investigate the protective effect of various genes against different neurological insults. The general experimental scheme is illustrated in Fig. 7. With this paradigm, neuronal cultures are infected with the bicistronic vector expressing the gene of interest (e.g., vα4GTβ-gal or vα22β-galα4GT) or with a control vector which expresses β-gal alone. The cultures are later challenged with neurological insults and the number of β-gal-expressing cells in each group is quantified. A representative photomicrograph of β-gal-expressing neurons is shown in Fig. 8. We have generated four different types of β-gal-expressing control vectors. For the first type, the vector contains a single transcriptional unit containing *lacZ* under the α4 or α22/α47 promoter (e.g., vα4β-gal or vα22β-gal). For the second, the vector also contains only the *lacZ* gene, but the IRES signal is placed 5′ of and in frame with the *lacZ* coding sequences (e.g., vα4EMCβ-gal). The

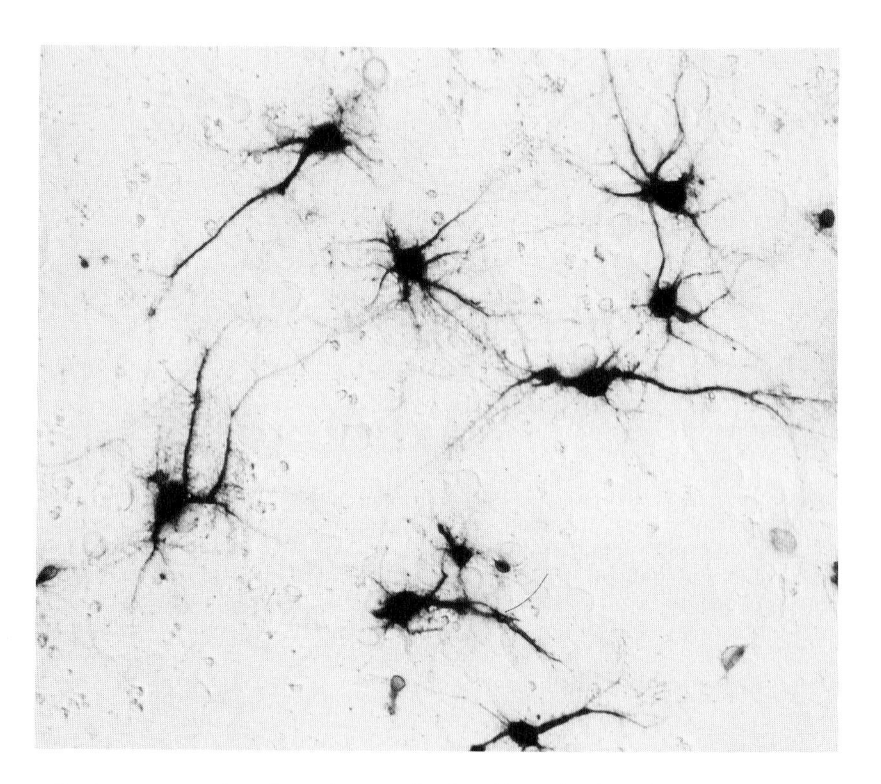

Figure 8 A representative photomicrograph of vα22β-galα4GT-infected hippocampal culture demonstrating β-gal-positive neurons. A 10-day-old hippocampal culture was infected with the bicistronic vector vα22β-galα4GT (m.o.i. $\approx$ 0.1). Despite the presence of glial cells in the culture (neuron/astrocytes ratio $\approx$ 50 : 50%), the majority of β-gal-expressing cells are neuronal.

third type contains both the gene of interest and *lacZ*, but the gene of interest is placed in a reverse orientation, such that antisense transcripts would be made (e.g., vα4antiCBβ-gal; see below). The fourth type contains both *lacZ* and the gene of interest in the correct orientation, but translation of the latter is terminated by insertion of stop codons into the coding sequences (e.g., vα4gtsβ-gal or vα22β-galα4gts); thus, only a truncated, nonfunctional protein product would be made. The *lacZ* gene can be expressed through a separate promoter (vα22β-galα4gts) or through the IRES signal (vα4gtsβ-gal). Since the size of an amplicon will affect its replication efficiency as well as the number of its copies packaged into the virion, the third and the fourth types, which are of the same size as the test vectors, will be better controls. However, given that antisense expression of a certain gene may affect the endogenous level of that gene product, we believe that control vectors with stop codons inserted into the gene of interest would be most appropiate.

Using these bicistronic vectors, we have demonstrated protection of neuronal cultures against hypoglycemia and various types of toxins (Ho *et al.*, 1995). For instance, vα22β-galα4GT infection can protect hippocampal neurons against hypoglycemia and glutamate toxicity (Fig. 9); vα4GTβ-gal protects

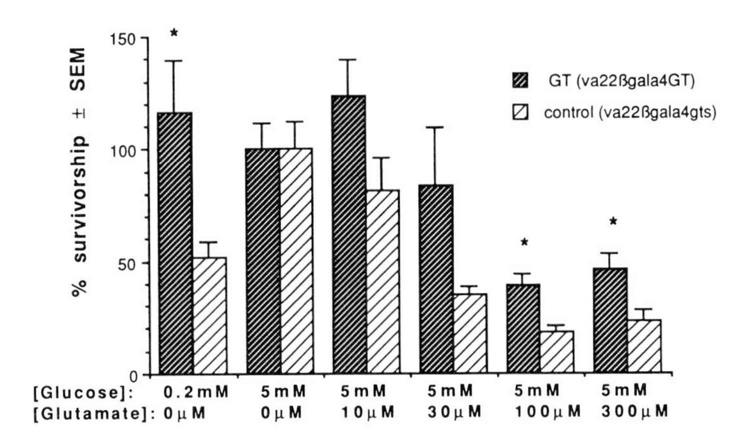

Figure 9 vα22β-galα4GT infection protects neurons against hypoglycemia and glutamate toxicity. Eleven-day-old mixed hippocampal cultures in 48-well plates were infected with vα22β-galα4GT or vα22β-galα4gts. Eight hours after infection, the cultures were rinsed once with Cl^- free medium (80 m*M* $NaSO_4$/4 m*M* K_2SO_4/1m*M* $MgSO_4$/1.5 m*M* $CaCl_2$/15 m*M* dextrose/5 m*M* glucose/27 m*M* $NaHCO_3$) and were incubated with the Cl^- medium containing various amount of glutamate, as indicated in the figure. Thirty minutes later, the cultures were rinsed once with DMEM containing 0.2 m*M* or 5 m*M* glucose, then maintained in DMEM/0.2 or 5 m*M* glucose for 16 hr. At the end of the incubation, the cells were fixed and reacted with X-gal. The number of blue-staining neurons were counted and were expressed as the percentage mean values of the control groups which did not receive any glutamate or hypoglycemia challenge (i.e., 0 m*M* glutamate/5 m*M* glucose). Asterisks represent significant differences between the two groups where $P < 0.01$. *n* averaged 6.

striatal cultures against the toxicity of 3-nitropropionic acid (Fig. 10), an inhibitor of oxidative metabolism which selectively damages striatum in animals and has been used to model Huntington's disease-like damage (Beal *et al.*, 1993).

This bicistronic approach has also aided us in identifying vector-infected cells in the brain. In the previous section, we have described the protective function of vα4GT against kainic acid-induced seizure in the hippocampus. Using the bicistronic vector vα22β-galα4GT, we have been able to correlate the number of infected neurons with the extent of protection. As mentioned prevously, kainic acid-induced damage is localized in the CA3 cell field; however, the majority of β-gal-expressing cells (which also expressed GT) are dentate gyrus neurons. We observe a significant relationship between the number of β-gal-expressing cells in the dentate gyrus and the degree of protection in CA3 against kainic acid (Lawrence *et al.*, 1994). This observation agrees with our current understanding of the action of kainic acid and the circuitry of the hippocampus. Neurotoxicity depends upon intact dentate gyrus projections to CA3 neurons because kainic acid triggers release of EAAs by dentate axon terminals (Gannon and Terrian, 1991; Pasturszko *et al.*, 1984) and/or disrupts EAA uptake at such terminals (Johnston *et al.*, 1979; Pasturszko *et al.*, 1984). Energy failure leads to membrane depolarization, calcium-dependent

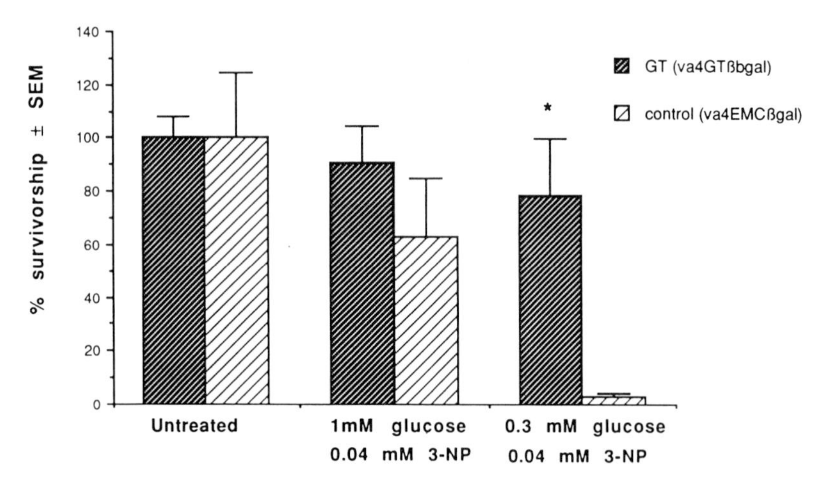

Figure 10 vα4GTβ-gal protects striatal neurons from hypoglycemia and 3-nitropropionic acid challenge. Nineteen-day-old striatal culures in 48-well plates were infected with vα4GTβ-gal or vα4EMCβ-gal. Seven hours later, medium was aspirated off and replaced with DMEM containing 0.4 m*M* 3-nitropropionic acid with either 1 or 0.3 m*M* glucose. Thirteen hours later, cells were fixed and reacted with X-gal. The number of blue-staining neurons in each well was counted. Those from the experimental groups were expressed as the percentage mean values of the control groups which did not receive any medium change or toxins. Signifiant difference between the two groups was seen with the 0.3 m*M* glucose/ 0.04 m*M* 3-NP condition ($P < 0.005$). $n = 6$ for all groups.

release of EAAs, and ultimately the reversal of EAA/Na^+ cotransporter (Kauppinen *et al.*, 1988). This results in both increased EAA release and decreased reuptake. Hence, bolstering dentate neurons energetically with vα22β-galα4GT may allow them to contain these imbalances, leading to increased neuronal survivorship in CA3.

IV. Additional Neuroprotective Strategies Using HSV Vectors

Besides increasing the level of glucose uptake, we have also attempted to deliver other genes into neurons which may block the cascade that leads to cell death. For instance, the increase in intracellular calcium after excitotoxic insults has been implicated as a mechanism involved in neuron death (reviewed in Choi, 1988; Landfield *et al.*, 1992) In an attempt to buffer the rise of

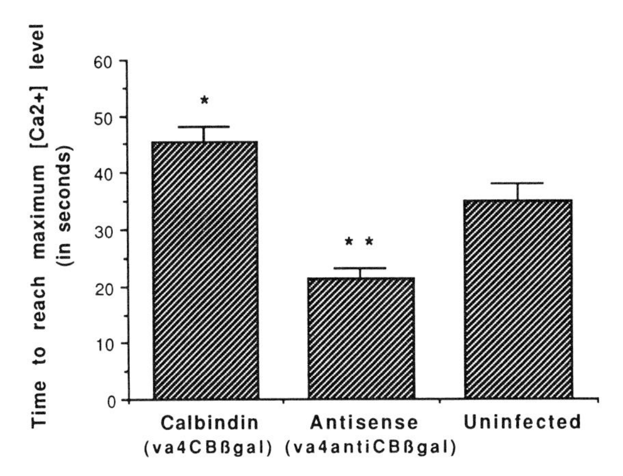

Figure 11 vα4CBβ-gal infection delays the rise of intracellular Ca^{2+} after glutamate challenge. Twelve- or thirteen-day-old mixed hippocampal cultures were mock infected or infected with vα4CBβ-gal or vα4antiCBβ-gal (antisense control) (m.o.i. = 0.2 for vector; m.o.i. = 1 for helper virus). The cultures were loaded with fura-2 AM-ester 10 to 12 hr postinfection and intracellular Ca^{2+} levels of single neurons were calculated by fura-2 ratio imaging microfluorimetry at 340 and 380 nm. Resting intracellular Ca^{2+} levels were recorded for 90 sec, followed by introduction of 1 ml of 1 m*M* glutamate in Hepes-buffered Hank's balanced salt solution into the superfusion chamber of the microscope. Fresh buffer without glutamate continued to flow through the chamber during the treatment. Neurons from vα4CBβ-gal-infected cultures exhibited significantly longer times to reach peak Ca^{2+} levels than those from mock-infected or vα4antiCBβ-gal-infected cultures. Expression of antisense calbindin transcripts (i. e., vα4antiCBβ-gal infection) also significantly shortened the time to reach peak Ca^{2+} levels. Asterisks represent significant differences from mock-infected control where $^{**}P < 0.001$ and $^{*}P < 0.01$. *n* averaged 500.

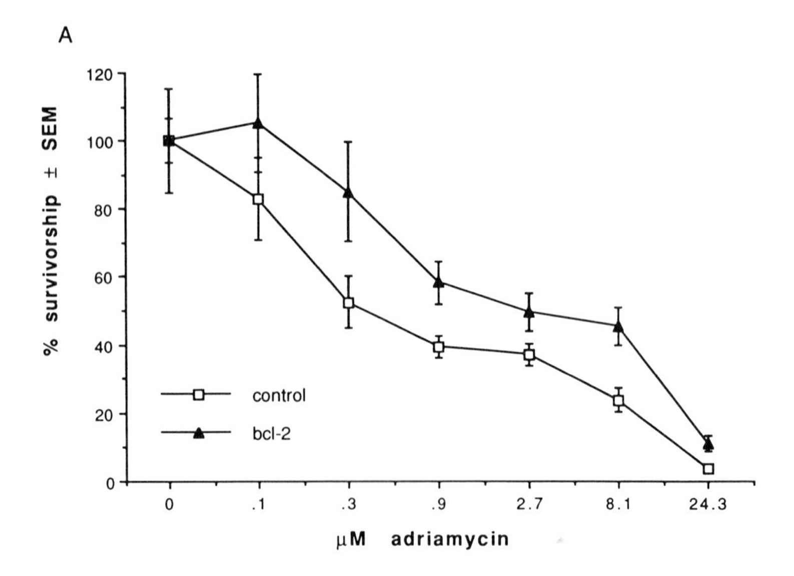

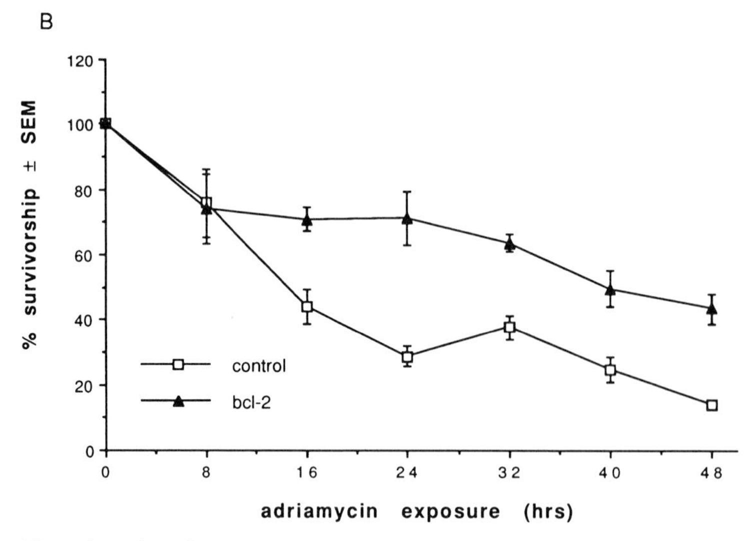

Figure 12 vα22β-galα4Bcl-2 infection protects hippocampal neurons against an oxidative insult. Eleven-day-old mixed hippocampal cultures were infected with vα22β-galα4Bcl-2 or vα22β-galα4bst (vector titer: 6.0×10^5 infectious particles/ml; helper virus titer: 2.2×10^6 PFU/ml). Eight hours afterward cells were exposed to an oxidative challenge by the addition of various dilutions of adriamycin to the culture medium to achieve adriamycin concentrations of 0, 0.1, 0.3, 0.9, 2.7, 8.1, and 24.3 m*M*. Cells were fixed and stained with X-gal 24 hr later and the number of blue-staining neurons were expressed as the percentage mean values of the control groups ± SEM, which did not receive adriamycin (A). Signifcant differences were observed between the two vectors at 0.9 ($P < 0.05$), 2.7 ($P < 0.05$), 8.1

intracellular calcium, we have constructed a bicistronic vector expressing the calcium-binding protein calbindin D-28K (vα4CBβ-gal). Using calcium imaging techniques, we have noted that this vector can significantly delay the rise of intracellular calcium after the cells are challenged with glutamate (Fig. 11).

We are also interested in the potential protective effects of a protooncogene *bcl-2* (Tsujimoto *et al.*, 1986). Although this gene was originally identified as an apoptosis inhibitor in lymphocytes (Tsujimoto *et al.*, 1986), later studies have indicated that it can also prevent apoptosis in sympathetic neurons deprived of nerve growth factor (Garcia *et al.*, 1992). We have constructed a bicistronic vector vα22β-galα4Bcl-2 expressing *bcl-2* and *lacZ*. Using the experimental scheme described in Fig. 7, we have observed protection of hippocampal cultures by vα22β-galα4Bcl-2 against an oxygen radical-generator adriamycin (Fig. 12), which strongly agrees with the proposed role of *bcl-2* as an antioxidant (Hockenbery *et al.*, 1993). Our preliminary data also suggest that vα22β-galα4Bcl-2 can also protect hippocampal neurons against glutamate toxicity, indicating that oxidative damage may be involved in the cascade of EAA-induced neuron death (Fig. 13).

V. Cytopathic Effects of HSV Vectors

Despite our successful gene expression and neuroprotection with HSV vectors, one of the obstacles that we face is their cytotoxicity. This problem is not unique to HSV vectors. Although most viral vector systems employ viral mutants incapable of replication in target cells, virion components, as well as residual gene expression from the viral genomes, may still have cytopathic effects. For HSV, a virion component termed virion host shutoff, can turn off host protein synthesis (Read and Frenkel, 1983; Read *et al.*, 1993; Schek and Bachenheimer, 1985; Smibert *et al.*, 1992); other constituents of the virions may also affect cellular physiology in as yet to be identified manners. In addition to the virion components, the immediate-early genes of HSV have also been shown to be cytotoxic (Johnson *et al.*, 1992a).

Several groups reported that after microinfusion of viral vectors into adult rat brain, little or no adverse pathologic effects were observed in the animals (Chiocca *et al.*, 1990; Kaplitt *et al.*, 1991; Le Gal La Salle *et al.*, 1993). Such conclusions were drawn based on the absence of behavioral or

($P < 0.005$), and 24.3 m*M* ($P < 0.005$) adriamycin concentrations. In a further time course study infected mixed hippocampal cultures were fixed and stained 8, 16, 24, 32, 40, and 48 hr after the addition of 0.9 m*M* adriamycin (B). Signifcant differences were observed between the two vectors at all time points ($P < 0.005$) except at 8 hr. *n* averaged 10 for each data point.

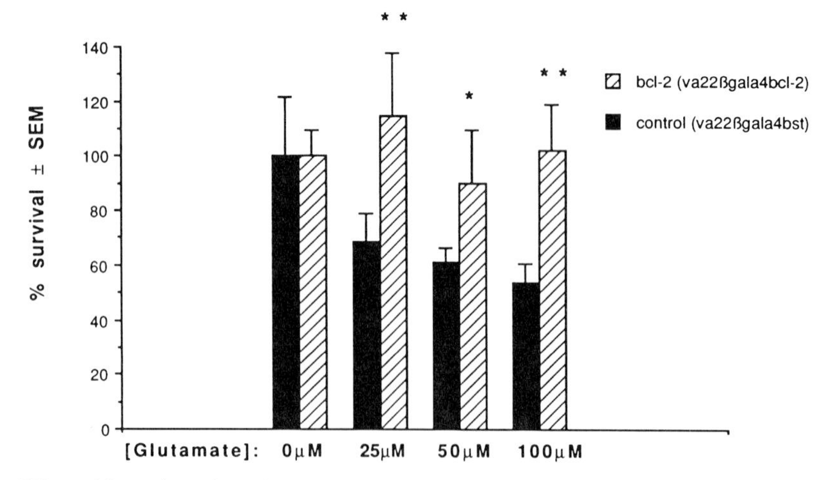

Figure 13 vα22β-galα4Bcl-2 infection protects hippocampal neurons against glutamate toxicity. Hippocampal cultures were infected with vα22β-galα4Bcl-2 or vα22β-galα4bst as described in the legend to Fig. 12 and were treated with glutamate as described in the legend to Fig. 9. Asterisks represent significant differences between the two vectors where $^{**}P < 0.02$ and $^{*}P < 0.05$. *n* averaged 10.

neurological abnormalities and lack of enlargement of the ventricles or other gross anatomical changes in infected animals. However, for HSV (or any other viral vectors) to be used for gene therapy, its cytopathicity must be more carefully examined. We have examined the cytopathic effects of two HSV-1 α4 mutants, ts756 and d120, in adult rat hippocampus and striatum and in hippocampal neurons in culture. α4 mutants have been commonly used as helpers to propagate amplicon-based vectors (Geller and Breakefield, 1988; Ho *et al.*, 1993; Kaplitt *et al.*, 1991) and have also been the backbones of many recombinant vectors (Chiocca *et al.*, 1990; Dobson *et al.*, 1990; Johnson *et al.*, 1992b; Miyanohara *et al.*, 1992; Roemer *et al.*, 1991; Weir and Elkins, 1993). Given that α4 mutants cannot replicate but only express the immediate-early gene products (and the large subunit of ribonucleotide reductase) (DeLuca *et al.*, 1985; Paterson and Everett, 1990), they are certainly among the most benign mutants available.

We have not observed any gross behavioral or anatomical changes in the animals after microinfusion of ts756 or d120 into rat hippocampus or striatum. However, when we assessed damage by stringent counting of surviving neurons surrounding the injection sites, we observed that ts756, but not d120, produces a significant amount of damage in the CA4 cell field and dentate gyrus of the hippocampus; neither ts756 nor d120 infection produces any observable damage in other hippocampal cell fields or in the striatum (Ho *et al.*, 1994). These findings are encouraging in that the cytotoxicity of d120 (and probably other $\alpha4$ deletion mutants) is negligible in our system. However, they also caution that crude examination is insufficient to detect subtle but potentially meaningful degrees of neuron loss due to the vectors. Such loss may be of functional significance. Furthermore, while the endpoint of our assay is neuron loss, the vectors might also affect cellular physiology in other subtle ways.

For hippocampal neurons in culture, we have observed that infection with d120 induces two phases of cell death: acute cytotoxicity, which occurs within the first 24 hr postinfection, and delayed cytotoxicity, occuring after 48 hr (Ho *et al.*, 1994). The acute cytotoxicity is also observed by adding uninfected cell lysates to the neurons; thus, such cytotoxicity is probably not contributed by the virus per se, but arises from the host cells used for viral propagation. Furthermore, the acute cytotoxicity can be blocked by a glutamate antagonist, suggesting that activation of glutamate receptors is involved (Ho *et al.*, 1994). It is conceivable that the plentiful amounts of glutamate and aspartate utilized by the host cells for protein synthesis can generate sufficiently high concentrations in the cell lysates or viral inoculum so as to damage neurons by excitotoxic mechanisms. We have found that partially purifying the virus from the crude cell lysates can significantly decrease the amount of acute cytotoxicity (Ho *et al.*, 1994).

In contrast, the delayed cytotoxicity pertains to the virus itself. Similar to the finding of Johnson et al. (1992a), we observed that such cytotoxicity can be blocked by uv irradiating the virus (data not shown), suggesting viral gene expression is involved. Since d120, similar to other $\alpha4$ mutants, can express only the other four immediate-early proteins and the large subunit of ribonucleotide reductase, one or more of these genes may contribute to the cytotoxicity; however, the exact candidate(s) have not yet been identified.

VI. Conclusions

Our understanding of the basic EAA/calcium cascade by which neurons die during necrotic insults is now at a stage where one can conceive of interventions designed to lessen the neurotoxicity. The general strategy of using HSV vectors for such interventions is hampered by numerous technical limitations, chief among them being the limited spread of vectors from an injection site. Nonetheless, we feel that the initial successes in the use of such vectors suggest the neuroprotective potential of this approach, and we anticipate considerable progress in this nascent field.

Acknowledgments

This work was supported by the American Paralysis Association and the Stanford Office of Technology Licensing (R.M.S.), a Huntington's Disease Society of America fellowship (D.Y.H.), a MSTP grant from National Institute of General Medical Sciences (S.L.F.), and a Howard Hughes predoctoral fellowship (M.S.L.).

References

Beal, M. (1992). Does impairment of energy metabolism result in excitotoxic neuronal death in neurodegenerative illnesses? *Ann. Neurol.* **31,** 119–130.

Beal, M. F., Brouillet, E., Jenkins, B. G., Ferrante, R. J., Kowall, N. W., Miller, J. M., Storey, E., Srivastava, R., Rosen, B. R., and Hyman, B. T. (1993). Neurochemical and histologic characterization of striatal excitotoxic lesions produced by the mitochondrial toxin 3-nitropropionic acid. *J. Neurosci.* **13,** 4181–4192.

Becky, A. S., and Sapolsky, R. M. (1988). Chemical adrenalectomy reduces hippocampal damage induced by kainic acid. *Brain Res.* **473,** 175–180.

Ben-Ari, Y. (1985). Limbic seizure and brain damage produced by kainic acid: Mechanisms and relevance to human temporal lobe epilepsy. *Neuroscience* **14,** 375–403.

Birnbaum, M. J., Haspel, H. C., and Rosen, O. M. (1986). Cloning and characterization of a cDNA encoding the rat brain glucose-transporter protein. *Proc. Natl. Acad. Sci. USA* **83,** 5784–5788.

Chiocca, E. A., Choi, B. B., Weizhong, N. A., DeLuca, N. A., Schaffer, P. A., DeFiglia, M., Breakefield, X. A., and Martuza, R. L. (1990). Transfer and expression of the *lacZ* gene in rat brain neurons mediated by herpes simplex virus mutants. *New. Biol.* **2,** 739–746.

Choi, D. W. (1990). Cerebral hypoxia: Some new approaches and unanswered questions. *J. Neurosci.* **10,** 2493–2498.

Choi, D. W. (1988). Calcium-mediated neurotoxicity: Relationships to specific channel types and role in ischemic damage. *Trends Neurosci.* **11,** 465–469.

Collins, G., Anson, J., and Surtees, L. (1983). Presynaptic kainate and N-methyl-D-aspartate receptors regulate excitatory amino acids release in the olfactory cortex. *Brain Res.* **265,** 157–159.

Cox, J., Lysko, P., and Henneberry, R. (1989). Excitatory amino acid neurotoxicity at the NMDA receptor in cultured neurons: Role of the voltage-dependent magnesium block. *Brain Res.* **499,** 267–273.

DeLuca, N. A., McCarthy, A. M., and Schaffer, P. A. (1985). Isolation and characterization of deletion mutants of herpes simplex virus type 1 in the gene encoding immediate-early regulatory protein ICP4. *J. Virol.* **56,** 558–570.

Dobson, A. T., Margolis, T. P., Sedarati, F., Stevens, J. G., and Feldman, L. T. (1990). A latent, nonpathogenic HSV-1-derived vector stably expresses β-galactosidase in mouse neurons. *Neuron* **5,** 353–360.

Duke, G. M., Hoffman, M. A., and Palmenberg, A. C. (1992). Sequence and structural elements that contribute to efficient encephalomyocarditis virus RNA translation. *J. Virol.* **66,** 1602–1609.

Dumuis, A., Sebben, M., Haynes, L., Pin, J., and Bockaert, J. (1988). NMDA receptors activate the arachidonic acid cascade system in striatal neurons. *Nature* **336,** 68–72.

Frenkel, N. (1981). Defective interfering herpesviruses. In "The Human Herpesviruses–An Interdisciplinary Prospective" (A. J. Nahmias, W. R. Dowdle, and R. F. Schinazi, Eds.), pp. 91–120. Elsevier/North-Holland, New York.

Gannon, R., and Terrian, D. (1991). Presynaptic modulation of glutamate and dynorphin release by excitatory amino acids in the guinea-pig hippocampus. *Neuroscience* **41,** 401–406.

Garcia, I., Matinou, I., Tsuijimoto, Y., and Martinou, J. C. (1992). The prevention of programmed cell death of sympathetic neurons by the *bcl*-2 protooncogene. *Science* **258,** 302–304.

Geller, A. I., and Breakefield, X. O. (1988). A defective HSV-1 vector expresses *Escherichia coli* β-galactosidase in cultured peripheral neurons. *Science* **241,** 1667–1669.
Ho, D. Y. (1994). Amplicon-based herpes simplex virus vectors. *Methods Cell Biol.* **43,** 191–210.
Ho, D. Y., Fink, S. L., Lawrence, M. S., Meier, T. J., Saydam, T. C., Dash, R., and Sapolsky, R. M. (1994). Herpes simplex virus vector system: Analysis of its in vivo and in vitro cytopathic effects. *J. Neurosci. Methods,* (in press).
Ho, D. Y., Mocarski, E. S., and Sapolsky, R. M. (1993). Altering central nervous system physiology with a defective herpes simplex virus vector expressing the glucose transporter gene. *Proc. Natl. Acad. Sci. USA* **90,** 3655–3659.
Ho, D. Y., Saydam, T. C., Fink, S. L., Lawrence, M. S., and Sapolsky, R. M. (1995). Defective herpes simplex virus vectors expressing the rat brain glucose transporter protect cultured neurons from necrotic insults. *J. Neurochem.* (in press).
Hockenbery, D. M., Oltvai, Z. N., Yin, X. M., Milliman, C. L., and Korsmeyer, S. J. (1993). *Bcl-2* functions in an antioxidant pathway to prevent apoptosis. *Cell* **75,** 241–251.
Hughes, R. G., Jr., and Munyon, W. H. (1975). Temperature-sensitive mutants of herpes simplex virus type 1 defective in lysis but not in transformation. *J. Virol.* **16,** 275–283.
Jang, S. K., Kräusslich, H., Nicklin, M. J. H., Duke, G. M., Palmenberg, A. C., and Wimmer, E. (1988). A segment of the 5′ nontranslated region of encephalomyocarditis virus RNA directs internal entry of ribosomes during in vitro translation. *J. Virol.* **62,** 2636–2643.
Jang, S. K., Davies, M. V., Kaufman, R. J., and Wimmer, E. (1989). Initiation of protein synthesis by internal entry of ribosomes into the 5′ nontranslatd region of encephalomyocarditis virus RNA in vivo. *J. Virol.* **63,** 1651–1660.
Johnson, P. A., and Friedmann, T. (1994). Replication-defective recombinant herpes simplex virus vectors. *Methods Cell Biol.* **43,** 211–232.
Johnson, P. A., Miyanohara, A., Levine, F., Cahill, T., and Friedmann, T. (1992a). Cytotoxicity of a replication-defective mutant of herpes simplex virus type 1. *J. Virol.* **66,** 2952–2965.
Johnson, P. A., Yoshida, K., Gage, F. H., and Friedmann, T. (1992b). Effects of gene transfer into cultured CNS neurons with a replication-defective herpes simplex virus type 1 vector. *Mol. Brain Res.* **12,** 95–102.
Johnston, G., Kennedy, S., and Twitchin, B. (1979). Action of the neurotoxin kainic acid on high affinity uptake of L-glutamic acid in rat brain slices. *J. Neurochem.* **32,** 121–127.
Kaplitt, M. G., Pfaus, J. G., Kleopoulos, S. P., Hanlon, B. A., Rabkin, S. D., and Pfaff, D. W. (1991). Expression of a functional foreign gene in adult mammalian brain following in vivo transfer via a herpes simplex virus type 1 defective viral vector. *Mol. Cell. Neurosci.* **2,** 320–330.
Kauppinen, R., McMahon, H., and Nicholls, D. (1988). Calcium-dependent and calcium-independent glutamate release, energy status and cytosolic free calcium concentration in isolated nerve terminals following metabolic inhibition: Possible relevance to hypoglycemia and anoxia. *Neuroscience* **27,** 175–182.
Landfield, P. W., Thibault, O., Mazzanti, M. L., Porter, N. M., and Kerr, D. S. (1992). Mechanisms of neuronal death in brain aging and Alzheimer's Diseases: Role of endocrine-mediated calcium dyshomeostasis. *J. Neurobiol.* **23,** 1247–1260.
Lawrence, M. S., Ho, D. Y., Dash, R., and Sapolsky, R. M. (1995). Herpes simplex virus vectors overexpressing the glucose transporter gene protect against excitotoxic seizures. Submitted for publication.
Le Gal La Salle, G., Robert, J. J., Berrard, S., Ridoux, V., Stratford-Perricaudet, L. D., Perricaudet, M., and Mallet, J. (1993). An adenovirus vector for gene transfer into neurons and glia in the brain. *Science* **259,** 988–990.
Lipton, S., Choi, Y., Pan, Z., Lei, S., Chen, H., Sucher, N., Loscalzo, J., Singel, D., and Stamler, J. (1993). A redox-based mechanism for the neuroprotective and neurodestructive effects of nitric oxide and related nitroso-compounds. *Nature* **364,** 626–630.

Lysko, P., Cox, J., Givano, M. A., and Henneberry, R. (1989). Excitatory amino aicd neurotoxicity at the NMDA receptor in cultures neurons: Pharmacological characterization. *Brain Res.* **499,** 258–262.

McBurney, R., and Neering, I. (1987). Neuronal calcium homeostasis. *Trends Neurosci.* **10,** 164–169.

McConnell, H. M., Owicki, J. C., Parce, J. W., Miller, D. L., Baxter, G. T., Wada, H. G., and Pitchford, S. (1992). The cytosensor microphysiometer: Biological applications of silicon technology. *Science* **257,** 1906–1912.

McCord, J. (1985). Oxygen-derived free radicals in post-ischemic tissue injury. *N. Engl. J. Med.* **312,** 159–167.

Miyanohara, A., Johnson, P. A., Elam, R. L., Dai, Y., Witstum, J. L., Verma, I. M., and Friedmann, T. (1992). Direct gene transfer to the liver with herpes simplex virus type 1 vectors: Transient production of physiologically relevant levels of circulating factor IX. *New Biol.* **4,** 238–246.

Novelli, A., Reilly, J., Lysko, P., and Henneberry, R. (1988). Glutamate becomes neurotoxic via the NMDA receptor when intracellular energy levels are reduced. *Brain Res.* **451,** 205–211.

Pardridge, W. M., Boado, R. J., and Farrel, C. R. (1990). Brain-type glucose transporter (GLUT-1) is selectively localized to the blood–brain barrier. *J. Biol. Chem.* **265,** 18035–18040.

Pasturszko, A., Wilson, D., and Erecinska, F. (1984). Effects of kainic acid in rat brain synaptosome: The involvement of calcium. *J. Neurochem.* **43,** 747–752.

Paterson, T., and Everett, R. D. (1990). A prominent serine-rich region in Vmw175, the major regulatory protein of herpes simplex virus type 1 is not essential for virus growth in tissue culture. *J. Gen. Virol.* **71,** 1775–1783.

Raley-Susman, K. M., Miller, K. R., Owicki, J. O., and Sapolsky, R. M. (1992). Effects of excitotoxin exposure on metabolic rate of primary hippocampal cultures: Application of silicon microphysiometry to neurobiology. *J. Neurosci.* **12,** 773–780.

Read, G. S., and Frenkel, N. (1983). Herpes simplex virus mutants defective in the virion-associated shutoff of host polypeptide synthesis and exhibiting abnormal synthesis of alpha (immediate early) viral polypeptides. *J. Virol.* **46,** 498–512.

Read, G. S., Karr, B. M., and Kinght, K. (1993). Isolation of a herpes simplex virus type-1 mutant with a deletion in the virion host shutoff gene and identification of multiple forms of vhs (UL41) polypeptides. *J. Virol.* **67,** 7149–7160.

Roemer, K., Johnson, P. A., and Friedmann, T. (1991). Activity of the simian virus 40 early promoter–enhancer in herpes simplex virus type 1 vectors is dependent on its position, the infected cell type, and the presence of Vmw175. *J. Virol.* **65,** 6900–6912.

Roizman, B., and Sears, A. E. (1990). Herpesviruses and their replication. In "Virology" (B. N. Fields and D. M. Knipe, Eds.), pp. 1795–1841. Raven Press, New York.

Sapolsky, R. (1992). "Stress, the Aging Brain, and the Mechanisms of Neuron Death." MIT Press, Cambridge, MA.

Sapolsky, R., and Stein, B. (1988). Status epilepticus-induced hippocampal damage is modulated by glucose availability. *Neursoci. Lett.* **97,** 157–162.

Schek, N., and Bachenheimer, S. L. (1985). Degradation of cellular mRNAs induced by a virion-associated factor during herpes simplex virus infection of vero cells. *J. Virol.* **55,** 601–610.

Smibert, C. A., Johnson, D. C., and Smiley, J. R. (1992). Identification and characterization of the virion-induced host shutoff product of herpes simplex virus gene UL41. *J. Gen. Virol.* **73,** 467–470.

Spaete, R. R., and Frenkel, N. (1982). The herpes simplex virus amplicon: A new eucaryotic defective-virus cloning-amplifying vector. *Cell* **30,** 295–304.

Spear, P. G., Shieh, M. T., Herold, B. C., WuDunn, D., and Koshy, T. I. (1992). Heparan sulfate glycosaminoglycans as primary cell surface receptors for herpes simplex virus. *Adv. Exp. Med. Biol.* **313,** 341–353.

Tsujimoto, Y., Finger, L. R., Yunis, J., Nowell, P. C., and Croce, C. M. (1986). Analysis of the structure, transcripts and protein products of Bcl-2, the gene involved in human follicular lymphoma. *Proc. Natl. Acad. Sci. USA* **83,** 5214–5218.

Weir, J. P., and Elkins, K. L. (1993). Replication-incompetent herpesvirus vector delivery of an interferon α gene inhibits human immunodeficiency virus replication in human monocytes. *Proc. Natl. Acad. Sci. USA* **90**, 9140–9144.
Zeevalk, G., and Nicklas, W. (1992). Evidence that the loss of the voltage-dependent magnesium block at the NMDA receptor underlies receptor activation during inhibition of neuronal metabolism. *J. Neurochem.* **59**, 1211–1218.
Zeevalk, G., and Nicklas, W. (1991). Mechanisms underlying initiation of excitotoxicity associated with metabolic inhibition. *J. Pharmacol. Exp. Ther.* **257**, 870–877.

CHAPTER 10

In Vivo Promoter Analysis in the Adult Central Nervous System Using Viral Vectors

Jun Yin
Michael G. Kaplitt
Donald W. Pfaff
Laboratory of Neurobiology and Behavior
The Rockefeller University
New York, New York

I. Introduction

Herpes-based viral vectors can be applied in modern neurobiology for at least three purposes: gene transfer into the brain, prolonged expression of antisense RNA, or *in vivo* promoter analysis. The usefulness of the latter application depends on the following reasoning. Assuming that nerve cells are different from those cell types traditionally used for transfection studies, at least by virtue of their synaptic inputs, there is the likelihood that the nuclear protein environment *in vivo* imposes special requirements for valid transcriptional analyses. If, as a consequence, typical transfection studies using kidney cells, skin cells, or other convenient cell types are not entirely satisfactory, there are two choices. The first choice, *in vitro* transcription methodology, attempts to treat RNA synthesis as a defined set of chemical reactions outside the context of the cell and is intriguing for the neurobiologist who wants to avoid the errors implicit in using the wrong cell types. We have developed an *in vitro* transcription assay using, in some cases, neuronal nuclear extracts supplemented by HeLa cell extract and in other cases only neuronal material and have used this assay to help prove transcriptional facilitation for the preproenkephalin gene by the hormone estradiol (Freidin *et al.*, 1994; and manuscript in preparation). However, this assay is finicky, is not truly quantitative, and will always be problematic for the neurobiologist who has small amounts of starting material. In addition, studies of gene regulation under

VIRAL VECTORS

varied physiological conditions are very limited with this technique. Therefore, the second choice, *in vivo* promoter analysis, holds much promise.

What types of processes can be studied using *in vivo* promoter analyses? As an example, at the messenger RNA level, three sorts of phenomena can be analyzed. First, specificity across brain regions is a prominent feature of gene expression for the progesterone receptor and for the opioid peptide proenkephalin (Harlan *et al.*, 1987; Romano *et al.*, 1989a). Second, the steroid hormone estradiol leads to a handsome induction both of the progesterone receptor and preproenkephalin in medial hypothalamic tissue (Romano *et al.*, 1988, 1989a). Finally, there are obvious sex differences in the magnitude of hormonal induction for both of these genes (Romano *et al.*, 1989b; Lauber *et al.*, 1991). In both cases the estrogen induction is weak or absent in the male, and the sex difference is not simply due to a requirement for testosterone, which also is ineffective. Neither is this ever due to a simple inability of the male hypothalamic cell to transcribe that gene, since for both genes basal expression in the male is robust. Rather, the hypothalamic neuronal nucleus in the genetic male lacks specific nuclear proteins required for the estrogenic facilitation or contains additional squelching or other inhibitory nuclear proteins. The promoter of the gene for the opioid peptide enkephalin has been used as an example in the studies quoted below to show how well-designed viral vectors can be used for this facet of molecular neurobiology.

The defective HSV vector possesses many features advantageous for *in vivo* promoter analysis. This system is plasmid-based, which should allow adaptation to a wide variety of research applications. This also should permit mutagenesis of promoters in standard *in vitro* systems. The resulting plasmids may then be used to rapidly generate viruses for further *in vivo* studies. Potential confounding *cis*-acting influences from other regulatory sequences are minimized with this approach. Moreover, promoters from any mammalian species can be studied within the same species. Since the local microenvironment of an organ can strongly influence promoter function (Clayton *et al.*, 1985a,b; Friedmann *et al.*, 1986), the defective HSV vector provides an opportunity for *in vivo* analysis of the function of an isolated promoter within desired cell types *in situ*.

II. Regional Differences in Expression of the Preproenkephalin Gene in the Brain

Direct gene transfer by microinjection of a viral DNA vector was used for the analysis of the preproenkephalin promoter (Kaplitt *et al.*, 1994). The generation, use, and biology of this vector has been described elsewhere in this volume (Kwong *et al.*). The defective viral vector dvHENK was generated from the amplicon pHENK. pHENK contains the bacterial *lacZ* gene under the control of a 2.7-kb fragment of the rat PPE promoter. This transcription unit was inserted into the HSV amplicon pSRa-ori (Kaplitt *et al.*, 1991), which contains the HSV origin of DNA replication (HSV ori) and a cleavage/packaging sequence (HSV a).

dvHENK was stereotaxically microinjected into regions of the rat brain which express endogenous PPE mRNA (Khachaturian *et al.*, 1983; Harlan *et al.*, 1987), including the caudate nucleus, piriform cortex, and amygdala. Numerous cells containing β-galactosidase activity were observed in each of these regions. This is not only comparable to the known pattern of PPE expression, but local β-galactosidase production within injected sites appears to be restricted to a nonrandom pattern dictated by cell group boundaries. For example, positive cells were found dispersed in the caudate nucleus after vector application there, which is expected since cells which express endogenous PPE are distributed throughout this region (Khachaturian *et al.*, 1983; Harlan *et al.*, 1987). In the piriform cortex, the majority of positive cells were observed within the pyramidal cell layer, consistent with a specific pattern of endogenous PPE expression characteristic of cells within this region (Khachaturian *et al.*, 1983; Harlan *et al.*, 1987).

To further analyze PPE promoter function, dvHENK was injected into the dorsolateral thalamic region and dorsolateral neocortex, two regions which do not contain cells which express normal PPE transcripts (Khachaturian *et al.*, 1983; Harlan *et al.*, 1987). Some positive cells were observed in the dorsolateral thalamus. By contrast, positive cells were not observed in the dorsolateral neocortex, as predicted from the lack of endogenous PPE transcripts in this region. As a positive control, a defective vector containing the human cytomegalovirus (HCMV) immediate–early (IE) promoter yielded positive cells within the dorsolateral neocortex (Kaplitt *et al.*, 1991), showing that defective HSV vectors are capable of transferring and expressing the *lacZ* gene in this region. Despite the presence of a needle track within the substance of the neocortex, it was still possible that a negative result was due to physical loss of defective viral particles during or after surgery. To address this question, coverslips were floated off stained sections and "nested" PCR analysis was performed. Nested PCR involves two successive reactions, with the second reaction using a sample of DNA from the first reaction. The primers from the second reaction are internal to the first, which increases the specificity of the final amplified product, since it is extremely unlikely that a nonspecific product from reaction A would be reamplified with the internal primers from reaction B. The results show that dvHENK DNA was present in both the amygdala and the dorsolateral cortex, even though *lacZ* expression was only observed in the amygdala. No DNA was present in an uninjected region of the same brain, as expected. Since dvHENK DNA was present in the dorsolateral neocortex, the absence of positive cells must have been due to a lack of transcription in this region (Kaplitt *et al.*, 1994).

Since the PPE promoter is an endogenous promoter, we also wanted to determine whether this would permit long-term expression of a foreign gene in the adult brain. Rats were injected with dvHENK in the amygdala and the ventromedial hypothalamus. Two months following injection, positive cells were observed in both regions. We then extended the histochemical assay with PCR analysis of dvHENK DNA, as described earlier. Bands of the predicted size were observed in PCR products from three different *lacZ*-expressing sections, thereby demonstrating that defective HSV DNA is maintained stably *in vivo* (Kaplitt *et al.*, 1994). It should be noted that although fewer positive cells were observed in these sections when compared with the short-term promoter analysis study (above), a vector stock with a far lower titer was used in this instance. Therefore, this result should be considered as a qualitative demonstration of long-term expression, and not as a quantitative analysis of this property.

A recent study examined a 200-bp fragment of human PPE promoter in transgenic mice (Donovan *et al.*, 1992). Significant differences were noted between the gross regional patterns of reporter gene and endogenous PPE expression within the brain, and the low level of reporter gene expression within the brain prevented highly localized regional analyses. When promoter function was studied with a defective HSV vector, a far larger PPE promoter fragment was employed, and the rat promoter was studied within the rat brain (Kaplitt *et al.*, 1994). Both factors may have contributed to our observations of restricted expression consistent with endogenous PPE distribution.

Positive cells were unexpectedly observed in the dorsolateral thalamic region (Kaplitt *et al.*, 1994), suggesting that this region contains factors which permit transcription from the PPE promoter. It is possible that an element which silences transcription in thalamus is present beyond the 2.7-kb sequence used in this study, and complex combinations of enhancers and silencers have been found in regulatory regions of other genes expressed in brain (Mori *et al.*, 1990). In addition, recent observations of PPE RNA in brain cell types previously thought to lack such transcripts, including the reticular nucleus of the thalamus, indicate that previous hybridization studies may not have completely characterized endogenous PPE expression (Kilpatrick *et al.*, 1985; Vilijn *et al.*, 1988; La Gamma *et al.*, 1992; P. J. Brooks, personal communication). By contrast, the lack of transcription in the dorsolateral neocortex indicates that this region is incapable of supporting PPE promoter activity, due either to a lack of a necessary transcription factor(s) or to the presence of a highly specific silencer within the 2.7-kb promoter fragment. Further studies with both larger and deleted promoter fragments, as well as *in vitro* binding assays, may provide additional insights into these phenomena.

Using the PPE promoter, we also reported (Kaplitt *et al.*, 1994) the first demonstration of long-term expression of a foreign gene in the adult rat brain following direct, *in vivo* transfer via an HSV defective viral vector. Earlier, we observed expression in rat brain of *lacZ* under the control of the HCMV IE promoter (Kaplitt *et al.*, 1991), which was limited to 2 weeks following viral injection. This was consistent with other *in vivo* studies using the HCMV IE promoter (Scharfmann *et al.*, 1991; Fink *et al.*, 1992). Although downregulation of promoter function would have explained this result, it was also possible that either the vector DNA was degraded with time or that cells containing the transferred gene may have died. Stable maintenance of defective HSV DNA has not previously been demonstrated, but PCR analysis of stained sections confirms that defective HSV vector DNA is stable within at least some cells of the rat brain. The replacement of a viral immediate–early promoter with a promoter for an endogenous cellular gene resulted in long-term expression, and it is possible that the continuous promoter activity contributed to the stability of the vector DNA, although this has not been tested.

In addition to novel findings with respect to PPE promoter activity, our results (Kaplitt *et al.*, 1994) have implications for the emerging field of direct gene transfer as a therapeutic modality. The ability to regulate and maintain expression of a foreign gene *in vivo* has been difficult and is central to the continued development of approaches to genetic therapy (Miller, 1992). The observation that defective HSV vector DNA is stably maintained *in vivo* suggests that such vectors are applicable for long-term expression studies. Locally restricted, long-term PPE promoter-driven expression in brain further indicates that certain promoters may allow for specifically targeted and stable production of a foreign gene product *in vivo*. The combination of defective HSV vectors and endogenous cellular promoters may provide both new approaches to understanding gene regulation *in vivo* and new opportunities for genetic therapy through stable modification of the physiology of nondividing, terminally differentiated cells.

III. *In Vivo* Deletion Analysis of the Preproenkephalin Promoter

There have been few studies to date examining regions of the rat PPE promoter which influence gene expression. Plasmids containing various fragments of the rat PPE promoter controlling expression of the chloramphenicol acetyltransferase (CAT) gene have been tested in cultured C6 glioma cells (Joshi and Sabol, 1991). CAT mRNA levels were very low when a 2.7-kb promoter fragment was utilized, but significant stimulation occurred following forskolin treatment, which was then inhibited by further addition of dexamethasone. This was in contrast to expression of endogenous PPE in C6 glioma cells, in which forskolin stimulation of expression is potentiated by glucocorticoids. The presence of possible negative regulatory elements was suggested by this study, but the relevance of these data to PPE regulation in normal neurons is unclear, since glucocorticoid treatment does not have major effects upon PPE mRNA levels in brain tissue of normal rats (Chao and McEwen, 1990).

More recently, transgenic mice have been created using small fragments (<200 bp) of the human proenkephalin promoter controlling the CAT gene (Donovan *et al.*, 1992). The pattern of enzymatic activity within brain regions did not parallel the distribution of endogenous PPE expression. Furthermore, although CAT activity in the spinal cord dorsal horn increased in response to primary afferent stimulation, as occurs with the endogenous PPE gene, CAT mRNA levels were so low in brain and spinal cord that *in situ* hybridization failed to produce positive signals (Donovan *et al.*, 1992; Takemura *et al.*, 1992). Since a single transgenic founder line was selected for the synaptic regulation studies, it is possible that a synaptic response element does reside within this region, but that basal promoter activity was influenced by sequences near the site of transgene insertion. In addition, differences between the human and murine promoters may have influenced expression as well. Another transgenic mouse study, however, used a 1.6-kb fragment of the rat PPE promoter with the CAT gene as a reporter, and again CAT mRNA levels in the brain were barely detectable in numerous founder lines (Zinn *et al.*, 1991), although expression in the testis was quite high. These studies suggest that numerous regulatory sequences may reside at various sites near to and distal to the transcriptional start site for PPE, but sequences influencing the normal pattern of rat PPE expression in brain have not been identified.

The purpose of our study was to utilize the defective HSV vector in order to identify regions of the rat PPE promoter which are required for expression within regions of the rat brain. Using a 2.7-kb fragment of the rat PPE promoter controlling expression of the bacterial *lacZ* gene, we have confirmed and expanded upon our earlier observation that this construct yields a pattern of expression in rat brain which is analogous to the distribution of PPE mRNA as determined by *in situ* hybridization (Harlan *et al.*, 1987). Deletions within the rat PPE promoter were created, and *in vivo* analysis of these constructs revealed the presence of a element in the region between −1431 and −2700 bp upstream of the PPE transcriptional start site which positively influences expression in the amygdala/piriform cortex and caudate nucleus (Kaplitt, 1993). Finally, previous observations of long-term expression in rat brain using the 2.7-kb PPE promoter fragment were confirmed, although there was a significant decrease in the number of positive cells observed over time.

In order to expand upon earlier observations, the pattern of *lacZ* expression from dvHENK was compared with *in situ* hybridization for endogenous PPE transcripts. dvHENK was infused into the piriform cortex region, and the resulting pattern of expression 3 days following injection was largely limited to the neuronal layer of the piriform cortex. This pattern is highly comparable to the endogenous pattern of PPE expression, as demonstrated by *in situ* hybridization, and these results are consistent with earlier observations. In addition, dvHENK was infused into the ventromedial nucleus (VMN) of the hypothalamus. After 3 days, the majority of positive cells were limited to the VMN, with the characteristic oval shape evident near the third ventricle. Positive cells within the VMN appear to be concentrated in the ventrolateral and dorsomedial portions of the VMN, with fewer cells in the middle of this nucleus. This parallels the distribution of cells expressing PPE transcripts within the VMN. Thus, although an absolutely exclusive argument is not possible, in the VMN as well as in the piriform cortex, *lacZ* expression appears to be limited largely to cell groups which express endogenous PPE transcripts, following transfer with dvHENK.

It should also be noted that we have clearly observed retrograde transport of these defective viral particles for the first time. A grouping of cells was noted in the uninjected, contralateral amygdala of subjects injected with dvHENK in the amygdala/piriform cortex region. This was observed repeatedly, but not in all subjects. These cells would be the result of synaptic uptake of viral vector particles, followed by retrograde transport of vector capsids through commissural connections to cell bodies residing in the contralateral amygdala. No evidence of transport was noted in other regions, although this was not extensively analyzed, since this result was not a specific aim of the current study. Although significant numbers of cells were demonstrated in the opposite amygdala (100–400 positive cells/subject), this still represented a very minor portion of the total number of positive cells. This confirms earlier statements indicating that the majority of vector uptake and expression is local to the site of injection (Kaplitt *et al.*, 1991).

We have demonstrated above that a 2.7-kb fragment of the rat PPE promoter yields a pattern of expression analogous to that of the endogenous PPE transcript. In anticipation of deletion analysis, the viral vector dvHENK was renamed dvHENK.2700. Following restriction digestion of plasmid pHENK.2700, three new amplicons were generated. These were identical to the parental plasmid, with the exception that the PPE promoter was progressively deleted. Amplicons pHENK.1431, pHENK.841, and pHENK.500 contain fragments of the PPE promoter which are 1431, 841, and 500 bp upstream of the transcriptional start site, respectively. Viral vectors were then generated from these amplicons by cotransfection of the amplicons with DNA from the temperature-sensitive HSV1 mutant tsK. The resulting viral stocks were serially propagated until the defective vectors represented the majority of the viral stock.

Titers of the defective viral vectors were obtained from analysis of labeled viral DNA. Briefly, Vero cells were infected with a given passage in media without phosphate. Following several hours, radioactive orthosphosphate was added to specifically label viral DNA, which was harvested the following day. When the viral DNA was subjected to *Sal*I restriction digestion and electrophoresis, the resulting autoradiography revealed the ratio of defective vector DNA to helper virus DNA. At early passages, the majority of the DNA consisted of helper virus DNA. By passage 5, however, the defective vector represented the majority of the virus stock, since the defective vector band was clearly present following an overnight exposure, while the helper virus bands were still not evident. The titers of the defective vectors were then estimated from the titer of the helper virus, as determined by plaque assay, and the ratio of defective vector/helper virus DNA.

The vectors dvHENK.1431, dvHENK.841, and dvHENK.500 were each infused into the amygdala/piriform cortex or caudate nucleus of different female Sprague–Dawley rats. As a control, dvHENK.2700 was also infused into the amygdala/piriform cortex or caudate nucleus of additional subjects. Approximately 200,000 functional defective vector particles were infused into each region of each subject in a total volume of 2 μl, and all animals were sacrificed 3 days following vector injection.

A dramatic decrease in the number of positive cells was evident within the amygdala/piriform cortex and caudate nucleus with all three of the vectors containing deletions in the PPE promoter (Tables 1 and 2; Kaplitt, 1993). The results demonstrate that there is a 50- to 100-fold decrease in the number of positive cells observed in the amygdala/piriform cortex and caudate nucleus

Table 1
***lacZ* Expression in Brain Driven by Deleted Enkephalin Promoters. I. Amygdala/Piriform**

	No. positive cells
−2700	
Rat D131	>10,000
Rat D132	>10,000
−1431	
Rat D50	52
Rat D60	219
Rat D120	123
−841	
Rat D30	127
Rat D40	42
Rat D100	4
−500	
Rat D10	95
Rat D20	103
Rat D80	124

Table 2
***lacZ* Expression in Brain Driven by Deleted Enkephalin Promoters. II. Caudate**

	No. positive cells
−2700	
Rat D131	>10,000
Rat D132	>10,000
−1431	
Rat D50	60
Rat D110	72
−841	
Rat D30	44
Rat D90	20
−500	
Rat D10	91
Rat D70	166

when dvHENK.1431, dvHENK.841, or dvHENK.500 was utilized compared with the number of cells obtained with dvHENK.2700. This indicates that a regulatory element(s) exists between −1413 and −2700 bp upstream of the PPE transcriptional start site which positively influences PPE expression.

The identification of a transcriptional enhancer between −1431 and −2700 bp is consistent with data generated from a study utilizing transgenic mice (Zinn *et al.,* 1991). In that instance, a 1.6-kb fragment of the rat PPE promoter controlling expression of the CAT gene was used as the transgene. The purpose of that study was to examine expression of a PPE transcript with a testis-specific start site, and so extensive analysis of transgene expression in brain was not performed. However, the level of CAT mRNA in whole brain, testis and other tissues was examined. The distribution of CAT mRNA between organs was analogous to the endogenous PPE distribution. CAT mRNA levels were quite high in the testis of all founder males and F_1 male offspring of founder females, yet CAT mRNA was barely detectable or undetectable in brains of each of these animals. This is in direct contrast with endogenous PPE mRNA, which is normally higher in brain than in testis (Kilpatrick *et al.,* 1985, 1990). It is unlikely that insertional influences were significant, since unexpectedly low levels of CAT mRNA were demonstrated in each of several different founder animals. These data also suggest that a positive regulatory element, which is important for expression in brain, was absent from the 1.6kb promoter fragment used in the transgenic study. Since caudate nucleus and amygdala/piriform cortex contain a substantial fraction of PPE transcripts in the brain (Harlan *et al.,* 1987), it is likely that low CAT expression in whole brain reflects a low level of activity of the 1.6kb PPE promoter fragment in those regions. It is also of interest that a DNAse hypersensitivity site in brain has been identified near −2.0kb of the rat PPE promoter (Funabashi *et al.*, 1993). These sites are frequently associated with mechanisms important for transcription.

IV. Hormonal Induction of the Preproenkephalin Promoter

In vivo promoter analysis can also be used to look for regulated expression which, in the present case, is a hormonal effect on the preproenkephalin promoter (Yin *et al.*, 1994a). Estradiol *in vivo* causes an increase in PPE mRNA levels (Romano *et al.*, 1988), with a doubling of mRNA concentration within 1 hr (Romano *et al.*, 1989b). Further, gel shift assays reveal estrogen receptor-like binding to the PPE promoter in a region with an estrogen response element-like sequence (Zhu *et al.*, 1994, and manuscript submitted for publication). All of these facts suggest that estradiol induces transcription of PPE, a hypothesis we tested here using viral vector technology for gene transfer into adult brain (Kaplitt *et al.*, 1991). For this purpose, 2×10^5 particles of dvHENK.2700 were microinjected using stereotactic techniques into the brains of ovariectomized female rats. Two days after brain surgery, half of the rats received estradiol benzoate injections subcutaneously while the other half received oil vehicle control injections. After 2 additional days of hormone treatment, all rats were sacrificed by cardiac perfusion with 2% paraformaldehyde with EGTA. Sections stained for β-galactosidase were analyzed with the light microscope.

Estrogen treatments significantly increased the number of cells staining positive for the *lacZ* gene product (β-galactosidase) following viral vector injections into the hypothalamus (Table 3). As a control, dvHENK was injected

Table 3
Effect of Estradiol Benzoate (EB) on PPE Promoter Function *in Vivo*

	Mean no. of cells expressing B-gal transgene	
	Control, no EB	Systemic EB
Hypothalamic microinjection of viral vector		
Basomedial hypothalamus	82 ± 27	223 ± 50
Amygdala	187 ± 120	148 ± 80
Caudate microinjection of viral vector		
Caudate	193 ± 84	272 ± 108

into the caudate of the same animals. There was no difference in the number of positive cells seen in the estrogen-treated and untreated groups. This is not unexpected given an absence of estrogen induction of endogenous PPE in the caudate, and this may be due to the absence estrogen receptors in this region. Surprisingly, a large number of amygdaloid and piriform cortex neurons were stained following hypothalamic injections, and these are now being analyzed. In ovariectomized control animals which did not receive estrogen treatment, there were few β-galactosidase-positive cells following dvHENK injection into the ventromedial hypothalamus. PCR–*in situ* hybridization (described below) revealed, however, that ample viral vector DNA was present in these brain sections, suggesting that the paucity of positive cells was secondary to a lack of estrogen rather than variabilities in injections or transduction efficiency. Taken together, these results indicate that the 2700-base PPE promoter fragment contains *cis*-regulatory elements sufficient for a transcriptional effect of estradiol in rat hypothalamic neurons.

In turn, new data indicate that *in vivo* promoter analysis is necessary to reveal the hormonal sensitivity of the preproenkephalin promoter. A well-chosen neuronal cell line was not adequate to show this regulation (Attardi, unpublished data). The initial presumption would be that an ideal cell line for revealing a steroid hormone effect would be a cell line which expresses neuronal characteristics, which is derived from the basal forebrain and which has endocrine functions. The GT-1 cell line satisfies all of these requirements. The GT-1 cell line was produced by targeted oncogenesis using the GnRH promoter, expresses neuronal characteristics and was derived from the forebrain. Yet, when transfected with estrogen receptor and with 2700 bases of the preproenkephalin promoter no increase in the luciferase reporter gene was observed after estradiol (Attardi, unpublished observation). This result underscores the necessity of studying promoter function in the adult brain for the purposes of revealing at least a subset of transcriptional regulatory events.

V. *In Situ* PCR as a Control for Viral Vector-Mediated *in Vivo* Promoter Analysis

A technique is required to show that under different experimental conditions invasion or survival of viral vector DNA is not influenced in a manner which would confound interpretations of varying degrees of transgene expression. Histochemical staining for the presence of a transgene product, such as β-galactosidase, and *in situ* hybridization to detect mRNA are both straightforward. It is far more difficult, however, to reliably and easily document the presence of vector DNA within individual cells in a tissue section. *In situ* PCR has been used in cell suspensions and surgical tissues for detection of target DNAs (Haase *et al.*, 1990; Nuovo, 1992). Using *in situ* PCR, we can detect low-number copies of target DNA after amplification with specific primers, and we have used *in situ* PCR to detect viral vector DNA within individual cells in rat brain sections (Yin *et al.*, 1994b, and manuscript submitted for publication). In this manner, we can count the number of neurons transfected and thus test whether a given experimental condition has influenced viral vector invasion or survival. We have described above an example of the use of this approach to control for potentially confounding effects of variability in transduction efficiency which could influence the interpretation of promoter studies. Below we review the development and adaptation of this technique for use in mammalian brain tissue sections.

Brain sections prefixed with 2% paraformaldehyde were treated with PBS–detergent (0.01% DOC, 0.02% NP-40) before PCR amplification to increase the permeability of peptides and oligonucleotides across cellular barriers in brain tissue. Pretreatment with detergent retains better brain morphology than protease treatment (Table 4). The PCR mixture containing dNTPs, primers, digoxigenin (dig)–dUTP, and buffer was loaded onto each brain section. Slides containing brain sections were placed in an aluminum boat and then on

Table 4
***In Situ* PCR Applied to CNS**

1. Gradual $-20°C \rightarrow 4°C \rightarrow 20°C$.
2. 1 × PBS (0.01% DOC, 0.02% NP-40) 1 hr. Then 2 × PBS, 15 min.
3. Drain slides. Load PCR Rx mix, with primers and digoxygenin dUPT (200 μl/section). Coverslip, anchored by nail polish.
4. Slides → aluminum boat → PCR Thermal Cycler. Heat block → 82°C.
5. Lift coverslip, add 2 μl (4 units) DNA *Taq* polymerase. Mix using coverslip.
6. Overlay coverslip with mineral oil preheated to 82°C.
7. Denature 94°C, 3 min. Then 35 PCR cycles:

Anneal	55°C	2 min
Extend	72°C	2 min
Denature	92°C	2 min

8. Wash slides in xylene ≥ 15 min. Lift coverslip.
9. Rehydrate (100% ETOH, 95%, 70%, PBS).
10. Detect digoxygen by alkaline phosphatase-conjugated anti-dig. antibody and then BCIP/NBT.
    - ≥ 30 min blocking solution.
    - Wash in TBS 3 × 10 min
    - 200 μl/section anti-dig, 1 hr.
    - Wash in TBS 3 × 10 min.
    - Stain with BCIP/NBT.
11. Dry, counterstain, dehydrate, coverslip.

the block of a thermal cycler. Temperature was brought to 82°C before adding *Taq* polymerase (hot start method) followed by coverslipping and PCR amplification with 35 cycles at 55, 72, and 94°C, 2 min each. Dig-labeled PCR fragments were then detected by alkaline-phosphatase-linked anti-dig antibody. Positive signals were seen within the nucleus of transfected neurons, indicating presence of viral DNA, which then serves as an internal control for transgene presence during comparisons of experimental groups.

Additional control experiments were done to test the validity of the *in situ* PCR approach in rat brain (Yin *et al.*, 1994b, and manuscript submitted for publication). As expected, testing for transgene expression with uninjected brain tissue or with needle damage only yielded negative results. Leaving out the *Taq* enzyme also yielded negative results as, usually, did the microinjection with helper virus only. However, we sometimes saw positive results when a nonsense primer was used or when primers were left out. This led us to the hypothesis that primer-free amplification was possible, perhaps because of repair of damaged viral DNA. Since accurate detection of viral vector DNA could be confounded by this mechanism, additional control experiments were performed in which the digoxygenin label was not incorporated through the free nucleotide bases but instead was attached to the primers themselves. Under these circumstances the positively labeled cells for the transgene not only required the *Taq* polymerase enzyme but also required priming by the proper sequences, ensuring a valid measure of viral vector in brain cells.

VI. Summary

Viral vectors represent powerful tools in the emerging field of gene therapy of central nervous system disease. However, these vehicles are also revolutionizing the manner in which the basic functioning and development of the brain is analyzed. This technique is likely to be of particular importance to the study of regulation of gene expression in the living mammalian brain. For example, the ability to rapidly generate and test constructs *in vivo* is very attractive. Equally advantageous is the necessity of direct injection of episomal vectors into the fully developed adult brain. This limits the effect of development or site of insertion upon the properties of a given promoter under study, as may occur in transgenic animals. In addition, both control and experimental vectors can be tested within the same animal by unilateral injection of each vector into a given brain region.

This chapter has summarized our experience using the defective herpes simplex viral vector for analysis of the rat preproenkephalin promoter. We have demonstrated fidelity of regional expression and hormonal inducibility and documented sites of potential importance via deletion analysis. All of the data generated have been consistent with previous observations which to date were not completely interpretable. Finally, we have applied novel techniques to enhance detection and normalization of data to limit the effect of variables which could potentially confound the interpretation of results obtained from *in vivo* promoter analysis. While the field of *in vivo* promoter analysis using defective viral vectors is relatively new, the observations and techniques summarized here should be broadly applicable to the study of a wide range of promoters under various physiological conditions. In combination with more conventional techniques, this approach may provide new insight into the complex and highly specific regulation of gene expression within cells of the mammalian brain.

References

Chao, H. M., and McEwen, B. S. (1990). Glucocorticoid regulation of preproenkephalin mRNA in the rat striatum. *Endocrinology* **126**, 3124–3130.

Clayton, D. F., and Darnell, J. E., Jr. (1985a). *Mol. Cell. Biol.* **5**, 2623–2632.

Clayton, D. F., and Darnell, J. E., Jr. (1985b). *Mol. Cell. Biol.* **5**, 2633–2641.

Donovan, D. M., Takemura, M., O'Hara, B. F., Brannock, M. T., and Uhl, G. R. (1992). Preproenkephalin promoter "cassette" confers brain expression and synaptic regulation in transgenic mice. *Proc. Natl. Acad. Sci. USA* **89**, 2345–2349.

Fink, D. J., Sternberg, L., Mata, M., and Glorioso, J. C. (1992). HSV-mediated gene transfer to the PNS. *Neurology* **42**(4, Suppl. 3), 145.

Freidin, M., Priest, C., and Pfaff, D. W. (1994). Nuclear extracts from rat brain regulate PPE promoter activity in an *in vitro* transcription assay. *Soc. Neurosci.* **20**, 53.

Friedman, J. M., Babiss, L. W., Clayton, D. F., and Darnell, J. E., Jr. (1986). Cellular promoters incorporated into the adenovirus genome: Cell specificity of albumin and immunoglobulin expression. *Mol. Cell. Biol.* **6**, 3791–3797.

Funabashi, T., Brooks, P. J., Mobbs, C. V., and Pfaff, D. W. (1993). DNA methylation and DNase-hypersensitive sites in the 5′ flanking and transcribed regions of the rat preproenkephalin gene: Studies of mediobasal hypothalamus. *Mol. Cell. Neurosci.* **4**, 499–509.

Haase, A. T., Retzel, E. E., and Staskus, K. A. (1990). Amplification and detection of antiviral DNA inside cells. *Proc. Natl. Acad. Sci. USA* **87**, 4971–4975.

Harlan, R. E., Shivers, B. D., Romano, G. J., Howells, R. D., and Pfaff, D. W. (1987). Localization of preproenkephalin mRNA in the rat brain and spinal cord by *in situ* hybridization. *J. Comp. Neurol.* **258**, 159–184.

Joshi, J., and Sabol, S. L. (1991). Proenkephalin gene expression in C6 rat glioma cells: Potentiation of cyclic AMP transcription by glucocorticoids. *Mol. Endocrinol.* **6**, 1069–1080.

Kaplitt, M. G., Pfaus, J. G., Kleopoulos, S. P., Hanlon, B. A., Rabkin, S. D., and Pfaff, D. W. (1991). Expression of a functional foreign gene in adult mammalian brain following *in vivo* transfer via a herpes simplex virus type 1 defective viral vector. *Mol. Cell. Neurosci.* **2**, 320–330.

Kaplitt, M. G. (1993). "Development and Application of Herpes Simplex Viral Vectors for in Vivo Manipulation of Gene Expression in Mammalian Brain." Doctoral Thesis, The Rockefeller University.

Kaplitt, M. G., Kwong, A. D., Kleopoulos, S. P., Mobbs, C. V., Rabkin, S. D., and Pfaff, D. W. (1994). Preproenkephalin promoter yields region-specific and long-term expression in adult brain following direct *in vivo* gene transfer via a defective herpes simplex viral vector. *Proc. Natl. Acad. Sci. USA* **91**, 8979–8983.

Khachaturian, H., Lewis, M. E., and Watson, S. J. (1983). Enkephalin systems in diencephalon and brainstem of the rat. *J. Neurosci.* **3**, 844–855.

Kilpatrick, D. L., Howells, R. D., Noe, M., Bailey, L. C., and Undefriend, S. (1985). Expression of preproenkephalin-like mRNA and its peptide products in mammalian testes and ovary. *Proc. Natl. Acad. Sci. USA* **83**, 7467–7469.

Kilpatrick, D. L., Zinn, S. A., Fitzgerald, M., Higuchi, H., Sabol, S. L., and Meyerhardt, J. (1990). Transcription of the rat and mouse proenkephalin genes is initiated at distinct sites in spermatogenic and somatic cells. *Mol. Cell. Biol.* **10**, 3717–3726.

La Gamma, E. F., Agarwal, B. F., and DeCristofaro, J. D. (1992). Regulation of adrenomedullary preproenkephalin mRNA: Effects of hypoglycemia during development. *Mol. Brain Res.* **13**, 189–197.

Lauber, A. H., Romano, G. J., and Pfaff, D. W. (1991). Sex difference in estradiol regulation of progestin receptor mRNA in rat mediobasal hypothalamus as demonstrated by *in situ* hybridization. *Neuroendocrinology* **53**, 608–613.

Miller, A. D. (1992). Human gene therapy comes of age. *Nature* **357**, 455–460.

Nuovo, G. J. (1992). "PCR In Situ Hybridization: Protocols and Applications." Raven Press, New York.

Mori, N., Stein, R., Sigmund, O., and Anderson, D. J. (1990). A cell type-preferred silencer element that controls the neural specific expression of the SCG10 gene. *Neuron* **4**, 583–594.

Romano, G. J., Harlan, R. E., Shivers, B. D., Howells, R. D., and Pfaff, D. W. (1988). Estrogen increases proenkephalin messenger ribonucleic acid levels in the ventromedial hypothalamus of the rat. *Mol. Endocrinol.* **2**, 1320–1328.

Romano, G. J., Krust, A., and Pfaff, D. W. (1989a). Expression and estrogen regulation of progesterone receptor mRNA in neurons of the mediobasal hypothalamus: An *in situ* hybridization study. *Mol. Endocrinol.* **3**, 1295–1300.

Romano, G. J., Mobbs, C. V., Howells, R. D., and Pfaff, D. W. (1989b). Estrogen regulation of proenkephalin gene expression in the ventromedial hypothalamus of the rat: Temporal qualities and synergism with progesterone. *Mol. Brain Res.* **5**, 51–58.

Scharfmann, R., Axelrod, J. H., and Verma, I. M. (1991). Long-term *in vivo* expression of retrovirus-mediated gene transfer in mouse fibroblast implants. *Proc. Natl. Acad. Sci. USA* **88**, 4626–4630.

Takemura, *et al.* (1992). Primary different stimulation acts through a 193 base pair promoter region to upregulate preproenkephalin expression in dorsal horn of transgenic mice. *Mol. Brain Res.* **13**, 207–212.

Vilijn, M. H., Vaysse, P. J.-J., Zukin, R. S., and Kessler, J. A. (1988). Expression of proenkephalin mRNA by cultured astrocytes and neurons. *Proc. Natl. Acad. Sci. USA* **85**, 6551–6555.

Yin, J., Kaplitt, M. G., and Pfaff, D. W. (1994a). *In vivo* promoter analysis for detecting an estrogen effect on preproenkephalin (PPE) transcription in hypothalamic neurons. In "Proceedings, the Endocrine Society 76th Annual Meeting," pp. 318. [Abstract]

Yin, J., Kaplitt, M., Kwong, A. D., and Pfaff, D. W. (1994b). *In situ* PCR for *in vivo* detection of foreign gene transfer in rat brain. In "Proceedings, 24th Annual Meeting of Society for Neuroscience, Miami Beach, Florida," Vol. 20, pp. 220.

Zhu, Y.-S., *et al.* (1994). DNA binding of hypothalamic and pituitary nuclear protein on ERE and proenkethalin (PENK) promoter. *Soc. Neurosci. Abst.* **20**(1), 53.

Zinn, S. A., Ebert, K. A., Mehta, N. D., Joshi, J., and Kilpatrick, D. L. (1991). Selective transcription of rat proenkephalin fusion genes from the spermatogenic cell-specific promoter in testis of transgenic mice. *J. Biol. Chem.* **266**, 23850–23855.

CHAPTER 

Adenoviral-Mediated Gene Transfer: Potential Therapeutic Applications

Beverly L. Davidson
Department of Internal Medicine
University of Iowa

Blake J. Roessler
Department of Internal Medicine
University of Michigan

Effective methods for gene transfer to the central nervous system (CNS) will find applications in the treatment of neurologic deficits characteristic of many inborn errors of metabolism, as well as the ablation or growth inhibition of CNS neoplasms. This chapter provides a summary of a portion of the ongoing work in our laboratory to determine the capacity of recombinant adenovirus to correct enzyme defects *in vivo,* and to ameliorate existing and/or ongoing neuropathology in animal models of disease.

I. Application to Inborn Errors with Neurologic Involvement

Many inborn errors have as a sequelae of the disorder, a severe and often devastating neurologic component. Representative of such disorders are deficiencies of purine metabolism and the mucopolysaccharidoses (MPS), or more specifically, the Lesch–Nyhan syndrome (LNS; Lesch and Nyhan, 1964) and β-glucuronidase deficiency (MPS VII; Sly syndrome; Sly *et al.,* 1973), respectively. Both disorders result in a very poor quality of life and shortened life span, resulting from numerous movement and intellectual deficits that often require specialized care.

Therapies directed toward correction of the MPS or LNS usually have either no effect (direct enzyme replacement, fibroblast transplantation, or amnion membrane transplantation) or questionable effects (bone marrow transplantation with wild-type or genetically modified deficient cells) on the CNS involvement (Yatziv *et al.,* 1982; Slavin and Yatziv, 1980; Hoogerbrugge *et*

VIRAL VECTORS

al., 1987; Birkenmeier *et al.*, 1991; Birkenmeier, 1991; Sands *et al.*, 1993). This may in part be due to difficulties in delivering enzymes or cells across the blood/brain barrier (BBB). Vascular delivery of purified recombinant β-glucurondidase to neonates does result in detectable levels of activity in the CNS (Vocler *et al.*, 1993), with detectable levels in the meninges, vessels, and ganglia neurons of the peripheral nervous system. Importantly, the enzyme is cleared from the serum within an hour, and intracellular activity diminishes by 4 to 5 days with little to no detectable replacement to CNS neurons. Enzyme replacement to the CNS may therefore be more effectively approached by the direct delivery of transgenes, or transvascular delivery following BBB disruption.

Recently, direct gene replacement to the brain has involved the use of recombinant herpes simplex virus vectors (HSV-I; Palella *et al.*, 1989; Huang *et al.*, 1992; Andersen *et al.*, 1992; Wolfe *et al.*, 1992; Fink *et al.*, 1992), adenoassociated virus (Kaplitt *et al.*, 1994), liposome/plasmids (Davidson and Roessler, 1994; Jiao *et al.*, 1992; Ono *et al.*, 1990), and adenoviruses (Davidson *et al.*, 1993; Bajocchi *et al.*, 1993; Akli *et al.*, 1993; Le Gal La Salle *et al.*, 1993; Davidson *et al.*, 1994a). This chapter describes the results of a series of experiments that demonstrate the effectiveness of recombinant adenoviruses to mediate gene transfer to the CNS using either direct or global delivery methods.

A. Metabolic Correction of HPRT Deficiency *in Vitro*

Recent publications have focused on the ability of recombinant adenoviruses containing the reporter gene for *Escherichia coli* β-galactosidase to efficiently transduce both neuronal and glial cells in culture or following direct *in vivo* application to striatum, neocortex, ventricles, and the eye (Davidson *et al.*, 1993; Akli *et al.*, 1993; Le Gal La Salle *et al.*, 1993; Bajocchi *et al.*, 1993; Li *et al.*, 1994). The description of these model systems has led to the development of recombinant adenoviruses containing therapeutically relevant transgenes, such as the HPRT gene, with experiments designed to test their efficacy *in vitro* prior to *in vivo* applications.

In vitro analysis of recombinant adenovirus containing the rat cDNA was done to determine if HPRT activity could be measured in HPRT$^-$ rat neuroblastoma cells and if the enzyme could salvage radiolabeled purine ([^{3}H]hypoxanthine) substrate to IMP and its catabolites (Davidson *et al.*, 1994). HPRT$^-$ B103-4C cells were grown to near confluence and infected with Ad. RSV*rHPRT* at multiplicities of infection (pfu/cell) ranging from 10^2 to 10^4. AdRSV*rHPRT* contains the rat HPRT cDNA flanked by the Rous sarcoma virus (RSV) LTR for promoter activity and a polyadenylation signal from SV40. Infection was done in serum-free media for 4 hr to overnight, and cells were incubated in complete media until time of harvest for enzyme assay or

Southern analysis. Cell passage was required prior to harvest for later time points. Southern blotting detected transgene DNA out to 8 days, with the copy number per cell slowly diminishing from approximately 5 to 0.5 (Fig. 1). Endonuclease restriction using an enzyme which cuts once within the rHPRT sequence revealed that the majority of the transgene persisted as episomal (single fragment of 2.2 kb) DNA.

Enzyme activity persisted out to 28 days, but was highest at Day 15 (Table 1). The fact that HPRT is a relatively stable protein may account for the temporal discrepancy between detectable levels of DNA (by Southern) and detectable enzyme activity (by assay).

Purine salvage was demonstrated following infection of subconfluent B103-4C cells. Cells were infected with Ad.RSV*rHPRT* at a multiplicity of infection (m.o.i.) of 10^3 for 4 hr in serum-free media, and serum was added to a final concentration of 2% overnight. Cells were then washed with PBS and incubated in complete media for 40 hr. Following a PBS wash, cells were incubated in fresh media containing [^{3}H]hypoxanthine for 2 hr. Cells were harvested and nucleotides extracted and quantified using HPLC (Sidi and Mitchell, 1985). As seen in Table 2, cells transduced with Ad.RSV*rHPRT* were able to salvage [^{3}H]hypoxanthine in contrast to uninfected or mock-infected (Ad.RSV*lacZ*) controls. Radiolabeled pool levels increased in a dose-dependent fashion; [^{3}H]ATP pools at a m.o.i. of 10^4 were greater than 50-fold higher

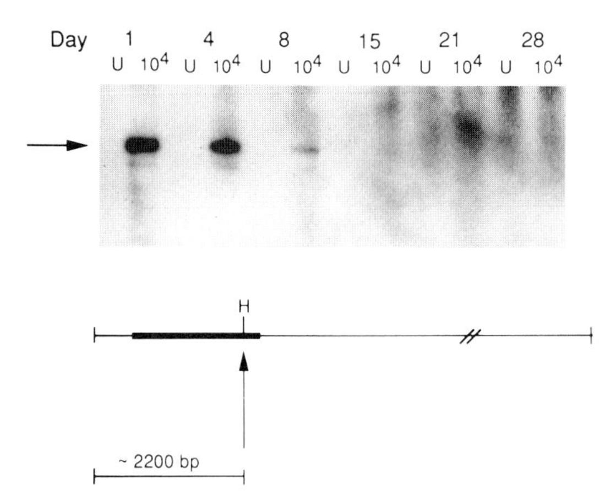

Figure 1 Southern analysis of DNA isolated from B103-4C cells following Ad.RSV*rHPRT* infection. Cells were infected at a m.o.i. of 10^4 and harvested at the times indicated. DNA was isolated and digested with *Hpa*I. This enzyme cuts once within the rHPRT sequence in the left half of the full-length recombinant adenoviral genome. Following fractionation on a 0.8% agarose gel, DNA was transferred to nitrocellulose and blots were hybridized to radiolabeled probes specific for rat HPRT. U, uninfected; H, *Hpa*I. The arrow denotes the 2.2-kb fragment predicted following *Hpa*I digestion.

Table 1
HPRT Enzyme Activity in Ad.RSVrHPRT-Infected B103-4C Cells

Day	Multiplicity of infection			
	Un[a]	10^2	10^3	10^4
1	1.1[b]	1.0	1.1	7.4
5	1.2	1.6	3.4	12.2
8	1.8	2.8	17.1	102.0
15	1.4	14.2	46.2	104.8

[a] Uninfected.
[b] CPM/mg total protein $\times 10^{-3}$.

than those seen at a m.o.i. of 1 (Table 2). This data, in conjunction with the Southern and enzyme activity data, indicate that the enzyme is functionally active *in vitro*.

B. Metabolic Correction of HPRT Deficiency *in Vivo*

Short-term studies designed to examine the ability of Ad.RSV*rHPRT* to modulate purine pool levels or to salvage radiolabeled purines *in vivo* were done in mice deficient in HPRT (Davidson *et al.*, 1994). Direct inoculations into the right striatum were done using 5 μl of purified adenovirus (experimental or control) at a titer of 1×10^{10} pfu/ml. Five days postinoculation, animals were sacrificed, and brain sections surrounding the injection site analyzed for nucleotide pool levels. As shown in Table 3, there was evidence of increased pool levels in the right, infected hemisphere compared to the contralateral controls.

Table 2
[^{3}H]-Nucleotide Pool Levels Following Transduction of B103-4C Neuroblastoma Cells

Sample	AMP[a]	ADP	ATP	GDP	GTP
Uninfected	—		11.0		
Ad.RSV*lacZ*(10^3)[b]	—		7.0		
Ad.RSV*rHPRT*(10^2)	—	182.3	105.2	644.9	90.0
Ad.RSV*rHPRT*(10^3)	600.8	1140.8	822.1	10420	328.5
Ad.RSV*rHPRT*(10^4)	2232.8	9900.2	5066.5	7252	9334.6

[a] Counts per minute/nmole.
[b] Multiplicity of infection.

Table 3
Nucleotide Pool Levels in Mouse Brain Extracts[a]

Hemisphere	AMP	GMP	ADP	GDP	ATP	GTP
Left	7.16	0.54	2.28	1.16	4.32	1.01
Right	8.53	0.59	3.89	1.49	5.46	1.35

[a] nmol/100 μl.

Transgene expression was also assayed *ex vivo* using an *in situ* tissue assay. Sections (2 mm) were overlaid with an enzyme assay mix containing [^{14}C]hypoxanthine (100 m*M*, 1 mCi/mmol), 6 m*M* $MgCl_2$, and 10 m*M* PRPP, and incubated for 1 hr at 37°C in a humidified atmosphere. Following incubation, sections were homogenized and nucleotides extracted for analysis of [^{14}C]hypoxanthine metabolites. Increased uptake of radiolabeled [^{14}C]hypoxanthine was noted in the experimental (Ad.RSV*rHPRT*) vs the contralateral control left hemisphere (Table 4). Animals injected with Ad.RSV*lacZ* did not demonstrate any increases in radiolabeled purine pools compared to the contralateral hemisphere (data not shown), indicating that increased salvage is due to increased expression of transgenic HPRT.

C. Safety Issues in Nonhuman Primates

We have performed a series of experiments to examine the clinical safety following injection of recombinant adenovirus containing a xenogeneic transgene, the gene for rat HPRT, into primate brain (Doran *et al.*, 1994a). Four rhesus macaques (*Mucaca mulatta*) weighing approximately 3 kg each were injected with adenovirus and followed clinically until the time of sacrifice. Three of the four animals were examined by positron emission tomography (PET), and two of these three by T_1 and T_2 weighted magnetic resonance imaging (MRI).

Mucaca mulatta 1 (MM1) was injected with 200 μl of Ad.RSV*rHPRT* into the left caudate nucleus (1.6×10^{11} pfu/ml). The opposite hemisphere

Table 4
[^{14}C]hx Metabolites in Murine Brain Sections

Hemisphere	IMP	GMP	ADP	GDP
Left	954[a]	1938	90	111
Right	11524	14556	212	156

[a] cpm/section.

was injected with 200 μl of Ad.RSV*lacZ* (1.6×10^{11} pfu/ml). The animal experienced no perioperative morbidity and throughout the 1-week observation period showed no clinical signs of cerebritis, meningitis, or encephalitis. The animal groomed and fed normally. One week postsurgery, the animal was anesthetized, euthenized, and the brain was perfused with 120 cc of ice-cold saline. The brain was removed and the caudate nucleus blocked into four equal sections. One section was taken for *in situ* RNA hybridization, one for enzyme activity, and two for histochemistry and immunohistochemistry. Enzyme assay and *in situ* RNA analyses were positive for transgene expression (Davidson *et al.*, 1994a). Although clinical signs of disease were not evident, routine H&E revealed a cellular infiltrate consisting of foamy macrophages and lymphocytes. The infiltrate radiated approximately 3 or 4 mm from the injection site, with the number of inflammatory cells diminishing as distance from the injection site increased. Postoperative complete blood counts (CBC) were all within normal limits.

A second long-term nonhuman primate study was also done. MM2 was injected with 200 μl of Ad.RSV*rHPRT* into the right caudate nucleus. Clinically the animal recovered well, and throughout the 1-year observation period has shown no signs of cerebritis, meningitis, or encephalitis. The animal continues to groom and feed normally. MM2 was studied using the TCC 4600A PET. Cerebral glucose metabolism was studied using bolus injection of 2-[^{18}F]fluoro-2-deoxy-D-glucose ([^{18}F]FDG), with imaging at 40–60 min after injection. Dopamine receptors in the striatum were studied using infusion of [^{11}C]raclopride to a steady state and imaging from 40–60 min. In all studies, there were no significant differences in left vs right hemisphere comparisons of either [^{18}F]FDG or [^{11}C]raclopride. Comparison of pre- vs postoperative scans were also unremarkable. Serum drawn from the animal 1.5 months postadministration of virus was analyzed for the presence of a cytotoxic T-cell response against Ad.RSV*rHPRT*-infected syngeneic fibroblasts and was negative. Postoperative CBCs have remained within normal limits.

Two animals, MM3 and MM4, were followed by both PET and MRI to clinically evaluate the host response to Ad infection in the CNS. MM3 and MM4 animals were injected with 200 μl of Ad.RSV*rHPRT* into the right caudate nucleus [total particles (pt) of 10^9 and 10^{10}]. The animals recovered well clinically, and throughout the nearly 2-month observation period have shown no signs of cerebritis, meningitis, or encephalitis. Serial MRI and PET scans were initiated 1 week postoperative [^{18}F]FDG and [^{11}C]raclopride scanning revealed no gross metabolic differences between the injected and uninjected hemisphere. Preoperative MR T_1 and T_2 imaging (GE Signa, 1.5T; transmit–receive Knee Coil) were normal. T_1 (500/18/4; TR/TE/nex) and T_2-weighted (2800/92/4) images were obtained with 3-mm slice thickness and 0.5-mm skip. Matrix was 256×256. MRI scans at 1 week postoperative in MM3, the animal receiving a dose of 10^{10} pt, demonstrated only minimal increased signal intensity on T_2 images and minimal enhancement on T_1 images along the needle tract. At 4 weeks postoperative there was only minimal increased signal intensity in white matter adjacent to the injection tract on T_2 images with no distinct enhancement or altered signal intensity in brain parenchyma on T_1 images. PET remained unremarkable at 4 weeks postoperative. At 8 weeks the region of increased signal intensity on T_2 images was even less evident and there was no abnormal enhancement. MM4, which received 10^9 pt, showed changes on T_2 weighted MRI at 1 week which were consistent with a $6 \times 4 \times 9$-mm region of edema at the site of virus administration. T_1 weighted MR images demonstrated slightly decreased signal intensity in a $1 \times 3 \times 3$-mm area with no enhancement following gadopentetate dimeglumide administration. At 4 weeks postoperative T_2 images demonstrated this region of apparent edema to be smaller ($4 \times 3 \times 9$ mm) and less intense. However, T_1 images without contrast showed increased signal intensity in the same distribution as the T_2 signal changes seen at 1 week. With gadopentetate dimeglumide administration there was minimal enhancement in the same region. At 8 weeks the region of increased signal intensity on T_2 images was distinctively smaller and there was no abnormal enhancement. The animal remains clinically well. There was minimal enhancement of the meninges over the frontal regions in both postoperative studies in both animals. Serial CBCs have remained within normal limits for both MM3 and MM4 during the experimental period.

These experiments indicate that although the direct administration of recombinant adenovirus to brain induces a notable immune response, the response resolves over time. The consequence of redosing in the same animal, which is an important issue as no current generation of viruses can direct transgene expression indefinitely, remains to be determined.

D. Alternate Delivery Strategies

A major obstacle to the delivery of recombinant viruses to the CNS is the presence of a BBB formed by the brain capillary endothelial cells and a blood–CSF barrier formed by the choroid plexus epithelial cells and the meninges (Rapoport and Robinson, 1986). These barriers greatly impede the passage of most polar molecules from the blood to the brain. Although the interface between the CSF and brain is lined by ependymal cells in the cerebral ventricles and by pial cells in the subarachnoid space, these cells do not form barriers. Several methods have been developed to reversibly and transiently open the BBB including intracarotid infusion of hyperosmotic solutions (Neuwelt *et al.*, 1979), polycations (Hardebo and Kahrstrom, 1985; Strausbaugh, 1987; Westergren and Johansson, 1993), oleic acid (Sztriha and Betz, 1991), and histamine (Gross *et al.*, 1982).

Osmotic BBB disruption requires intracarotid infusion of extremely hypertonic solutions (e.g., 1.6 *M* arabinose) for 20 sec or more. The barrier remains open for up to 2 hr (Rapoport *et al.*, 1980). The mechanism of barrier opening appears to involve separation of tight junctions (Dorovini-Zis *et al.*, 1984) perhaps due to shrinkage of the endothelial cells (Dorovini-Zis *et al.*, 1983). This technique has been successfully adapted for the delivery of chemotherapeutic agents in humans (Neuwelt *et al.*, 1986). A possible limitation to this approach for delivery of viruses is that the barrier may not open enough to accommodate the size of adenovirus (Ad5 has a diameter of about 70 nm). However, Neuwelt *et al.* (1991) showed increased uptake of ultraviolet-inactivated HSV-1 (native size of 120 nm in size) following osmotic disruption of the BBB and intravascular administration of virus. Thus, preliminary experiments were done to determine if recombinant adenovirus can cross the BBB (Doran *et al.*, 1994b).

Male Sprague–Dawley rats weighing 250 gm were anesthesized with isoflourane (5% induction, 2% maintenance) and the right carotid artery was exposed through a ventral neck incision. A catheter filled with heparinized saline was introduced into the right external carotid artery and advanced retrograde to the bifurcation of the common carotid artery. Evans blue solution (2 ml/kg of 2 g/100 ml solution) was given intravenously 5 min prior to blood–brain barrier disruption to serve as a marker of altered BBB integrity. Blood–brain barrier disruption was then performed as previously described (Neuwelt *et al.*, 1991). In six animals, mannitol (25%, 37°C) was infused cephalad into the internal carotid artery at a rate of 0.12 ml/sec for 30 sec. Control animals were infused with 0.9% saline, 37°C, into the internal carotid artery at the same rate and duration. After waiting 60 sec, 5×10^{11} particles of Ad.RSV*lacZ* in 1 ml of 0.9% saline were infused in all animals over a period of 60 sec. Animals were then recovered and housed in a biocontainment facility.

The degree of blood–brain barrier disruption based on Evans blue staining was graded from 0 to 3+, as previously described (Rapoport *et al.*, 1980). The Evans blue–albumin complex does not cross the normal BBB and the degree of staining in the osmotically disrupted brain is a measure of BBB opening. Of the six rats receiving intracarotid mannitol, three had 2+ Evans blue staining and three had 1+ staining of the ipsilateral cerebral hemisphere (Fig. 2). As these animals were sacrificed 4 days after barrier disruption, and as Evans blue is gradually cleared from the disrupted brain, it is likely that the degree of barrier disruption was greater at the actual time of adenoviral infusion. No Evans blue staining was found in brains of animals receiving intracarotid saline.

Following X-gal staining, brain sections from rats that had undergone mannitol-induced BBB disruption followed by intracarotid Ad.RSV*lacZ* injection were all *E. coli lacZ* positive in pericapillary astrocyte. (Fig. 2). No intracerebral *E. coli lacZ* activity was found in control animals receiving intracarotid saline followed by Ad.RSV*lacZ*. In animals receiving intracarotid mannitol, X-gal-stained cells were found almost exclusively in the disrupted right hemisphere. Rare positive cells were found in the extreme medial aspect of the opposite hemisphere, in the so-called "watershed" areas of vascular anastamoses. The highest concentration of X-gal-stained cells was found in the cerebral cortex, followed by the deep gray nuclei. The number of positive cells correlated with the degree of Evans blue staining: hemispheres with 2+ Evans blue staining typically had 5–10 *lacZ*-positive cortical cells per low-powered field, while hemispheres with 1+ Evans blue staining had 0–2 cortical cells per low-powered field.

Escherichia coli lacZ activity was found only in astrocytes as shown morphologically by X-gal staining. X-gal staining was not seen in neurons or ependymal cells. X-gal-positive astrocytes were found primarily adjacent to capillaries (Fig. 2). TEM and immunohistochemistry were also done to confirm the identity of the transduced cells of the cerebral cortex based on ultrastructural criteria (data not shown). Histochemistry, immunohistochemistry, and TEM failed to detect transduced neurons. The most likely explanation for this finding is the normal anatomic relationship between cerebral capillaries and astrocytes. Surrounding the basement membranes of the capillary endothelium are the foot processes of astrocytes. After intracarotid infusion of mannitol, the capillary endothelial cells presumably shrink and the tight junctions are temporarily opened, allowing the recombinant adenovirus to leave the cerebral vasculature and enter the perivascular space. The first structure then encountered would be the foot processes of astrocytes. The capacity of astrocyte transduction following BBB -disruption to globally correct the neuropathology in animal models of metabolic disease is currently under study.

II. Application to the Treatment of CNS Neoplasms

In addition to the application for the treatment of inherited metabolic disorders, recombinant adenoviruses may be useful in the treatment of CNS neoplasms. Over 17,000 new cases of primary intracranial tumors are diagnosed annually. The incidence of all primary brain tumors in the United States increases with increasing age, extending from 2.3 per 100,000 during childhood to a peak rate of 20.4 per 100,000 in the 55- to 75-year-old age group. Gliomas are the most frequently encountered brain tumor in all age groups, accounting for approximately 57% of all primary intracranial brain neoplasms (New, 1993; Levin *et al.*, 1993). They encompass a range of tumors including astrocytomas, glioblastoma, oligodendrogliomas, and ependymal tumors. These tumors are thought to arise from the neoplastic transformation of differentiated cells at loci of glial cell development. Gliomas are usually nonencapsulated and infiltrate into the surrounding brain parenchyma.

Because of the extremely poor prognosis, alternate management strategies are continually being developed. Gene therapy has recently been investigated as an alternative for brain tumor treatment. Viral or synthetic transducing agents containing a "suicide gene," usually thymidine kinase (TK) from HSV-I (Borrelli *et al.*, 1988; Barba *et al.*, 1993; Moolten, 1986), are used to deliver the gene to the tumor (Moolten and Wells, 1990; Ezzeddine *et al.*, 1991; Culver *et al.*, 1992; Ram *et al.*, 1993; Plautz *et al.*, 1991). The expressed gene is not in itself toxic to the cell. However, administration of an antiviral 2'-deoxyguanosine analog drug, ganciclovir (Cytovene; GCV), results in the formation of toxic phosphorylated guanosine intermediates in cells expressing the transduced TK (Shewach *et al.*, 1994). Cell death occurs only in actively dividing cells, presumably as a result of the incorporation of the toxic nucleoside intermediates into DNA. Because the cell must be actively dividing to respond to the harmful effects of the drug, the *tk* gene is a good candidate for gene therapy treatment of gliomas (Moolten and Wells, 1990; Ezzeddine *et al.*, 1991; Culver *et al.*, 1992; Ram *et al.*, 1993). Importantly, surrounding glial and neuronal tissue have an extremely slow mitotic rate imparting a protective effect from the phosphorylated analog.

A variety of transducing agents has been used to transfer DNA to the tumor mass, including liposomes (Yoshida *et al.*, 1992) and retroviral vectors (Chu *et al.*, 1993; Culver *et al.*, 1992; Oldfield *et al.*, 1993; Ram *et al.*, 1993, 1992; Short *et al.*, 1990; Takamiya *et al.*, 1992, 1993). Gene transfer with retroviral vectors is restricted to actively dividing cells, which is an attractive feature when targeting malignant cells amidst a background of quiescent, non-proliferating cells. A major drawback, however, is that low efficiency of gene transfer and inability to obtain high titers necessitates the *in situ* introduction of producer cell lines to obtain optimal levels of gene transfer. Also, the introduction of mouse fibroblast (producer cells) will likely confound readministration of producer cells due to host immune response. In contrast, recombinant adenoviruses possess a number of characteristics which make them attractive candidates for gene therapy (Kozarsky and Wilson, 1993). Adenoviruses can be grown to very high titers, 10^{13} plaque-forming units (pfu) vs $10^{5\text{-}7}$ pfu for retroviruses, and have no requirement for cell division for transgene expression (Mulligan, 1993). This is an important feature as only approximately 10% of the tumor is actively dividing at any one time, limiting the level of transduction that can occur with retroviral producer cells.

Ad.RSV*tk*, a recombinant adenovirus designed to deliver *tk* from HSV-I, has been shown to infect brain tumor cells *in vitro* and *in vivo* and render them susceptible to ganciclovir toxicity (Shewach *et al.*, 1994; Smythe *et al.*, 1994, 1995; Ram, 1994; Chen *et al.*, 1994; Davidson *et al.*, 1994b). When the virus is injected stereotactically into the tumor, the risk of infecting nontumor tissue is minimal. Also, Ad.RSV*tk*/GCV therapy in nontumor-bearing tissue in rodents does not appear to induce damage; animals are clinically normal and there is no gross evidence of disease. Similar studies are currently being carried out in nonhuman primates.

A. Transduction of Rodent Tumor Cell Lines *in Vitro*

Initial experiments have been done to determine the ability of Ad.RSV*tk* to confer GCV toxicity to glioma cells in culture. TK-deficient C6BU1 cells, which are derived from C6 rat glioma cells, were infected at varying m.o.i.s (pfu/cell) (Shewach *et al.*, 1994; Takamiya *et al.*, 1992). As seen in Fig. 3, a linear relationship exists between m.o.i. and HSV-TK activity over a range of 2 to 10,000 m.o.i. without apparent saturation. Transduction with Ad.RSV*tk* resulted in HSV-TK activity levels that were approximately 600-fold higher than those with retroviral-mediated HSV-TK transfer into C6BU1 cells (Shewach *et al.*, 1994).

The high levels of HSV-TK activity in the C6BU1 cells would be expected to lead to greater sensitivity to the cytotoxic effects of GCV. As shown in Fig. 4, cytotoxicity was enhanced with increasing m.o.i. In addition, the IC_{50} values for the drug decreased more than 20-fold with increasing m.o.i. from 0.077 μM at an m.o.i. of 10 to 0.0037 μM at an m.o.i. of 1000. A similar dose-dependent decrease in IC_{50} is seen in 9L glioma cells (data not shown). This allows for the use of lower systemic doses of GCV in animals following Ad.RSV*tk* gene transfer. The results of these experiments are described in detail in Shewach *et al.* (1994).

Mixing experiments with Ad.RSV*tk*-infected mesothelioma cells have demonstrated that as little as 10% Ad.RSV*tk*-infected cells are capable of producing a tumoricidal effect (Smythe *et al.*, 1995). These experiments have

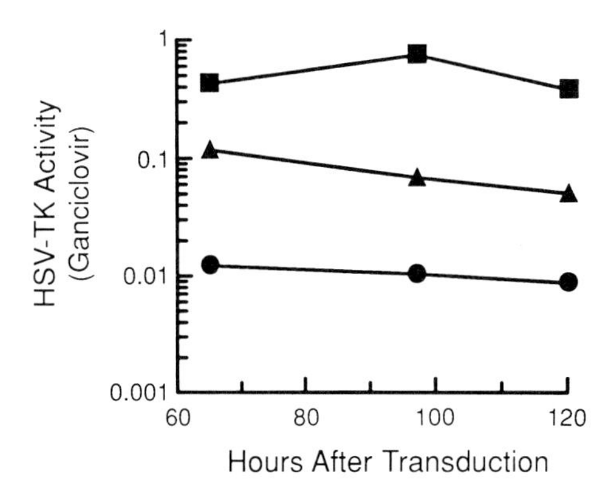

Figure 3 Relationship between m.o.i. and HSV-TK activity. C6BU1 cells were transduced with Ad.RSV*tk* 3 days prior to harvesting for measurement of HSV-TK activity. HSV-TK activity is expressed as nmol/hr per 10^6 cells.

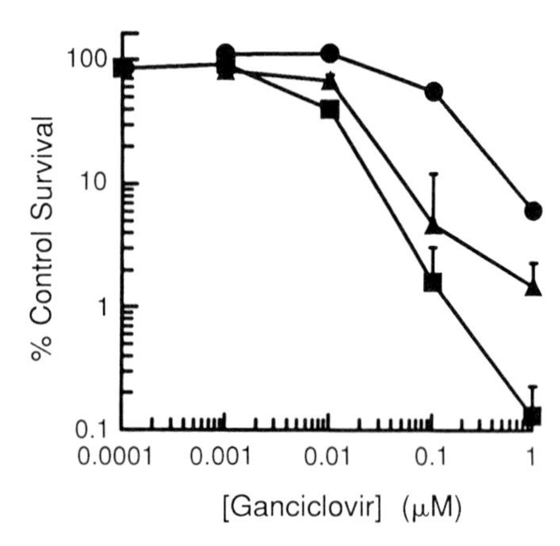

Figure 4 Effect of increasing m.o.i.s on the cytotoxicity of GCV. Cells were transduced with Ad.RSV*tk* at m.o.i.s of (●) 10, (▲), 100, or (■) 1000. Clonogenic cell survival was determined at each m.o.i. (1).

also been done in rat glioblastoma cell line C6 with similar results (Chen *et al.*, 1994).

B. *In Vivo* Studies in Rodent Models

1. HSV-TK Enzyme Activity and Effects on Growth Kinetics

The ability of Ad.RSV*tk* to confer susceptibility to GCV *in vivo* was demonstrated in an animal model of glioblastoma (Ross *et al.*, 1995). 9L glioma cells were grown as monolayers and 10^5 cells injected into the right striatum of Fischer 344 rats 3 mm from the cortical surface. Initial studies were done to determine HSV-I TK activity on Days 1–3 following Ad.RSV*tk* injection into tumors. Enzyme assay of glioma tissue removed from animals sacrificed at Day 1 postadministration of virus was 0.514 ± 0.26 nmol phosphorylated GCV/gram tumor tissue ($n = 5$). By Day 2, this level had increased to 0.9851 ± 0.28 ($n = 5$). Levels at Day 3 were similar ($n = 3$). Based on this information, twice daily injections of GCV (15 mg/kg) were initiated 24 to 30 hr postvirus administration in all animal studies.

Growth kinetics were determined by MRI scans that were initiated between 8 and 10 days postimplantation and repeated every other day to obtain volume measurements of tumors. At approximately Day 16, tumors were injected with 2.5×10^8 pfu of freshly prepared Ad.RSV*tk* in PBS. GCV treatment (15 mg/kg b.i.d. for 10 days) was administered ip with the first injection occurring 24 hr post-Ad.RSV*tk* infection. Note that 15 mg/kg was effective, compared to 150 mg/kg in other studies (Moolten and Wells, 1990; Ezzeddine *et al.*, 1991; Culver *et al.*, 1992; Ram *et al.*, 1993; Plautz *et al.*, 1991).

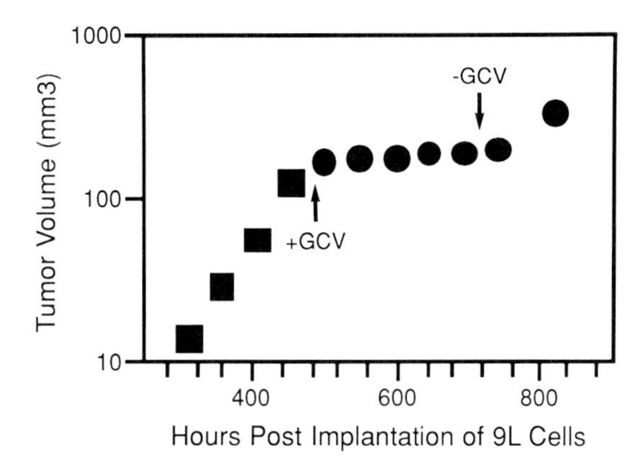

Figure 5 A representative growth curve for an intracranial 9L glioma treated with Ad.RSV*tk* and GCV. Note the marked growth retardation during the treatment period.

A representative curve demonstrating the growth kinetics of the tumor prior, during, and post-treatment is shown in Fig. 5. Marked growth retardation represented by an increase in tumor doubling time (T_d) from 53 ± 2.5 to 403 ± 111 hr (range 102 to 964 hr) during GCV administration was observed in 9 of 11 animals. Controls receiving Ad.RSV*lacZ*/GCV showed no change in T_d. Data summarized from these 9 animals are shown in Table 5.

Table 5
Tumor Doubling Time Pre- and Post-Ad.RSV*tk* and GCV Treatment

Rat No.	T_d (pre)[a]	T_d (post)
1	44	964
2	59	276
3	51	206
4	59	∞[b]
5	43	213
6	37	140
7	65	102
8	56	192
9	55	183
Mean ± SE	53 ± 2.5	403 ± 111

[a] T_d (tumor doubling time) is given in hours.
[b] Tumor volume decreased. T_d of 1000 hours used for statistical analysis.

2. Survival Study

Standard survival studies were done on rats implanted with 9L glioma cells (Ross *et al.*, 1995). Two cohorts of animals consisting of eight animals each were implanted with 9L cells. A third cohort of eight animals remained tumor free. All animals were inoculated on Day 7 with 20 μl of Ad.RSV*tk* (eight tumor-bearing and eight nontumor bearing animals; 1.2×10^{11} pfu/ml, 4.6×10^{12} pt/ml) or 20 μl of Ad.RSV*lacZ* (tumor bearing; 1.6×10^{11} pfu/ml, 5.5×10^{12} pt/ml). GCV was given twice daily, 5 mg/kg, with the first dose given 24 hr following Ad infection. GCV treatment was continued for 14 days. Figure 6 demonstrates that (i) animals receiving Ad.RSV*tk*/GCV survived significantly longer than those receiving control virus (Ad.RSV*lacZ*/GCV), and (ii) nontumor bearing animals receiving Ad.RSV*tk*/GCV therapy were clinically normal and did not succumb to any untoward side effects of infecting normal brain tissue with the HSVtk gene, and subsequent conversion of GCV to the toxic triphosphate form. To summarize, rats receiving Ad.RSV*tk* and GCV treatment ($n = 8$) survived significantly longer (50% increase in survival time) than animals receiving control adenovirus (Ad.RSV*lacZ*; $n = 8$). Mean survival ± SEM was 2.5 ± 1.15 days in Ad.RSV*lacZ*/GCV rats compared to 38 ± 4.49 days in Ad.RSV*tk*/GCV-treated animals. The effect of Ad.RSV*tk*/GCV on normal brain was unremarkable ($n = 8$).

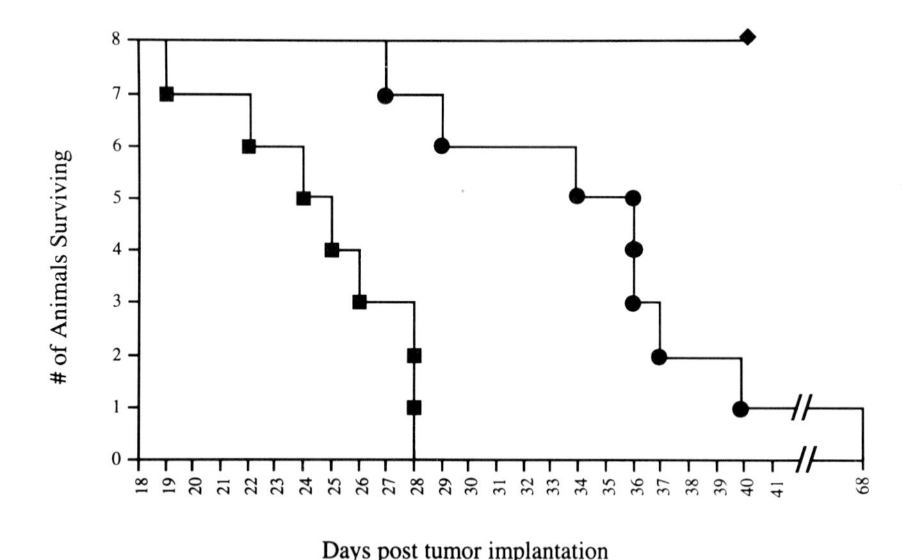

Figure 6 Survival study of rats following Ad.RSV*lacZ* and Ad.RSV*tk* injections. See text for a description of the experiments. ■, Ad.RSV*lacZ*/GCV; ●, Ad.RSV*tk*/GCV; ◆, Ad.RSV*tk*/GCV (no tumor).

3. *In Vivo* Spectroscopy Measurements

MRS can be used to study tumor metabolism *in vivo* and, as such, is a powerful tool to assess treatment efficacies (Ross *et al.,* 1992). The metabolic parameters which can be analyzed using MRS differ substantially and are therefore complementary to those provided by PET. Thus, MRS offers us an additional tool to determine the *in vivo* effects of the therapy. In addition to the detection of lipid and lactate, which are indicators of necrosis, MRS can localize neuronal-specific metabolites (e.g., *N*-acetyl aspartate). Because spectroscopic changes are usually observed from hours to a day posttherapy, rather than days to weeks with MRI, this technique should provide evidence for therapeutic efficacy which will precede morphologic changes such as decreased tumor volume. This data could also yield valuable insights into the duration of therapeutic effects which would assist in the optimization of redosing strategies.

A preliminary study was designed using MRS to determine if spectroscopic changes occurred following Ad.RSV*tk*/GCV treatment prior to notable radiographic changes. In these studies, spatially localized ^{1}H MRS studies were accomplished on Ad.RSV*tk*/GCV-treated gliomas during the time course of the treatment protocol. Localized *in vivo* ^{1}H spectra were acquired from 20-μl tissue volumes using adiabatic pulses combined with one-dimensional spectroscopic imaging along a column selected by two-dimensional ISIS (Ross *et al.,* 1992). Metabolites which are notably different between glioma tissue and normal brain are diminished concentrations of *N*-acetyl-aspartate and creatine and increased levels of choline (Ross *et al.,* 1995). A representative spectra from a tumor-bearing and AdRSV*tk*/GCV-treated animal is shown in Fig. 7.

III. Future Directions—Vector Development

There are several requirements which must be met before replication-deficient adenovirus can be considered a viable vehicle for the delivery of genes into the brain parenchyma for correction of inherited disorders of metabolism. In addition to its ability to infect and express in multiple cell types relatively efficiently, transgene expression must persist for long periods of time and the adenovirus must be amenable to redosing when transgene expression diminishes. The most significant problem associated with persistence of transgene expression may be directly related to the immunotoxicity of the virus. Akli and others (Akli *et al.,* 1993; Bajocchi *et al.,* 1993; Davidson *et al.,* 1993) demonstrated an acute toxicity following direct injection into brain, particularly at high viral titers. This acute response also occurs in lungs and is unrelated

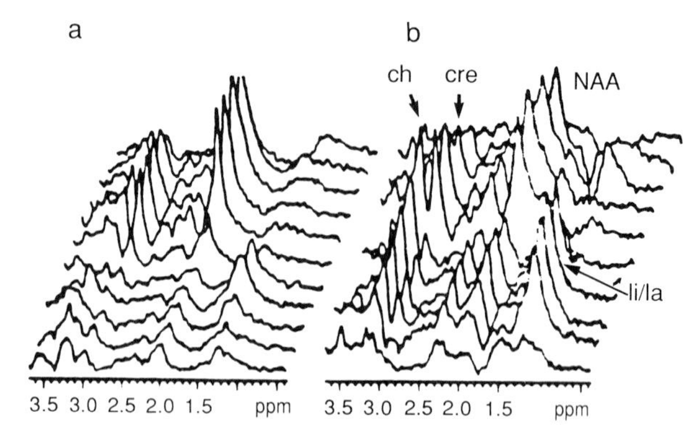

Figure 7 Each figure displays a series of spectra from 25-μl voxels along a two-dimensional column of the rat. Spectra of untreated gliomas (a) is distinctive, with an absence of *N*-acetyl aspartate and reduced levels of creatine in the right hemisphere (tumor bearing; first five spectra in both panels from front to back) compared to the contralateral (nontumor-bearing; last six spectra) hemisphere. The lipid/lactate resonance from untreated gliomas and healthy brain is barely visible. In the GCV–Ad.RSV*tk*-treated glioma (b), ^{1}H spectra revealed an increase in the lipid/lactate resonance and a decrease in creatine levels. li, lipid; la, lactate; NAA, *N*-acetyl aspartate; ch, choline; cre, creatine.

to viral or transgene expression (McCoy *et al.*, 1994). At all titers, transgene expression diminishes over time. This is due in part to a CD8-dependent host response which can be tempered following backbone modifications to further cripple expression of immunogenic viral proteins (Engelhardt *et al.*, 1994a,b; Yang *et al.*, 1994). Whether or not such modifications result in prolonged expression in the CNS as a result of diminished immune response remains to be determined.

Vector improvements which result in persistence are less an obstacle for the use of adenovirus in anticancer therapies. However, the efficient delivery of suicide genes to effect a complete tumoridical response will no doubt require redosing of the virus. Whether or not vector modifications can be made to overcome the obstacles inherent in the current generation of adenoviral vectors is an area of intense research interest.

In summary, adenoviral vectors will undoubtedly lead the field in applications directed at the study of CNS biochemistry, physiology, and neuropharmacology because of their relative ease in generation, manipulation, and application (high efficiency of infection with low toxicity) compared to other viral vectors. The direct application to CNS disease in humans, however, awaits further development and thorough safety and efficacy studies in animal models of disease.

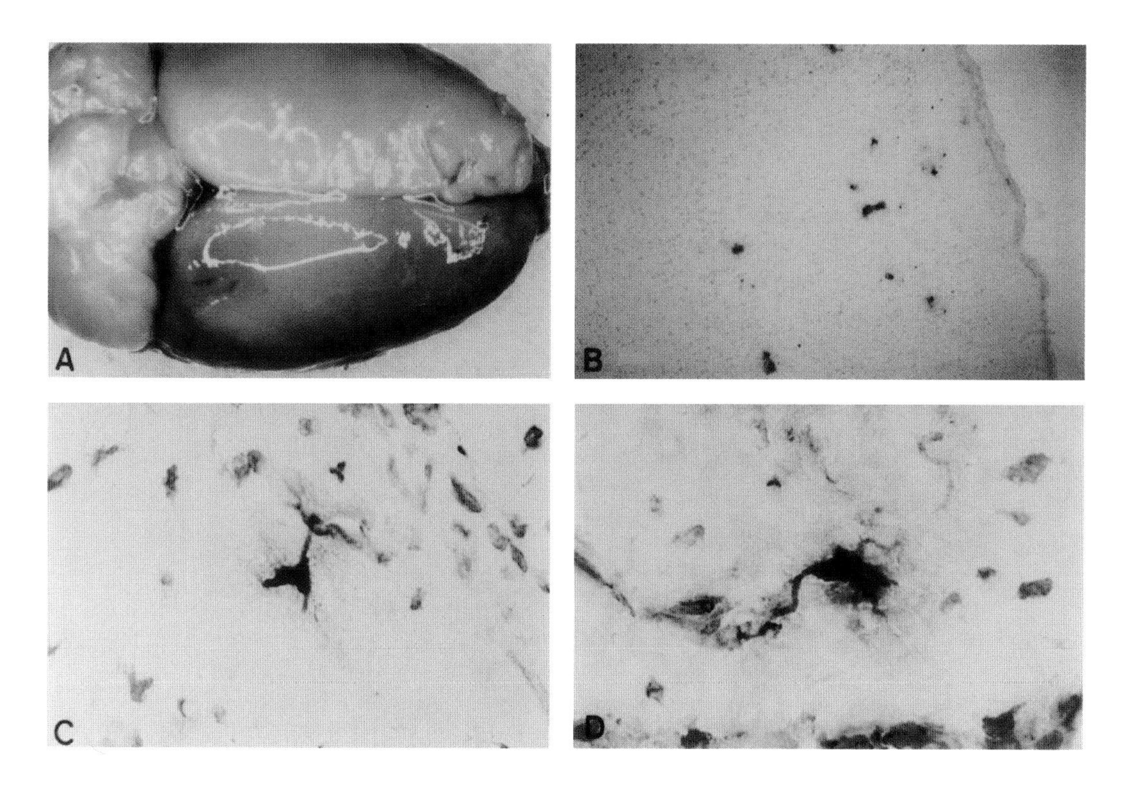

Figure 2 (A) Evans blue staining (graded 2+) of the right cerebral hemisphere following osmotic disruption with intracarotid mannitol. (B) Low-powered view (100×) of a X-gal-stained rat cerebral cortex following BBB disruption and intracarotid infusion of Ad.RSV*lacZ*. Note the blue-stained cells scattered throughout the cerebral cortex, indicating transgene expression. (C and D) High-powered view (400×) of astrocytes from the cerebral cortex expressing transgenic *lacZ*. In D, the foot process of the astrocyte is in direct contact with an adjacent capillary.

References

Akli, S., Caillaud, C., Vigne, E., Stratford-Perricaudet, L. D., Perricaudet, M., and Peschankski, M. R. (1993). Transfer of a foreign gene into the brain using adenovirus vectors. *Nature Genet.* **3**, 224–228.

Andersen, J. K., Garber, D. A., Meaney, C. A., and Breakefield, X. O. (1992). Gene transfer into mammalian central nervous system using herpes virus vectors: Extended expression of bacterial lacZ in neurons using the neuron-specific enolase promoter. *Hum. Gene Ther.* **3**, 487–499.

Bajocchi, G., Feldman, S. H., Crystal, R. G., and Mastrangeli, A. (1993). Direct *in vivo* gene transfer to ependymal cells in the central nervous system using recombinant adenovirus vectors. *Nature Genet.* **3**, 229–234.

Barba, D., Hardin, J., Ray, J., and Gage, F. H. (1993). Thymidine kinase-mediated killing of rat brain tumors. *J. Neurosurg.* **79**, 729–735.

Birkenmeier, E. H. (1991). Correction of murine mucopolysaccharidosis type VII (MPS VII) by bone marrow transplantation and gene transfer therapy. *Hum. Gene Ther.* **2**, 113.

Birkenmeier, E. H., Barker, J. E., Vogler, C. A., Kyle, J. W., Sly, W. S., Gwynn, B., Levy, B., and Pegors, C. (1991). Increased life span and correction of metabolic defects in murine mucopolysaccharidosis type VII after syngeneic bone marrow transplantation. *Blood* **78**, 3081–3092.

Borrelli, E., Heyman, R., Hsi, M., and Evans, R. M. (1988). Targeting of an inducible toxic phenotype in animal cells. *Proc. Natl. Acad. Sci. USA* **85**, 7572–7576.

Chen, S-H., Shine, H. D., Goodman, J. C., Grossman, R. G., and Woo, S. L. C. (1994). Gene therapy for brain tumors: Regression of experimental gliomas by adenovirus-mediated gene transfer *in vivo*. *Proc. Natl. Acad. Sci. USA* **91**, 3054–3057.

Chu, M. Y., Lipsky, M. H., Whartenby, K. A., Freeman, S., Chen, T. M., Epstein, J., Forman, E. N., and Calabresi, P. (1993). *In vivo* assessment of therapy on human carcinomas transduced with STK gene *Proc. Annu Meeting Am. Assoc. Cancer Res.* **34**, A2008. [Meeting abstract]

Culver, K. W., Ram, Z., Wallbridge, S., Ishii, H., Oldfield, E. H., and Blaese, R. M. (1992). *In vivo* gene transfer with retroviral vector-producer cells for treatment of experimental brain tumors. *Science* **256**, 1550–1552.

Davidson, B. L., Allen, E. D., Kozarsky, K. F., Wilson, J. M., and Roessler, B. J. (1993). A model system for *in vivo* gene transfer into the CNS using an adenoviral vector. *Nature Genet.* **3**, 219–223.

Davidson, B. L., Doran, S. E., Shewach, D. S., Latta, J. M., Hartman, J. W., and Roessler, B. J. (1994a). Expression of *Escherichia coli* beta-galactosidase and rat HPRT in the CNS of *Macaca mulatta* following adenoviral mediated gene transfer. *Exp. Neurol.* **125**, 258–267.

Davidson, B. L., and Roessler, B. J. (1994). Direct plasmid mediated transfection of adult murine brain cells *in vivo* using cationic liposomes. *Neurosci. Lett.* **167**, 5–10.

Davidson, B. L., Roessler, B. J., and Shewach, D. S. (1995). Evidence of hypoxanthine salvage in HPRT deficient neuronal cells following gene transfer. Submitted for publication.

Doran, S. E., Brunberg, J., Kilbourn, M., Davidson, B. L., and Roessler, B. J. (1995). MRI and PET evaluation following administration of recombinant adenovirus to primate brain. Submitted for publication.

Doran, S. E., Ren, X. D., Betz, A. L., Pagel, M. A., Neuwelt, E. A., Roessler, B. J., and Davidson, B. L. (1994). Gene expression from recombinant viral vectors in the CNS following blood–brain barrier disruption. Submitted for publication.

Dorovini-Zis, K., Sato, M., Goping, G., Rapoport, S., and Brightman, M. (1983). Ionic lanthanum passage across cerebral endothelium exposed to hyperosmotic arabinose. *Acta Neuropathol (Berlin)* **60**, 49–60.

Dorovini-Zis, K., Bowman, P. D., Betz, A. L., and Goldstein, G. W. (1984). Hyperosmotic arabinose solutions open the tight junctions between brain capillary endothelial cells in tissue culture. *Brain Res.* **302**, 383–386.

Engelhardt, J. F., Litzky, L., and Wilson, J. M. (1994a). Prolonged transgene expression in cotton rat lung with recombinant adenoviruses defective in E2a. *Hum. Gene Ther.* (in press).

Engelhardt, J. F., Ye, X., Doranz, B., and Wilson, J. M. (1994b). Ablation of E2a in recombinant adenoviruses improves transgene persistence and decreases inflammatory response in mouse liver. *Proc. Natl. Acad. Sci. USA* **91**, 6196–6200.

Ezzeddine, Z. D., Martuza, R. L., Platika, D., Short, M. P., Malick, A., Choi, B., and Breakefield, X. O. (1991). Selective killing of glioma cells in culture and in vivo by retrovirus transfer of the herpes simplex virus thymidine kinase gene. *New Biol.* **3**, 608–614.

Fink, D. J., Sternberg, L. R., Weber, P. C., Mata, M., Goins, W. F., and Glorioso, J. C. (1992). *In vivo* expression of beta-galactosidase in hippocampal neurons by HSV-mediated gene transfer. *Hum. Gene Ther.* **3**, 11–19.

Gross, P. M., Teasdale, G. M., Graham, D. I., Angerson, W. J., and Harper, A. M. (1982). Intra-arterial histamine increases blood–brain transport in rats. *Am. J. Physiol.* **243**, H307–H317.

Hardebo, J. E., and Kahrstrom, J. (1985). Endothelial negative surface charge areas and blood–brain barrier function. *Acta Physiol. Scand.* **125**, 495–499.

Hoogerbrugge, P. M., Poorthuis, B. J., Mulder, A. H., Wagemaker, G., Doren, L. J., Vossen, J. M., and van Bekkum, D. W. (1987). Correction of lysosomal enzyme deficiency in various organs of beta-glucuronidase-deficient mice by allogenic bone marrow transplantation. *Transplantation* **43**, 609–614.

Huang, Q., Vonsattel, J. P., Schaffer, P. A., Martuza, R. L., Breakefield, X. O., and DiFiglia, M. (1992). Introduction of a foreign gene (*Escherichia coli lacZ*) into rat neostriatal neurons using herpes simplex virus mutants: A light and electron microscopic study. *Exp. Neurol.* **115**, 303–316.

Jiao, S., Acsadi, G., Jani, A., Felgner, P. L., and Wolff, J. A. (1992). Persistence of plasmid DNA and expression in rat brain cells *in vivo*. *Exp. Neurol.* **115**, 400–413.

Kaplitt, M. G., Leone, P., Samulski, R. J., Xiao, X., Pfaff, D. W., O'Malley, K. L., and During, M. J. (1994). Long-term gene expression and phenotypic correction using adeno-associated virus vectors in the mammalian brain. *Nature Genet.* **8**, 148–154.

Kozarsky, K. F., and Wilson, J. M. (1993). Gene therapy: Adenovirus vectors. *Curr. Opin. Genet. Dev.* **3**, 499–503.

Le Gal La Salle, G., Robert, J. J., Berrard, S., Ridoux, V., Stratford-Perricaudet, L. D., Perricaudet, M., and Mallet, J. (1993). An adenovirus vector for gene transfer into neurons and glia in the brain. *Science* **259**, 988–990.

Lesch, M., and Nyhan, W. L. (1964). A familial disorder of uric acid metabolism and central nervous system function. *Am. J. Med.* **36**, 561–570.

Levin, V. A., Gutin, P. H., and Leibel, S. (1993). Neoplasms of the central nervous system. In "Cancer: Principles and Practice of Oncology" (V. T. DeVita, S. Hellman, and S. A. Rosenberg, Eds.), pp. 1679–1737. J. B. Lippincott, Philadelphia.

Li, T., Adamian, M., Roof, D. J., Berson, E. L., Dryja, T. P., Roessler, B. J., and Davidson, B. L. (1994). *In vivo* transfer of a reporter gene to the retina mediated by an adenoviral vector. *Invest. Ophthalmol. Vis. Sci.* **35**, 2543–2549.

McCoy, R. D., Davidson, B. L., Roessler, B. J., Huffnagle, G. B., and Simon, R. H. (1995). Expression of human interleukin-1 receptor in mouse lungs using a recombinant adenovirus: Effects on vector-induced inflammation. *Gene Ther. J.* (in press).

Moolten, F. L. (1986). Tumor chemosensitivity conferred by inserted herpes thymidine kinase genes: Paradigm for a prospective cancer control strategy. *Cancer Res.* **46**, 5276–5281.

Moolten, F. L., and Wells, J. M. (1990). Curability of tumors bearing herpes thymidine kinase genes transferred by retroviral vectors. *J. Natl. Cancer Inst.* **82**, 297–300.

Mulligan, R. C. (1993). The basic science of gene therapy. *Science* **260**, 926–932.

Neuwelt, E. A., Maravilla, K. R., Frenkel, E. P., Rapoport, S. I., Hill, S. A., and Barnett, P. A. (1979). Osmotic blood–brain barrier disruption. *J. Clin. Invest.* **64**, 684–688.

Neuwelt, E. A., Howieson, J., Frenkel, E. P., Specht, H. D., Weigel, R., Buchan, C. G., and Hill, S. A. (1986). Therapeutic efficacy of multiagent chemotherapy with drug delivery enhancement by blood–brain barrier modification in glioblastoma. *Neurosurgery* **19**, 573–582.

Neuwelt, E. A., Pagel, M. A., and Dix, R. D. (1991). Delivery of ultraviolet-inactivated 35S-herpesvirus across an osmotically modified blood–brain barrier. *J. Neurosurg.* **74**, 475–479.

New, P. Z. (1993). Central nervous system cancer. In "Clinical Oncology" (G. R. Weiss, Ed.), pp. 146–167. Appleton & Lange, Norwalk, CT.

Oldfield, E. H., Ram, Z., Culver, K. W., Blaese, R. M., DeVroom, H. L., and Anderson, W. F. (1993). Gene therapy for the treatment of brain tumors using intra-tumoral transduction with the thymidine kinase gene and intravenous ganciclovir. *Hum. Gene Ther.* **4**, 39–69.

Ono, T., Fujino, Y., Tsuchiya, T., and Tsuda, M. (1990). Plasmid DNAs directly injected into mouse brain with lipofectin can be incorporated and expressed by brain cells. *Neurosci. Lett.* **117**, 259–263.

Palella, T. D., Hidaka, Y., Silverman, L. J., Levine, M., Glorioso, J., and Kelley, W. N. (1989). Expression of human HPRT mRNA in brains of mice infected with a recombinant herpes simplex virus-1 vector. *Gene* **80**, 137–144.

Plautz, G., Nabel, E. G., and Nabel, G. J. (1991). Selective elimination of recombinant genes *in vivo* with a suicide retroviral vector. *New Biol.* **3**, 709–715.

Ram, Z., Culver, K. W., Walbridge, S., Blaese, R. M., and Oldfield, E. H. (1992). Retroviral mediated thymidine kinase gene transfer for the treatment of malignant brain tumors. *Hum. Gene Ther.* **3**, 615–610. [Abstract]

Ram, Z., Culver, K. W., Walbridge, S., Blaese, R. M., and Oldfield, E. H. (1993). *In situ* retroviral-mediated gene transfer for the treatment of brain tumors in rats. *Cancer Res.* **53**, 83–88.

Ram, Z. (1994). Adenovirus-mediated gene transfer into experimental solid brain tumors and leptomeningeal cancer. *J. Neurosurg.* (in press).

Rapoport, S. I., Fredericks, W. R., Ohno, K., and Pettigrew, K. D. (1990). Quantitative aspects of reversible osmotic opening of the blood–brain barrier. *Am. J. Physiol.* **238**, R421–R431.

Rapoport, S. I., and Robinson, P. J. (1988). Tight-junctional modification as the basis of osmotic opening of the blood–brain barrier. *Ann. N.Y. Acad. Sci.* **481**, 250–267.

Ross, B. D., Kim, B., and Davidson, B. L. (1995). MRI and ^{1}H MRS assessment of ganciclovir toxicity to experimental intracranial gliomas following recombinant adenoviral mediated gene transfer of the herpes simplex virus thymidine kinase gene. *Clin. Cancer Res.* (in press).

Ross, B. D., Merkle, H., Hendrich, K., Staewen, R. S., and Garwood, M. (1992). Spatially localized *in vivo* 1H magnetic resonance spectroscopy of an intracerebral rat glioma. *Magn. Reson. Med.* **23**, 96–108.

Sands, M. S., Barker, J. E., Vogler, C., Levy, B., Gwynn, B., Galvin, N., Sly, W. S., and Birkenmeier, E. (1993). Treatment of murine mucopolysaccharidosis type VII by syngeneic bone marrow transplantation in neonates. *Lab. Invest.* **68**, 676–686.

Shewach, D. S., Zerbe, L. K., Hughes, T. L., Roessler, B. J., Breakefield, X. O., and Davidson, B. L. (1994). Enhanced cytotoxicity of antiviral drugs mediated by adenoviral directed transfer of the herpes simplex virus thymidine kinase gene in rat glioma cells. *Cancer Gene Ther.* **1**(2), 107–112.

Short, M. P., Choi, B. C., Lee, J. K., Malick, A., Breakefield, X. O., and Martuza, R. L. (1990). Gene delivery to glioma cells in rat brain by grafting of a retrovirus packaging cell line. *J. Neurosci. Res.* **27**, 427–439.

Sidi, Y., and Mitchell, B. S. (1985). Z-nucleotide accumulation in erthrocytes from Lesch–Nyhan patients. *J. Clin. Invest.* **76**, 2416–2419.

Slavin, S., and Yatziv, S. (1980). Correction of enzyme deficiency in mice by allogeneic bone marrow transplantation with total lymphoid irradiation. *Science* **210,** 1150–1152.

Sly, W. S., Quinton, B. A., McAlister, W. H., and Rimoin, D. L. (1973). Beta glucuronidase deficiency: Report of clinical, radiologic, and biochemical features of a new mucopolysaccharidosis. *J. Pediatr.* **82,** 249–257.

Smythe, W. R., Hwang, H. C., Amin, K. M., Eck, S. J., Davidson, B. L., Wilson, J. M., Kaiser, L. R., and Albelda, S. M. (1994). Use of recombinant adenovirus to transfer the HSV–thymidine kinase gene to thoracic neoplasms: An effective *in vitro* drug sensitization system. *Cancer Res.* **54,** 2055–2059.

Smythe, W. R., Hwang, H. C., Amin, K. M., Eck, S. L., Davidson, B. L., Wilson, J. M., Kaiser, L. R., and Albelda, S. M. (1995). Successful treatment of experimental human mesothelioma using adenovirus transfer of the herpes simplex–thymidine kinase gene. *Ann. Surg.* (in press).

Strausbaugh, L. J. (1987). Intracarotid infusions of protamine sulfate disrupt the blood–brain barrier of rabbits. *Brain Res.* **409,** 221–226.

Sztriha, L., and Betz, A. L. (1991). Oleic acid reversibly opens the blood–brain barrier. *Brain Res.* **550,** 257–262.

Takamiya, Y., Short, M. P., Ezzeddine, Z. D., Moolten, F. L., Breakefield, X. O., and Martuza, R. L. (1992). Gene therapy of malignant brain tumors: A rat glioma line bearing the herpes simplex virus type 1-thymidine kinase gene and wild type retrovirus kills other tumor cells. *J. Neurosci. Res.* **33,** 493–503.

Takamiya, Y., Short, M. P., Moolten, F. L., Fleet, C., Mineta, T., Breakefield, X. O., and Martuza, R. L. (1993). An experimental model of retrovirus gene therapy for malignant brain tumors. *J. Neurosurg.* **79,** 104–110.

Vocler, C., Sands, M., Higgins, A., Levy, B., Grubb, J., Birkenmeier, E. H., and Sly, W. S. (1993). Enzyme replacement with recombinant b-glucuronidase in the newborn mucopolysaccharidosis type VII mouse. *Pediatr. Res.* **34,** 837–840.

Westergren, I., and Johansson, B. B. (1993). Altering the blood–brain barrier in the rat by intracarotid infusion of polycations: A comparison between protamine, poly-L-lysine and poly-L-arginine. *Acta Physiol. Scand.* **149,** 99–104.

Wolfe, J. H., Deshmane, S. L., and Fraser, N. W. (1992). Herpesvirus vector gene transfer and expression of beta-glucuronidase in the central nervous system of MPS VII mice. *Nature Genet.* **1,** 379–384.

Yang, Y., Nunes, F. A., Berencsi, K., Gonczol, E., Engelhardt, J. F., and Wilson, J. M. (1994). Inactivation of E2a in recombinant adenoviruses limits cellular immunity and improves the prospect for gene therapy of cystic fibrosis. *Nature Genet.* **7,** 362–369.

Yatziv, S., Weiss, L., Morecki, S., Fuks, Z., and Slavin, S. (1982). Long-term enzyme replacement therapy in beta-glucuronidase-deficient mice by allogeneic bone marrow transplantation. *J. Lab. Clin. Med.* **99,** 792–797.

Yoshida, J., Mizuno, M., and Yagi, K. (1992). A prelude to interferon gene therapy for brain tumors. *Proc. Annu Meeting Am. Assoc. Cancer Res.* **33,** A1428. [Meeting abstract]

CHAPTER

Transfer and Expression of Potentially Therapeutic Genes into the Mammalian Central Nervous System in Vivo Using Adeno-Associated Viral Vectors

Michael G. Kaplitt
Division of Neurosurgery
Department of Surgery
New York Hospital
Cornell University Medical College
and Laboratory of Neurobiology and Behavior
The Rockefeller University
New York, New York

Matthew J. During
Department of Neurosurgery
Yale University School of Medicine
New Haven, Connecticut

I. Introduction

Modification of mammalian cellular gene expression through direct, *in vivo* gene transfer represents one of the most exciting and powerful developments in modern biology. In no field does this approach present greater opportunities as well as potential pitfalls than neurobiology. The use of viral vectors for *in vivo* gene transfer into the fully developed, living mammalian brain provides previously unimaginable opportunities to study basic neurobiological functions. Furthermore, gene therapy may eventually have the greatest impact upon the treatment of human neurological disease, due to the many unique properties of the central nervous system (CNS) which often prevent intervention by conventional means. This chapter will concentrate upon the relevant biology and current applications of adeno-associated virus (AAV) vectors to the study and treatment of the fully developed, adult mammalian central nervous system. (For a more complete overview of the biology of AAV and AAV vectors, see the chapter by Bartlett, *et al.*)

It is clear from the numerous contributions to this volume that there is an ever-increasing array of viral vectors which can be considered for application

VIRAL VECTORS

to the nervous system. Retroviral vectors are among the earliest systems developed for gene transfer. Since they absolutely require active cell division (Miller *et al.*, 1990) their use in the CNS has been limited to the developmental studies of actively dividing neuronal precursors, *in vivo* treatment of dividing brain tumor cells, or *ex vivo* modification of dividing cells in tissue culture prior to transplantation. Vectors based upon DNA viruses are, therefore, more appropriate for transduction of postmitotic cells within the brain. Recombinant herpes simplex virus (HSV) and adenovirus vectors have been used for gene transfer into neurons in tissue culture and *in vivo* (Ho and Mocarski, 1988; Breakefield and Deluca, 1991; Fink *et al.*, 1992; Le Gal La Salle *et al.*, 1993). Typically, these vectors carry deletions in viral genes which render them replication defective. However, they retain numerous functional viral genes (Roizman and Jenkins, 1985; Johnson *et al.*, 1992; Akli *et al.*, 1993; Davidson *et al.*, 1993), which may be cytotoxic to the recipient cells, induce an immune response against the transduced cell, or potentially reactivate latent viruses which exist within most adults (Johnson *et al.*, 1992; Neve, 1993; Shenk, this volume).

In order to eliminate viral gene expression, defective viral vectors have been developed. The first example of a defective DNA viral vector was the defective HSV vector (Spaete and Frenkel, 1982). This plasmid-based system contains an HSV origin of DNA replication and a cleavage/packaging signal, which are simply recognition sequences. Therefore, all viral sequences which code for protein products are eliminated. In the presence of a helper virus, the plasmid is replicated and packaged into an HSV particle, creating a defective viral vector (Spaete and Frenkel, 1982; Kwong and Frenkel, 1985; Kaplitt *et al.*, 1991; Kwong *et al.*, this volume). While the vector contains no viral genes, questions of safety remain since helper virus cannot be completely eliminated from the final vector stock. Even a few particles of residual wild-type virus could be sufficient to cause severe encephalitis and possibly death. Although the use of HSV mutants as helper viruses has substantially limited this possibility (Geller *et al.*, 1988, 1990; Kaplitt *et al.*, 1991, 1994a), issues of pathogenicity and reversion to wild-type still remain (During *et al.*, 1994a). This vector remains attractive as a tool for many neurobiological studies due to certain unique properties (discussed in this volume by Kwong *et al.*, During *et al.*, and Yin *et al.*). At present, however, the inability to completely eliminate contaminating helper virus remains concerning and will likely continue to limit consideration of this vector for human therapeutic applications.

II. Biology of Adeno-Associated Virus Vectors

The AAV vector system is a new approach to gene transfer in the CNS. Wild-type AAV is a nonpathogenic parvovirus which integrates into specific sites within the host cell chromosome (Berns and Hauswirth, 1979; Muzyczka, 1992; Bartlett *et al.,* this volume). A productive infection requires coinfection by a non-AAV helper virus, usually adenovirus, which provides proteins necessary for AAV replication and packaging. In the AAV vector system, 96% of the parental genome has been deleted such that only the terminal repeats remain, containing only recognition signals for DNA replication and packaging. AAV structural proteins are provided *in trans* by cotransfection of the AAV vector with a helper plasmid containing the missing AAV genes but lacking replication/packaging signals (Samulski *et al.,* 1987, 1989). Following infection with adenovirus, two populations of particles are obtained: progeny helper adenovirus and AAV vectors encoding foreign genes. The helper plasmid lacking AAV terminal repeats is not packaged. Since the AAV coat proteins are structurally distinct from the helper adenovirus, contaminating adenovirus particles can be completely removed. The AAV vector is thus unique among DNA viral vectors, as it contains only the gene of interest with no viral genes and is completely free of contaminating helper virus. An important basis for pathogenicity is thereby eliminated, rendering this system particularly suited to human gene therapy.

Although wild-type AAV integrates within a specific site within human chromosome 19, the fate of AAV vector DNA within transduced cells is controversial. The preponderance of evidence currently suggests that the AAV vector does integrate within the chromosome of the host cell. At least two studies have directly demonstrated integration of AAV vector genomes within host chromosomes (Walsh *et al.,* 1992; Russell *et al.,* 1994). These studies used Southern blots to demonstrate that cells stably transduced with AAV vectors contain genomes integrated within the host chromosome. While Russell *et al.* demonstrated integration in 15/17 clonal cell lines tested, Walsh *et al.* observed integration in all samples tested. Both studies demonstrated that the integration of AAV vectors occurred in single-copy fashion. This indicates that AAV proteins are not absolutely required for chromosomal integration. While others have suggested that AAV vectors do not integrate (Flotte *et al.,* 1994), the positive data from earlier studies remain compelling, and several possible explanations exist for a lack of detection of integrated DNA. In particular, since wild-type AAV tends to integrate in multiple copies but the above studies demonstrate single-copy integration of AAV vectors, it is possible that detection of vector integration may be subject to greater variability between studies. Regard less of localization, however, all groups have demonstrated the persistence of vector DNA within transduced cells. This supported the idea that AAV vectors should not only be safe, but would likely yield long-term maintenance of vector DNA within CNS cells, thereby presenting the possibility for stable gene expression *in vivo.*

III. Development of an AAV Vector for Efficient, Safe, and Stable Gene Transfer into the Mammalian Brain

An AAV plasmid encoding the bacterial *lacZ* gene was created by subcloning the human cytomegalovirus (CMV) immediate-early promoter, *lacZ*, and an SV40 polyadenylation signal between the terminal repeats of the AAV genome in plasmid psub201 (Samulski *et al.*, 1989). These termini contain the recognition signals necessary for cleavage and packaging into an AAV vector. Cells were cotransfected with pAAVlac and pAd/AAV, which provide AAV structural proteins but lack AAV termini and thus cannot package into virus (Samulski *et al.*, 1989). Cells were then infected with adenovirus type 5 to provide remaining functions necessary for replication and packaging. Residual helper adenovirus was then eliminated by heating the stock at 56°C for 30 min. Gross debris was eliminated by low-speed centrifugation. AAV vectors were then purified by one of two methods. At first, ultracentrifugation of vectors over a cushion of 25% sucrose in PBS was used. This approach was fairly quick and simple, but did not result in absolute purification of the crude stock despite vector concentration. For later studies (particularly for primate studies), a highly purified stock of virus was generated by ammonium sulfate precipitation followed by centrifugation on a cesium chloride gradient and removal of cesium by dialysis of the vector band (see Samulski *et al.*, this volume, for more details).

AAV vectors were titered by infection of cultured 293 cells followed by histochemical staining for β-galactosidase expression and counting of the resulting blue cells. There was no difference in the number of cells observed at 1 and 5 days following infection, demonstrating an absence of vector replication and spread. When the process was repeated using a *lacZ* plasmid without the AAV recognition signals, no positive cells were observed following infection with the resulting stock. The complete elimination of adenovirus was confirmed by the inability to detect any viral plaques in cultured cells 1 week following infection with this viral stock. This indicates that the *lacZ* gene was packaged into an AAV virus which was incapable of autonomous replication while residual adenovirus was completely eliminated.

AAVlac was stereotaxically microinjected into various regions of the adult rat brain, including caudate nucleus, amygdala, striatum, and hippocampus. Animals were initially sacrificed between 1 and 3 days following injection, and sections were processed for β-galactosidase expression via X-gal histochemistry. Positive cells were demonstrated within each region (Kaplitt *et al.*, 1994b). The efficiency of gene transfer into the brain appeared to be at least equivalent to that observed previously with HSV or adenovirus vectors (Ho and Mocarski, 1988; Breakefield and Deluca, 1991; Kaplitt *et al.*, 1991, 1994a; Akli *et al.*, 1993; Davidson *et al.*, 1993; Ho *et al.*, 1993; Le Gal La Salle *et al.*, 1993).

In order to analyze the long-term stability of AAV gene transfer and expression within the mammalian brain, animals were injected in the caudate nucleus with AAVlac and sacrificed 2–3 months following surgery. Sections were processed for X-gal histochemistry, using a technique which completely eliminates background staining (Kaplitt *et al.*, 1991), in order to identify cells containing functional β-galactosidase. Positive cells were identified within injected regions of the caudate nucleus up to 3 months following vector injection (Kaplitt *et al.*, 1994b). Additional animals were then sacrificed and tissue sections from these animals were examined using the polymerase chain reaction, which was modified to permit amplification and visualization of viral vector DNA *in situ*. (Nuovo *et al.*, 1991, 1993; Yin *et al.*, this volume). Numerous cells within the brain were detected which retained the bacterial *lacZ* gene after 2 months (Kaplitt *et al.*, 1994b). There was no staining on the contralateral side, in sections processed without *Taq* polymerase or in sections from brains injected with adenovirus alone. At no time were behavioral or physiological abnormalities detected within the animal subjects, and the brain sections showed no evidence of pathology resulting from the AAV gene transfer.

IV. Genetic Therapy of a Rodent Model of Parkinson's Disease Using an AAV Vector

One CNS disorder which may be amenable to gene transfer techniques is Parkinson's disease (PD). PD is characterized by loss of the nigrostriatal pathway and is responsive to treatments which facilitate dopaminergic transmission in the caudate–putamen (Yahr *et al.*, 1969; Yahr and Bergman, 1987). In rodent models of PD, mesencephalic fetal cells or genetically modified cells expressing tyrosine hydroxylase (TH) synthesize dihydroxyphenylalanine (L-dopa), which induces behavioral recovery (Freed *et al.*, 1987; Wolff *et al.*, 1989; Horrelou *et al.*, 1990; Jiao *et al.*, 1993). Moreover, direct gene transfer into the denervated striatum of lesioned animals with defective HSV vectors encoding TH can also be successful (During *et al.*, 1994b).

In order to generate vector AAVth, which may have therapeutic utility in human PD patients, the bacterial *lacZ* gene was replaced with a human TH cDNA (form II) (O'Malley *et al.*, 1987). AAVth was packaged and helper virus was eliminated as described above for AAVlac. Following injection of AAVth into the denervated striatum of rats with substantia nigra lesions , animals were analyzed for TH expression at times ranging from 24 hr to 4 months (Fig.1). Expression of TH from the AAV vector was detected using immunocytochemistry with a mouse monoclonal anti-TH antibody. Although this antibody does not distinguish between the rat and human protein, TH is not expressed within either the intrinsic neurons or glia of the rat striatum (Dubach *et al.,* 1990). Furthermore, endogenous TH immunoreactivity (TH-IR) is limited to the dopaminergic afferent fibers within the striatum of unlesioned animals and is therefore completely absent in the fully denervated striatum. In both control, uninjected rats and AAVlac-injected rats there was no striatal TH-IR on the denervated side (Fig. 1A). In contrast, in the denervated striata injected with AAVth, numerous TH-IR cells were clustered around the injection site (Figs. 1B and 1D) and extending to 2 mm away from the injection (Fig. 1C). The majority of cells within the striatum appeared to be neurons morphologically, and double-labeling with both the anti-TH monoclonal antibody and an anti-neurofilament antibody confirmed that a substantial number of intrinsic striatal neurons expressed immunoreactive TH *de novo* (Figs. 1E and 1F). Additional sections were double labeled with an antibody to glial fibrillary acidic protein, which is a marker of astrocytes and oligodendrocytes, and TH-positive glial cells were also identified. Thus, following gene transfer into the rat striatum via an AAV vector, the majority of TH-IR cells in the striatum were neurons, while a small percentage of TH-IR cells were glial cells (Kaplitt *et al.,* 1994b).

The titer of the AAVth stock used for these *in vivo* studies was 5×10^6 vector particles/ml. Therefore a single injection of 2 ml would result in 10,000 positive cells if the efficiency of infection was 100% and each particle infected a different cell. In the AAVth-injected animals, the total number of striatal cells containing TH-IR consistently exceeded 1000 for each of the 2-ml injections, suggesting a minimum of 10% *in vivo* efficiency. This is in fact significantly greater than our previous observations using defective HSV-1 vectors (Kaplitt *et al.,* 1991, 1994a). Since previous infection of AAV does not prevent subsequent reinfection or multiple particles infecting the same cell (Lebkowski *et al.,* 1988), the actual efficiency may be higher, since some cells may have been infected with multiple vector particles. In addition, only the striatum was examined histologically; however, AAVth might also infect axons and terminals and, following retrograde transport, yield additional TH expression in distal regions containing cell bodies of these striatal projections.

Striatal TH expression was also examined at times ranging from 3 days to 7 months following injection of AAVth. Expression persisted throughout this 7-month period (Figs. 1B and 1C), although the level of expression diminished by approximately 50%. Furthermore, gene transfer appeared to be completely safe in the experimental subjects, as there were no signs of cytopathic effects in any animal (AAVth or AAVlac) at any time. The only changes observed in the short-term AAVth animals (examined less than 1 week following injection) was a slight needle injury at the injection site, which was similar in PBS-injected and AAVlac-injected animals. In the long-term animals (greater than 2 months), the residual needle track was not consistently visible and there was no evidence of any neuronal injury or reactive gliosis. There were also no behavioral or gross pathological signs of brain damage in any subject. Long-term expression of a foreign gene was therefore demonstrated with no evidence of pathological changes to the brain tissue (Kaplitt *et al.*, 1994b).

Although the number of positive cells declined during the 7-month period, a greater percentage of positive cells was retained long-term compared with previous reports using other vectors (Kaplitt *et al.*, 1991, 1994a; Akli *et al.*, 1993; Davidson *et al.*, 1993; Le Gal La Salle *et al.*, 1993). While it is possible that vector genomes were lost from some cells or that some positive cells died, it is more likely that long-term expression was limited due to the use of the CMV IE promoter. Previous studies have indicated that long-term expression may be limited when using viral transcriptional control elements such as the CMV promoter (Scharfmann *et al.*, 1991; Kaplitt *et al.*, 1991, 1994a). Therefore, endogenous cellular promoters may provide more uniform long-term expression. Cellular promoters may also confer regional or cell-type specific gene expression (Kaplitt *et al.*, 1994a). At present, however, most of these promoters are extremely large fragments of genomic DNA. Given the 5-kb insert size limit for AAV vectors, current cellular promoters thus would not be compatible with many medium to large size cDNAs. This may be facilitated in the future by continued identification of control elements which confer *in vivo* specificity, thereby permitting creation of smaller promoters which would yield satisfactory long-term and/or restricted expression within the mammalian brain.

Unilateral 6-hydroxydopamine lesions of the substantia nigra have been used to generate an established rodent model of PD. In this model, elimination of nigral dopaminergic inputs results in upregulation of dopamine receptors in the lesioned striatum, while the striatal dopamine receptor density on the unlesioned side remains unchanged. The asymmetry caused by the resulting differential postsynaptic receptor sensitivities between the denervated and intact striatum results in rotational behavior (contralateral to the side of the lesion) following systemic administration of dopaminergic agents, such as the direct-acting agonist, apomorphine (Hefti *et al.*, 1980). The rate of assymetrical rotation is directly related to the severity of the striatal dopamine deficit. As a result, this model has predictive ability in defining treatments which may have therapeutic efficacy in PD (Freed *et al.*, 1987; Hargraves and Freed, 1987).

The goal of viral-based therapy was to restore dopamine synthesis locally and induce a downregulation of dopamine receptors on the treated side, thereby equalizing striatal dopamine sensitivities. Lesioned rats were tested for apomorphine-induced rotation every 2 weeks on a mimumum of three occasions, and animals that satisfied behavioral criteria of >90% lesion efficacy were identified (Hefti *et al.*, 1980). AAVth or AAVlac virus was delivered by stereotactic injection into the denervated striatum. Animals were tested for apomorphine-induced assymetrical rotation at 1 and 2 months postinjection. The rotational behavior of the AAVlac injected animals was unaltered from baseline. In contrast, AAVth-injected animals demonstrated significant behavioral recovery, compared to the AAVlac-injected group. The average behavioral recovery caused by AAVth was 31 ± 6% at 1 month and was maintained at 32 ± 3% at 9 weeks ($P < 0.01$; repeated measures ANOVA) after injection (Kaplitt *et al.*, 1994b). It should be noted that although expression has been demonstrated for up to 7 months postinjection, behavioral analyses have as of now only been performed at 1 and 2 months after treatment since an inadequate number of animals remained beyond this period, thereby preventing statistically significant conclusions; however, larger long-term studies are currently in progress.

V. Expression of Two Genes from a Single AAV Vector Results in *de Novo* Synthesis of the Neurotransmitter Dopamine

Dopamine synthesis is catalyzed by two enzymes, TH and aromatic acid decarboxylase (AADC). The reaction catalyzed by TH results in the synthesis of L-dopa, and this is the rate-limiting step in the synthesis of dopamine. Dopamine then results from conversion of L-dopa by AADC. Although striatum does not contain cells which endogenously produce TH, there is a small percentage of striatal cells which produce AADC. Therefore, behavioral recovery in animals treated with AAVth (or other approaches using TH alone) presumably occurs secondary to conversion of the resulting L-dopa to dopamine by endogenous striatal AADC. Since a limited number of cells produce AADC, however, it is possible that synthesis of dopamine could be enhanced by expression of both TH and AADC in every transduced cell. In this manner, any target cell would become an autonomous dopamine-producing cell following gene transfer. Recent evidence, in fact, suggests that expression of both genes in the denervated striatum may be superior to expression of TH alone (Kang *et al.*, 1993). Furthermore, the most substantial behavioral recovery following cell transplantation occurred when TH-expressing muscle cells were utilized (Jiao *et al.*, 1993), and unlike fibroblasts from earlier studies, muscle cells express endogenous AADC activity. This suggested that creation of an AAV vector containing both TH and AADC would be valuable.

Due to the limitation on insert sizes in AAV vectors, several modifications were required in order to create a vector containing both genes. First, the TH gene was truncated, eliminating the 5′ end. Truncation of the TH gene has actually been shown to increase enzymatic activity due to removal of an amino terminal regulatory domain (Walker *et al.*, 1994). Therefore, this served a functional purpose as well as increasing the space available for other genetic elements. In addition, a small synthetic oligonucleotide, encoding a novel epitope, was at-

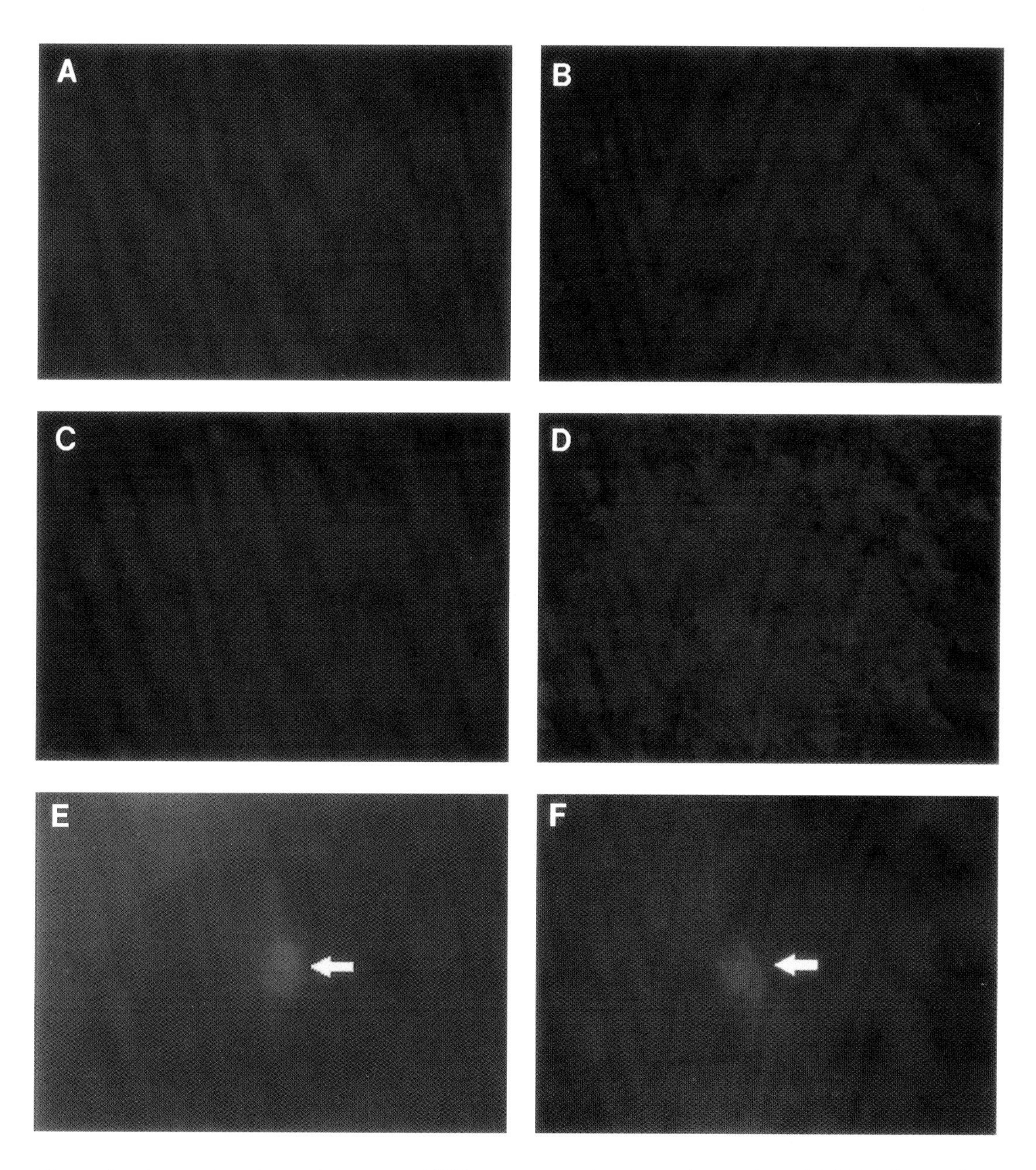

Figure 1 Immunohistochemical detection of hTH expression within the caudate nucleus of 6-OHDA-lesioned rats following injection of AAVth. (A) Absence of immunostaining in caudate following injection of AAVlac. No staining was ever observed in AAVlac animals, and staining was also always absent from the uninjected caudate from AAVth animals. (B,C) TH expression in cells of the caudate nucleus 4 months following injection of AAVth. These sections were 30 μm in thickness, which prevented morphological identification of positive cells. Approximately 30 cells are seen at the site of injection (B) and cells are also seen 2 mm away from the injection site (C), although fewer cells are present at 2 mm. This observation was repeated twice at 4 months following injection, while comparable results were obtained from three animals at 2 months and two animals at 1 month following injection. (D) TH expression

in caudate 1 week following AAVth injection. This section was 7 μm in thickness, revealing the neuronal appearance of the majority of positive cells. Fifty positive cells can be seen in this section, which is representative of approximately 50 consecutively positive sections obtained from each short-term animal. Fewer cells were observed as far as 280 sections (2 mm) away from the injection site. This result was repeated twice at 1 week following injection, and comparable results were obtained from 9 animals at 48 hr and 9 animals at 24 hr postinjection. (E,F) Double-label immunocytochemistry demonstrating neuronal TH expression. (E) TH expression in a caudate cell (arrow) was revealed using a FITC-labeled secondary antibody. (F) Neuronal identification of the TH-expressing cell (arrow) was obtained by sequentially staining the same section with an anti-neurofilament antibody and visualization with a Texas red-conjugated secondary antibody. Magnification: A–D, 400×; E, F, 630×. (Reprinted by permission from Kaplitt *et al.*, 1994.)

tached to the 5′ end of the truncated TH. This novel epitope, termed "Flag," is recognized by a commercially available monoclonal antibody; this provides an independent and unambiguous marker for expression of AAV-transduced TH *in vivo*.

After modifying TH, the AADC gene was inserted into the vector. Creation of two independent expression units, with two promoters and two polyadenylation signals, would have resulted in an insert size so large as to be incompatible with the constraints of the AAV vector. Therefore, an internal ribosome entry site (IRES) element was inserted between the Flag-TH and AADC cDNAs. Most eukaryotic mRNAs are monocistronic; they contain a single-open reading frame, and when translation of a peptide is stopped and the ribosome falls off of the transcript, additional downstream translational start sites cannot be utilized. When the IRES element is present on an mRNA downstream of a translational stop codon, it directs ribosomal re-entry (Ghattas *et al.*, 1991), which permits initiation of translation at a the start of a second open reading frame. In this manner, a eukaryotic bicistronic mRNA can be created which allows translation of two distinct peptides from a single mRNA (Fig. 2). Thus, with only a single promoter (CMV) and a single mRNA polyadenylation signal (SV40) directing expression of a single transcript, translation of both the Flag-TH and AADC proteins could occur within a single cell transduced with a single AAV vector.

Following creation of the plasmid AAV Flag-TH/AADC, each of the independent expression parameters was tested in culture. The plasmid was transfected into 293T cells, and then the following day the substrate tyrosine and an essential cofactor (tetrahydrobiopterin) were added to the tissue culture medium of some of these cultures. For comparison, additional cells were transfected with the plasmid AAVlac or were mock-transfected. Samples of medium were obtained at 30 and 60 min after addition of cofactors (or mock treatment), and these were analyzed for the presence of dopamine by high-performance liquid chromatography (HPLC). As indicated in Fig. 3, very high levels of dopamine were produced in 293T cells tranfected with AAV Flag-TH/AADC in the presence of both cofactors. In similarly transfected cells lacking the cofactors, barely detectable amounts of dopamine were produced, while AAVlac-transfected or mock-transfected cells yielded absolutely no dopamine synthesis even in the presence of adequate cofactors. This indicated that 293T cells were incapable of endogenously directing dopamine synthesis; however, introduction of the bicistronic vector AAV Flag-TH/AADC converted these cells into high-level, cofactor-dependent producers of dopamine. Finally, it should be noted that these cells were then fixed and stained with the anti-Flag monoclonal antibody, and this revealed highly specific histochemical detection of the Flag epitope with no background.

The specificity and function of the bicistronic AAV vector was further analyzed in cultured 293T cells. Despite the above data, it was still possible that 293T cells contained endogenous AADC activity. If this were true, then expression of Flag-TH alone would have yielded similar data without achieving translation of the second (AADC) open-reading frame. In order to test this, an additional vector was created. AAVFlag-TH contains a monocistronic insert with the Flag-TH open reading frame but lacking both the IRES sequence and the AADC open reading frame. 293T cells were then transfected with AAV Flag-TH/AADC, AAVFlag-TH, AAVlac or no plasmid. Both cofactors were added to all cultures the following day, and then samples of the medium were tested for both L-dopa and dopamine by HPLC (Table 1). Cells which were transfected with no plasmid or AAVlac could not synthesize any detectable level of either L-dopa or dopamine. The lack of L-dopa demonstrated that 293T cells do not posess any endogenous TH activity. Furthermore, cells

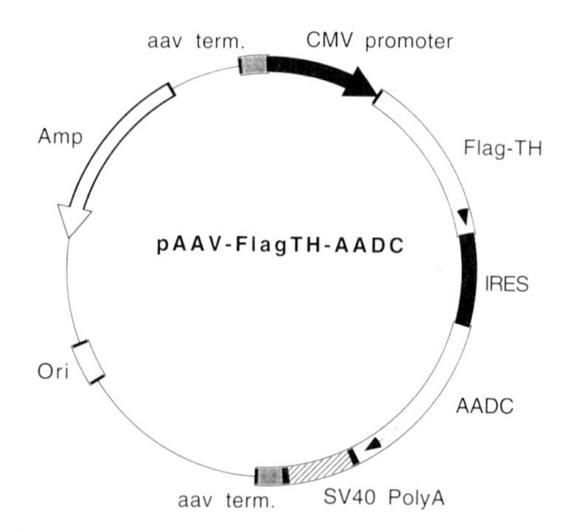

Figure 2 Plasmid pAAV-FlagTH-AADC. This bicistronic construct contains the bicistronic construct with open reading frames for truncated tyrosine hydroxylase containing the N-terminal Flag epitope (Flag-TH) and aromatic amino acid decarboxylase (AADC). TH converts tyrosine to L-dopa, and then AADC converts L-dopa to dopamine. Between the two open reading frames is a sequence allowing ribosome re-entry and initiation of translation of a second open reading frame downstream from a translational stop codon. This is the internal ribosome entry site (IRES). These are transcribed as a single messenger RNA from the human cytomegalovirus immediate early gene promoter (CMV promoter). At the 3′ end of the insert is a signal for polyadenylation of the mRNA derived from the SV40 virus (SV40 polyA). The entire insert is flanked by terminal repeats from the adeno-associated virus (AAV term.), which permits replication, excision, and packaging of the insert in the presence of proteins provided by the helper plasmid pAAV/Ad and helper adenovirus. The plasmid also contains standard plasmid sequences which permit replication and amplification of the DNA inside a bacterium (Ori) and selection of bacterial colonies harboring the plasmid through resistance to ampicillin (Amp). One of the several unique features of the AAV vector is that unlike other defective viral vectors, these plasmid sequences are lost when the DNA between the AAV termini is packaged.

transfected with AAVFlag-TH yielded very high levels of L-dopa, but undetectable amounts of dopamine. This demonstrates that 293T cells do not possess any AADC activity either. Furthermore, this indicates that the truncated TH is highly active and the addition of the 5′ Flag sequence did not adversely influence enzymatic activity. Finally, cells transfected with AAV Flag-TH/AADC produced significant amounts of L-dopa but very high levels of dopamine. Presumably the lower level of L-dopa in these cells compared with those transfected with AAVFlag-TH was due to the efficient conversion of L-dopa to dopamine. Thus two genes can be placed into a single AAV vector, and techniques such as insertion of an intervening IRES sequence can result in translation of both protein products. These data also indicate that AAV vectors can yield expression of multiple, functionally active proteins which can synergize in the production of a single, biologically active neurotransmitter. The Flag epitope was also shown to be a specific, independent marker of AAV-derived TH protein production without adversely influencing TH enzymatic activity.

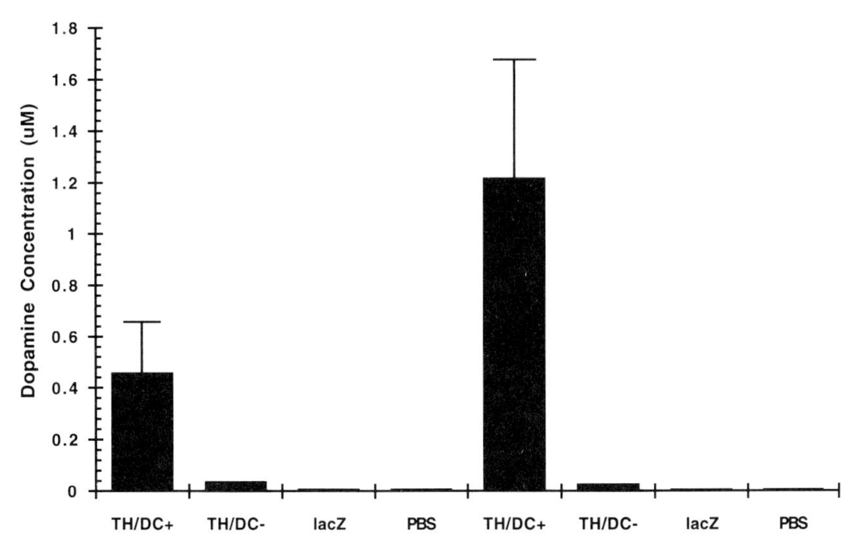

Figure 3 Dopamine release into culture medium following plasmid transfection in 293T cells. The first group of four samples represents 30 min following addition of tyrosine and tetrahydrobiopterin, while the second four samples were taken after 60 min. Dopamine release was significant at 30 min, and even higher at 60 min, in cells transfected with pAAV-FlgTH-AADC and given tyrosine and tetrahydrobiopterin (TH/DC+). Cells transfected with this plasmid but not given the substrate and co-factor (TH/DC-) synthesized negligible amounts of dopamine at both time points. Controls transfected with pAAVlac, expressing the bacterial *lacZ* gene, or mock-transfected with phosphate-buffered saline (PBS) produced no detectable amount of dopamine.

VI. AAV Vector-Mediated Gene Therapy of a Primate Model of Parkinson's Disease

The great potential of the bicistronic vector as a therapeutic agent for Parkinson's disease has led to the rapid initiation of primate studies. The primate model of Parkinson's disease is considered to be the gold-standard model for evaluation of potential therapies prior to entering human clinical trials. This model was originally developed from the observation in the early 1980s that groups of younger people were developing a neurodegenerative disorder strikingly similar to idiopathic Parkinson's disease. The source of this disorder was traced to the use of a street drug, and the specific causative agent was found to be 1-methyl-4-phenyl-1,2,3,6-tetrahydropyridine (MPTP) (Langston, 1985). When MPTP was then given to primates, the animals developed a parkinsonian disorder that has become the principle model for testing anti-parkinsonian agents. Peripherally administered MPTP will cross the blood–brain barrier, whereupon it is converted to MPP+ by monoamine oxidase B. This compound is then selectively concentrated within the dopamin-

Table 1
Release of L-Dopa and Dopamine into the Culture Medium of 293T Cells Following Plasmid Transfection

	L-Dopa (pg/ml)	Dopamine (pg/ml)
Blank	<40	<40
LacZ	<40	<40
Flag-TH	8200	<40
TH-AADC	800	4050

Controls which were transfected with pAAVlac or mock-transfected with PBS did not produce any detectable level of either L-dopa or dopamine, and therefore at the very least these cells did not contain TH activity. Following transfection with pAAV-FlagTH, which expresses only tyrosine hydroxylase, significant amounts of L-dopa were produced, but none was converted to dopamine. This demonstrated that the truncated TH with the N-terminal Flag epitope was enzymatically active, yet 293T cells contain no endogenous AADC activity, hence the absence of conversion to dopamine. By contrast, cells transfected with the bicistronic vector pAAV-FlagTH-AADC yielded significant levels of L-dopa, but far higher levels of dopamine. This demonstrated that the Flag-TH was fully active, synthesizing L-dopa, but that functional AADC was also translated from the same mRNA, and this converted much of the L-dopa to dopamine. Therefore, two enzymes were expressed from a single vector, thereby converting a cell with no endogenous TH or AADC activity into a dopamine-producing cell.

ergic neurons of the substania nigra via an energy-dependent, presynaptic uptake mechanism. This may be enhanced by the ability of neuromelanin, found within nigral neurons, to bind MPP+ (D'Amato *et al.,* 1986). MPP+ is a potent neurotoxin which eventually causes the degeneration of nigral dopaminergic neurons and loss of the nigro–striatal dopamine pathway, as is seen in Parkinson's disease (Redmond *et al.,* 1993; Tipton and Singer, 1993).

Early studies initiated in MPTP primates have been designed to test the safety of the AAV system in primates and to obtain information regarding the potential therapeutic efficacy of AAV Flag-TH/AADC for Parkinson's disease. The initial study employed a small number of animals with only moderate nigral lesions and was designed to determine whether AAV vectors can transfer genes in the adult primate brain and whether dopamine transmission could be increased in the striatum using the bicistronic vector. Purified vector was sterotaxically injected unilaterally into the striatum of MPTP-treated primates, and subjects were then sacrificed either 10 days or 4.5 months after injection. Tissue sections were analyzed for Flag immunoreactivity, and numerous positive cells were demonstrated in several sections from the injected striatum in both short- and long-term subjects, while sections from the uninjected side were completely negative. The majority of positive cells appeared morphologically to be neurons. This demonstrated for the first time that AAV vectors could successfully transfer genes into the primate brain (During *et al.,* 1994b).

Biochemical analysis of tissue samples from treated primates further indicated that the vector did cause an increase in striatal dopamine (During *et al.,* 1994b). For example, in one subject sacrificed at 10 days following treatment, the level of dopamine from a striatal tissue sample near the site of AAV injection was 18.93 ng/mg protein. An equivalent tissue sample from the uninjected, contralateral striatum yielded a dopamine level of 7.97 ng/mg protein. Tissue samples from distal sites on the injected and uninjected sides resulted in dopamine levels of 2.48 and 2.27 ng/mg, respectively. Since peripherally administered MPTP should result in roughly equal lesions to the substantia nigra bilaterally, the approximately 140% increase in dopamine levels in the injected striatum compared with the untreated side suggests that the AAV vector resulted in expression of functionally active enzymes.

A second study emplyed more severely lesioned primates in order to determine whether there is a therapeutic potential for AAV Flag-TH/AADC. Subjects were divided into two groups, with the treated group receiving AAV Flag-TH/AADC and controls receiving AAVlac. All animals received bilateral stereotaxic injections, with the same virus infused into the striatum on both sides of the brain. Subjects were then followed for 2.5 months after surgery. Although immunocytochemistry, tissue biochemistry, and formal, blinded behavioral analysis of video tapes are currently in progress, anecdotal observations suggests that the bicistronic vector resulted in sustained improvement in parkinsonian behavior (During *et al.,* 1994b). Monthly assessments of control and treated animals by blinded caretakers reported virtually no change in the behavior of animals which were subsequently determined to have been controls, while the response in treated subjects varied from modest improvement to substantial recovery of function. Most of the animals began the study spending much of their time facedown and requiring assistance in order to feed and groom themselves. Reports indicate that improvements in treated animals resulted in some cases in decreased time spent facedown and recovery of the ability to feed and groom themselves. While statistical analysis of data obtained from formal scoring of video tapes, these blinded observations are encouraging and suggest that AAV vectors may result in behavioral recovery of parkinsonian primates. It should also be noted that in both primate studies, there was no behavioral or histological evidence of toxicity due to the AAV vector. Further studies will be necessary to determine the optimal promoter, titers, and number of injections for maximum therapeutic efficacy. In addition, a direct comparison in rodents and primates of AAV vectors containing only TH and those containing TH and AADC will be very important. However, all of the above data indicate that safe, long-term improvement of human neurological diseases may be possible via genetic modification of adult brain cells *in vivo* using AAV vectors.

VII. Expression of a Growth Factor from an AAV Vector Can Yield Recovery of Function Following Neuronal Lesions

An additional AAV vector has been developed as an alternative approach to the treatment of Parkinson's disease. To date, the majority of therapeutic strategies for PD have concentrated upon increasing striatal dopamine levels. Although behavioral recovery in animal models has been repeatedly demonstrated, this is not a cure for the disease but rather symptomatic palliation. Neuronal degeneration in the substantia nigra is the pathological result of the disease process, and progression of neurodegeneration is not altered by increasing striatal dopamine. Recently, however, several reports have determined that growth factors such as glial-derived neurotrophic factor (GDNF) can be protective of and trophic for neurons of the substantia nigra (Lin *et al.*, 1993). Therefore, an AAV vector was created containing the cDNA for GDNF under the control of the CMV promoter.

Rats were lesioned with 6-OHDA and subsequently received injections AAVgdnf, AAVlac, or saline into the lesioned substantia nigra (During *et al.*,1994b). After several weeks, dopamine release into the striatum on the lesioned side was determined using intracerebral microdialysis. This technique permits sampling of local neurotransmitter release within a specific brain region of living animals (During and Spencer, 1993). Baseline dopamine levels were sampled three times and there was no difference between groups. Animals were then treated with potassium which induces release of dopamine from presynaptic terminals. Neither the AAVlac- nor saline-treated animals showed any variation in dopamine release from baseline, indicating that there were few dopaminergic terminals present within the striatum. The group treated with AAVgdnf, however, yielded a significant increase in dopamine release of 200% ($P < 0.05$). Since the AAV vector only contained the gene for a growth factor, the restoration of potassium-induced dopamine release into the striatum suggests that GDNF expression either protected or promoted regrowth of dopaminergic neurons in the substantia nigra following 6-OHDA treatment.

These results were further supported by subsequent administration of nomifensine to animal subjects after dopamine levels in the AAVgdnf group returned to baseline. Nomifensine is a drug which increases synaptic dopamine levels by inhibiting dopamine reuptake. Again both control groups showed no change in dopamine levels in response to nomifensin, while striatal dopamine increased 150% ($P < 0.05$) in the group treated with AAVgdnf. Together these data demonstrate that AAV-mediated transfer of a growth factor gene can either protect or restore dopaminergic inputs to the striatum. Thus gene therapy can be useful for both palliation of PD through striatal expression of synthetic enzymes for dopamine as well as for treatment of the underlying disease process by expressing growth factors which may protect or regenerate dopaminergic neurons.

VIII. Summary

Adeno-associated virus has been considered as a potentially useful gene transfer vehicle since the original demonstration that foreign DNA flanked by AAV terminal repeats can be packaged into an AAV coat in the presence of necessary protein factors (Samulski *et al.*, 1987, 1989). Until recently, the majority of studies have concentrated upon the use of AAV to transfer genes into dividing cells in tissue culture. Within the past year, however, it has become clear that AAV may also be a very useful vector for gene transfer into tissues composed largely of nondividing cells, particularly the brain (Kaplitt *et al.*, 1994b; During *et al.*, 1994b). This chapter has reviewed these recent developments, with a concentration upon application of this system towards genetic therapy of Parkinson's disease. Many of the features of AAV as a vector are particularly attractive for those interested in pursuing clinical applicability of viral vectors for genetic therapy of human disease. To date there have been no clinical trials approved for genetic therapy of brain disorders other than almost uniformly terminal brain tumors such as gliomas. The rapid advances in using AAV for the treatment of Parkinson's disease in both rodent and primate models suggests, however, that application to human clinical disease may be feasible in the near future. It should be clear from this chapter, however, that the benefits of the AAV vector may extend well beyond treatment of Parkinson's disease. As demonstrated by the varied contributions to this volume, numerous diseases of the nervous system are amenable to potential treatment by genetic intervention. Although continued research into viral vectors will undoubtedly continue to yield improved vector systems, at present the safe, stable, and versatile AAV vector may represent the most promising opportunity available for gene therapy of devastating but nonterminal neurological diseases.

References

Akli, S., Caillaud, *et al.* (1993). Transfer of a foreign into the brain using adenovirus vectors. *Nature Genet.* 3,224–228.

Berns, K. I., and Hauswirth, W. W. (1979). Adeno-associated viruses. *Adv. Virus Res.* **25**, 407–409.

Breakefield, X. O., and Deluca, N. A. (1991). Herpes simplex virus for gene delivery to neurons. *New Biol.* **3**, 203–218.

D'Amato, R. J., Lipman, Z. P., and Snyder, S. H. (1989). Selectivity of the parkinsonian neurotoxin MPTP: Toxic metabolite MPP+ binds to neuromelanin. *Science* **231**, 987.

Davidson, B. L., *et al.* (1993). A model system for in vivo gene transfer into the central nervous system using an adenoviral vector. *Nature Genet.* **3**, 219–223.

Dubach, M., *et al.* (1990). Primate neostriatal neurons containing tyrosine hydroxylase: Immunohistochemical evidence. *Neurosc. Lett.* **75**, 205–210.

During, M. J., *et al.* (1992). Biochemical and behavioral recovery in a rodent model of Parkinson's Disease following stereotactic implantation of dopamine-containing liposomes. *Exp. Neurol.* **115**, 193–199.

During, M. J., and Spencer, D. D. (1993). Extracellular hippocampal glutamate and spontaneous seizures in the conscious human brain. *Lancet* **341**, 1607–1610.

During, M. J., *et al.* (1994a). Long-term behavioral recovery in parkinsonian rats by an HSV vector expressing tyrosine hydroxylase. *Science* **266**, 1399–1403.

During, M. J., Kaplitt, M. G., O'Malley, K.L, *et al.* (1994b). AAV-mediated transfer of tyrosine hydroxylase and aromatic amino acid decarboxylase gene into the primate brain: A direct gene therapy approach to Parkinson's disease. *Abstr. Soc. Neurosci.* **20**, 1465.

Fink, D. J., Sternberg, L. R., Weber, P. C., Mata, M., Goins, W. F., and Glorioso, J. C. (1992). In vivo expression of beta-galactosidase in hippocampal neurons by HSV-mediated gene transfer. *Hum. Gene Ther.* 3(1), 9–11.

Flotte, T. R., *et al.* (1993). Stable in vivo expression of the cystic fibrosis transmembrane conductance regulator with an adeno-associated virus vector. *Proc. Natl. Acad. Sci. USA* 90, 10613–10617.

Flotte, T. R., Afione, S. A., and Zeitlin, P. L. (1994). Adeno-associated virus vector gene expression occurs in non-dividing cells in the absence of vector DNA integration. *Am. J. Respir. Cell Mol. Biol.* **11**, 517–521.

Freed, W. J., *et al.* (1987). Resoration of dopaminergic function by grafting of fetal rat substantia nigra to the caudate nucleus: Long-term behavioral, biochemical, and histochemical studies. *Ann. Neurol.* **8**, 510–519.

Ghattas, I. R., Sanes, J. R., and Majors, J. E. (1991). The encephalomyelitis virus internal ribosome entry site allow efficient coepression of two genes from a recombinant provirus in cultured cells and in embryos. *Mol. Cell. Biol.* **11**, 5848–5959.

Geller, A. I., Keyomarski, K., Bryan, J., and Pardee, A. B. (1990). An efficient deletion mutant packaging system for defective HSV-1 vectors: Potential applications to neuronal physiology and human gene therapy. *Proc. Natl. Acad. Sci. USA* **87**, 8950–8954.

Graham, F. L., and van der Eb, A. J. (1973). A new technique for the assay of infectivity of human adenovirus 5 DNA. *Virology* **52**, 456–467.

Graham, F. L., Smiley, J., Russell, W. C., and Nairn, R. (1977). Characterization of a human cell line transformed by DNA from human adenovirus type 5. *J. Gen. Virol.* **36**, 59–74.

Hargraves, R., and Freed, W. J. (1987). Chronic intrastriatal dopamine infusions in rats with unilateral lesions of the substantia nigra. *Life Sci.* **40**, 959–966.

Hefti, F., Melamed, E., and Wurtman, R. J. (1980). Partial lesions of the dopaminergic nigrostriatal system in rat brain: Biochemical characterization. *Brain Res.* **195**, 123–127.

Ho, D. Y., and Mocarski, E. S. (1988). β-Galactosidase as a marker in the peripheral and neural tissues of the herpes-simplex virus-infected mouse. *Virology* **167**, 279–283.

Ho, D. Y., Mocarski, E. S., and Sapolsky, R. M. (1993). Altering central nervous system physiology with a defective herpes simplex virus vector expressing the glucose transporter gene. *Proc. Natl. Acad. Sci. USA* **90**, 3655–3659.

Horrelou, P., *et al.* (1990). In vivo release of dopa and dopamine from genetically engineered cells grafted to the denervated rat striatum. *Neuron* **5**, 393–402.

Jiao, S., Gurevich, V., and Wolff, J. A. (1993). Long term correction of rat model of Parkinson's Disease by gene therapy. *Nature* **262**, 450.

Johnson, P. A., *et al.* (1992). Cytotoxicity of a replication defective mutant of herpes simplex virus 1. *J. Virol.* **66**, 2952–2955.

Jones, N. C., and Shenk, T. S. (1978). Isolation of deletion and substitution mutants of adenovirus type 5. *Cell* **13**, 181–188.

Kang, U. J., Fisher, L. J., Joh, T. H., *et al.* (1993). Regulation of dopamine production by genetically modified primary fibroblasts. *J. Neurosci.* **13**, 5203–5211.

Kaplitt, M. G., *et al.* (1991). Expression of a functional foreign gene in adult mammalian brain following in vivo transfer via a herpes simplex virus type 1 defective viral vector. *Mol. Cell. Neurosci.* **2**, 320–330.

Kaplitt, M. G., *et al.* (1994a). Preproenkephalin promoter yields region-specific and long-term expression in adult brain following direct in vivo gene transfer via a defective herpes simplex viral vector. *Proc. Natl. Acad. Sci. USA* **91**, 8979–8983.

Kaplitt, M. G., *et al.* (1994b). Long-term gene expression and phenotypic correction using adeno-associated virus vectors in the mammalian brain. *Nature Genet.* **8**, 148–153.

Kwong, A. D., and Frenkel, N. (1985). The herpes simplex virus amplicon IV: Effecient expression of a chimeric chicken ovalbumin gene amplified within defective virus genomes. *Virology* **142**, 421–425.

Langston, J. W. (1985). Mechanisms of MPTP toxicity: More answers, more questions. *Trends Pharmacol. Sci.* **6**, 375–378.

Lebkowski, J. S., McNally, M. M., Okarma, T. B., and Lerch, L. B. (1988). Adeno-associated virus: A vector system for efficient introduction and integration of DNA into a variety of mammalian cell types. *Mol. Cell. Biol.* **8**, 3988–3996.

Le Gal La Salle, G., *et al.* (1993). An adenovirus vector for gene transfer into neruons and glia in the brain. *Science* **259**, 988–990.

Lin, L. F., *et al.* (1993). GDNF: A glial cell line-derived neurotrophic factor for midbrain dopaminergic neurons. *Science* **260**, 1130–1132.

McLaughlin, S. K., *et al.* (1988). Adeno-associated virus general transduction vectors: Analysis of proviral structures. *J. Virol.* **62**, 1963–1973.

Miller, D. G., Adam, M. A., and Miller, A. D. (1990). Gene transfer by retrovirus vectors occurs only in cells that are actively replicating at the time of infection. *Mol. Cell. Biol.* **10,** 4329–4242.

Muzyczka, N. (1992). Use of adeno-associated virus as a general transduction vector for mammalian cells. *Curr. Topics Microbiol. Immunol.* **158,** 97–129.

Neve, R. L. (1993). Adenovirus vectors enter the brain. *TINS* **16,** 251–253.

Nuovo, G. J., *et al.* (1991). An improved technique for the in situ detection of DNA after polymerase chain reaction amplification. *Am. J. Pathol.* **139,** 1239–1244.

Nuovo, G. J., *et al.* (1993). Importance of different variables for enhancing in situ detection of PCR-amplified DNA. *PCR Methods Appl.* **2,** 305–312.

O'Malley, K. L., Anhalt, M. J., Martin, B. M., Kelsoe, J. R., Winfield, S. L., and Ginns, E. I. (1987). Isolation and characterization of the human tyrosine hydroxylase gene: Identification of 5′ alternative splice sites responsible for multiple mRNAs. *Biochemistry* **26,** 6910–6914.

Paxinos, G., and Watson, C. (1992). "The Rat Brain in Stereotaxic Coordinates." Academic Press, Sydney, Australia.

Perese, D. A., Ulman, J., Viola, J., Ewing, S. E., Bankiewicz, K. S. (1989). A 6-hydroxydopamine-induced selective parkinsonian rat model. *Brain Res.* **494,** 285–293.

Poole, S., *et al.* Gene transfer into hematopoietic stem cells using AAV vectors: Targeted integration into chromosome 19. *Blood,* in press.

Redmond, D. E., Jr., Roth, R. H., Spencer, D. D., *et al.* (1993). Neural transplantation for neurodegenerative diseases: Past, present and future. *Ann. N.Y. Acad. Sci.* **695,** 258–266.

Roizman, B., and Jenkins, F. J. (1985). Genetic engineering of novel genomes of large DNA viruses. *Science* **229,** 1208–1214.

Russell, D. W., Miller, A. D., and Alexander, I. E. (1994). Adeno-associated virus vectors preferentially transduce cells in S phase. *Proc. Natl. Acad. Sci. USA* **91** ,8915–8919.

Samulski, R. J., Chang, L. S., and Shenk, T. (1987). A recombinant plasmid from which an infectious adeno-associated virus genome can be excised in vitro and its use to study viral replication. *J. Virol.* **61,** 3096–3101.

Samulski, R. J., Chang, L-S., and Shenk, T. (1989). Helper-free stocks of adeno-associated viruses: Normal integration does not require viral gene expression. J. Virol. **63,** 3822–3828.

Samulski, R. J., *et al.* (1991). Targeted integration of adeno-associated virus (AAV) into human chromosome 19. *EMBO J.* **10,** 3941–3950.

Scharfmann, R., Axelrod, J. H., and Verma, I. M. (1991). Long-term in vivo expression of retrovirus-mediated gene transfer in mouse fibroblast implants. *Proc. Natl. Acad. Sci. USA* **88,** 4626–4630.

Spaete, R. R., and Frenkel, N. (1982). The herpes simplex virus amplicon: A new eucaryotic defective-virus cloning-amplifying vector. *Cell* **30,** 295–304.

Tipton, K. F., and Singer, T. P. (1993). Advances in our understanding of the mechanisms of the neurotoxicity of MPTP and related compounds. *J. Neurochem.* **61,** 1191–1206.

Walker, S. J., *et al.* (1994). Catalytic core of rat tyrosine hydroxylase: terminal deletion analysis of bacterially expressed enzyme. *Biochem. Biophys. Acta* **1206,** 113–119.

Walsh, C. E., Liu, J. M., Xiao, X., *et al.* (1992). Regulated high level expression of a human gamma-globin gene introduced into erythroid cells by an adeno-associated virus vector. *Proc. Natl. Acad. Sci. USA* **89,** 7257–7261.

Wolff, J. A., *et al.* (1989). Grafting fibroblasts genetically modified to produce L-dopa in a rat model of Parkinson's Disease. *Proc. Natl. Acad. Sci. USA* **86,** 9011–9014.

Yahr, M. D., *et al.* (1969). Treatment of parkinsonism with levadopa. *Arch. Neurol.* **21,** 343–354.

Yahr, M. D., and Bergmann, K. J. (Eds.) (1987). "Parkinson's Disease." Raven Press, New York.

CHAPTER 13

Genetic Modification of Cells with Retrovirus Vectors for Grafting into the Central Nervous System

Un Jung Kang
Department of Neurology
The University of Chicago
Chicago, Illinois

I. Introduction

Advances in molecular genetics have led to a new understanding of the genetic defects in human disorders. At the same time, the development of recombinant DNA technology has also produced means of introducing DNAs to mammalian cells. Gene therapy, as an experimental tool and therapeutic modality (Mulligan, 1993; Miller, 1992), holds great promise for the future, particularly in central nervous system (CNS) disorders whose treatment has been hampered by the lack of an efficient delivery system across the blood–brain barrier (Gage *et al.,* 1987, 1991b; Breakefield, 1993).

The goal of gene therapy is not restricted to the removal of the specific defective gene and/or replacement with the correct gene into the cells of patients. Molecular biological studies also help us to define the roles of specific gene products and the pathophysiology of various disorders. Therefore, gene therapy can be used to supplement multiple other steps in the metabolic pathway that may compensate for the primary genetic defect. Good examples of this approach are dopamine replacement therapy in Parkinson's disease (PD) and neurotrophic factor therapy in neurodegenerative disorders. The etiology of PD is unknown and there is no specific genetic abnormality that has been identified. Nevertheless, gene therapy with dopamine-producing genes has the potential for restoration of most of the major motor disabilities. Neurotrophic factors that can promote sprouting of neurons may also not be related to the primary genetic defect, but are potentially very effective in restoring the function of a diseased CNS.

Gene therapy can be targeted to germ line cells or somatic cells. Gene transfer to germ cells is employed in transgenic animal experiments. In the methods of transgenic animals, the entire developmental strategies of the organism interact with the transgene. On the other hand, somatic gene transfer provides a simpler experimental approach with which a single gene's effect in an adult organism can be studied. Two general approaches have been employed for somatic gene therapy: the *ex vivo* approach using *in vitro* modification of cells, followed by transplantation of the genetically modified cells (Gage *et al.*, 1991b), and the *in vivo* approach using direct gene transfer with viral vectors capable of introducing genes into somatic cells *in situ*, including nondividing neuronal cells (Fig. 1) (Breakefield, 1993; Le Gal La Salle *et al.*, 1993; Geller *et al.*, 1991). For *ex vivo* gene transfer, genetic modification of cells is carried out in culture, and the genetically modified cells are introduced into the subject. This approach allows us to control and monitor the gene transfer process before the cells are placed into subjects. Use of genetically modified cells allows flexibility and precise control of factors delivered to the host brain. *In vivo* gene transfers by adenoviruses, adenoassociated viruses, and herpes viruses are discussed in other chapters.

Introduction of genetic materials into somatic cells can be achieved by use of physical, chemical, or biological methods (Felgner and Rhodes, 1991;

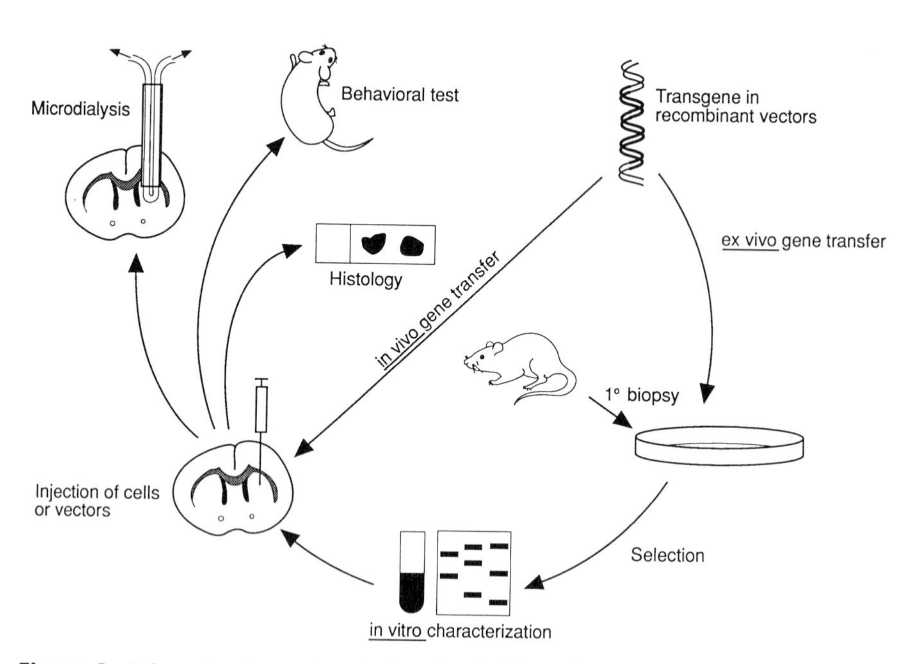

Figure 1 Schematic of gene transfer into the CNS by either *ex vivo* or *in vivo* approach.

Davis *et al.*, 1993; Curiel, 1994). Most of these methods are not very efficient and are most suitable for experimental studies with established cell lines. Cell lines are usually derived from tumors and may not be suitable for gene therapy in humans unless the growth is controlled and immunological responses are curtailed. Primary cells obtained from the individual patient provide better alternatives, although primary cells grow more slowly and have limited life span in culture. Therefore, more efficient methods of gene transfer are desirable. Various viral vectors derived from replication-defective recombinant viruses are most efficient in transducing the genes into primary cells. Retroviruses have been most widely used and extensively developed over the past decade. Most of the methods utilizing retrovirus vectors involve *ex vivo* gene transfer, since retrovirus vectors require actively dividing cells for successful transduction (Miller *et al.*, 1990). The exception to this generalization is gene transfer to tumor cells by retroviral vectors *in vivo* (discussed in another chapter in this volume). This chapter will focus on *ex vivo* approaches for CNS utilizing retrovirus vectors.

II. Retrovirus Vectors

The most widely used virus vectors for *ex vivo* gene transfer are disabled murine and avian retroviruses (Majors, 1992; Boris-Lawrie and Temin, 1993). More detailed discussions on the retrovirus vectors are found in other chapters in this volume and only the most relevant summary is presented in this chapter. Retroviruses consist of two copies of polyadenylated RNA genome packaged in a viral core which, in turn, is encapsidated inside a multiunit protein coat. The genome contains three major areas: the group-specific antigen gene (*gag*) encodes the viral core capsid protein and matrix protein. RNA-dependent DNA polymerase is encoded in the *pol* region, and the *env* region encodes the glycosylated envelope protein that determine the nature of the host infected by the viruses. Amphotropic viruses infect cells across a broad range of vertebrates including primates, rodents, and birds. The ecotropic viruses infect only murine or other rodent species. Retroviruses attach to receptors in the plasma membrane of host cells and gain entry into the cells by endocytosis. Ecotropic receptor has been recently identified to be a ubiquitously expressed cell surface transporter for charged amino acids (Kim *et al.*, 1991). After endocytosis, the contents of the viral capsids are released in the cytoplasm and the RNA genome is transcribed into double-stranded DNA by reverse transcriptase *pol.* The synthesized DNA is integrated into the host chromosome during S phase of the cell cycle, dependent on active DNA synthesis (Miller *et al.*, 1990). The integration site is apparently random and only one copy is integrated. The viral RNA is transcribed from the integrated proviral DNA and is packaged into a viral core particle along with the other viral components. The complete virus particles exit the cells by budding, taking plasma membrane fragments with envelope glycoproteins inserted in the membrane. The cells are left intact without lysis or morphological changes.

Many generations of modified vectors have been developed for safer and more efficient use (Majors, 1992; Boris-Lawrie and Temin, 1993). The basic principle lies in production of replication-defective vectors by deleting essential viral elements. Infective viral particles are produced by use of packaging cell lines with stably transfected packaging-deficient helper virus genome. Retroviral plasmid vectors contain packaging signals and the transgene with appropriate promoters, but do not have any genes that can produce viral proteins. Infectivity is conferred by the packaging cell line that contains the retroviral genes necessary to provide the infective viral particles but lacks the packaging signal. When the retroviral vector plasmid is transfected into the packaging cells, the cells will package the recombinant retrovirus containing the transgene into infectious viral particles. Once these viruses infect the host cells, the plasmid gene is integrated into the chromosomes of the host cell. Further spread of the plasmid gene is not possible since the plasmid gene does not produce any viral proteins.

A. Packaging Cell Lines

The first generation of retrovirus packaging cell lines consisted of the packaging-deficient retroviral genome stably transfected in fibroblast cells such as NIH 3T3 cells. ψ2 cells (Mann *et al.*, 1983) contain a stably transfected mutant virus genome, pMOVψ-, which was derived from Moloney murine leukemia virus (MoMuLV) by deleting 350 base pairs (bp) from the 5′ splice donor site of envelope message to AUG of the *gag* gene on order to remove the packaging signal. An amphotropic variant, ψAM cells (Cone and Mulligan, 1984) contain the *env* gene from amphotropic virus 4070A (pMAVψ-) to widen the host range. Wild-type viruses can still be generated from these packaging cells at a low frequency, from recombination between the vector and the packaging system. Further safety measures were incorporated into the second generation of packaging cells, such as PE501 and PA317, which have further deletions in the 5′ long-term repeat (LTR) and 3′ LTR regions (Miller and Buttimore, 1986). Such deletion of overlapping regions reduces the chance for homologous recombination. Most recent types of packaging cells have separated the viral genome into two plasmids to further reduce the chance for wild-type virus generation. Ecotropic ψ-CRE and amphotropic ψ-CRIP cells have two separate plasmids containing the retroviral genome: one with pMOVψ- with an additional insertion mutation in the *env* gene, and the other with intact *env* gene only (Danos and Mulligan, 1988). GP+E86 (Markowitz *et al.*, 1988b) and GP+AM12 (Markowitz *et al.*, 1988a) have deletion of the entire *env* gene on one plasmid and intact *env* gene on the other.

B. Plasmid Vectors

There are three basic designs for expressing multiple genes from a retrovirus vector. The prototype retroviral vector, pZIPNEOSV(X), includes a splicer site for transcription of two genes from a single LTR promoter, producing genomic and subgenomic transcripts by differential splicing (Cepko *et al.*, 1984). Since it is difficult to predict the effect of specific sequences to direct differential splicing, more consistent expression of two genes is obtained from directional orientation vectors which contain separate promoters for expression of each gene. However, selection for either promoter may suppress the second promoter activity, depending on the type of backbone retrovirus used (Emerman and Temin, 1984, 1986). The amount of suppression is not affected by distance, reverse orientation, or viral promoter activity (Emerman and Temin, 1986; Soriano *et al.*, 1991). Therefore, efforts were made to overcome this problem. More recent attempts include polycistronic vectors with an internal ribosomal entry site (IRES) preceding the second gene instead of a promoter. IRES from picornaviruses allows entry of ribosomes independent of the noncapped 5′ end of the mRNA and allows efficient translation of the second coding region from a single transcript (Jang *et al.*, 1988; Adam *et al.*, 1991; Ghattas *et al.*, 1991).

The discovery that a more efficient packaging signal includes a portion of *gag* gene lead to development of vectors with extended packaging signals, such as N2, with 10-fold increase in vector titer (Bender *et al.*, 1987). In addition, the start codon in the portion of the *gag* region that was included in N2 vector was mutated to a stop codon, and further deletions were made to reduce the possibilities for wild-type generation through homologous recombination (Miller and Rosman, 1989). Other precautions, such as elimination of polyadenylation signal from the foreign gene, were designed to increase the viral titer, since a premature termination of genomic retroviral transcript could be prevented (Miller *et al.*, 1983).

C. Advantages and Disadvantages of Retrovirus Vectors

Advantages of retrovirus vectors include efficiency of gene transfer and stable integration of provirus into the host genome as described above. The retrovirus vectors have also been extensively studied for the past decade and undergone many modifications through several generations for increased safety. Disadvantages of retroviral vectors include relatively low viral titers, usually less than 10^7 plaque forming units per milliliter. The size of the gene that can be inserted is limited to 8 to 10 kbp. The integration of provirus into the chromosome is stable and precise, but the location is apparently random and, therefore, could produce an insertional mutation by disrupting a normal gene. Proviral deletion/rearrangement occurs, albeit at a low frequency of 10^{-5}.

Homologous recombination of the retroviral vectors with the retroviral genome in the packaging cells could generate the wild-type virus, but the current generation of packaging cells has been modified to minimize the possibility of homologous recombination, as noted in previous sections. Retroviruses also require dividing cells with active replication and DNA synthesis for the provirus integration to occur (Miller *et al.*, 1990). Therefore, retroviruses are not useful for *in vivo* gene transfer into nondividing cells and are most suitable for *ex vivo* therapies. However, this property can be advantageous for developmental studies marking precursor cells *in vivo* or for CNS tumor therapy as described in other chapters in this volume.

III. Cell Types for *ex Vivo* Gene Therapy Carriers

A. Graft Properties for *ex Vivo* Therapy

To be a useful vehicle for transduction of genes, the cells should be easily obtainable and readily cultured, be able to express the transgene, and be able to undergo a selection process used to enrich the population of cells. The ideal characteristics of genetically modified cells for CNS disorders should approximate those of the normal neuronal cells that we are attempting to replace. The more neuronal features they have, the more useful they would be. For PD, the production of dopamine or 3,4-dihydroxy-L-phenylalanine (L-DOPA) may be the minimal neuronal feature necessary. Nonneuronal cells could be genetically modified to secrete the neurotransmitter and serve as a biological minipump at a localized site in the brain. The cells should be nononcogenic, immunologically compatible with the recipient, and survive well in the brain with minimal disruption of the host anatomy and physiology. Most of the established cell lines derive from tumor lines and are not suitable for long-term grafts. Therefore, primary cells have been extensively pursued as graft donor cells.

B. Primary Cells (Nonneuronal)

1. Primary Fibroblast Cells

Primary skin fibroblast cells taken from adult animals can be easily cultured and genetically modified by various methods including retrovirus vectors. Primary fibroblast cells stop dividing after reaching confluence. These cells can be maintained for a few weeks at the confluent state in culture. Autologous or syngeneic fibroblast cells are also contact inhibited and survive on a long-term basis without the formation of tumors or immune rejection after transplantation into the CNS (Kawaja *et al.*, 1991; Kawaja and Gage, 1992) as well as in the peripheral tissue (Palmer *et al.*, 1991). Initially they form cell-dense globular areas with finger-like projection of fibroblast cells into the host neuropil (Fig. 2). By 2 weeks, the graft reorganizes into cell-dense cores and collagen-rich areas surrounding the core (Fig. 3). Fibroblasts are seen both in the core and in the collagen-rich surroundings. The collagen-rich area has fibroblasts that are oriented along the

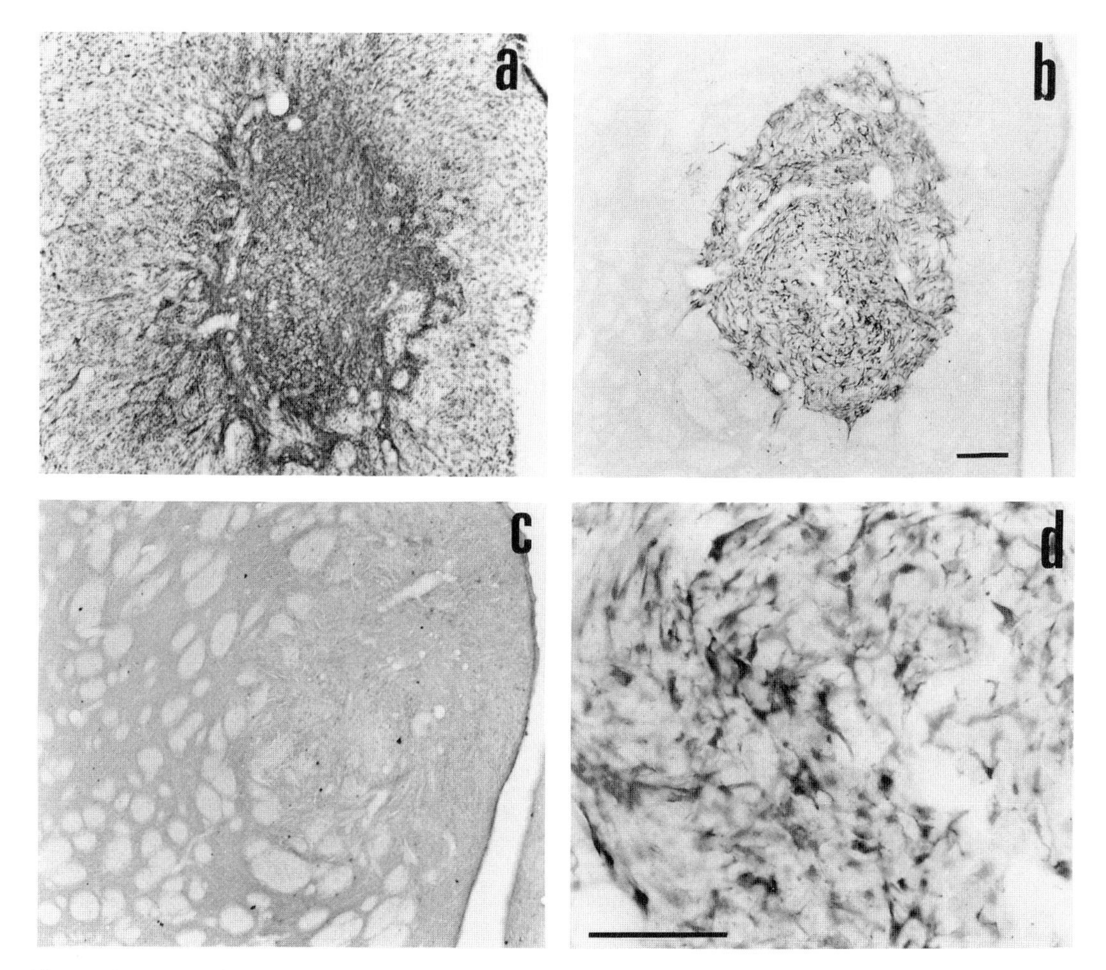

Figure 2 Primary fibroblast cells *in vivo* at 1 week. (a) Nissl stain at 1 week after grafting. The implanted primary fibroblasts are forming a dense globular area with some finger-like projections into the host brain. (b) The same graft was immunostained for the transgene, AADC. AADC immunoreactivity is robust within the fibroblasts. (c) Control transduced fibroblasts show no positive immunoreactivity. (d) High-power view of b. The scale bars represent 100 μm.

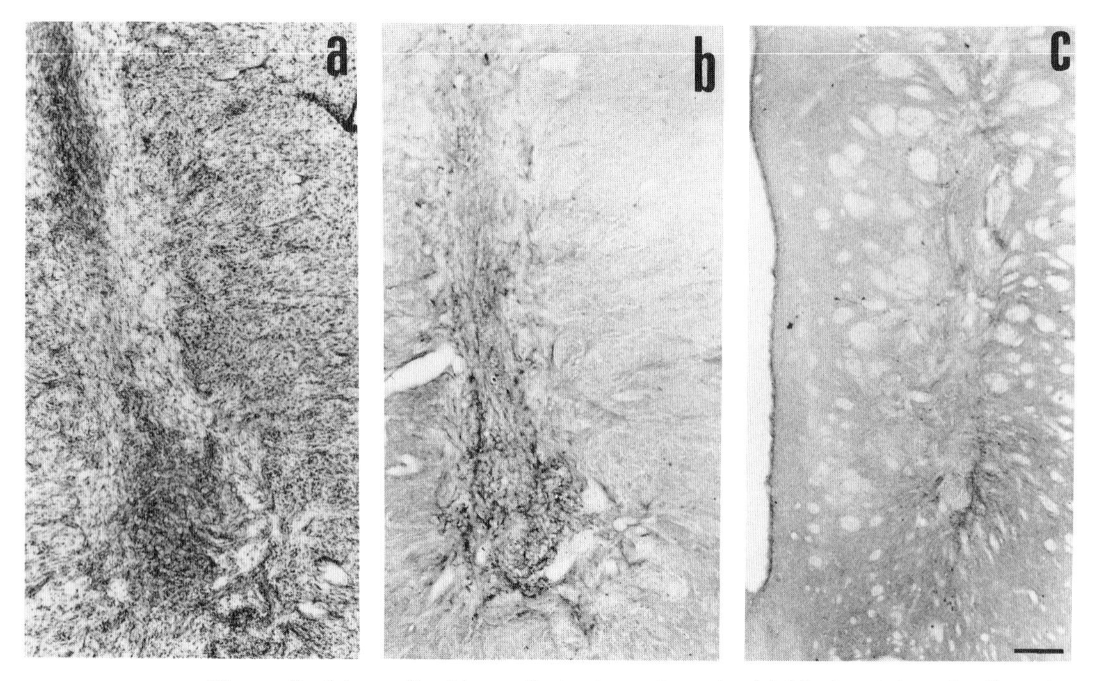

Figure 3 Primary fibroblast cells *in vivo* at 2 weeks. (a) Nissl staining of cells at 2 weeks postimplantation. The graft is organizing to cell-dense cores and collagen-rich areas surrounding the core. (b) AADC immunocytochemistry of same graft as that in a; AADC expression is noted both in the core and in the collagen-rich areas. (c) AADC immunocytochemistry of control fibroblast graft. The scale bars represent 100 μM.

collagen bundles in lower densities than are seen in the core region. The graft volume remains relatively constant from 2 to 10 weeks. Immunolabeling for the transgene is strongly positive at 1 and 2 weeks, with numerous transgene-expressing fibroblasts in both the core and the collagen-rich surroundings (Fig. 3). Transgene expression can still be noted at 10 weeks after grafting, although immunopositive cells appear sparser within the graft at this time than at an earlier time point (Fig. 4). The morphological features of grafted fibroblasts cells are similar to those normally found in the skin, including an elliptical nucleus with condensed chromatin, elongated mitochondria, and few Golgi apparatus on electron microscopic examination (Kawaja *et al.*, 1991). Viability of grafts has been also demonstrated by production of collagen stained with picrofuchsin and by abundant fibronectin production within the borders of the graft noted by immunostaining (Lucidi-Phillipi *et al.*, 1995). Reactive, glial fibrillary acidic protein (GFAP)-positive, astrocytes are also present in the border of grafts. Astrocytic processes penetrate the graft and support cholinergic axonal ingrowth when nerve growth factors are produced by the graft (Kawaja and Gage, 1991). Mild

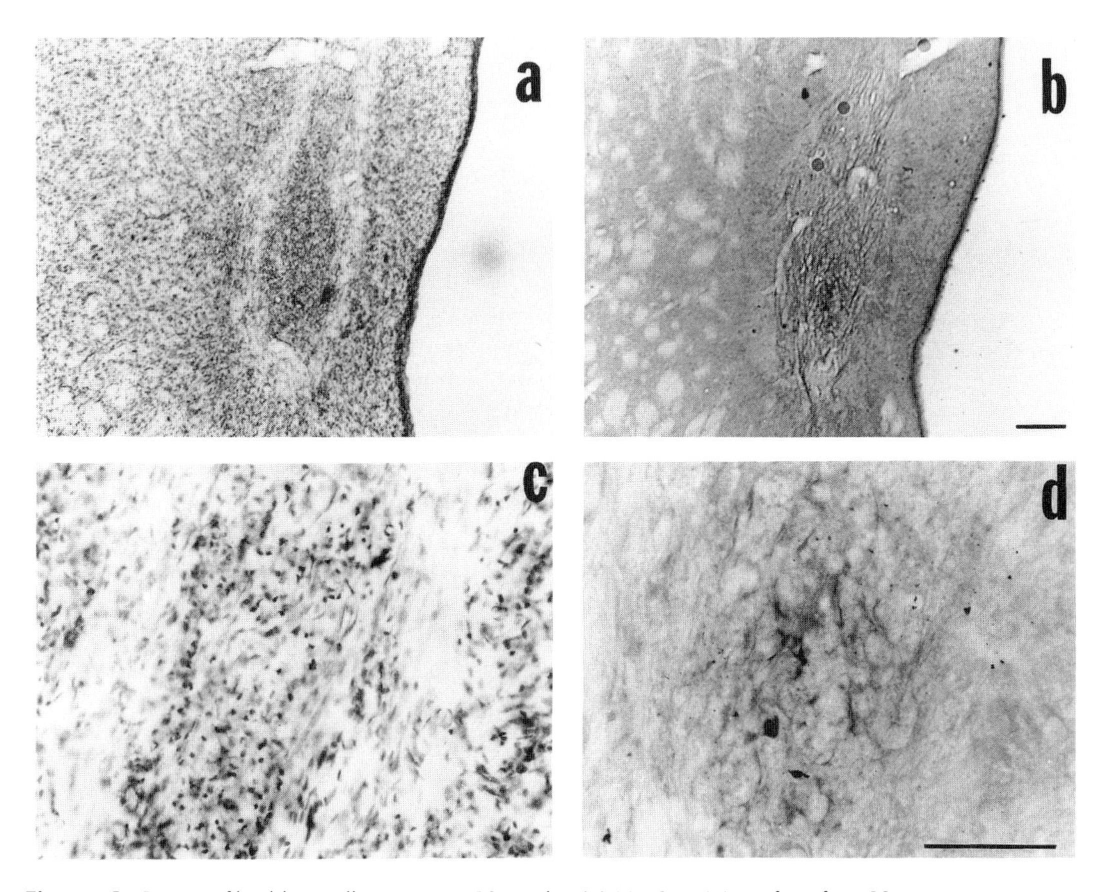

Figure 4 Primary fibroblast cells *in vivo* at 10 weeks. (a) Nissl staining of graft at 10 weeks postimplantation. The graft size remains the same as those at 2 or 5 weeks postimplantation. (b) The high-power photomicrograph of a shows the cell-dense cores with some cells with phagocytic materials and the collagen-rich surroundings with fibroblasts imbedded in them. (c) AADC immunocytochemistry shows that a small number of AADC-immunopositive cells remain at this time point. (d) High-power view of c. The scale bars represent 100 μm.

macrophage infiltration was also observed as noted by OX-42 immunoreactivity (Kawaja *et al.*, 1991). The graft is also well vascularized with nonfenestrated endothelial cells and the blood–brain barrier is restored within several weeks (Kawaja and Gage, 1992).

Fibroblast cells have been most extensively studied as a potential autologous donor cell type; these can be taken from adult animals and undergo genetic modifications and selection processes. Use of fibroblasts could be easily applied to patients: fibroblasts from their own skin biopsy could be modified into customized immunocompatible donor cells.

2. Muscle Cells

Primary myoblasts are amenable to high-efficiency transfection by plasmid DNAs using calcium phosphate precipitation or lipofection methods in addition to retroviral infection (Jiao *et al.*, 1992). Intramuscular injection of naked plasmid DNA can result in expression of the gene in the muscles, although at a very low efficiency (Wolff *et al.*, 1990). The mechanism underlying the relative ease of gene transfer into myoblast cells is not well understood. Some have suggested that nonviral plasmids allow more stable long-term expression than retrovirus vectors in muscle cells (Jiao *et al.*, 1993). Muscle cells have the unique mixture of myotubes, which are the differentiated muscle cells, and myoblasts, which retain the capacity to multiply efficiently. Cultured myoblasts, when injected intramuscularly, fuse into adjacent normal muscle fibers (Rando and Blau, 1994). Autologous muscles and cultured primary muscle cells from heterologous donors have also been shown to survive well in the brain after transplantation (Jiao and Wolff, 1992; Jiao *et al.*, 1992, 1993).

All of the studies mentioned were done with fetal or neonatal muscles, and autologous tissue may be difficult to obtain. The feasibility of myoblast growth from adult mice was also suggested and needs to be explored further (Naffakh *et al.*, 1993).

3. Glial Cells

The astrocyte is an attractive cell type for grafting studies in the CNS due to its intrinsic supportive role in the CNS. Fetal and neonatal astrocytes migrate extensively after grafting into adult cerebral cortex (Goldberg and Bernstein, 1988), although in some areas, such as hippocampus and hypothalamus, they tend to stay within the organ of origin (Hatton *et al.*, 1992). Primary type I astrocytes from neonatal rats have been transduced with retroviral vector expressing nerve growth factor (NGF) and selected for their expression of aminoglycoside phosphotransferase by growing in G418. These astrocytes were grafted and survived up to 2 weeks in the brain (Cunningham *et al.*, 1991). Astrocytes have been transduced with tyrosine hydroxylase (TH) gene to express the dopaminergic phenotype that is necessary for the grafting studies in animal models of PD (Lundberg *et al.*, 1991; La Gamma *et al.*, 1993). However, the use of astrocytes has been limited to fetal or neonatal astrocytes and hampered by their slow growth rate which makes expansion and gene transfer difficult. Obtaining autologous astrocytes would not be as benign as obtaining skin fibroblast cells or myoblasts either. Use of adenovirus vectors that does not require active cell division may overcome some of the technical difficulties of gene transfer into astrocytes (Ridoux *et al.*, 1994).

Other glial cell types include the oligodendrocyte type 2 astrocyte progenitor cells that develop into oligodendrocytes after transplantation into the neonatal rat brain (Espinosa de los Monteros *et al.*, 1993). The precursor cells have been immortalized with temperature-sensitive mutant of simian virus 40 (SV40) large T antigen and shown to express GFAP immunoreactivity (Whittemore *et al.*, 1994). Neonatal Schwann cells have also been transduced with TH gene in culture (Owens *et al.*, 1991). Again, obtaining autologous cells may be the major hurdle to the use of the glial cells.

4. Endocrine Cells

Chromaffin cells from adrenal medulla have been extensively studied as graft donor cells in animal models and patients with Parkinson's disease since chromaffin cells constitutively produce catecholamines without any genetic modification. However, production of dopamine is at a relatively low level compared to that of other catecholamines. Moreover, chromaffin cells have been shown to survive poorly in animal and in humans (Freed *et al.*, 1990). Cografting with peripheral nerves enhances survivability (Kordower *et al.*, 1990). This may be mediated by NGF, since supplementation of NGF has been shown to enhance their survival (Cunningham *et al.*, 1991). Some have suggested that chromaffin cells may survive well in the brain after grafting, if they are purified to remove other contaminating cell types (Schueler *et al.*, 1993).

Chromaffin cells may be useful if they are genetically modified with either NGF or other genes to enhance their survival and dopamine production. However, obtaining autologous chromaffin cells involves a major abdominal surgical procedure which accounts for the majority of morbidity in autologous adrenal transplantation into the striatum of the PD patients.

C. Neuronal Cells

The ideal cells for CNS somatic gene therapy would be cells of CNS origin that exhibit neuronal secretory pathways, secondary messengers, and signal transduction. Most cell lines are tumors which are transformed and proliferating cells such as PC12 cells and neuroblastoma cells. Recently, embryonic carcinoma cells such as human NT2N cells (Pleasure *et al.*, 1992) and murine PCC7-S-aza-R1009 cells (Wojcik *et al.*, 1993) or P19 cells (Morassutti *et al.*, 1994) have been shown to be capable of differentiation with proper stimulation. NT2N cells were derived from a human teratocarcinoma line, NTera 2 (NT2) cells, *in vitro* by retinoic acid treatment. NT2N extend long axons staining for phosphorylated neurofilaments and simplified dendrites containing microtubule-associated protein 2 after transplantation into the rodent brain (Trojanowski *et al.*, 1993). PCC7-S-aza-R1009 cells express a catecholaminergic phenotype as noted by TH immunoreactivity (Wojcik *et al.*, 1993).

However, when transplanted into the brain, these cells require immunosuppression for survival.

Immortalization of CNS fetal neurons has been extensively pursued as a tool to study their characteristics and to obtain readily available donor cells with neuronal phenotypes (Noble *et al.*, 1992). Recent techniques of conditional immortalization by retrovirus-mediated transduction of temperature-sensitive oncogenes have increased the safety and potential value of these cells as transplant donor cells (Eves *et al.*, 1992). Temperature-sensitive oncogene, SV40 large T antigen, would allow oncogenic growth of the cells at permissive temperatures (usually at 33°C) in culture but would degrade at body temperature (39°C), reverting the cells back to their neuronal features. When these cells are grafted isochronically, they survive and proliferate in a controlled fashion, then differentiate into phenotypes appropriate for the host site and integrate into the host architecture (Renfranz *et al.*, 1991; Snyder *et al.*, 1992). When these cells are grafted into adult CNS, some differentiation is observed depending on the host sites (Onifer *et al.*, 1993). These cells are obviously a very effective tool for developmental studies and could also be an ideal vehicle for gene transfer. Although cell lines driven from embryonic rat striatum have been shown to express gamma-aminobutyric acid (GABA) and GABA uptake ability (Giordano *et al.*, 1993), most of these conditionally immortalized cells tend to be progenitor cells that do not express the desired neurotransmitter phenotype. Therefore, additional gene transfers with other genes of interest such as neurotransmitter synthesizing genes, have been pursued (Anton *et al.*, 1994).

Primary neurons can be cultured and grafted, but their capacity to undergo genetic modification is limited. Propagation of progenitor cells by use of epidermal growth factor (Reynolds *et al.*, 1992) or fibroblast growth factor (Ray *et al.*, 1993) may make stable genetic modification by retroviruses feasible. *In vivo* gene transfer methods with adenovirus or herpes virus vectors may also make the primary neurons amenable to gene modification to augment their properties.

These approaches have the potential to produce universal donor cells with many of the desirable neuronal features. Long-term survival of these cells in the brain, the safety of temperature-sensitive oncogenes, and immune responses to the cells are some of the issues that need to be resolved before clinical applications can be contemplated.

IV. Grafting Genetically Modified Cells in Animal Models of Neurodegenerative Disorders

A. Neurotransmitter Replacement

1. Current Therapy of Neurodegenerative Disorders

a. Pharmacological Therapy The current treatments for most neurodegenerative disorders are unsatisfactory. First of all, the etiology and pathophysiology of most neurodegenerative disorders are unknown. Even with some knowledge of the CNS pathophysiology, the task of delivering the proper pharmacological agents into CNS structures is not trivial. The best examples are PD and Alzheimer's disease (AD). Though the etiology of these disorders is unknown, neurotransmitter deficits due to the degeneration of specific neuronal groups are thought to be responsible for their major symptoms. The loss of neurons in the nucleus basalis and reduction of cholinergic neurotransmitters are found in AD and are thought to be responsible for the memory loss (Bartus *et al.*, 1982). In PD, there is a selective loss of dopaminergic neurons in the SN and a corresponding decrease of dopamine in the target area, the striatum (Hornykiewicz and Kish, 1986). The extent of motor dysfunction in PD correlates well with the degree of dopaminergic loss (Bernheimer *et al.*, 1973).

The strategy of replacing the missing neurotransmitter dopamine by oral administration of the precursor, L-DOPA, or of other dopamine agonists attenuates many of the motor symptoms, at least in the short term (Cotzias *et al.*, 1969). Long-term treatment of PD with L-DOPA is, however, frequently complicated by loss of response, erratic responses ("wearing-off" and "on-off"), dyskinetic movements, and psychosis (Marsden and Parkes, 1977; Sweet *et al.*, 1976). Continuous delivery of dopamine agonists using duodenal (Kurlan *et al.*, 1988) or intravenous infusions (Quinn *et al.*, 1986) has demonstrated that some of the problem is due to the fluctuating levels of L-DOPA from oral administration.

However, any type of systemic delivery cannot avoid stimulation of the dopaminergic system in limbic areas, which often leads to the development of untoward symptoms such as psychosis. Continuous delivery of dopamine localized to the target area of the striatum may serve to alleviate and prevent many of the long-term complications of the presently available pharmacological treatments. Investigators are attempting to develop ligands that stimulate dopamine receptors in the striatum selectively; however, despite advances in the understanding of dopamine receptor subtypes (Civelli *et al.*, 1991), such selectivity has not yet been achieved. Moreover, most of the dopamine agonists are less potent than the natural compound L-DOPA and are best used as adjunct therapies to L-DOPA (Factor and Weiner, 1993). Therefore, other approaches for delivering L-DOPA or dopamine continuously and locally have been explored; namely, the transplantation of cells and tissues producing dopamine either endogenously or after genetic modification.

In AD, the lack of good cholinergic medication has hindered the test of the cholinergic hypothesis for memory loss and therapeutic development. Recently, however, a choline esterase inhibitor, tacrine, which increases brain acetylcholine levels, improved memory in Alzheimer's disease patients (Davis *et al.*, 1992). Neurotransmitter replacement for Huntington's disease (HD) has been even less successful. The selective vulnerability of striatal neurons in HD involves medium spiny neurons expressing GABA and enkephalin (Reiner *et al.*, 1994). GABA agonists have not helped the patients with HD. Since GABA is present in many different populations of neurons projecting to various structures (Albin *et al.*, 1989), activation of one subgroup of neurons with GABA receptors could counteract the activation of another group of neurons. The reason for the lack of response to neurotransmitter replacement is not clear and may be due to the inadequate level of neurotransmitter replacement or due to the indiscriminate delivery of the neurotransmitters. Localized delivery by genetically modified cells should be able to address this point, as will be detailed further in following sections.

b. CNS Transplantation The most successful dopaminergic transplant in the CNS has been achieved by use of fetal mesencephalic neurons in rat models of PD (Dunnett *et al.*, 1988). These successes have led to clinical trials of fetal tissue transplants in patients with PD. A limited number of clinical trials have been performed, and the results appear promising (Freed *et al.*, 1992; Lindvall *et al.*, 1992). Intrastriatal transplantation of fetal striatal cells has been shown to reverse abnormal movements in an animal model of HD (Hantraye *et al.*, 1992). However, there are many potential problems associated with using human fetal tissue. In addition to ethical and political considerations, practical issues of the availability and quality control of fetal tissue remain formidable obstacles to generalization of this approach beyond research studies. Estimates based on rat data and from clinical studies suggest that several fetuses are necessary for each side of the striatum for optimal dopamine restoration. Since the proper fetal age and fresh preparation are critical in successful graft survival (Brundin *et al.*, 1988), obtaining such materials from several fetuses for each patient presents logistical challenges and high cost. The use of cell cultures and autologous donor tissues has begun to be explored as an alternative. The adrenal medulla produces catecholamines and can be obtained for autografts; however, the survival of adrenal medullary chromaffin cells has been uniformly poor both in experimental animals (Freed *et al.*, 1990) and in human patients (Hirsh *et al.*, 1990). Although more recent studies with purified chromaffin cells have shown better survival (Schueler *et al.*, 1993), chromaffin cells produce mainly noradrenalin rather than dopamine. Combining cellular transplant and gene therapy may produce better customized donor cells (Gage *et al.*, 1991a,b) and this *ex vivo* approach will be discussed in more detail in the following sections.

2. Neurotransmitter Replacement by Gene Therapy

One of the first experiments applying gene therapy in animal models of CNS disorders was the use of the gene for TH in the animal model of PD (Wolff *et al.*, 1989). Initial studies were done with immortalized rat fibroblast 208F cells. Subsequently, the use of primary fibroblast cells (Fisher *et al.*, 1991) or myoblast cells (Jiao *et al.*, 1993) has been shown to produce long-term graft survival. Although the initial studies showed that this approach is feasible, the details regarding biochemical pathways necessary for the full effects of genetically modified cells are still controversial. Dopamine is synthesized in a series of enzymatic steps starting from the precursor L-tyrosine, an essential amino acid present in all types of cells. The first step is catalyzed by TH and is thought to be the rate-limiting step (Fig. 5). TH requires cofactors and is critically dependent on tetrahydrobiopterin (BH_4) for its activity. Although genetically modified fibroblast cells and myoblasts transduced with TH require addition of BH_4 for production of L-DOPA in culture, these grafts have been shown to attenuate abnormal rotational behavior of rats with experimental dopaminergic lesions (Wolff *et al.*, 1989; Horellou *et al.*, 1990b; Jiao *et al.*, 1993). These findings suggest that the levels of BH_4 available in these cells *in vivo* are sufficient to allow some dopamine synthesis. On the other hand, others have noted that their fibroblast cell lines do not express sufficient amounts of cofactor BH_4 to permit production of L-DOPA either *in vitro* or *in vivo*. Addition of BH_4 to these cells was necessary for L-DOPA production and behavioral reversal (Uchida *et al.*, 1992). It seems most likely that the supply of BH_4 in the grafted cells will be important for efficient production of L-DOPA. This could be achieved by the use of cells that can produce the cofactor endogenously or by genetically modifying the cells to produce BH_4 (Bencsics *et al.*, 1994).

The second step in dopamine synthesis involves the decarboxylation of L-DOPA by aromatic L-amino acid decarboxylase (AADC). Despite the loss of the majority of dopaminergic neurons, which are the major source of AADC, L-DOPA is still converted to dopamine within the CNS of patients with PD and in the rat model of PD with 6-hydroxydopamine (6-OHDA) lesions of the SN (Hefti *et al.*, 1981). The molecular basis of this conversion *in vivo* remains unclear. There is no significant nonenzymatic decarboxylation *in vivo*, nor are there other decarboxylating enzymes that have significant substrate specificity for L-DOPA (Kang *et al.*, 1992). The source and site of this remaining AADC in dopamine-depleted animals are not known (Melamed *et al.*, 1980, 1981). More recent studies suggest the presence of AADC message and protein in glial cells in culture (Juorio *et al.*, 1993; Li *et al.*, 1992). Although genetically modified fibroblast cell lines (Horellou *et al.*, 1990a) or myoblasts (Jiao *et al.*, 1993) transfected with TH cDNA alone have been shown to increase dopamine

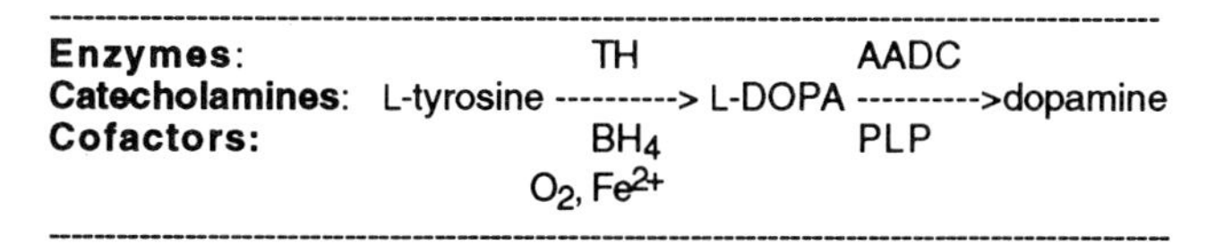

Figure 5 Dopamine biosynthetic pathway

in the denervated striatum after grafting, supplemental AADC expression by gene transfer might increase the efficiency of dopamine production (Kang *et al.*, 1993a,b). In addition, use of AADC-expressing cells in conjunction with precursor (L-DOPA) administration has been proposed as a method to regulate the final level of dopamine delivery (Kang *et al.*, 1993a). Placement of an additional AADC source in the striatum may allow use of lower doses of L-DOPA, which would have minimal effect on the other areas of brain and provide a high enough concentration of dopamine in the striatum. Secretion of both L-DOPA and dopamine from fibroblast cells is efficient and not affected by high potassium concentration, which is consistent with the lack of a vesicular storage system in fibroblast cells (Kang *et al.*, 1993a).

A similar approach can be taken for treatment of AD. The meager effect of cholinergic pharmacological therapy could be due to the lack of efficient delivery of acetylcholine to the target neurons. Genetically modified primary fibroblast cells that have been transduced with choline acetyltransferase produce acetylcholine. The level of acetylcholine was regulatable with administration of the precursor choline both *in vitro* (Schinstine *et al.*, 1992) and *in vivo* (Fisher *et al.*, 1993). Release of acetylcholine was independent of potassium and calcium concentrations (Misawa *et al.*, 1994). These cells would provide an excellent tool for neurotransmitter delivery to AD brains. Glutamic acid decarboxylase (GAD) is the rate-limiting step of GABA synthesis. Rat-1 fibroblast cells that were transduced with GAD express enzymatically active GAD and secrete GABA into the media (Ruppert *et al.*, 1993), raising the possibility of GABA replacement therapy for HD.

B. Neurotrophic Factor Delivery

The role of neurotrophic factors in neural development, maintenance of adult neurons, and degeneration of neurons in aging and disease is described in detail in other reviews (Hefti *et al.*, 1989; Barde, 1989; Thoenen, 1991; Snider, 1994). The use of neurotrophic factors to promote the sprouting of remaining neurons and to prevent further degeneration has been suggested (Rosenberg *et al.*, 1988). NGF is the prototype neurotrophic factor that has been shown to be effective in central cholinergic neurons. A family of related neurotrophic molecules has been discovered since then: brain-derived neurotrophic factor (BDNF) (Barde *et al.*, 1982; Leibrock *et al.*, 1989), neurotrophin-3 (Maisonpierre *et al.*, 1990; Hohn *et al.*, 1990), and neurotrophin-4/5 (Berkemeier *et al.*, 1991; Ip *et al.*, 1992). Since neurotrophic factors are big molecules that do not cross the blood–brain barrier, direct CNS delivery is difficult with conventional pharmacological techniques. Methods of neurotrophic factor application *in vivo* involve infusing the protein intraventricularly or intraparenchymally. In experimental paradigms, this approach has been especially problematic for BDNF. An abundance of truncated forms of the neurotrophin receptor, *trk*B (Middlemas *et al.*, 1991; Klein *et al.*, 1990; Rudge *et al.*, 1994) limits the diffusion of BDNF *in vivo,* since BDNF binds with an exceptionally high affinity to the ependymal cells lining the ventricular system and glial cells. In order to circumvent this problem, gene therapy techniques can be used for efficient delivery into the CNS.

NGF has been introduced to the brain by both genetically modified fibroblast cell lines (Rosenberg *et al.*, 1988) and primary fibroblasts (Kawaja *et al.*, 1992) and has been successful in rescuing cholinergic neurons following fimbria–fornix transection. Moreover, intracerebral grafting of genetically modified fibroblasts also affords the unique opportunity of studying the neurite extension properties of trophic molecules *in vivo*. Grafts of NGF-producing fibroblasts implanted into the septum or striatum encourage the ingrowth of cholinergic fibers along astrocytic processes penetrating the graft (Kawaja and Gage, 1991) and promote the regeneration of acetyl cholinesterase-positive axons when grafted into the lesioned fimbria–fornix cavity in a collagen matrix (Kawaja *et al.*, 1992). With this approach, one can produce trophic molecules *in vivo* and provide substrates for axonal growth along the astrocytic processes as well. Similar induction of dopaminergic fiber ingrowth has been demonstrated in BDNF-transduced fibroblasts implanted into the midbrain (Fig. 6) (Lucidi-Phillipi *et al.*, 1995). Genetically modified fibroblast cells secreting neurotrophic factors provide the proper environment for the axonal growth as well as providing trophic molecules.

Neurotrophic factor-secreting grafts may also have a neuroprotective effect. Intrastriatal transplants of NGF-secreting fibroblasts also reduce the area of degeneration found following quinolinic acid infusion (Frim *et al.*, 1993) and diminish the degeneration of striatal cholinergic interneurons and other striatal neurons (Schumacher *et al.*, 1991; Frim *et al.*, 1993). This may have relevance in treating or preventing progression of neuronal loss in HD. BDNF promotes the survival of fetal dopaminergic neurons in culture and protects them against a toxic metabolite of 1-methyl-4-phenyl-1,2,3,6-tetrahydropyridine, 1-methyl-4-phenylpyridinium (MPP^+) (Hyman *et al.*, 1991; Beck *et al.*, 1992; Spina *et al.*, 1992), 6-OHDA (Spina *et al.*, 1992), and 6-hydroxyDOPA (Skaper *et al.*, 1993) *in vitro*. Experiments with BDNF infusions *in vivo* have not shown any neuroprotective effect after axotomy (Lapchak *et al.*, 1993) or 6-OHDA lesions of the median forebrain bundle (Altar *et al.*, 1992) that result in near complete dopaminergic depletion, but they have noted functional effect of enhancing amphetamine-induced rotations and increased dopamine turnover. The lack of histological effect may be due to the devastating nature of the lesions and difficulty of achieving high concentration of BDNF at the proper site. Indeed, genetically modified cells expressing BDNF protect dopaminergic neurons *in vivo* from partial lesions produced by intrastriatal MPP^+ (Frim *et al.*, 1994) or 6-OHDA (Kang et al., unpublished) infusions.

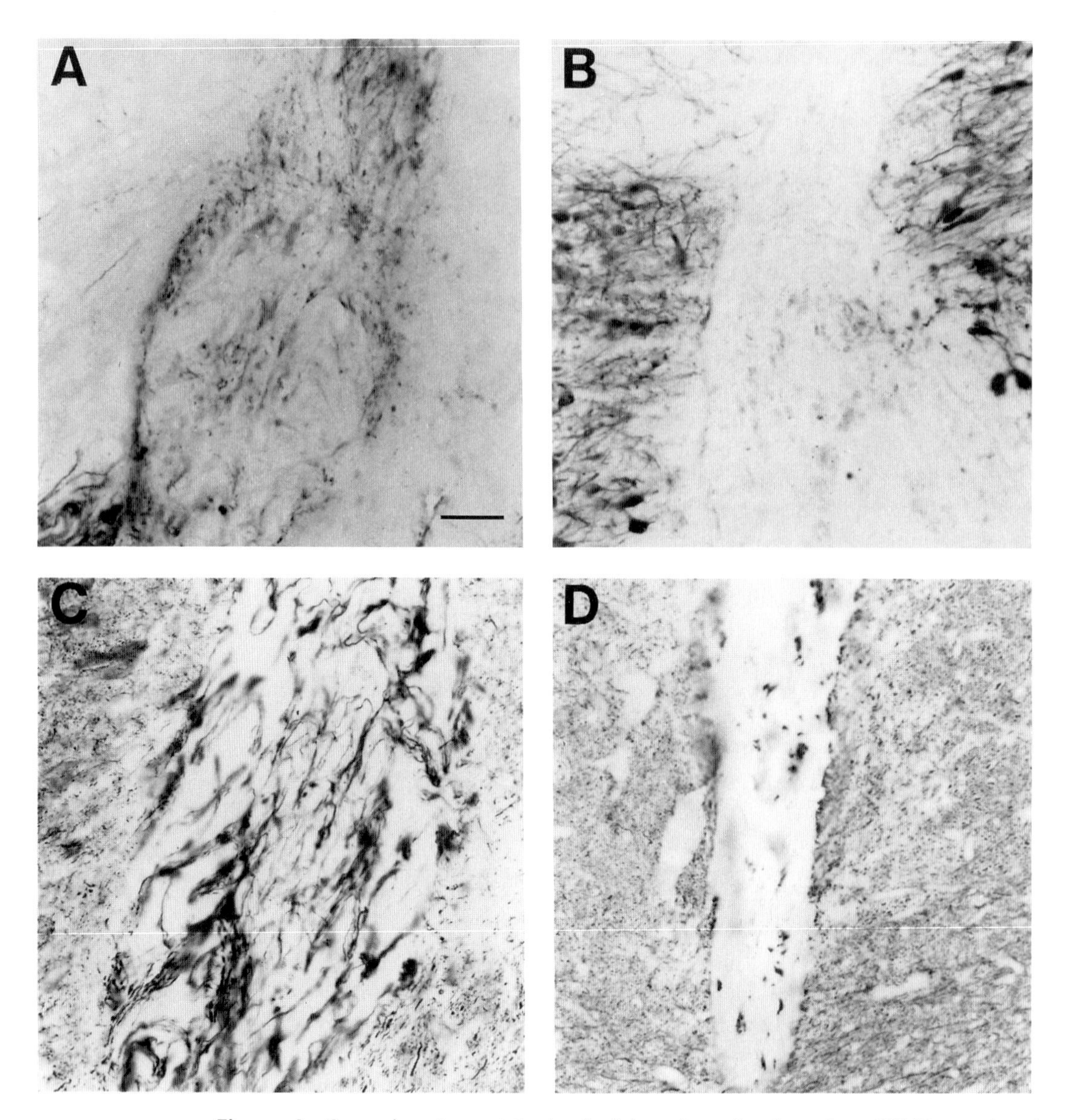

Figure 6 Coronal sections at the level of the substantia nigra show BDNF-transduced primary fibroblasts (A, C) and control fibroblasts transduced with the same retrovirus vector expressing β-galactosidase (B, D) grafted into the ventral mesencephalon. Six weeks following implantation, coronal sections were immunostained for TH (A, B) and neurofilament (C, D). TH-immunoreactive fibers (A) as well as neurofilament-immunoreactive profiles (C) are evident in BDNF-producing grafts but not in the control grafts (B,D). Scale bar = 50 mm. (reprinted with permission from Lucidi-Phillipi *et al.*, 1995)

V. Current Problems and Future Directions

The use of genetically modified cells has opened new possibilities of delivering large molecules directly into target sites in the CNS. Before these methods can be applied to human neurodegenerative disorders, however, several issues need to be resolved. First are the safety issues, concerning the virus vectors used and transgenes expressed. Retrovirus vectors have been used extensively and several generations of vectors have been developed that incorporate features that make the recombination event leading to the wild type virus generation less likely. Containing the retroviruses in the cells and characterizing them in culture allows the opportunity to screen the cells for possible transformation, the presence of wild-type viruses, and other deleterious effects. The other safety issue is the nature of the transgenes introduced into CNS. A deleterious effect of ectopic transgene expression may be difficult to predict. Whether exposure to the transgene will lead to the immune response is unknown, since many of the transgenes introduced are normally intracellular.

A second problem concerns the long-term expression of the transgene. Most of the vectors and methods mentioned above have suffered from short-lived expression of the transgenes (Palmer *et al.*, 1991). The reason for the transient nature of the transgene expression is not clear. Most of the gene therapy vectors use viral promoters to express the transgenes, and their transcriptional activity may be suppressed in somatic cells. Constitutive cellular promoters (Scharfmann *et al.*, 1991) or cell type-specific promoters (Dai *et al.*, 1992) have had some initial success. The effect of the promoter may depend on the cell type and vectors used (Roemer *et al.*, 1991). Stability of the transgene protein is an important variable as well that has not been studied as extensively. TH protein was noted to be much more stable in PC-12 cells than in fibroblast cells suggesting cell-specific environment for protein stability (Wu and Cepko, 1994).

Most current gene therapy methods do not allow regulation of transgene expression. Optimal levels are critical for the physiological function of many substances, and the ability to regulate the transgene expression is important. The design of promoters that can be regulated externally or have a built-in feedback mechanism will be an important advance for the future. The ability to accommodate larger genes will be important for many of the neuronal genes. We also need to better understand the function of the involved genes to devise rational strategies of gene therapy. Genetically modified cells will serve as biological minipumps that deliver the peptides that can be secreted or that express enzymatic machinery to produce and secrete neurotransmitters. For some other types of molecules, intracellular delivery by use of virus vectors capable of transducing neurons may be essential for their function.

In conclusion, genetically modified cells have become a useful experimental tool to examine the function of a single gene in the CNS of adult animals *in vivo*. The success of these approaches also has profound implications for therapy of neurodegenerative disorders that have defied conventional treatments.

Acknowledgments

The author gratefully acknowledges the support by USPHS Grant NS 32080, Joint Junior Faculty Award from Parkinson Disease Foundation & United Parkinson Foundation, National Parkinson Foundation, and Brain Research Foundation. I thank William Lytton and Anthony Reder for their critical comments on the manuscript.

References

Adam, M. A., Ramesh, N., Miller, A. D., and Osborne, W. R. A. (1991). Internal initiation of translation in retroviral vectors carrying picornavirus 5′ nontranslated regions. *J. Virol.* **65**, 4985–4990.

Albin, R. L., Young, A. B., and Penney, J. B. (1989). The functional anatomy of basal ganglia disorders. *Trends Neurosci.* **12**, 366–375.

Altar, C. A., Boylan, C. B., Jackson, C., Hershenson, S., Miller, J., Wiegand, S. J., Lindsay, R. M., and Hyman, C. (1992). Brain-derived neurotrophic factor augments rotational behavior and nigrostriatal dopamine turnover *in vivo*. *Proc. Natl. Acad. Sci. USA* **89**, 11347–11351.

Anton, R., Kordower, J. H., Maidment, N. T., Manaster, J. S., Kane, D. J., Rabizadeh, S., Schueller, S. B., Yang, J., Edwards, R. H., Markham, C. H., and Bredesen, D. E. (1994). Neural-targeted gene therapy for rodent and primate hemiparkinsonism. *Exp. Neurol.* **127**, 207–218.

Barde, Y.-A., Edgar, D., and Thoenen, H. (1982). Purification of a new neurotrophic factor from mammalian brain. *EMBO J.* **1**, 549–553.

Barde, Y. A. (1989). Trophic factors and neuronal survival. *Neuron* **2**, 1525–1534.

Bartus, R. T., Dean, R.L., III, Beer, B., and Lippa, A. S. (1982). The cholinergic hypothesis of geriatric memory dysfunction. *Science* **217**, 408–414.

Beck, K. D., Knüsel, B., Winslow, J. W., Rosenthal, A., Burton, L. E., Nikolics, K., and Hefti, F. (1992). Pretreatment of dopaminergic neurons in culture with brain-derived neurotrophic factor attenuates toxicity of 1-methyl-4-phenylpyridinium. *Neurodegeneration* **1**, 27–36.

Bencsics, C., Hatakeyama, K., Milstien, S., Cahill, A. L., and Kang, U. J. (1994). Regulation of L-DOPA production by 5,6,7,8-tetrahydro-L-biopterin in genetically modified cells. *Soc. Neurosci. Abstr.* **20**, 176. [Abstract]

Bender, M. A., Palmer, T. D., Gelinas, R. E., and Miller, A. D. (1987). Evidence that the packaging signal of Moloney murine leukemia virus extends into the *gag* region. *J. Virol.* **61**, 1639–1646.

Berkemeier, L. R., Winslow, J. W., Kaplan, D. R., Nikolics, K., Goeddel, D. V., and Rosenthal, A. (1991). Neurotrophin-5: A novel neurotrophic factor that activates trk and trkB. *Neuron* **7**, 857–866.

Bernheimer, H., Birkmayer, W., Hornykiewicz, O., Jellinger, K., and Seitelberger, F. (1973). Brain dopamine and the syndromes of Parkinson and Huntington: Clinical, morphological and neurochemical correlations. *J. Neurol. Sci.* **20**, 415–455.

Boris-Lawrie, K. A. and Temin, H. M. (1993). Recent advances in retrovirus vector technology. *Curr. Opin. Genet. Dev.* **3**, 102–109.

Breakefield, X. O. (1993). Gene delivery into the brain using virus vectors. *Nature Genet.* **3**, 187–189.

Brundin, P., Barbin, G., Strecker, R. E., Isacson, O., Prochiantz, A., and Björklund, A. (1988). Survival and function of dissociated rat dopamine neurones grafted at different developmental stages or after being cultured in vitro. *Dev. Brain Res.* **39**, 233–243.

Cepko, C. P., Roberts, B. E., and Mulligan, R. C. (1984). Construction and application of highly transmissible murine retrovirus shuttle vector. *Cell* **37**, 1053–1062.

Civelli, O., Bunzow, J. R., Grandy, D. K., Zhou, Q. Y., and Vantol, H.H.M. (1991). Molecular biology of the dopamine receptors. *Eur. J. Pharmacol-Molec. Pharm.* **207**, 277–286.

Cone, R., and Mulligan, R. (1984). High-efficiency gene transfer into mammalian cells: Generation of helper-free recombinant retrovirus with broad mammalian host range. *Proc. Natl. Acad. Sci. USA* **81**, 6349–6353.

Cotzias, G. C., Papavasiliou, P. S., and Gellene, R. (1969). Modification of parkinsonism: Chronic treatment with L-dopa. *N. Engl. J. Med.* **280**, 337–345.

Cunningham, L. A., Hansen, J. T., Short, M. P., and Bohn, M. C. (1991). The use of genetically altered astrocytes to provide nerve growth factor to adrenal chromaffin cells grafted into the striatum. *Brain Res.* **561**, 192–202.

Curiel, D. T. (1994). High-efficiency gene transfer employing adenovirus–polylysine–DNA complexes. *Nat. Immun.* **13**, 141–164.

Dai, Y., Roman, M., Naviaux, R. K., and Verma, I. M. (1992). Gene therapy via primary myoblasts: long-term expression of factor IX protein following transplantation in vivo. *Proc. Natl. Acad. Sci. USA* **89**, 10892–10895.

Danos, O., and Mulligan, R. C. (1988). Safe and efficient generation of recombinant retroviruses with amphotropic and ecotropic host ranges. *Proc. Natl. Acad. Sci. USA* **85**, 6460–6464.

Davis, H. L., Demeneix, B. A., Quantin, B., Coulombe, J., and Whalen, R. G. (1993). Plasmid DNA is superior to viral vectors for direct gene transfer into adult mouse skeletal muscle. *Hum. Gene Ther.* **4**, 733–740.

Davis, K. L., Thal, L. J., Gamzu, E. R., Davis, C. S., Woolson, R. F., Gracon, S. I., Drachman, D. A., Schneider, L. S., Whitehouse, P. J., Hoover, T. M., Morris, J. C., Kawas, C. H., Knopman, D. S., Earl, N. L., Kumar, V., Doody, R. S., and The Tacrine Collaborative Study Group (1992). A double-blind, placebo-controlled multicenter study of tacrine for Alzheimer's disease. *N. Engl. J. Med.* **327**, 1253–1259.

Dunnett, S. B., Hernandez, T. D., Summerfield, A., Jones, G. H., and Arbuthnott, G. (1988). Graft-derived recovery from 6-OHDA lesions: Specificity of ventral mesencephalic graft tissues. *Exp. Brain Res.* **71**, 411–424.

Emerman, M., and Temin, H. M. (1984). Genes with promoters in retrovirus vectors can be independently suppressed by an epigenetic mechanism. *Cell* **39**, 459–467.

Emerman, M., and Temin, H. M. (1986). Comparison of promoter suppression in avian and murine retrovirus vectors. *Nucleic Acids Res.* **14**, 9381–9396.

Espinosa de los Monteros, A., Zhang, M., and De Vellis, J. (1993). O2A progenitor cells transplanted into the neonatal rat brain develop into oligodendrocytes but not astrocytes. *Proc. Natl. Acad. Sci. USA* **90**, 50–54.

Eves, E. M., Tucker, M. S., Roback, J. D., Downen, M., Rosner, M. R., and Wainer, B. H. (1992). Immortal rat hippocampal cell lines exhibit neuronal and glial lineages and neurotrophin gene expression. *Proc. Natl. Acad. Sci. USA* **89**, 4373–4377.

Factor, S. A., and Weiner, W. J. (1993). Viewpoint: Early combination therapy with bromocriptine and levodopa in Parkinson's disease. *Movement Disorders* **8**, 257–262.

Felgner, P. L., and Rhodes, G. (1991). Gene therapeutics. *Nature* **349**, 351–352.

Fisher, L. J., Jinnah, H. A., Kale, L. C., Higgins, G. A., and Gage, F. H. (1991). Survival and function of intrastriatally grafted primary fibroblasts genetically modified to produce L-DOPA. *Neuron* **6**, 371–380.

Fisher, L. J., Schinstine, M., Salvaterra, P., Dekker, A. J., Thal, L., and Gage, F. H. (1993). *In vivo* production and release of acetylcholine from primary fibroblasts genetically modified to express choline acetyltransferase. *J. Neurochem.* **61**, 1323–1332.

Freed, C. R., Breeze, R. E., Rosenberg, N. L., Schneck, S. A., Kriek, E., Qi, J.-X., Lone, T., Zhang, Y.-B., Snyder, J. A., Wells, T. H., Ramig, L. O., Thompson, L., Mazziotta, J. C., Huang, S. C., Grafton, S. T., Brooks, D., Sawle, G., Schroter, G., and Ansari, A. A. (1992). Survival of implanted fetal dopamine cells and neurologic improvement 12 to 46 months after transplantation for Parkinson's disease. *N. Engl. J. Med.* **327**, 1549–1555.

Freed, W. J., Poltorak, M., and Becker, J. B. (1990). Intracerebral adrenal medulla grafts: a review. *Exp. Neurol.* **110**, 139–166.

Frim, D. M., Short, M. P., Rosenberg, W. S., Simpson, J., Breakefield, X. O., and Isacson, O. (1993). Local protective effects of nerve growth factor-secreting fibroblasts against excitotoxic lesions in the rat striatum. *J. Neurosurg.* **78**, 267–273.

Frim, D. M., Uhler, T. A., Galpern, W. R., Beal, M. F., Breakefield, X. O., and Isacson, O. (1994). Implanted fibroblasts genetically engineered to produce brain-derived neurotrophic factor prevent 1-methyl-4-phenylpyridinium toxicity to dopaminergic neurons in the rat. *Proc. Natl. Acad. Sci. USA* **91**, 5104–5108.

Gage, F. H., Wolff, J. A., Rosenberg, M. B., Xu, L., Yee, J.-K., Shults, C., and Friedmann, T. (1987). Grafting genetically modified cells to the brain: possibilities for the future. *Neuroscience* **23**, 795–807.

Gage, F. H., Kang, U. J., and Fisher, L. J. (1991a). Intracerebral grafting in the dopaminergic system: Issues and controversy. *Curr. Opin. Neurobiol.* **1**, 414–419.

Gage, F. H., Kawaja, M. D., and Fisher, L. J. (1991b). Genetically modified cells: applications for intracerebral grafting. *Trends Neurosci.* **14**, 328–333.

Geller, A. I., During, M. J., and Neve, R. L. (1991). Molecular analysis of neuronal physiology by gene transfer into neurons with herpes simplex virus vectors. *Trends Neurosci.* **14**, 428–432.

Ghattas, I. R., Sanes, J. R., and Majors, J. E. (1991). The encephalomyocarditis virus internal ribosome entry site allows efficient coexpression of two genes from a recombinant provirus in cultured cells and in embryos. *Mol. Cell. Biol.* **11**, 5848–5859.

Giordano, M., Takashima, H., Herranz, A., Poltorak, M., Geller, H. M., Marone, M., and Freed, W. J. (1993). Immortalized GABAergic cell lines derived from rat striatum using a temperature-sensitive allele of the SV40 large T antigen. *Exp. Neurol.* **124**, 395–400.

Goldberg, W. J., and Bernstein, J. J. (1988). Fetal cortical astrocytes migrate from cortical homografts throughout the host brain and over the glia limitans. *J. Neurosci. Res.* **20**, 38–45.

Hantraye, P., Riche, D., Maziere, M., and Isacson, O. (1992). Intrastriatal transplantation of cross-species fetal striatal cells reduces abnormal movements in a primate model of Huntington disease. *Proc. Natl. Acad. Sci. USA* **89**, 4187–4191.

Hatton, J. D., Garcia, R., and Sang U. H. (1992). Migration of grafted rat astrocytes: Dependence on source/target organ. *Glia* **5**, 251–258.

Hefti, F., Melamed, E., and Wurtman, R. J. (1981). The site of dopamine formation in rat striatum after L-dopa administration. *J Pharmacol. Exp. Ther.* **217**, 189–197.

Hefti, F., Hartikka, J., and Knüsel, B. (1989). Function of neurotrophic factors in the adult and aging brain and their possible use in the treatment of neurodegenerative diseases. *Neurobiol. Aging* **10**, 515–533.

Hirsh, E. C., Duyckaerts, C., Javoy-Agid, F., Hauw, J.-J., and Agid, Y. (1990). Does adrenal graft enhance recovery of dopaminergic neurons in Parkinson's disease? *Ann. Neurol.* **27**, 676–682.

Hohn, A., Leibrock, J., Bailey, K., and Barde, Y. A. (1990). Identification and characterization of a novel member of the nerve growth factor/brain-derived neurotrophic factor family. *Nature* **344**, 339–341.

Horellou, P., Brundin, P., Kalen, P., Mallet, J., and Björklund, A. (1990a). *In vivo* release of DOPA and dopamine from genetically engineered cells grafted to the denervated rat striatum. *Neuron* **5**, 393–402.

Horellou, P., Marlier, L., Privat, A., and Mallet, J. (1990b). Behavioral effect of engineered cells that synthesize L-DOPA or dopamine after grafting into the rat neostriatum. *Eur. J. Neurosci.* **2**, 116–119.

Hornykiewicz, O., and Kish, S. J. (1986). Biochemical pathophysiology of Parkinson's disease. *Adv. Neurol.* **45**, 19–34.

Hyman, C., Hofer, M., Barde, Y.-A., Juhasz, M., Yancopoulos, G. D., Squinto, S. P., and Lindsay, R. M. (1991). BDNF is a neurotrophic factor for dopaminergic neurons of the substantia nigra. *Nature* **350**, 230–232.

Ip, N. Y., Ibanez, C. F., Nye, S. H., Mcclain, J., Jones, P. F., Gies, D. R., Belluscio, L., Lebeau, M. M., Espinosa, R., Squinto, S. P., Persson, H., and Yancopoulos, G. D. (1992). Mammalian neurotrophin-4: Structure, chromosomal localization, tissue distribution, and receptor specificity. *Proc. Natl. Acad. Sci. USA* **89**, 3060–3064.

Jang, S. K., Kräusslich, H.-G., Nicklin, M. J. H., Duke, G. M., Palmenberg, A. C., and Wimmer, E. (1988). A segment of the 5′ nontranslated region of encephalomyocarditis virus RNA directs internal entry of ribosomes during in vitro translation. *J. Virol.* **62**, 2636–2643.

Jiao, S., Schultz, E., and Wolff, J. A. (1992). Intracerebral transplants of primary muscle cells—A potential platform for transgeneexpression in the brain. *Brain Res.* **575**, 143–147.

Jiao, S., Gurevich, V., and Wolff, J. A. (1993). Long-term correction of rat model of Parkinson's disease by gene therapy. *Nature* **362**, 450–453.

Jiao, S., and Wolff, J. A. (1992). Long-term survival of autologous muscle grafts in rat brain. *Neurosci. Lett.* **137**, 207–210.

Juorio, A. V., Li, X.-M., Walz, W., and Paterson, I. A. (1993). Decarboxylation of L-Dopa by cultured mouse astrocytes. *Brain Res.* **626**, 306–309.

Kang, U. J., Park, D. H., Wessel, T., Baker, H., and Joh, T. H. (1992). DOPA-decarboxylation in the striata of rats with unilateral substantia nigra lesions. *Neurosci. Lett.* **147**, 53–57.

Kang, U. J., Fisher, L. J., Joh, T. H., O'Malley, K. L., and Gage, F. H. (1993a). Regulation of dopamine production by genetically modified primary fibroblasts. *J. Neurosci.* **13**, 5203–5211.

Kang, U. J., Fisher, L. J., Kuczenski, R., Jinnah, H. A., Joh, T. H. and Gage, F. H. (1993b). Grafting genetically modified cells in a rat model of Parkinson's disease. *Neurology* **43** (Suppl 2), A222. [Abstract]

Kawaja, M. D., Fagan, A. M., Firestein, B. L., and Gage, F. H. (1991). Intracerebral grafting of cultured autologous skin fibroblasts into the rat striatum: An assessment of graft size and ultrastructure. *J. Comp. Neurol.* **307**, 695–706.

Kawaja, M. D., Rosenberg, M. B., Yoshida, K., and Gage, F. H. (1992). Somatic gene transfer of nerve growth factor promotes the survival of axotomized septal neurons and the regeneration of their axons in adult rats. *J. Neurosci.* **12**, 2849–2864.

Kawaja, M. D., and Gage, F. H. (1991). Reactive astrocytes are substrates for the growth of adult CNS axons in the presence of elevated levels of nerve growth factor. *Neuron* **7**, 1019–1030.

Kawaja, M. D., and Gage, F. H. (1992). Morphological and neurochemical features of cultured primary skin fibroblasts of Fischer-344 rats following striatal implantation. *J. Comp. Neurol.* **317**, 102–116.

Kim, J. W., Closs, E. I., Albritton, L. M., and Cunningham, J. M. (1991). Transport of cationic amino-acids by the mouse ectropic retrovirus recptor. *Nature* **352**, 725–728.

Klein, R., Martin-Zanca, D., Barbacid, M., and Parada, L. F. (1990). Expression of the tyrosine kinase receptor gene trkB is confined to the murine embryonic and adult nervous system. *Development* **109**, 845–850.

Kordower, J. H., Fiandaca, M. S., Notter, M. F. D., Hansen, J. T. and Gash, D. M. (1990). NGF-like trophic support from peripheral nerve for grafted Rhesus adrenal chromaffin cells. *J. Neurosurg.* 73, 418. [Abstract]

Kurlan, R., Nutt, J. G., Woodward, W. R., Rothfield, K., Lichter, D., Miller, C., Carter, J. H., and Shoulson, I. (1988). Duodenal and gastric delivery of levodopa in parkinsonism. *Ann. Neurol.* **23,** 589–595.

La Gamma, E. F., Weisinger, G., Lenn, N. J. and Strecker, R. E. (1993). Genetically modified primary astrocytes as cellular vehicles for gene therapy in the brain. *Cell Transplant.* **2,** 207–214.

Lapchak, P. A., Beck, K. D., Araujo, D. M., Irwin, I., Langston, J. W., and Hefti, F. (1993). Chronic intranigral administration of brain-derived neurotrophic factor produces striatal dopaminergic hypofunction in unlesioned adult rats and fails to attenuate the decline of striatal dopaminergic function following medial forebrain bundle transection. *Neuroscience* **53,** 639–650.

Le Gal La Salle, G., Robert, J. J., Berrard, S., Ridoux, V., Stratford-Perricaudet, L. D., Perricaudet, M., and Mallet, J. (1993). An adenovirus vector for gene transfer into neurons and glia in the brain. *Science* **259,** 988–990.

Leibrock, J., Lottspeich, F., Hohn, A., Hofer, M., Hengerer, B., Masiakowski, P., Thoenen, H., and Barde, Y.-A. (1989). Molecular cloning and expression of brain-derived neurotrophic factor. *Nature* **341,** 149–152.

Li, X.-M., Juorio, A. V., Paterson, I. A., Walz, W., Zhu, M.-Y., and Boulton, A. A. (1992). Gene expression of aromatic L-amino acid decarboxylase in cultured rat glial cells. *J. Neurochem.* **59,** 1172–1175.

Lindvall, O., Widner, H., Rehncrona, S., Brundin, P., Odin, P., Gustavii, B., Frackowiak, R., Leenders, K. L., Sawle, G., Rothwell, J. C., Björklund, A., and Marsden, C. D. (1992). Transplantation of fetal dopamine neurons in Parkinson's disease—One-year clinical and neurophysiological observations in two patients with putaminal ismplants. *Ann. Neurol.* **31,** 155–165.

Lucidi-Phillipi, C. A., Gage, F. H., Shults, C. W., Jones, K. R., Reichardt, L. F., and Kang, U. J. (1995). BDNF-transduced fibroblasts: Production of BDNF and effects of grafting to the adult rat brain. *J. Comp. Neurol.* **354,** 361–376.

Lundberg, C., Horellou, P., Brundin, P., Wictorin, K., Kalen, P., Colin, P., Julien, J. F., Björklund, A., and Mallet, J. (1991). Transplantation of primary glial cells that produce DOPA after retroviral TH gene transfer in the rat model of Parkinson's disease. *Soc. Neurosci. Abstr.* **17,** 570. [Abstract]

Maisonpierre, P. C., Belluscio, L., Squinto, S., Ip, N. Y., Furth, M. E., Lindsay, R. M., and Yancopoulos, G. D. (1990). Neurotrophin-3: A neurotrophic factor related to NGF and BDNF. *Science* **247,** 1446–1451.

Majors, J. E. (1992). Retroviral vectors—Strategies and applications. *Semin. Virol.* **3,** 285–295.

Mann, R., Mulligan, R. C., and Baltimore, D. (1983). Construction of a retrovirus packaging mutant and its use to produce helper-free defective retrovirus. *Cell* **33,** 153–159.

Markowitz, D., Goff, S., and Bank, A. (1988a). Construction and use of a safe and efficient amphotropic packaging cell line. *Virology* **167,** 400–406.

Markowitz, D., Goff, S., and Bank, A. (1988b). A safe packaging line for gene transfer: Separating viral genes on two different plasmids. *J. Virol.* **62,** 1120–1124.

Marsden, C. D., and Parkes, J. D. (1977). Success and problems of long-term levodopa therapy in Parkinson's disease. *Lancet* **1,** 345–349.

Melamed, E., Hefti, F., Liebman, J., Schlosberg, A. J., and Wurtman, R. J. (1980). Serotonergic neurons are not involved in action of L-DOPA in Parkinson's disease. *Nature* **283,** 772–774.

Melamed, E., Hefti, F., Pettibone, D. J., Liebman, J., and Wurtman, R. J. (1981). Aromatic L-amino acid decarboxylase in rat corpus striatum: Implications for action of L-DOPA in parkinsonism. *Neurology* **31,** 651–655.

Middlemas, D. S., Lindberg, R. A., and Hunter, T. (1991). trkB, a neural receptor protein-tyrosine kinase: evidence for a full-length and two truncated receptors. *Mol. Cell. Biol.* **11,** 143–153.

Miller, A. D., Jolly, D. J., Friedmann, T., and Verma, I. M. (1983). A transmissible retrovirus expressing human hypoxanthine phosphoribosyltransferase (HPRT): Gene transfer into cells obtained from humans deficient in HPRT. *Proc. Natl. Acad. Sci. USA* **80**, 4709–4713.

Miller, A. D., and Buttimore, C. (1986). Redesign of retrovirus packaging cell lines to avoid recombination leading to helper virus production. *Mol. Cell. Biol.* **6**, 2895–2902.

Miller, A. D., and Rosman, G. J. (1989). Improved retroviral vectors for gene transfer and expression. *BioTechniques* **7**, 980–989.

Miller, A. D. (1992). Human gene therapy comes of age. *Nature* **357**, 455–460.

Miller, D. G., Adam, M. A., and Miller, A. D. (1990). Gene transfer by retrovirus vectors occurs only in cells that are actively replicating at the time of infection. *Mol. Cell. Biol.* **10**, 4239–4242.

Misawa, H., Takahashi, R., and Deguchi, T. (1994). Calcium-independent release of acetylcholine from stable cell lines expressing mouse choline acetyltransferase cDNA. *J. Neurochem.* **62**, 465–470.

Morassutti, D. J., Staines, W. A., Magnuson, D. S. K., Marshall, K. C., and McBurney, M. W. (1994). Murine embryonal carcinoma-derived neurons survive and mature following transplantation into adult rat striatum. *Neuroscience* **58**, 753–763.

Mulligan, R. C. (1993). The basic science of gene therapy. *Science* **260**, 926–932.

Naffakh, N., Pinset, C., Montarras, D., Pastoret, C., Danos, O., and Heard, J. M. (1993). Transplantation of adult-derived myoblasts in mice following gene transfer. *Neuromuscular Disorders* **3**, 413–417.

Noble, M., Groves, A. K., Ataliotis, P., and Jat, P. S. (1992). From chance to choice in the generation of neural cell lines. *Brain Pathol.* **2**, 39–46.

Onifer, S. M., Whittemore, S. R., and Holets, V. R. (1993). Variable morphological differentiation of a raphé-derived neuronal cell line following transplantation into the adult rat CNS. *Exp. Neurol.* **122**, 30–42.

Owens, G. C., Johnson, R., Bunge, R. P., and O'Malley, K. L. (1991). L-3,4-dihydroxyphenylalanine synthesis by genetically modified Schwann cells. *J. Neurochem.* **56**, 1030–1036.

Palmer, T. D., Rosman, G. J., Osborne, W. R. A., and Miller, A. D. (1991). Genetically modified skin fibroblasts persist long after transplantation but gradually inactivate introduced genes. *Proc. Natl. Acad. Sci. USA* **88**, 1330–1334.

Pleasure, S. J., Page, C., and Lee, M.-Y. (1992). Pure, postmitotic, polarized human neurons derived from NTera 2 cells provide a system for expressing exogenous proteins in terminally differentiated neurons. *J. Neurosci.* **12**, 1802–1815.

Quinn, N., Parkes, J. D., and Marsden, C. D. (1986). Control of on/off phenomenon by continuous intravenous infusion of levodopa. *Neurology* **34**, 1131–1136.

Rando, T. A., and Blau, H. M. (1994). Primary mouse myoblast purification, characterization, and transplantation for cell-mediated gene therapy. *J. Cell Biol.* **125**, 1275–1287.

Ray, J., Peterson, D. A., Schinstine, M., and Gage, F. H. (1993). Proliferation, differentiation, and long-term culture of primary hippocampal neurons. *Proc. Natl. Acad. Sci. USA* **90**, 3602–3606.

Reiner, A., Albin, R. L., Anderson, K. D., D'Amato, C. J., Penney, J. B., and Young, A. B. (1994). Differential loss of striatal projection neurons in Huntington's disease. *Proc. Natl. Acad. Sci. USA* **85**, 5733–5737.

Renfranz, P. J., Cunningham, M. G., and McKay, D. G. (1991). Region-specific differentiation of the hippocampal stem cell line HiB5 upon implantation into the developing mammalian brain. *Cell* **66**, 713–729.

Reynolds, B. A., Tetzlaff, W., and Weiss, S. (1992). A multipotent EGF-responsive striatal embryonic progenitor cell produces neurons and astrocytes. *J. Neurosci.* **12**, 4565–4574.

Ridoux, V., Robert, J. J., Zhang, X., Perricaudet, M., Mallet, J., and Le Gal La Salle, G. (1994). The use of adenovirus vectors for intracerebral grafting of transfected nervous cells. *NeuroReport* **5**, 801–804.

Roemer, K., Johnson, P. A., and Friedmann, T. (1991). Activity of the simian virus 40 early promoter–enhancer in herpes simplex virus type 1 vectors is dependent on its position, the infected cell type, and the presence of Vmw175. *J. Virol.* **65**, 6900–6912.

Rosenberg, M. B., Friedmann, T., Robertson, R. C., Tuszynski, M., Wolff, J. A., Breakefield, X. O., and Gage, F. H. (1988). Grafting genetically modified cells to the damaged brain: restorative effects of NGF expression. *Science* **242**, 1575–1578.

Rudge, J. S., Li, Y., Pasnikowski, E. M., Mattsson, K., Pan, L., Yancopoulos, G. D., Wiegand, S. J., Lindsay, R. M., and Ip, N. Y. (1994). Neurotrophic factor receptors and their signal transduction capabilities in rat astrocytes. *Eur. J. Neurosci.* **6**, 693–705.

Ruppert, C., Sandrasagra, A., Anton, B., Evans, C., Schweitzer, E. S., and Tobin, A. J. (1993). Rat-1 fibroblasts engineered with GAD65 and GAD67 cDNAs in retroviral vectors produce and release GABA. *J. Neurochem.* **61**, 768–771.

Scharfmann, R., Axelrod, J. H., and Verma, I.M. (1991). Long-term in vivo expression of retrovirus-mediated gene transfer in mouse fibroblast implants. *Proc. Natl. Acad. Sci. USA* **88**, 4626–4630.

Schinstine, M., Rosenberg, M. B., Routledgeward, C., Friedmann, T., and Gage, F. H. (1992). Effects of choline and quiescence on Drosophila choline acetyltransferase expression and acetylcholine production by transduced rat fibroblasts. *J. Neurochem.* **58**, 2019–2029.

Schueler, S. B., Ortega, J. D., Sagen, J., and Kordower, J. H. (1993). Robust survival of isolated bovine adrenal chromaffin cells following intrastriatal transplantation: A novel hypothesis of adrenal graft viability. *J. Neurosci.* **13**, 4496–4510.

Schumacher, J. M., Short, M. P., Hyman, B. T., Breakefield, X. O., and Isacson, O. (1991). Intracerebral implantation of nerve growth factor-producing fibroblasts protects striatum against neurotoxic levels of excitatory amino acids. *Neuroscience* **45**, 561–570.

Skaper, S. D., Negro, A., Facci, L., and Dal Toso, R. (1993). Brain-derived neurotrophic factor selectively rescues mesencephalic dopaminergic neurons from 2,4,5-trihydroxyphenylalanine-induced injury. *J. Neurosci. Res.* **34**, 478–487.

Snider, W. D. (1994). Functions of the neurotrophins during nervous system development: What the knockouts are teaching us. *Cell* **77**, 627–638.

Snyder, E. Y., Deitcher, D. L., Walsh, C., Arnold-Aldea, S., Hartwieg, E. A., and Cepko, C. L. (1992). Multipotent neural cell lines can engraft and participate in development of mouse cerebellum. *Cell* **68**, 33–51.

Soriano, P., Friedrich, G., and Lawinger, P. (1991). Promoter interactions in retrovirus vectors introduced into fibroblasts and embryonic stem cells. *J. Virol.* **65**, 2314–2319.

Spina, M. B., Squinto, S. P., Miller, J., Lindsay, R. M., and Hyman, C. (1992). Brain-derived neurotrophic factor protects dopamine neurons against 6-hydroxydopamine and *N*-methyl-4-phenylpyridinium ion toxicity: involvement of the glutathione system. *J. Neurochem.* **59**, 99–106.

Sweet, R. D., McDowell, F. H., Feigenson, J. S., Loranger, A. W., and Goodell, H. (1976). Mental symptoms in Parkinson's disease during chronic treatment with levodopa. *Neurology* **26**, 305–310.

Thoenen, H. (1991). The changing scene of neurotrophic factors. *Trends Neurosci.* **14**, 165–170.

Trojanowski, J. Q., Mantione, J. R., Lee, J. H., Seid, D. P., You, T., Inge, L. J., and Lee V. M.-Y. (1993). Neurons derived from a human teratocarcinoma cell line establish molecular and structural polarity following transplantation into the rodent brain. *Exp. Neurol.* **122**, 283–294.

Uchida, K., Tsuzaki, N., Nagatsu, T., and Kohsaka, S. (1992). Tetrahydrobiopterin-dependent functional recovery in 6-hydroxydopamine-treated rats by intracerebral grafting of fibroblasts transfected with tyrosine hydroxylase cDNA. *Dev. Neurosci.* **14**, 173–180.

Whittemore, S. R., Neary, J. T., Kleitman, N., Sanon, H. R., Benigno, A., Donahue, R. P., and Norenberg, M. D. (1994). Isolation and characterization of conditionally immortalized astrocyte cell lines derived from adult human spinal cord. *Glia* **10**, 211–226.

Wojcik, B. E., Nothias, F., Lazar, M., Jouin, H., Nicolas, J.-F., and Peschanski, M. (1993). Catecholaminergic neurons result from intracerebral implantation of embryonal carcinoma cells. *Proc. Natl. Acad. Sci. USA* **90**, 1305–1309.

Wolff, J. A., Fisher, L. J., Jinnah, H. A., Langlais, P. J., Iuvone, P. M., O'Malley, K. L., Rosenberg, M. B., Shimohama, S., Friedmann, T., and Gage, F. H. (1989). Grafting fibroblasts genetically modified to produce L-dopa in a rat model of Parkinson disease. *Proc. Natl. Acad. Sci. USA* **86**, 9011–9014.

Wolff, J. A., Malone, R. W., Williams, P., Chong, W., Acsadi, G., Jani, A., and Felgner, P. L. (1990). Direct gene transfer into mouse muscle in vivo. *Science* **247**, 1465–1468.

Wu, D. K., and Cepko, C. L. (1994). The stability of endogenous tyrosine hydroxylase protein in PC-12 cells differs from that expressed in mouse fibroblasts by gene transfer. *J. Neurochem.* **62**, 863–872.

CHAPTER 14

Virus Vector-Mediated Transfer of Drug-Sensitivity Genes for Experimental Brain Tumor Therapy

Ming X. Wei
Takashi Tamiya
Xandra O. Breakefield
E. Antonio Chiocca
Neurosurgery Service
Department of Surgery and Molecular Neurogenetics Laboratory
Department of Neurology
Massachusetts General Hospital East
Harvard Medical School
Charlestown, Massachussetts

I. Introduction

Brain tumors are estimated to cause 100,000 deaths in the United States every year. Malignant gliomas are the most common form of brain tumor and represent about 40% of all primary brain tumors (Schoenberg, 1983). They are considered incurable and even multimodal approaches, including surgery, radiation therapy, and chemotherapy, prolong the survival of patients by only a few months (Mahaley *et al.*, 1989). This resistance to surgery is due in part to glioma cell migration and phenotypic heterogeneity, as well as to the poor diffusion of most chemotherapeutic agents across the blood–brain barrier (BBB). Therefore, innovative approaches for the treatment of these tumors should be investigated.

With the arrival of recombinant DNA techniques and the subsequent identification of genes shown to have the potential to kill tumor cells, it is now possible to introduce, by viral vectors and other gene-transfer methods, genes that can kill tumor cells (Mulligan, 1993; Chiocca *et al.*, 1994). Clinical trials, using retrovirus-mediated delivery of the herpes simplex virus thymidine kinase

(HSVtk) gene to confer ganciclovir sensitivity to tumor cells, have been initiated (Oldfield *et al.*, 1993).

Herein, we will summarize current gene therapy strategies that employ virus vectors to deliver drug-sensitivity genes into brain tumor cells.

II. Viral Vectors for Gene Transfer into Brain Tumor Cells

Drug-sensitivity genes encode enzymes that activate prodrugs into their toxic metabolites. In order to transfer these genes into tumor cells *in vivo*, modified viruses and other agents can be employed as vectors (Table 1). Three types of viral vectors have been reported so far to be efficient vectors when applied to experimental brain tumors. They are retrovirus, herpes virus, and adenovirus.

A. Retrovirus Vectors

One major difference between retrovirus, herpes simplex virus, and adenovirus vectors is that with retroviruses, gene expression requires integration into the host cell chromosome and this integration occurs only in dividing cells (Miller and Buttimore, 1986). This property provides for selectivity in the delivery of drug-sensitivity genes into dividing tumor cells since most endogenous cells in the brain are postmitotic. The relative instability of retroviral particles *in vivo* necessitates direct grafting of the mouse fibroblast packaging

Table 1
Gene-Transfer Vehicles

Virus vectors
DNA virus
Herpes simplex virus
Adenovirus
Adenoassociated virus
Vaccinia virus
Epstein–Barr virus
Papova virus
RNA virus
Retrovirus
Nonvirus vectors
Liposome
DNA–polylysine complexes
Lipofection
Direct injection

cells which continually produce retrovirus vectors (Short *et al.*, 1990). Issues of safety have been met by genetically modifying and designing Moloney murine leukemia virus-based (MoMLV) vectors and mouse fibroblast packaging cells (ψ CRE/CRIP) to minimize the risk of recombination after infection, which could result in regeneration of pathogenic replication-competent retrovirus (Miller and Buttimore, 1986; Cepko, 1988; Danos and Mulligan, 1988). Injection of HSVtk retrovirus vector-producing cells into the brain of rats and monkeys has not been associated with significant toxicity to the brain or to remote organs (Ram *et al.*, 1993b). Therefore, retrovirus vectors are relatively safe and effective vehicles for delivery of drug-sensitivity genes for the treatment of brain tumors.

Retrovirus-mediated delivery of a reporter gene, *Escherichia coli* LacZ, into intracerebral gliomas in the brain can be obtained by direct intratumoral injection of retroviral particles, albeit achieving a very low efficiency of gene delivery (only 0.1% of tumor cells). However, it was recently reported that a retrovirus vector, in which the envelope glycoprotein is completely replaced by the G protein of vesicular stomatitis virus, can be concentrated by ultracentrifugation to higher titers ($>10^9$ cfu/ml) (Burns *et al.*, 1993). Intratumoral grafting of the mouse fibroblast packaging cells which continually produce and release the retrovirus vectors bearing the LacZ gene significantly increased the *in vivo* gene-transfer efficiency into tumor cells (Short *et al.*, 1990; Ram *et al.*, 1993a) (Fig. 1). In these studies, the grafted mouse fibroblast cells did not appear to survive more than 1 or 2 weeks in the brain of rats probably because they were rejected. Recently, our laboratory has expanded these studies: grafting of retroviral packaging CRIP cells bearing the LacZ gene in the MFG retrovirus vector (Dranoff *et al.*, 1993) resulted in 30–40% of LacZ-positive C6 glioma cells grafted in the brains of athymic mice (Tamiya *et al.*, 1995). The highest percentage of LacZ-positive tumor cells could be measured 2 or 3 weeks after grafting packaging cells in the brain. It is likely that the LacZ-positive cells found throughout the tumor were daughter cells of tumor cells that were originally infected by retrovirus and that they were migrating. To determine the fate of retrovirus packaging cells and the involvement of the retrovirus, we engineered "packaging" cells derived from CRIP fibroblasts, which expressed the LacZ gene themselves, but did not release retrovirus particles bearing the LacZ gene. The number of LacZ-positive cells in the tumor was only 5%, indicating that retrovirus-mediated gene delivery was needed to achieve a higher percentage of LacZ positivity in the tumor. Retrovirus packaging cells did not appear to significantly migrate in the normal surrounding brain parenchyma or within the tumor. The life span of the mouse fibroblast packaging cells was estimated to be about 3 weeks in the brain of athymic mice.

Retrovirus-mediated transfer of a drug-sensitivity gene, HSVtk, into tumor cells has been reported by several laboratories (Moolten, 1986; Moolten

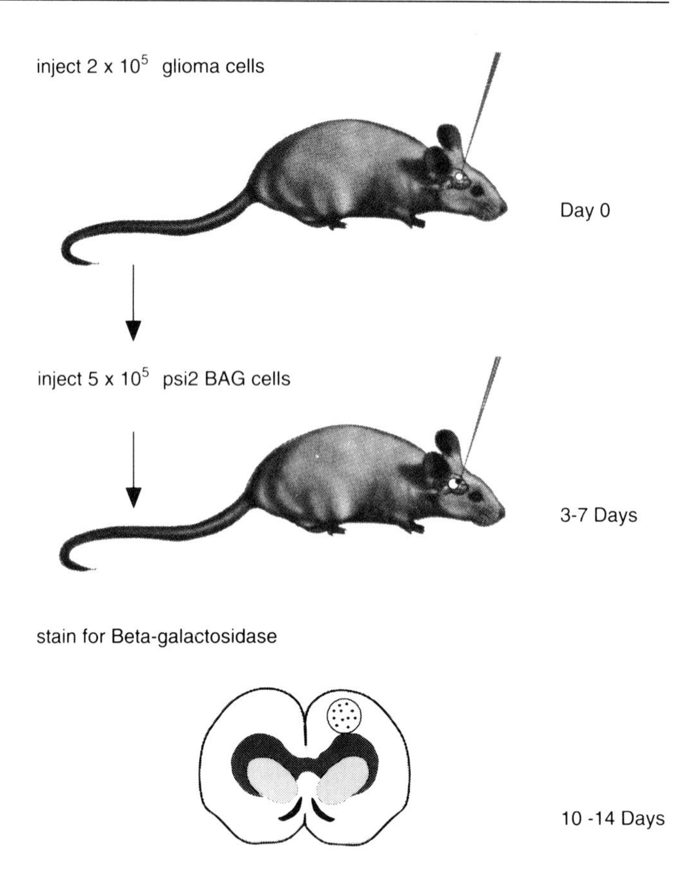

Figure 1 Experimental protocol for gene transfer *in vivo* using retrovirus vectors. Glioma cells are first inoculated stereotactically into the right frontal lobe of rat brains. Three to 5 days later when tumor is established in the brain, retrovirus-producing cells are then stereotactically grafted into the tumor bed. Seven to 10 days later, animals are either sacrificed for LacZ staining or treated with drugs.

et al., 1990; Ezzeddine *et al.*, 1991; Culver *et al.*, 1992; Takamiya *et al.*, 1992; Ram *et al.*, 1993a; Barba *et al.*, 1994). HSVtk phosphorylates nucleoside analogs, such as ganciclovir (GCV) and acyclovir, leading to their incorporation into DNA during its replication (Elion, 1983). The incorporated nucleoside analogs cause DNA strand breaks, resulting in cell death. The first brain tumor model shown to possess therapeutic efficacy was a rat C6 glioma. Rat C6 glioma cells, engineered to express HSVtk, exhibited increased sensitivity to the prodrug, ganciclovir, compared to parental tumor cells. This increased drug-sensitivity was further confirmed in tumor models in which daily intraperitoneal injections were given to mice bearing subcutaneous tumors (Ezzeddine *et al.*, 1991). In addition, a packaging cell line bearing this HSVtk gene also rendered

C6 tumors highly sensitive to ganciclovir *in vivo* (Takamiya *et al.*, 1992). Subsequently, rat 9L gliosarcoma cells were also shown to have acquired increased sensitivity to ganciclovir (Culver *et al.*, 1992; Ram *et al.*, 1993; Barba *et al.*, 1994). Intratumoral grafting of packaging cells releasing a retrovirus vector bearing the HSVtk gene, followed by daily intraperitoneal injections of ganciclovir, resulted in significant tumor regression and prolonged survival of rats bearing intracerebral 9L tumors. Two selective mechanisms of tumor therapy are operative in this model. One consists of selective targeting of replicating DNA strands by ganciclovir metabolites. The other consists of selective retrovirus-mediated gene integration and expression in dividing cells (Miller and Buttimore, 1986). Both of these mechanisms exclude normal nondividing glia and neurons from retrovirus-mediated gene expression and ganciclovir toxicity.

B. Adenovirus Vectors

Although retrovirus vectors have some advantages for treating brain tumors, they also have two disadvantages: (1) virus titers are low, since it is difficult to obtain supernatants which contain more than 10^5 or 10^6 cfu/ml and the actual release *in vivo* can not be easily quantitated, and (2) malignant gliomas have a high percentage of resting cells which are in the G_0 phase of the cell cycle and are thus not infectable by retrovirus (Yoshii *et al.*, 1986; Hoshino *et al.*, 1986). Therefore, other virus vectors which can be grown to high titers and which can also infect nondividing tumor cells might provide a useful alternative.

Replication-defective adenovirus vectors have been employed to deliver the reporter gene, LacZ, into endogenous neural cells, as well as into tumor cells in the brain (Akli *et al.*, 1993; Davidson *et al.*, 1993; Bajocchi *et al.*, 1993; La Gal LaSalle *et al.*, 1993; Boviatsis *et al.*, 1994a). Unlike retrovirus vectors, adenovirus vectors infect all cells in the brain (neurons, astrocytes, ependymal cells, and tumor cells). This vector can be grown to high titers and does not integrate into the host cell genome. Gene expression in cells is transient because the vector's genes are contained in extrachromosomal elements. Selective expression of the drug-sensitive phenotype in tumor cells in the brain requires stereotactic delivery to the tumor bed or use of a tumor-specific promoter. At a low multiplicity of infection, LacZ gene transfer *in vitro* with adenovirus seems lower than that with retrovirus or herpes virus vectors (Boviatsis *et al.*, 1994a). This indicates that apparently lower titers of herpes virus or retrovirus are needed for gene delivery into tumor cells compared to adenovirus.

The issue of safety with replication-defective adenovirus vectors remains to be addressed since these viruses have some neurotoxicity (Breakefield, 1993). Nevertheless, promising results have been obtained for delivery of the HSVtk gene to treat experimental rat gliomas (Chen *et al.*, 1994). In culture, tumor cells can be efficiently infected by a high-titer preparation of a replication-defective recombinant adenovirus bearing the HSVtk gene (Chen *et al.*, 1994; Smythe *et al.*, 1994). Infected tumor cells showed enhanced sensitivity to ganciclovir in a dose-dependent manner. Tumor regression was also achieved in an intracerebral C6 tumor model in athymic mice. Brain tumor volume was reduced by more than 500-fold in ganciclovir-treated compared to untreated animals. Recombinant adenovirus vectors are being used in a number of human clinical trials and may prove useful for the treatment of brain tumors.

C. Herpes Virus Vectors

Two types of herpes virus vectors have been developed for gene delivery to the central nervous system: replication-defective recombinant viruses and plasmid-derived amplicons (Breakefield and DeLuca, 1991; Chiocca *et al.*, 1990, 1994). Replication of HSV results in direct cell toxicity and death. While this property is deleterious for transgene delivery to neurons, it could be exploited for the killing of tumor cells. Therefore, the objective of an effective HSV vector for brain tumor therapy consists of rendering its replication selective for neoplastic cells.

HSV, modified through a deletion in the viral tk gene (TK^-), can only replicate in dividing cells which possess high amounts of the mammalian endogenous TK (Coen *et al.*, 1989). Direct injection of such a mutant virus into intracerebral brain tumors leads to improved survival of animals (Martuza *et al.*, 1991). Postmitotic brain cells, such as neurons and glia, have extremely low levels of endogenous TK making them nonpermissive for replication of TK^- HSV. The destructive ability toward rat 9L gliosarcoma cells was tested by comparing mutants of HSV which have an inactivating insertion of the LacZ gene into the gene loci for ICP0, ICP4 (viral transcription factors), TK, and ribonucleotide reductase (RR), as well as a deletion mutant in gamma 34.5 (neurovirulence factor) (Boviatsis *et al.*, 1994b). HSV vectors defective in TK or RR were more efficient in killing tumor cells than the other mutants. In an intracerebral tumor model, injection of the TK or RR virus mutants into tumor beds resulted in selective expression of the LacZ gene in tumor cells and in tumor necrosis (Boviatsis *et al.*, 1994b; Mineta *et al.*, 1994). Rats, treated with either of these two mutants, lived significantly longer than control animals.

The HSV RR^- mutant retains the ability to confer ganciclovir sensitivity on tumor cells since the viral TK gene is intact. In fact, ganciclovir treatment results in potentiation of the vector's cytotoxic action against 9L gliosarcoma tumors. Approximately 50% of rats treated with the HSV RR^- vector plus ganciclovir survived more than 90 days compared to 20% survival in rats treated with HSV RR^- alone (Boviatsis *et al.*, 1994).

In conclusion, each of the aforementioned virus vectors possess different characteristics which can be exploited in experimental therapeutic strategies. Retroviral vectors are relatively safe and very attractive due to their selective gene delivery to tumor cells in the brain. However, their use as gene-transfer vehicles is limited primarily by their low titer. Adenovirus vectors can partially circumvent this limitation due to their relatively high titers, but they do not possess tumor selectivity. Herpes virus vectors can be grown to high titers and can replicate selectively in brain tumor cells, but issues regarding their safety still need to be resolved (Table 2).

III. Drug-Sensitivity Genes

Several types of genes have been shown to possess antitumor effectiveness: (1) drug-enhancing genes encode enzymes that catalyze the conversion of prodrug into active antineoplastic metabolites; (2) immune-response enhancer genes, which encode factors, such as interleukin-2 (Rosenberg, 1988), interleukin-4 (Yu *et al.*, 1993; Wei *et al.*, 1995), and GM-CSF (Dranoff *et al.*, 1993), increase an immune reaction against tumors; (3) tumor-suppressor genes such as p53 (Cai *et al.*, 1993; Fujiwara *et al.*, 1994), retinoblastoma (Weinberg, 1991) and neurofibromatosis type-2 (Trofatter *et al.*, 1993); and (4) antisense RNAs such as that to insulin-like growth factor I (Trojan *et al.*, 1993).

As mentioned in previous sections, the basic principle in the genetic treatment of experimental gliomas in rodents consists of the intratumoral (*in situ*) grafting of a cell line that produces a retrovirus vector or direct injection of viral vectors that bears a therapeutic gene. The first example of such a gene was HSVtk which encodes an enzyme that converts the prodrug ganciclovir into an active metabolite which can kill dividing cells. Three other genes encoding enzymes which convert prodrugs to drugs have also been shown to possess therapeutic usefulness in animal models of brain tumors (Table 3).

Table 2
A Comparison of Three Viral Vectors Used in Brain Tumor Therapy

Vectors	Particle stability	Packaging cells	Toxicity	Titer	Selectivity for tumor cells	Clinical trials
Retrovirus	−	+	−	Low	+	+
Herpes	+	−	+	High	+	−
Adenovirus	±	−	?	High	−	−

Table 3
Drug-Enhancing Genes

Genes	Origin	Prodrug	Nucleoside precursor[a]	Alkylating agent[a]	Bystander effect	Secretory effect[b]	Clinical trials
HSVtk	Virus	GCV	+	−	+	−	+
E. coli. gpt	Bacterium	6-TX	+	−	+	−	−
CD	Bacterium fungi	5-FC	+	−	?	−	−
P450 2B1	Rat liver	CPA	−	+	+	+	−

Note. HSVtk, herpes simplex virus thymidine kinase; *E. coli.* gpt, *Escherichia coli* guanine phosphoribosyltransferase; CD, cytosine deaminase; GCV, ganciclovir; 6-TX, 6-thioxanthine; 5-FC, 5-fluorocytosine; CPA, cyclophosphamide.
[a] Type of activated metabolites.
[b] In P450 2B1/CPA system, the killing of transgene-negative tumor cells is also mediated through the conditioned medium of transgene-positive tumor cells treated with CPA (Wei *et al.*, unpublished results).

A. HSVtk Gene

As discussed previously, this gene encodes an enzyme which converts ganciclovir into a phosphorylated nucleoside analogue that is then incorporated into DNA during its replication, ultimately resulting in mutations, genomic dysfunction, and cell death. A similar viral tk gene from Varicella Zoster has also been shown to confer ganciclovir susceptibility to tumor cells (Huber *et al.*, 1991).

The "bystander" killing of transgene-negative tumor cells grown in proximity to ganciclovir-exposed, transgene-positive tumor cells has been shown to be an important part of the antitumor potential of the HSVtk/ganciclovir system (Moolten, 1986; Moolten, 1990; Takamiya *et al.*, 1993; Culver *et al.*, 1992; Bi *et al.*, 1993; Freeman *et al.*, 1993). The presence of this bystander killing effect is important in achieving complete regression of brain tumors since none of the viral vectors described above are able to transduce 100% of tumor cells in the brain. Four potential mechanisms have been invoked to explain this phenomenon: (1) development of antitumor immunity (Chiocca *et al.*, 1994; Barba *et al.*, 1994). This immunity might be against viral antigens and/or against tumor-specific antigens; (2) transfer of phosphorylated ganciclovir moieties from HSVtk-positive cells to HSVtk-negative cells through gap junctions (Moolten, 1986; Bi *et al.*, 1993; Culver *et al.*, 1992); (3) direct cell-to-cell transfer of phosphorylated ganciclovir moieties through apoptotic vesicles (Freeman *et al.*, 1993); and (4) destruction of newly formed blood vessels in the tumor bed by retroviral infection of proliferating endogenous cells and their subsequent death through acquisition of ganciclovir susceptibility (Ram *et al.*, 1994).

It is likely that the *in vivo* mechanism of the bystander effect involves a combination of all of the above. Nevertheless, it was reported that development of antitumor immunity alone can effect the long-term survival of previously treated animals (Barba *et al.*, 1994). After intracerebral implantation of 9L gliosarcoma cells, the grafting of mouse fibroblasts producing retrovirus vectors that bear the HSVtk gene followed by ganciclovir treatment resulted in significant tumor regression in the brain and long-term survival of treated animals. These animals were then able to resist a second challenge with parental 9L tumor cells. Immunohistological analysis of the brain sections showed an infiltration of macrophages/microglia and CD8-positive T cells in and around residual tumor cells. However, it is not clear at present whether this protective immunity is specific to the HSVtk/ganciclovir-treated death of the 9L tumor cells or whether it can be replicated with other types of tumor cells and/or other types of therapy.

B. *Escherichia coli* Cytosine Deaminase (CD) Gene

CD is normally present in some bacteria and fungi, but is not present in normal mammalian cells. Its normal function is to deaminate cytosine into uracil. Based on this, a cytosine analog, 5-fluorocytosine (5-FC), was developed as a prodrug. 5-FC can be converted by CD into highly toxic 5-fluorouracil which causes nucleotide strand breaks (Mullen *et al.*, 1992, 1994; Huber *et al.*, 1993; Austin and Huber, 1993; Harris *et al.*, 1994). The transfer of the CD gene into mammalian cells renders them selectively sensitive to 5-FC. A variety of tumor cells (colorectal carcinoma, adenocarcinoma, and fibrosarcoma), engineered to express the CD gene, have been shown to acquire sensitivity to 5-FC treatment both *in vitro* and *in vivo*. It has been shown recently that animals whose CD-positive tumors were eliminated by 5-FC treatment developed resistance to further challenges with wild-type tumor cells, presumably through an immune mechanism (Mullen *et al.*, 1994). It is likely that the CD/5-FC paradigm will also be useful in the treatment of experimental brain tumors.

C. *Escherichia coli* GPT Gene

In a manner similar to the previous systems, this gene encodes an enzyme, xanthine–guanine phosphoribosyltransferase, which converts the xanthine analogue, 6-thioxanthine (6-TX) to a nucleoside precursor. This nucleoside can be incorporated into replicating DNA strands (Besnard *et al.*, 1987), subsequently leading to cell death. One interesting aspect of this enzyme is that drugs are available to select both for and against it. Therefore, the gpt gene can be employed both as a positive and as a negative selection marker. K3T3 sarcoma cells, transfected with the gpt gene, were selected for with a drug regimen containing mycophenolic acid and xanthine (Mulligan and Berg, 1980, 1981). At the same time, gpt-positive tumor cells were 18 to 86 times more sensitive to 6-TX than parental tumor cells *in vitro*. This increased drug sensitivity was also confirmed *in vivo* by using a subcutaneous tumor model (Mroz and Moolten, 1993).

This system, gpt/6-TX, has also been employed in our laboratory to treat C6 glioma tumor cells (T. Tamiya and E. A. Chiocca, unpublished results). A retrovirus bearing the xgpt gene was used to infect C6 tumor cells. After selecting with mycophenolic acid and xanthine, gpt-positive C6 cells (C6-gpt) became more sensitive to 6-TX *in vitro*. In a subcutaneous tumor model, C6-gpt tumor growth was significantly inhibited by the daily intraperitoneal injection of 6-TX. Based on these results, this treatment was further tested in the brain. Three days after injection of 1000 C6 or C6-gpt cells in the right frontal lobe of athymic mouse brains, 6-TX was given by daily intraperitoneal injection. A significant increase in the long-term survival of animals bearing C6-gpt tumors treated with 6-TX occurred when compared to saline-treated animals bearing C6-gpt tumors. In addition, animals bearing C6 tumors showed no significant difference in survival among those treated with 6-TX versus saline. These results indicated that 6-TX treatment can kill C6 glioma cells which express the gpt gene both *in vitro* and *in vivo*. Further experiments are under way to test viral vector delivery of the gpt gene for the treatment of experimental brain tumors.

One potential difficulty common to these three drug-sensitivity genes is that they convert prodrugs into nucleoside analogues which need to be incorporated into DNA during its replication, and thus exert their toxic effects only on cells that are dividing at the time of drug treatment. Since malignant gliomas possess a large percentage of resting cells in the G_0 phase of the cell cycle (Yoshii *et al.*, 1986; Hoshino *et al.*, 1986), regimens employing the above gene therapy paradigms might fail. Therefore, an alternative approach for malignant brain tumors might employ prodrugs which are able to target tumor cells in all phases of the cell cycle.

D. Cytochrome P450 2B1 Gene

With the above considerations in mind, we have focused our attention on one alkylating agent, phosphoramide mustard (PM), which is the active metabolite of the chemotherapeutic agent, cyclophosphamide (CPA) (Clarke and Waxman, 1989). Alkylating agents provoke deleterious interstrand covalent links in DNA in all phases of the cell cycle (Sladek, 1973, 1987; Colvin and Hilton, 1981).

CPA itself is a biologically inactive prodrug, which must be metabolized by a liver-specific enzyme (cytochrome P450 2B1) into its active metabolite (4-hydroxy CPA) to exert its antitumor effect (Clarke and Waxman, 1989; LeBlanc and Waxman, 1990). This metabolite is unstable and spontaneously undergoes ring opening to form aldophosphamide mustard which then spontaneously decomposes to give PM and acrolein. The former binds covalently to DNA and causes DNA interstrand cross-links which lead to cell death during DNA replication (Sladek, 1973, 1987; LeBlanc and Waxman, 1989; Dearfield *et al.*, 1986) (Fig. 2). The lipophilicity of 4-hydroxy CPA should also allow its diffusion into adjacent tumor cells, which provides a way to kill tumor cells that do not express the enzyme. CPA has proven to be effective for numerous neoplasms, but has displayed limited activity against gliomas (Genka *et al.*, 1990; Colvin and Hilton, 1981; Sladek *et al.*, 1984; Sladek, 1988; Dedrick and Morrison, 1992). Due to its increased hydrophilicity with respect to CPA,

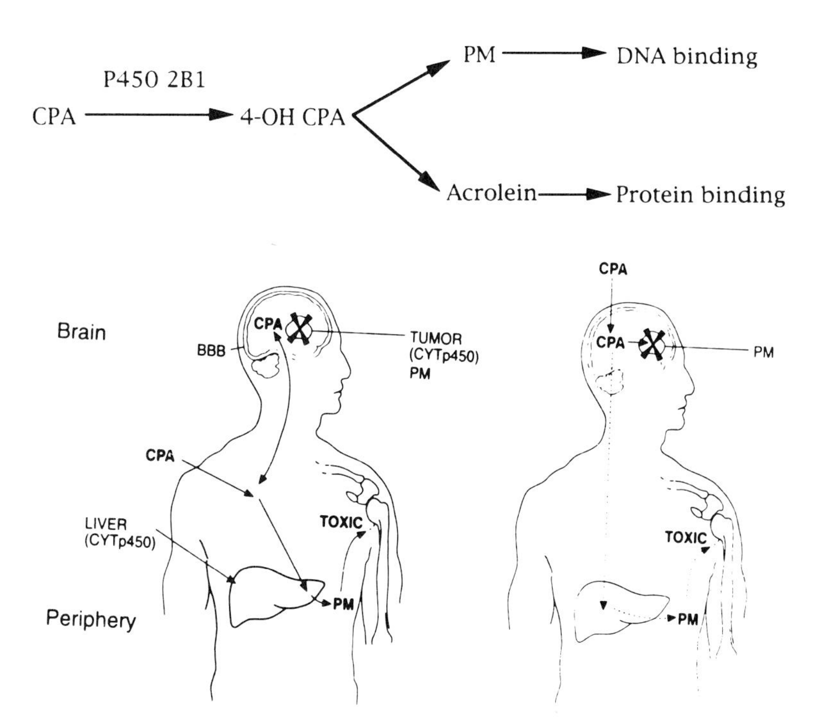

Figure 2 Gene therapy strategy using cytochrome P450 2B1. The top panel illustrates the biotransformation of the prodrug CPA into its active metabolites by the hepatic enzyme cytochrome P450 2B1. The bottom panel illustrates the strategy of using CPA to treat brain tumors. The hepatic enzyme cytochrome P450 2B1 was transduced into brain tumor cells by retrovirus vectors. The transduced tumor cells then become sensitive to CPA administered by intratumoral or intraperitoneal injection.

4-hydroxy CPA crosses the BBB at a very low rate and PM, which is very hydrophilic, is essentially blocked from entry (Fuchs *et al.*, 1990; Arndt *et al.*, 1987, 1988). It is not possible to augment the intravenous dosing of CPA in patients because of its systemic toxicity. We thus reasoned that glioma tumor cells engineered to express the hepatic-converting enzyme might become extremely sensitive to CPA through the *in situ* generation of toxic metabolites. The transfected glioma cells can also provide an intrathecal source of production of the hydroxylated active metabolite of CPA, which can destroy adjacent tumor cells even if they have not been transduced by retroviral vectors (Fig. 2). The cytochrome P450 2B1/CPA therapy has the advantage that tumor cells can be killed independently of the phase in the cell cycle during which the drug is administered. Even tumor cells which are not dividing at the time of CPA treatment are susceptible because PM provokes covalent links in DNA chains, disrupting DNA replication if the cell begins to divide. In this sense, this study represents an innovative gene therapy approach for the treatment of brain tumors.

A rat cytochrome P450 2B1 cDNA (Monier *et al.*, 1988), inserted into the pMFG retroviral vector (Dranoff *et al.*, 1993), was transfected into C6 glioma cells (Wei *et al.*, 1994). Stable clones were characterized for expression of cytochrome P450 2B1 by Western blot analysis and enzyme activity. CPA-mediated inhibition of proliferation in culture was then assayed for the clone that possessed the highest enzyme activity (named C450-8), for a clone that did not exhibit activity (named Cneo-1), and for parental tumor cells, C6. The number of actively proliferating C450-8 cells was reduced 50% by 27 μM CPA in 4 days, whereas this dose had no effect on Cneo-1 or on parental C6 cells even with higher concentrations (up to 1.0 mM) (Wei *et al.*, 1994) (Fig. 3). These results indicate that glioma cells transfected with the cytochrome P450 2B1 cDNA had acquired chemosensitivity to the prodrug CPA.

A bystander killing effect was present in the P450 2B1/CPA system as well. We found that when 10% of tumor cells in culture contained the P450 2B1 gene, there was a 75% inhibition of cell proliferation (Wei *et al.*, unpublished results).

The CPA susceptibility of tumors formed by C6 or C450-8 cells was assessed *in vivo* by subcutaneous growth in athymic mice. C6 or C450-8 cells were injected into the flanks of athymic mice. After tumors were established, animals were injected twice (Days 3 and 14) either with saline as a control or with 2 mg of CPA (intratumorally or intraperitoneally). Tumor growth was not significantly different for C6 or C450-8 cells when treated with saline. After 17 days, C450-8 tumors grew to an average volume of 0.45 cm^3, while C6 tumors grew to 0.75 cm^3 ($P > 0.1$). Treatment of animals with CPA significantly inhibited C450-8 tumor growth. C6 tumor volume was reduced by CPA 8.3-fold compared to untreated tumors at 17 days, whereas C450-8 tumor volume was reduced 22.5-fold following intramural injection ($P < 0.03$). When

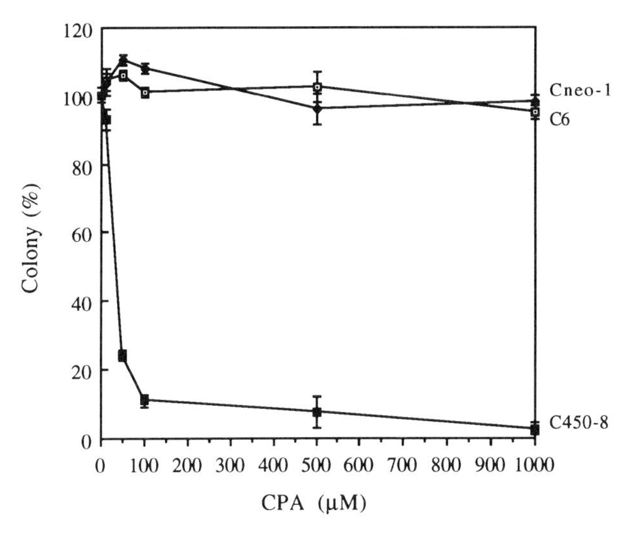

Figure 3 CPA sensitivity of rat C6 glioma cells. A rat glioma clone, C450-8, was derived from C6 by stable transfection with an expression plasmid containing the cytochrome P450 2B1 cDNA and the neoR gene. Clone Cneo-1 was used as a control since it was neomycin resistant but did not exhibit CPA susceptibility. Various concentrations of CPA were added to 1000 cells/10 cm dish the day after plating. The colony percentage represents the number of colonies in the dishes with a particular dose of CPA divided by the number in the dishes with no drug. Experiments were performed in triplicate and the mean ± SEM was shown.

CPA was administered intraperitoneally, the tumor volume of C6 was reduced 8.3-fold at 17 days, whereas that of C450-8 was reduced 45-fold ($P < 0.01$). These results indicated that peripheral C450-8 tumors had acquired additional chemosensitivity *in vivo* to CPA compared to C6 tumors (Wei *et al.*, 1994).

A producer cell line releasing retrovirus which bears the cytochrome P450 2B1 gene was generated from murine CRE fibroblasts (Danos and Mulligan, 1988). This line, named R450-2, was selected because of its chemosensitivity to CPA and its relatively elevated cytochrome P450 2B1 activity (Wei *et al.*, 1994). C6 glioma cells were inoculated stereotactically into athymic mouse brains, followed 3 days later by stereotactic inoculations of either 2.5×10^6 CRELacZ cells as a control (kindly provided by Dr. R. Mulligan, MIT) or 2.5×10^6 R450-2 cells (Fig. 1). Seven days later, CPA (0.3 mg in 1 μl) was injected three times every 4 days, through the same skull opening, into the tumor and meningeal space of mice from both CRELacZ and R450-2 groups. Ten days later, animals were sacrificed. Extensive tumor was found in the meningeal covering of the brains of eight/eight animals in the control group (Fig. 4A). In contrast, seven/eight animals in the R450-2 treatment group showed no evidence of meningeal tumor (Fig. 4B). Histological analysis of six brains from each group revealed minimal tumor necrosis in the parenchymal

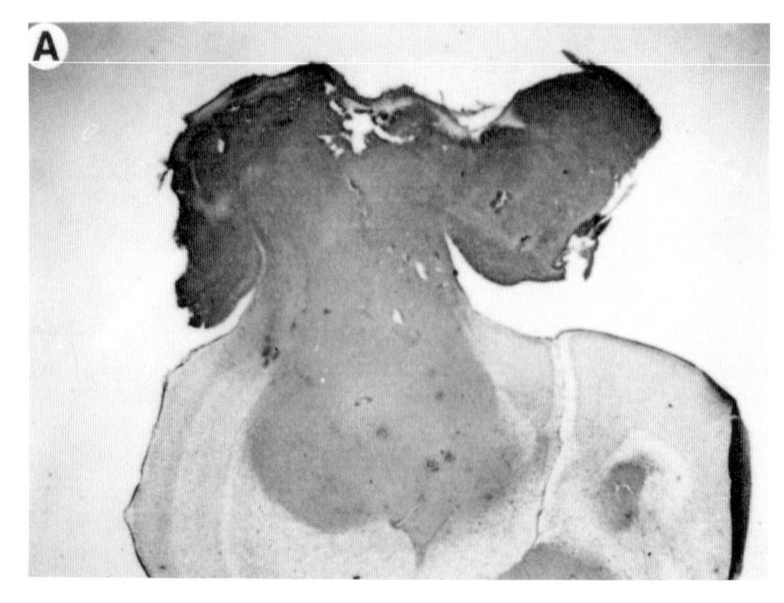

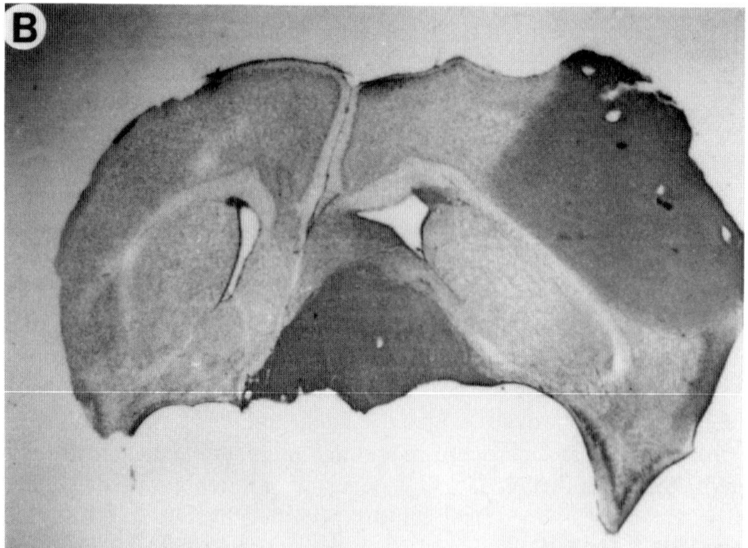

Figure 4 Inhibition of brain tumor growth. C6 glioma cells (1000) were injected into the right frontal lobe of nude mouse brains. Three days later, 2.5×10^6 LacZ (A) or cytochrome P450 2B1 (B) retrovirus-producer fibroblasts were injected into the tumor mass, with subsequent intratumoral administration of CPA. The histological coronal sections from nude mouse brains are shown. The extensive tumor tissue is marked by the dark area. (Reproduced with permission from Wei *et al.*, 1994.)

tumor mass of control animals and extensive tumor necrosis in two/six brains from the treatment group with minimal necrosis in the other four brains. We calculated the intracerebral tumor volumes from serial tissue sections using computerized image analysis. Brain tumor volumes in three/six animals in the treatment group were approximately one-twentieth, one-fifth, and one-half that of the average brain tumor volume in mice in the control group (Wei *et al.*, 1994). These promising results suggested that the cytochrome P450 2B1/CPA might constitute a novel gene therapy approach for brain tumors.

In conclusion, the four drug-sensitivity genes can be used in complementary fashion. The HSVtk, CD, and *E. coli* gpt genes target tumor cells in the S phase of the cell cycle. The rat cytochrome P450 2B1 gene targets tumor cells in all phases of the cell cycle. It is likely that a combination of these drug-sensitivity genes and immune-enhancing genes, such as interleukin-4 (Yu *et al.*, 1993; Wei *et al.*, 1995), could successfully treat brain tumors.

References

Akli, S., Caillaud, C., Vigne, E., Stratford-Perricaudet, L. D., Poenaru, L., Perricaudet, M., Kahn, A., and Peschanski, M. (1993). Transfer of a foreign gene into the brain using adenovirus vectors. *Nature Genet.* **3**, 224–228.

Arndt, C. A. S., Colvin, O. M., Balis, F. M., Lester, C. M., Johnson, G., and Poplack, D. G. (1987). Intrathecal administration of 4-hydroperoxycyclophosphamide in rhesus monkeys. *Cancer Res.* **47**, 5932–5934.

Arndt, C. A. S., Balis, F. M., McCully, C. L., Colvin, O. M., and Poplack, D. G. (1988). Cerebrospinal fluid penetration of active metabolites of cyclophosphamide and ifosfamide in rhesus monkeys. *Cancer Res.* **48**, 2113–2115.

Austin, E. A., and Huber, B. E. (1993). A first step in the development of gene therapy for colorectal carcinoma: Cloning, sequencing, and expression of *Escherichia coli* cytosine deaminase. *Mol. Pharmacol.* **43**, 380–387.

Bajocchi, G., Feldman, S. H., Crystal, R. G., and Mastrangeli, A. (1993). Direct *in vivo* gene transfer to ependymal cells in the central nervous system using recombinant adenovirus vectors. *Nature Genet.* **3**, 229–234.

Barba, D., Hardin, J., Sadelain, M., and Gage, F. H. (1994). Development of anti-tumor immunity following thymidine kinase-mediated killing of experimental brain tumors. *Proc. Natl. Acad. Sci. USA* **91**, 4348–4352.

Besnard, C., Monthioux, E., and Jami, J. (1987). Selection against expression of the *Escherichia coli* gene *gpt* in $hprt^+$ mouse teratocarcinoma and hybrid cells. *Mol. Cell. Biol.* **7**, 4139–4141.

Bi, W. L., Parysek, L. M., Warnick, R., and Stambrook, P. J. (1993). *In vitro* evidence that metabolic cooperation is responsible for the bystander effect observed with HSV tk retroviral gene therapy. *Hum. Gene Ther.* **4**, 725–731.

Boviatsis, E. J., Chase, M., Wei, M. X., Tamiya, T., Hurford, R. K., Jr., Kowall, N. W., Tepper, R. I., Breakefield, X. O., and Chiocca, E. A. (1994a). Gene transfer into experimental brain tumors mediated by adenovirus, herpes virus, and retrovirus vectors. *Hum. Gene Ther.* **5**, 183–191.

Boviatsis, E. J., Scharf, J. M., Chase, M., Harrington, K., Kowall, N. W., Breakefield, X. O., and Chiocca, E. A. (1994b). Antitumor activity and reporter gene transfer into rat brain neoplasms inoculated with herpes simplex virus vectors defective in thymidine kinase or ribonucleotide reductase. *Gene Ther.* **1**, 1–9.

Boviatsis, E. J., Park, J. S., Sena-Esteves, M., Kramm, C. K., Chase, M., Efird, J. T., Wei, M. X., Breakefield, X. O., and Chiocca, E. A. (1994c). Long-term survival of rats harboring brain neoplasms treated with ganciclovir and a herpes simplex virus vector that retains an intact thymidine kinase gene. *Cancer Res.* **54**, 5745–5751.

Breakefield, X. O., and Deluca, N. A. (1991). Herpes simplex virus for gene delivery to neurons. *New Biol.* **3**, 203–218.

Breakefield, X. O. (1993). Gene delivery in the brain using virus vectors. *Nature Genet.* **3**, 187–188.

Burns, J. C., Friedmann, T., Driever, W., Burrascano, M., and Yee, J. K. (1993). Vesicular stomatitis virus G glycoprotein pseudotyped retroviral vectors: Concentration to very high titer and efficient gene transfer into mammalian and nonmammalian cells. *Proc. Natl. Acad. Sci. USA* **90**, 8033–8037.

Cai, D. W., Mukhopadhyay, T., Lui, Y., Fujiwara, T., and Roth, J. A. (1993). Stable expression of the wild-type p53 gene in human lung cancer cells after retrovirus-mediated gene transfer. *Hum. Gene Ther.* **4**, 617–624.

Cepko, C. (1988). Retrovirus vectors and their applications in neurobiology. *Neuron* **1**, 345–353.

Chen, S. H., Shine, H. D., Goodman, J. C., Grossman, R. G., and Woo, S. L. C. (1994). Gene therapy for brain tumors: Regression of experimental gliomas by adenovirus-mediated gene transfer *in vivo*. *Proc. Natl. Acad. Sci. USA* **91**, 3054–3057.

Chiocca, E. A., Choi, B. B., Weizhong, C., Deluca, N. A., Schaffer, P. A., Difiglia, M., Breakefield, X. O., and Martuza, R. L. (1990). Transfer and expression of the LacZ gene in rat brain neurons mediated by herpes simplex virus mutants. *New Biol.* **2**, 739–746.

Chiocca, E. A., Anderson, J. K., Takamiya, Y., Martuza, R. L., and Breakefield, X. O. (1994). Virus-mediated genetic treatment of rodent gliomas. In "Gene Therapeutics" (J. A. Wolff, Ed.), pp. 245–262. Birkhauser, Boston.

Clarke, L., and Waxman, D. J. (1989). Oxidative metabolism of cyclophosphamide: Identification of the hepatic monooxygenase of drug activation. *Cancer Res.* **49**, 2344–2350.

Coen, D. M., Kosz-Vnenchak, M., Jacobson, J. G., Leib, D. A., Board, C. L., Schaffer, P. A., Tylor, K. L., and Knipe, D. M. (1989). Thymidine kinase-negative herpes simplex virus mutants establish latency in mouse trigeminal ganglia but do not reactivate. *Proc. Natl. Acad. Sci. USA* **86**, 4736–4740.

Colvin, M., and Hilton, J. (1981). Pharmacology of cyclophosphamide and metabolites. *Cancer Treatment Rep.* **65**, 89–95.

Culver, K. W., Ram, Z., Wallbridge, S., Ishii, H., Oldfield, E. H., and Blaese, R. M. (1992). In vivo gene transfer with retroviral vector-producer cells for treatment of experimental brain tumors. *Science* **256**, 1550–1552.

Danos, O., and Mulligan, R. C. (1988). Safe and efficient generation of recombinant retroviruse with amphotropic and ecotropic host ranges. *Proc. Natl. Acad. Sci. USA* **85**, 6460–6464.

Davidson, B. L., Allen, E. D., Kozarsky, K. F., Wilson, J. M., and Roessler, B. J. (1993). A model system for *in vivo* gene transfer into the central nervous system using an adenoviral vector. *Nature genetics* **3**, 219–223.

Dearfield, K. L., Jacobson-Kram, D., Huber, B. E., and Williams, J. R. (1986). Induction of sister chromatid exchanges in human and rat hepatoma cell lines by cyclophosphamide and phosphoramide mustard and the effects of cytochrome P-450 inhibitors. *Biochem. Pharmacol.* **35**, 2199–2205.

Dedrick, R. L., and Morrison, P. F. (1992). Carcinogenic potency of alkylating agents in rodents and humans. *Cancer Res.* **52**, 2464–2467.

Dranoff, G., Jaffee, E., Lazenby, A., Golumbek, P., Levitsky, H., Brose, K., Jackson, V., Hamada, H., Pardoll, D., and Mulligan, R. C. (1993). Vaccination with irradiated tumor cells engineered to secrete murine granulocyte-macrophage colony-stimulating factor stimulates potent, specific, and long-lasting anti-tumor immunity. *Proc. Natl. Acad. Sci. USA* **90**, 3539–3543.

Elion, G. B. (1983). The biochemistry and mechanism of action of acyclovir. *J. Antimicrobiol. Chemother.* **12**, 9–17.

Ezzeddine, Z. D., Martuza, R. L., Platika, D., Short, M. P., Malick, A., Choi, B., and Breakefield, X. O. (1991). Selective killing of glioma cells in culture and *in vivo* by retrovirus transfer of the herpes simplex virus thymidine kinase gene. *New Biol.* **3**, 608–614.

Freeman, S. M., Abboud, C. N., Whartenby, K. A., Packman, C. H., Koeplin, D. S., Moolten, F. L., and Abraham, G. N. (1993). The "bystander effect": Tumor regression when a fraction of the tumor mass is genetically modified. *Cancer Res.* **53**, 5247–5283.

Fuchs, H. E., Archer, G. E., Colvin, O. M., Bigner, S. H., Schuster, J. M., Fuller, G. N., Muhlbaier, L. H., Schold, S. C., Jr., Friedman, H. S., and Bigner, D. D. (1990). Activity of intrathecal 4-hydroperoxycyclophosphamide in a nude rat model of human neoplastic meningitis. *Cancer Res.* **50**, 1954–1959.

Fujiwara, T., Grimm, E. A., Mukhopadhyay, T., Zhang, W. W., Owen-Schaub, L. B., and Roth, J. A. (1994). Induction of chemosensitivity in human lung cancer cells *in vivo* by adenovirus-mediated transfer of the wild-type p53 gene. *Cancer Res.* **54**, 2287–2291.

Genka, S., Deutsch, J., Stahle, P. L., Shetty, U. H., John, V., Robinson, C., Rapoport, S. I., and Greig, N. H. (1990). Brain and plasma pharmacokinetics and anticancer activities of cyclophosphamide and phosphoramide mustard in the rat. *Cancer Chemother. Pharmacol.* **27**, 1–7.

Harris, J. D., Gutierrez, A. A., Hurst, H. C., Sikora, K., and Lemoine, N. R. (1994). Gene therapy for cancer using tumor-specific prodrug activation. *Gene Ther.* **1**, 170–175.

Hoshino, T., Nagashima, T., Cho, K. G., Murovic, J. A., Hodes, J. E., Wilson, C. B., Edwards, M. S., and Pitts, L. H. (1986). S-phase fraction of human brain tumors in situ measured by uptake of bromodeoxyuridine. *Int. J. Cancer* **38**, 369–174.

Huber, B. E., Richards, C. A., and Krenitsky, T. A. (1991). Retroviral-mediated gene therapy for the treatment of hepatocellular carcinoma: An innovative approach for cancer therapy. *Proc. Natl. Acad. Sci. USA* **88**, 8039–8043.

Huber, B. E., Austin, E. A., Good, S. S., Knick, V. C., Tibbels, S., and Richards, C. A. (1993). *In vivo* antitumor activity of 5-fluorocytosine on human colorectal carcinoma cells genetically modified to express cytosine deaminase. *Cancer Res.* **53**, 4619–4626.

LeBlanc, G. A., and Waxman, D. J. (1989). Interaction of anticancer drugs with hepatic monooxygenase enzymes. *Drug Metab. Rev.* **20**, 395–439.

LeBlanc, G. A., and Waxman, D. J. (1990). Mechanisms of cyclophosphamide action on hepatic P-450 expression. *Cancer Res.* **50**, 5720–5726.

Le Gal La Salle, G., Robert, J. J., Berrard, S., Ridoux, V., Stratford-Perricaudet, L. D., Perricaudet, M., and Mallet, J. (1993). An adenovirus vector for gene transfer into neurons and glia in the brain. *Science* **259**, 988–990.

Mahaley, M. S., Mettlin, C., Natarajan, N., Laws, E. R., and Peace, B. B. (1989). National survey of patterns of care for brain-tumor patients. *J. Neurosurg.* **71**, 826–836.

Martuza, R. L., Malick, A., Markert, J. M., Ruffner, K. I., and Coen, D. M. (1991). Experimental therapy of human glioma by means of a genetically engineered virus mutant. *Science* **252**, 854–856.

Miller, A. D., and Buttimore, C. (1986). Redesign of retrovirus packaging cell lines to avoid recombination leading to helper virus production. *Mol. Cell. Biol.* **6**, 2895–2902.

Mineta, T., Rabkin, S. D., and Martuza, R. L. (1994). Treatment of malignant gliomas using ganciclovir-herpersensitive, ribonucleotide reductase-deficient herpes simplex viral mutant. *Cancer Res.* **54**, 3963–3966.

Monier, S., Van Luc, P., Kreibich, G., Sabatini, D. D., and Adesnik, M. (1988). Signals for the incorporation and orientation of cytochrome P450 in the endoplasmic reticulum membrane. *J. Cell Biol.* **107**, 457–470.

Moolten, F. L. (1986). Tumor chemosensitivity conferred by inserted thymidine kinase genes: Paradigm for a prospective cancer control strategy. *Cancer Res.* **46**, 5276–5281.

Moolten, F. L., Wells, J. M., Heyman, R. A., and Evans, R. M. (1990). Lymphoma regression induced by ganciclovir in mice bearing a herpes thymidine kinase transgene. *Hum. Gene Ther.* **1**, 125–134.

Mroz, P. J., and Moolten, F. L. (1993). Retrovirally transduced *Escherichia coli gpt* genes combine selectability with chemosensitivity capable of mediating tumor eradiction. *Hum. Gene Ther.* **4**, 589–595.

Mullen, C. A., Kilstrup, M., and Blaese, R. M. (1992). Transfer of the bacterial gene for cytosine deaminase to mammalian cells confers lethal sensitivity to 5-fluorocytosine: A negative selection system. *Proc. Natl. Acad. Sci. USA* **9**, 33–37.

Mullen, C. A., Coale, M. M., Lowe, R., and Blaese, R. M. (1994). Tumors expressing the cytosine deaminase suicide gene can be eliminated *in vivo* with 5-fluorocytosine and induce protective immunity to wild type tumor. *Cancer Res.* **54**, 1503–1506.

Mulligan, R. C. (1993). The basic science of gene therapy. *Science* **260**, 926–931.

Mulligan, R. C., and Berg, P. (1980). Expression of a bacterial gene in mammalian cells. *Science* **209**, 1422–1427.

Mulligan, R. C., and Berg, P. (1981). Selection for animal cells that express the *Escherichia coli* gene encoding for xanthine–guanine phosphoribosyltransferase. *Proc. Natl. Acad. Sci. USA* **78**, 2072–2076.

Oldfield, E. H., Ram, Z., Culver, K. W., Blaese, R. M., DeVroom, H. L., and Anderson, W. F. (1993). Clinical Protocols: Gene therapy for the treatment of brain tumors using Intratumoral transduction with the thymidine kinase gene and intravenous ganciclovir. *Hum. Gene Ther.* **4**, 39–68.

Ram, Z., Culver, K. W., Walbridge, S., Blaese, R. M., and Oldfield, E. H. (1993a). *In situ* retroviral-mediated gene transfer for the treatment of brain tumors in rats. *Cancer Res.* **53**, 83–88.

Ram, Z., Culver, K. W., Walbridge, S., Frank, J. A., Blaese, R. M., and Oldfield, E. H. (1993b). Toxicity studies of retroviral-mediated gene transfer for the treatment of brain tumors. *J. Neurosurg.* **79**, 400–407.

Ram, Z., Walbridge, S., Shawker, T., Culver, K. W., Blaese, R. M., and Oldfield, E. H. (1994). The effect of thymidine kinase transduction and ganciclovir therapy on tumor vasculature and growth of 9L gliomas in rats. *J. Neurosurg.* **81**, 256–260.

Rosenberg, S. A. (1988). Immunotherapy of cancer using interleukin-2: Current status and future prospects. *Immunol. Today* **9**, 58–62.

Schoenberg, B. S. (1983). The epidemiology of nervous system tumors. In "Oncology of the Nervous System" (M. D. Walker, Ed.). Martinus Nijhoff, Boston.

Short, M. P., Choi, B. C., Lee, J. K., Malick, A., Breakefield, X. O., and Martuza, R. L. (1990). Gene delivery to glioma cells in rat brain by grafting of a retrovirus packaging cell line. *J. Neurosci Res.* **27**, 427–439.

Sladek, N. E. (1973). Bioassay and relative cytotoxic potency of cyclophosphamide metabolites generated in vitro and in vivo. *Cancer Res.* **33**, 1150–1158.

Sladek, N. E. (1987). Oxazaphosphorines In "Metabolism & Action of Anti-Cancer Drugs" (G. Powis and R. A. Prough, Eds.), pp. 48–90. Taylor & Francis, London.

Sladek, N. E. (1988). Metabolism of oxazaphosphorines. *Pharmacol. Ther.* **37**, 301–355.

Sladek, N. E., Doeden, D., Powers, J. F., and Krivit, W. (1984). Plasma concentrations of 4-hydroxycyclophosphamide and phosphoramide mustard in patients repeatedly given high doses of cyclophosphamide in preparation for bone marrow transplantation. *Cancer Treatment Rep.* **68**, 1247–1254.

Smythe, W. R., Hwang, H. C., Amin, K. M., Eck, S. L., Davidson, B. L., Wilson, J. M., Kaiser, L. R., and Albelda, S. M. (1994). Use of recombinant adenovirus to transfer the herpes simplex virus thymidine kinase (HSVtk) gene to thoracic neoplasms: An effective *in vitro* drug sensitization system. *Cancer Res.* **54**, 2055–2059.

Takamiya, Y., Short, M. P., Ezzeddine, Z. D., Moolten, F. L., Breakefield, X. O., and Martuza, R. L. (1992). Gene therapy of malignant brain tumors: A rat glioma line bearing the herpes simplex virus type 1-thymidine kinase gene and wild type retrovirus kills other tumor cells. *J. Neurosci. Res.* **33**, 493–503.

Takamiya, Y., Short, M. P., Moolten, F. L., Fleet, C., Mineta, T., Breakefield, X. O., and Martuza, R. L. (1993). An experimental model of retrovirus gene therapy for malignant brain tumors. *J. Neurosurg.* **79**, 104–110.

Tamiya, T., Wei, M. X., Chase, M., Breakefield, X. O., and Chiocca, E. A. (1995). Transgene inheritance and retroviral infection contribute to the efficiency of gene expression in solid tumors inoculated with retroviral vector producer cells. *Gene Ther.* **2**, (in press).

Trofatter, J. A., MacCollin, M. M., Rutter, J. L., Murrell, J. R., Duyao, M. P., Parry, D. M., Eldridge, R., Kley, N., Menon, A. G., Pulaski, K., Haase, V. H., Ambrose, C. M., Munroe, D., Bove, C., Haines, J. L., Martuza, R. L., MacDonald, M. E., Seizinger, B. R., Short, M. P., Buckler, A. J., and Gusella, J. F. (1993). A novel moesin-, ezrin-, radixin-like gene is a candidate for the neurofibromatosis 2 tumor suppressor. *Cell* **72**, 791–800.

Trojan, J., Johnson, T. R., Rudin, S. D., Ilan, J., Tykocinski, M. L., and Ilan, J. (1993). Treatment and prevention of rat glioblastoma by immunogenic C6 cells expressing antisense insulin-like growth factor I RNA. *Science* **259**, 94–97.

Wei, M. X., Tamiya, T., Chase, M., Boviatsis, E. J., Chang, T. K. H., Kowall, N. W., Hochberg, F. H., Waxman, D. J., Breakefield, X. O., and Chiocca, E. A. (1994). Experimental tumor therapy in mice using the cyclophosphamide-activating cytochrome P450 2B1 gene. *Hum. Gene Ther.* **5**, 969–978.

Wei, M. X., Tamiya, T., Hurford, R. K., Jr., Boviatsis, E. J., Tepper, R. I., and Chiocca, E. A. (1995). Enhancement of Interleukin 4-mediated tumor regression in athymic mice by *in situ* retroviral gene transfer. *Hum. Gene Ther.* **6**, (in press).

Weinberg, R. A. (1991). Tumor suppressor genes. *Science* **254**, 1138–1146.

Yoshii, Y., Maki, Y., Tsuboi, K., Tomono, Y., Nakagawa, K., and Hoshino, T. (1986). Estimation of growth fraction with bromodeoxyuridine in human central nervous system tumors. *J. Neurosurg.* **65**, 659–663.

Yu, J. S., Wei, M. X., Chiocca, E. A., Martuza, R. L., and Tepper, R. I. (1993). Treatment of glioma by engineered Interleukin 4-secreting cells. *Cancer Res.* **53**, 3125–3128.

CHAPTER

Brain Tumor Therapy Using Genetically Engineered Replication-Competent Virus

William Hunter
Samuel Rabkin
Robert Martuza
Molecular Neurosurgery Laboratory and Georgetown Brain Tumor Center
Department of Neurosurgery
Georgetown University Medical Center
Washington, DC

The most malignant and fastest growing tumor in the brain is glioblastoma multiforme. This tumor, even with modern technology using neurosurgical techniques, radiation, and chemotherapy, will allow a patient a median survival of about 1 year after diagnosis (Daumas-Duport *et al.*, 1988; Kim *et al.*, 1991; Mahaley *et al.*, 1989; Onoyama *et al.*, 1976; Salazar *et al.*, 1979; Shapiro *et al.*, 1989; Walker *et al.*, 1980).

Over the past few years, we have reintroduced an alternative therapy for killing tumor cells. Viruses have been tested in the past in an attempt to kill tumor cells (Austin and Boone, 1979; Burdick and Hawk, 1964; Cassel *et al.*, 1983; Lobayashi, 1979; Moore, 1960; Pearson and Lagerborg, 1956; Saueter *et al.*, 1978; Siegal *et al.*, 1955; Southam, 1960). The proposed therapeutic mechanisms included, in some cases, direct cell killing by the virus and, in others, the production of new antigens on the tumor cell surface to induce immunologic rejection. However, in prior studies, wild-type virus, passage-attenuated virus, spontaneous mutants, or infected cell preparations were used. Some of the problems of these earlier studies can now be overcome by techniques of genetic engineering which allow directed mutations of viruses to be made and tested. In this era of genetic engineering, viruses therefore offer an attractive alternative to conventional means of antitumor therapy.

Viruses are the most efficient means for getting foreign genes into cells. Various preliminary studies have suggested that the concept of viral tumor

VIRAL VECTORS

therapy is feasible: namely, that a virus can be genetically engineered to kill glioblastoma cells *in situ* with relative sparing of the surrounding brain. Moreover, these vectors are not necessarily limited to just nervous system tumors. With further engineering, including mutations of different viral genes, introduction of tumor-specific cytotoxic genes, or immune modulatory genes, and the use of cell-specific promoters, these vectors could be applied to tumors outside the nervous system as well. Our initial strategies have been relatively straightforward and have utilized various mutants of herpes simplex virus (HSV) known to have decreased neurovirulence and/or reduced ability to replicate in nondividing cells. The pathological consequences of HSV infection in the central nervous system (CNS) are governed by a large number of genes, mutations in which lead to diminished neurovirulence; ICP34.5 (Chou *et al.*, 1990), UL53 (gK) (Moyal *et al.*, 1992), UL50 (dUTPase) (Pyles *et al.*, 1992), UL46 or UL47 (modulate VP16-dependent transactivation) (Zhang *et al.*, 1991), UL39 (ribonucleotide reductase, ICP6) (Cameron *et al.*, 1988; Yamada *et al.*, 1991), UL23 (thymidine kinase) (Field and Wildy, 1978), US1 (Sears *et al.*, 1985), US3 (Fink *et al.*, 1992; Meignier *et al.*, 1988), and US7 (gE) (Meignier *et al.*, 1988; Rajcnai *et al.*, 1990). Of these genes, only UL53 is essential for growth in tissue culture. In glioblastoma multiforme, the tumor cells are rapidly proliferating, whereas the normal brain cells are quiescent and postmitotic. Therefore, the tumor cells may provide a selected target for the mutated virus to replicate, lyse these cells, and spread to the surrounding proliferating tumor cells.

Many steps are involved in viral antitumor therapy. The virus must distinguish between normal tissue and abnormal tumor cells. The virus must target these abnormal tumor cells and effectively eliminate them, while at the same time amplify this effect to surrounding tumor cells. Finally, this therapy must be safe to patients and the normal tissue.

I. Replication-Incompetent Vectors for Gene Transfer to Tumors

A number of different types of viral vectors are currently being explored for tumor therapy. These include HSV (Boviatsis *et al.*, 1994a,b; Jia *et al.*, 1994; Kaplitt *et al.*, 1994; Market *et al.*, 1992; Markert *et al.*, 1993; Martuza *et al.*, 1991; Mineta *et al.*, 1994), retroviral vectors (Boris-Lawrie and Temin, 1993; Culver *et al.*, 1992; Ezzeddine *et al.*, 1991; Oldfield *et al.*, 1993; Short *et al.*, 1990; Ram *et al.*, 1993; Takamiya *et al.*, 1992; Yamada *et al.*, 1992), and adenoviruses (Boviatsis *et al.*, 1994a; Chen *et al.*, 1994; Davidson *et al.*, 1994; Le Gal La Salle *et al.*, 1993; Rosenfeld *et al.*, 1992).

Prior to using the HSV model, we (Ezzenddine *et al.*, 1991; Short *et al.*, 1990; Takamiya *et al.*, 1992) and other investigators (Culver *et al.*, 1992;

Moolten and Wells, 1990; Yamada *et al.*, 1992) examined the use of retroviruses genetically engineered for gene therapy. This therapy is limited because each replication-incompetent retrovirus particle can only integrate and express the gene to be delivered in a single dividing cell and cannot spread to other dividing cells. The viral vector is genetically modified to the point that it is incapable of producing active viral particles in infected host cells (Boris-Lawrie and Temin, 1993). These vectors are not intrinsically cytotoxic, but cytotoxic genes or "suicide" genes whose expression leads to a product that is able to metabolically convert a prodrug into a toxic compound can be introduced into the vector's genome to achieve tumor cell destruction (Culver *et al.*, 1992; Ezzeddine *et al.*, 1991; Moolten, 1986; Moolten and Wells, 1990; Oldfield *et al.*, 1993; Ram *et al.*, 1993). These are useful in an *in vitro* setting or *in vivo* in the presence of small tumors, but the low infectivity and inability to spread may prove too inefficient *in vivo* for penetrating deep into a tumor and killing multiple cells.

Another replication-incompetent virus model in use is based upon adenovirus vector(s) (Chen *et al.*, 1994; Davidson *et al.*, 1994; Le Gal La Salle *et al.*, 1993; Rosenfeld *et al.*, 1992). It has been shown that adenovirus may be an alternative vector for gene delivery. Replication-incompetent adenovirus vectors are usually not toxic to cells at low titers. Adenovirus will infect most cells, not just proliferating ones. The major advantages over the retroviral vector are (i) higher titers to treat the tumor *in situ* and (ii) larger genes can be placed into the adenoviral genome (Boviatsis *et al.*, 1994a).

II. Replication-Competent Vectors for Gene Transfer to Tumors

Various viruses may be genetically engineered for use as replication-competent vectors for tumor therapy. Our current strategy utilizes HSV-I. HSV-I mutants may be constructed which are attenuated for growth in nondividing cells but replication-competent within the tumor. This allows the virus to enter one tumor cell, make multiple copies, kill the cell, and spread to additional tumor cells. The surrounding normal brain cells are nondividing and therefore unable to support replication of these vectors. Additional advantages to using this virus for possible therapy include: (i) HSV is known to infect a large variety of tumor cells with high efficiency; (ii) the HSV genome is well defined and large enough to allow for insertion of considerable genetic material while retaining replication competence; (iii) antiherpetic agents are available to abort any unforeseen infection if necessary; (iv) mutants have been identified which are attenuated for neurovirulence, thus minimizing possible damage to normal brain cells; and (v) the HSV genome does not integrate into the host genome.

III. HSV Biology

HSV type I is a human, neurotropic virus which can infect most vertebrate cells (Roizman and Sears, 1990). Natural infections follow either a lytic, replicative cycle or establishment of latency, usually in peripheral ganglia, where the DNA is maintained indefinitely in an episomal state. From a clinical perspective, HSV encephalitis is the most commonly reported viral infection of the CNS in the United States, with an estimated incidence of 2.3 cases per million population (Corey and Spear, 1986). HSV encephalitis is usually localized to the temporal lobe and the limbic system and histological examination of autopsy cases demonstrated viral antigen at these sites (Esiri, 1982). A number of drugs are available to control HSV infection including acyclovir (9-(2-hydroxyethoxymethyl) guanine, Zovirax) (Whitley *et al.,* 1990), ganciclovir (9(1,3-dehydroxy-2-propoxy)methylguanine, DHPG, 2′NDG, Cytovene) (DeArmond, 1991), adenine arabinoside (Vidarabine) (Whitley *et al.,* 1977), and foscarnet (phosphonoformic acid) (Oberg, 1983).

HSV-I contains a double-stranded, linear DNA genome, 153 kb in length which has been completely sequenced (McGeoch *et al.,* 1988). DNA replication and virion assembly occurs in the nucleus of infected cells. Late in infection, concatemeric viral DNA is cleaved into genome length molecules which are packaged into virions. In the CNS, HSV spreads transneuronally followed by intraaxonal transport, either retrograde or antegrade, to the nucleus where replication occurs (Ugolini *et al.,* 1989).

Mutations in HSV genes can be isolated after mutagenesis or constructed via recombination between the viral genome and genetically engineered sequences (Roizman and Jenkins, 1985). The high rate of recombination and the fact that transfected viral DNA is infectious means that genetic manipulation is very straightforward. These genetically altered viruses can then be used as the backbone for HSV-based vectors.

IV. Viral Tumor Therapy

Our initial strategy (Martuza *et al.,* 1991) involved the use of a thymidine kinase (tk)-deficient HSV-I mutant termed *dl*sptk. Malignant glioma cells are a dividing tumor cell population. In contrast, the surrounding normal brain is mostly composed of nondividing neurons and minimally dividing glia. Certain HSV mutants involved in nucleotide metabolism (including those that are deficient for the virus-encoded enzyme tk) can replicate in dividing cells, but are severely impaired for replication in nondividing cells (Field and Wildy, 1978; Jamieson *et al.,* 1974; Goldstein and Weller, 1988b) and for replication in the mammalian brain (Cameron *et al.,* 1988; Coen *et al.,* 1989; Field and

Darby, 1980; Field and Wildy, 1978; Jacobson *et al.*, 1989; Tenser *et al.*, 1979). We hypothesized that such HSV mutants might replicate in gliomas while sparing normal brain. HSV-I mutant, *dl*sptk (Coen *et al.*, 1989), completely lacks viral tk activity due to a 360-bp deletion within the tk gene. Because *dl*sptk has a deletion, it is unable to revert to wild type and prior studies have also shown that it cannot be reactivated from latency in the trigeminal ganglion (Coen *et al.*, 1989). In cell culture, even low multiplicities of infection of *dl*sptk sustained a spreading infection that destroyed the entire monolayer of malignant glioma cells (see Fig. 1) (Martuza *et al.*, 1991). Short-term primary glioma cultures can be established by explanting malignant human gliomas obtained at surgery. Three primary malignant gliomas (one anaplastic astrocytoma and two glioblastomas) were susceptible to *dl*sptk induced cytopathic effects (Martuza *et al.*, 1991).

The effects of *dl*sptk on human gliomas *in vivo* were studied in nude mice injected subcutaneously with tumor cells (U87-malignant human glioblastoma cell line). When tumors were palpable and growing (5 weeks later) they were inoculated intratumorally with *dl*sptk or medium alone (control). Within 1 month, the virus-treated tumors were significantly smaller than control tumors as determined by external caliper measurements (Martuza *et al.*, 1991).

The most representative model is intracerebral tumor growth (see Fig. 2). Nude mice were stereotactically inoculated in the right frontal lobe with a tumor cell suspension which caused 100% mortality within 6 weeks. Ten days after tumor cell implantation, various doses of *dl*sptk were inoculated at the stereotactic coordinates initially used to inject the tumor cells (Martuza *et al.*, 1991). By 7 weeks, all control animals were dead. At 19 weeks, 29% of the high-dose virus treatment group (10^5 pfu) were still alive, at which time they were sacrificed. These animals were still healthy and neurologically normal and pathological examination revealed no definite evidence of tumor.

This experimental therapy was applied to other nervous system tumors in addition to gliomas (Market *et al.*, 1992). Three human malignant meningiomas, one atypical meningioma, five neurofibrosarcomas, and two medulloblastomas were susceptible to killing by *dl*sptk. *In vivo* inhibition of tumor growth by *dl*sptk was demonstrated for one medulloblastoma and for one malignant meningioma in another animal model—implantation into the subrenal capsule (Market *et al.*, 1992).

V. Safety Issues

A concern for the therapeutic use of HSV against gliomas is the potential killing of other dividing cells such as the endothelium in brain vessels or at mucocutaneuous sites following systemic spread. However, clinical studies indicate that even wild-type HSV-I viruses generally do not spread far from the

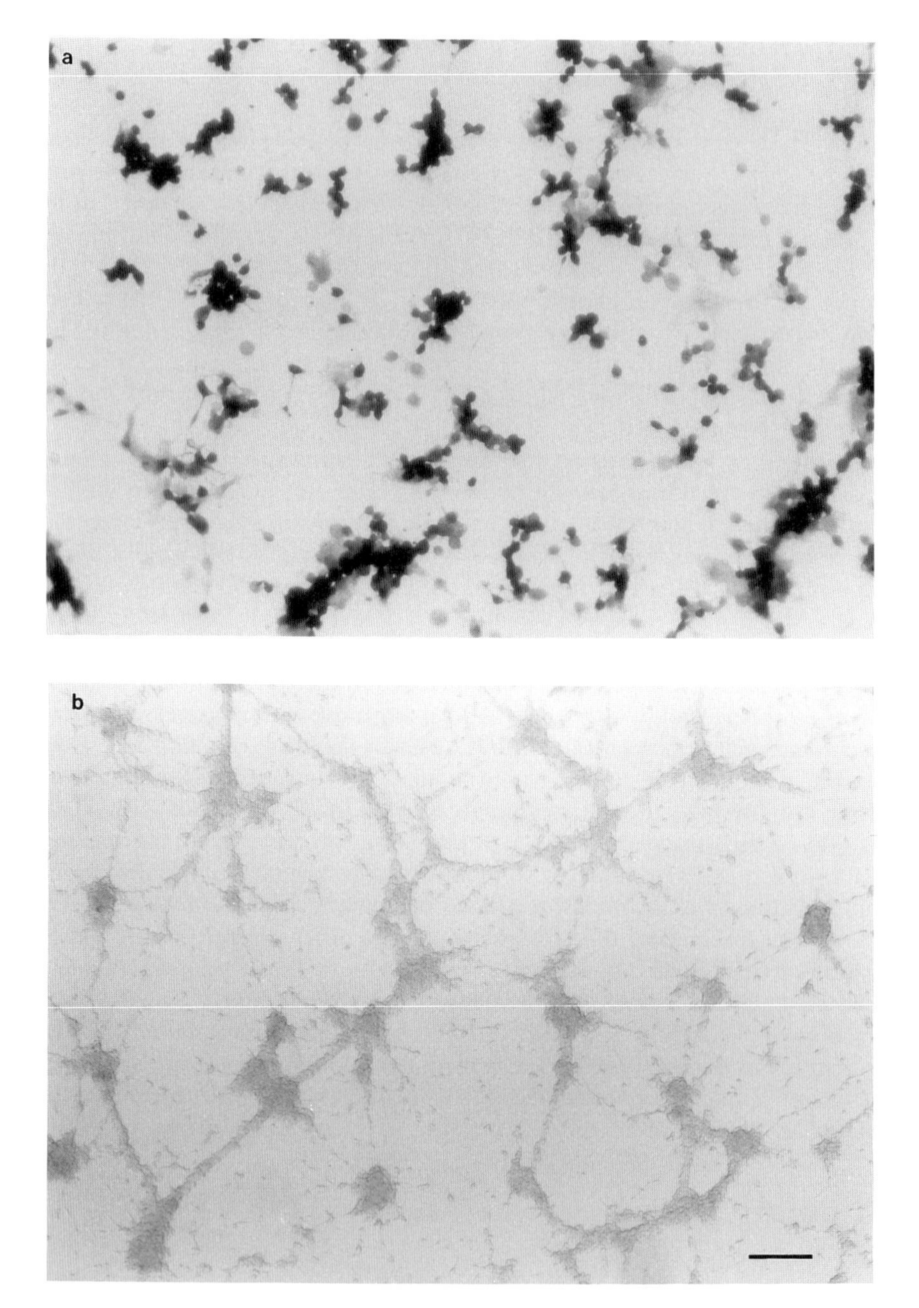

Figure 1 U 87 (human glioma) cells infected with hrR3 at a MOI = 0.1 (a) or mock-infected (b) were fixed 3 days postinfection and histochemically stained for the presence of β-galactosidase. Positive cells appear black and rounded up (hrR3-infected cells).

site of initial infection or cause serious systemic disease in immunocompetent individuals (Sacks *et al.,* 1989). Although *dl*sptk is relatively attenuated for neuropathogenicity (Coen *et al.,* 1989; Field and Darby, 1980; Field and Wildy, 1978; Tenser *et al.,* 1979), tk-deficient viruses have sometimes been associated with progressive disease in certain immunocompromised patients (Erlich *et al.,* 1989; Sacks *et al.,* 1989). *dl*sptk is resistant to nucleoside analogs, such as acyclovir, which require viral tk for their antiviral activity. It is sensitive to vidarabine and foscarnet, which act without requiring viral tk activity and have been used to treat HSV infections (Chatis *et al.,* 1989; Erlich *et al.,* 1989a; Hirsch and Schooley 1983a,b; Sacks *et al.,* 1989; Youle *et al.,* 1988).

In the nervous system, the major side effect of using replication-competent viruses as therapeutic agents is the possible production of encephalitis. Indeed, the virus (*dl*sptk)-treated animals had more early deaths in the intracranial study than the controls, despite the fact that the long-term results were significantly better in the *dl*sptk-treated group. The LD_{50} in immune-competent mice from subsequent studies is 10^6 pfu (Markert *et al.,* 1993); therefore, using *dl*sptk at 10^5 pfu intracranially with nude mice, we were working near the margin of toxicity. This early death rate probably reflects death due to viral encephalitis; in fact, some evidence of mild encephalitis was seen in the surviving animals despite normal neurological functioning at sacrifice.

VI. tk-Proficient HSV-I Mutants

In order to try to overcome these potential safety problem, yet still have effective tumor cell killing, other HSV mutants with decreased neurovirulence were explored. Such mutant(s) would enable the administration of higher doses of virus, rendering treatment more effective. Five different HSV mutants known to replicate well in cultured cells but which demonstrate decreased neurovirulence were investigated. $AraA^r9$ (Coen *et al.,* 1982; Field and Coen, 1986) and $AraA^r13$ (Coen *et al.,* 1982) contain point mutations in the gene-encoding HSV DNA polymerase. These mutants are hypothesized to be replication compromised in the brain due to a lower affinity for deoxynucleoside triphosphates, which are presumed to be at low concentrations in brain. Another mutant, RE6 (Thompson and Stevens, 1983; Thompson *et al.,* 1989), is an intertypic recombinant (HSV-I and HSV-2) with at least two lesions conferring neuroattenuation. It maps to the inverted repeat in the long segment of the HSV genome. The mutant, R3616, contains a 1000-bp deletion in both copies of the γ 34.5 gene in the inverted repeat of the long segment (Chou *et al.,* 1990; Dolan *et al.,* 1992). The γ 34.5 gene of HSV has been shown to play a vital role in viral neurovirulence (Bolivan *et al.,* 1994; Chou *et al.,* 1990; MacLean *et al.,* 1991; Whitley *et al.,* 1993). The final mutant, hrR3, is a $HSV\text{-}RR^-$ (ribonucleotide reductase deficient) mutant. HSV-I encodes its own ribonucleo-

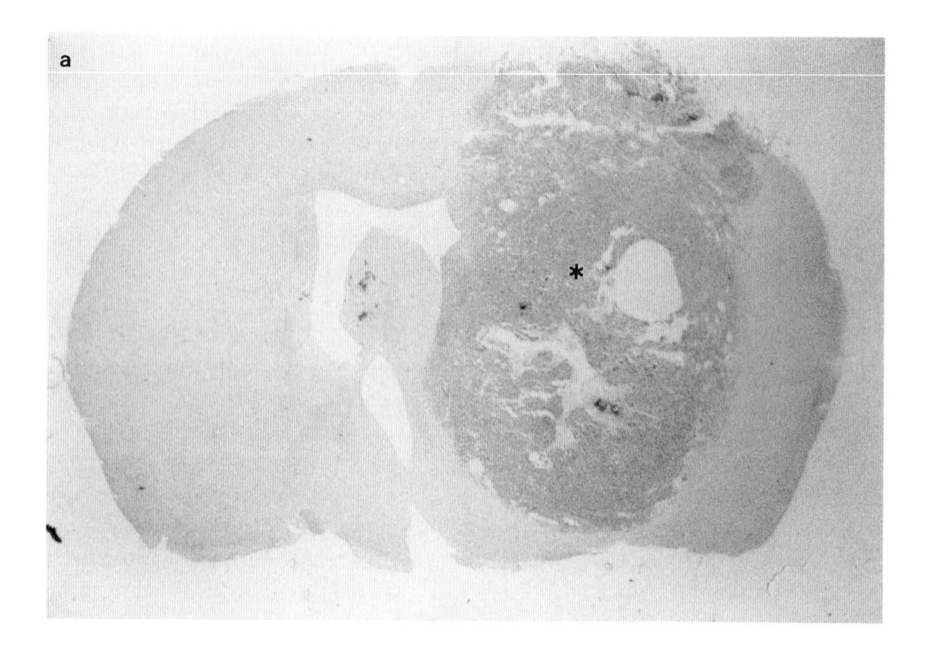

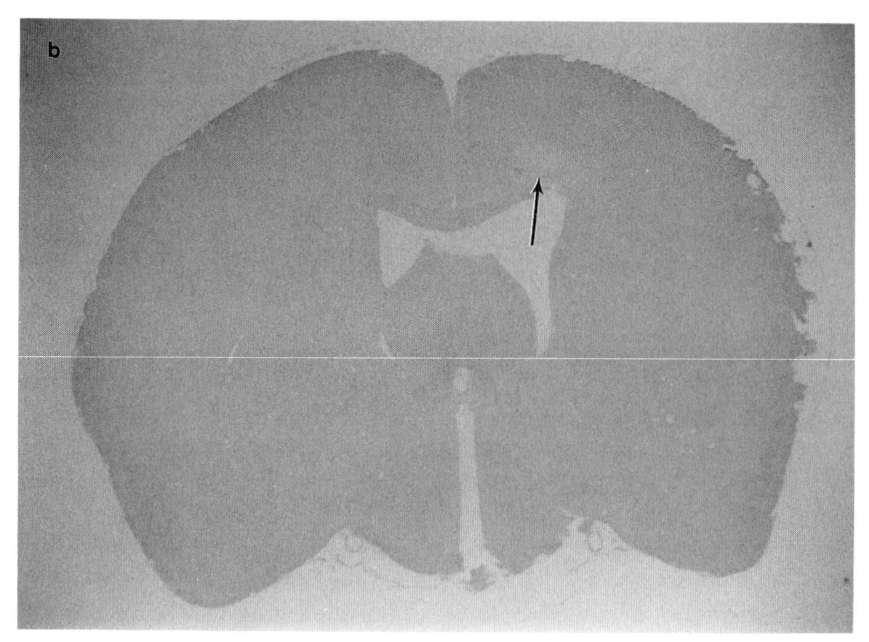

Figure 2 U87 (human glioma) cells inoculated intracranially (2×10^5 cells/5 μl) into nude mice were treated on Day 10 with (a) medium and sacrificed on Day 28, or (b) *dl*sptk (1×10^4 pfu/2 μl) and sacrificed on Day 80. Note the large necrotic tumor (*) in the media-injected control (a) and the small injection region (arrow) in virus-treated brain (b).

tide reductase, an enzyme involved in the *de novo* synthesis of DNA precursors (Averett *et al.*, 1983; Preston *et al.*, 1984). In hrR3, the *Escherichia coli lacZ* gene was inserted into the HSV ICP6 gene (encoding the large subunit of ribonucleotide reductase) (Goldstein and Weller, 1988b). HSV mutants without RR are unable to replicate in many rodent cell lines nor cause any pathology in infected animals (Cameron *et al.*, 1988; Goldstein and Weller, 1988a; Jacobson *et al.*, 1989; Yamada *et al.*, 1991). (see Table 1).

In cell culture, all the above described viral mutants were able to destroy a human glioma line, U87 (Markert *et al.*, 1993; Mineta *et al.*, 1994). Additionally, short-term glioma cultures, established by explanting human glioblastomas, were destroyed in 10 days or less.

In vivo studies were performed using the subcutaneous tumor model described earlier. At the highest inoculum available (5×10^6 pfu), AraA^{r}9 never demonstrated effective inhibition of tumor growth, whereas the other four mutants tested all showed significant tumor growth inhibition (Markert *et al.*, 1993; Mineta *et al.*, 1994).

Table I
HSV Mutants Tested for Antitumor Efficacy

HSV mutants	Mutations	Effects *in vitro*	Effects *in vivo*	Ref.
*dl*sptk	Thymidine kinase gene deletion	+	+	(Martuza *et al.*, 1991)
KOS-SB	Thymidine kinase gene deletion	+	+	(Jia *et al.*, 1994)
dl 8.36tk	Thymidine kinase gene deletion	+	+	(Kaplitt *et al.*, 1994)
RH105	*Lac Z* inserted in thymidine kinase	+	+	(Boviatsis *et al.*, 1994b)
AraA^{r}9	Point mutation of DNA polymerase	+	−	(Markert *et al.*, 1993)
AraA^{r}13	Point mutation of DNA polymerase	+	+/−	(Markert *et al.*, 1993)
R3616	Deletion of γ 34.5 genes	+	+	(Markert *et al.*, 1992)
hrR3	*Lac Z* gene inserted in ribonucleotide reductase	+	+	(Mineta *et al.*, 1994; Boviatsis *et al.*, 1994b)

Note. Mutants and their effects on *in vitro* or *in vivo* tumor cell growth are presented. +, The mutated virus demonstrated significant tumor cell destruction either *in vitro,* or *in vivo;* −, no statistically significant inhibitory effect on *in vivo* studies using intracranial inoculation (all doses were at the same pfu) or subcutaneous model; +/−, AraA^{r}13 has a significant effect in the subcutaneous model but not in the intracranial model.

Using the intracranial tumor model, mice treated with intracranial inoculations of AraA^{r}9 or AraA^{r}13 demonstrated a decrease in median days of survival when compared to controls, indicating a lack of efficacy and probable virally induced pathology. For treatment with RE6 and R3616, statistically significant increases in survival were seen in all treated groups when compared with controls (Markert *et al.*, 1993). In virus-treated groups no deaths occurred prior to those in the control group. All control animals were dead within 2 months and after 4 months the surviving virus-treated animals were sacrificed. These mice appeared neurologically normal at this time. Moreover, the groups showed a significant decrease in size of tumor and no evidence of widespread encephalitis was noted in any brains.

As demonstrated, genetically engineered viruses can be constructed to destroy glioma cells *in vitro* and *in vivo*. Inhibition of tumor growth was achieved with a tk-negative HSV mutant and mutants retaining tk proficiency. However, careful choice of mutant is essential for effective *in vivo* tumor inhibition and survival prolongation. Subcutaneous tumor growth was significantly inhibited by one of the DNA polymerase mutants (AraA^{r}13), but neither of the DNA polymerase mutants prolonged survival in the intracranial model. With proper selection of HSV mutants, early encephalitic deaths and chronic encephalitis can be eliminated. An additional safety advantage of the tk-proficient mutants is susceptibility to acyclovir. The hrR3 mutant is actually hypersensitive to acyclovir and ganciclovir (Coen *et al.*, 1989; Mineta *et al.*, 1994) and is temperature sensitive for viral growth (Goldstein and Weller, 1988a; Mineta *et al.*, 1994; Preston *et al.*, 1988).

These studies have been very promising; however, the effect of such treatment in immune-competent animal models is important. All our studies described in this chapter have utilized athymic nude mice. Using nude mice allows investigators to directly test human tumors (both primary and passaged cell lines) with the HSV mutants. Unfortunately, in the rodent models there are very few syngeneic glioma tumors available and some HSV mutants are specifically compromised for replication in rats. We and others (Boviatsis *et al.*, 1994a,b; Jia *et al.*, 1994; Kaplitt *et al.*, 1994) have begun to evaluate these HSV mutants in the presence of a competent immune system. In preliminary experiments, we have been able to show that in such an immune-competent system the virus-treated animals have prolonged survival (data not published). In the rat model, using 9L glioma cells, HSVtk$^-$ mutants were shown to effectively inhibit tumor growth in intracranial tumors (Boviatsis *et al.*, 1994b; Jia *et al.*, 1994) and in subcutaneous tumors of W256 cells (Kaplitt *et al.*, 1994).

These studies provide strong evidence that HSV mutants can effectively limit tumor growth. However, we have still not reached the ultimate goal—effective and safe human use. Current treatment of primary malignant glioma, consisting of gross surgical resection followed by radiation and/or chemotherapy, is cytoreductive in nature but has failed to substantially change the outcome of patients with glioblastoma (Walker *et al.*, 1978). This has led to the investigation of other modalities of treatment, including immunotherapy. Lymphocytic infiltrates have been described in primary brain tumors and were speculated to be a host-mediated response against the tumor (Ridley and Cavanaugh, 1971). The presence of lymphocytic infiltrates has been associated with improved prognosis and as an adjuvant treatment for brain tumors (Brooks *et al.*, 1976; Yu *et al.*, 1993). A number of cytokines have been identified which reduce tumorgenicity (Jacobs *et al.*, 1986; Kuppner *et al.*, 1988; Nitta *et al.*, 81994;Roszman*etal.*,1991;TepperandMule,1994;Tepper*etal.*,1989,1992).

Immune-mediated tumor therapy via gene delivery is an active area of research. An increasing number of human trials have been approved using retroviral vectors to transfer cytokine genes to cells *ex vivo* followed by subsequent transplantation of the transferred cells (Tepper and Mule, 1994). The viruses described in this chapter can be further engineered to produce cytokines to increase antitumor therapy. We recently demonstrated (Yu *et al.*, 1993) that interleukin-4 (IL-4) expressed within a glioma can cause tumor killing via an elicited eosinophilic response. The construction of HSV-expressing IL-4 and other cytokines is a reasonable avenue for further study. These possibilities are promising not only for the malignant brain tumors but also for other tumors.

References

Austin, F. C., and Boone, C. W. (1979). Virus augmentation of the antigenicity of tumor cell extracts. *Adv. Cancer Res.* **30**, 301–345.

Averett, D. R., Lubbers, C., Elion, G. B., and Spector, T. (1983). Ribonucleotide reductase induced by herpes simplex type 1 virus. Characterization of a distinct enzyme. *J. Biol. Chem.* **258**, 9831–9838.

Bolovan, C. A., Sawtell, N. M., and Thompson, R. L. (1994). ICP34.5 mutants of herpes simplex virus type 1 strain syn+ are attenuated for neurovirulence in mice and for replication in confluent primary mouse embryo cell cultures. *J. Virol.* **68**, 48–55.

Boris-Lawrie, K. A., and Temin, H. M. (1993). Recent advances in retrovirus vector technology. *Curr. Opin. Genet. Dev.* **3**, 102–109.

Boviatsis, E. J., Chase, M., Wei, M. X., Tamiya, T., Hurford, R. K., Kowall, N. W., Tepper, R. I., Breakefield, X. O., and Chiocca, E. A. (1994a). Gene transfer into experimental brain tumors mediated by adenovirus, herpes simplex virus, and retrovirus. *Hum. Gene Ther.* **5**, 183–191.

Boviatsis, E. J., Scharf, J. M., Chase, M., Harrington, K., Knowall, N. W., Breakefield, X. O., and Chiocca, E. A. (1994b). Antitumor activity and reporter gene transfer into rat brain neoplasms inoculated with herpes simplex virus vectors defective in thymidine kinase or ribonucleotide reductase. *Gene Ther.* **1**, 323–331.

Brooks, W. H., Rozman, T. L., and Rogers, A. S. (1976). Impairment of rosette-forming T lymphocytes in patients with primary intracranial tumors. *Cancer (Philadelphia)* **37**, 1869–1873.

Burdick, K. H., and Kawk, W. A. (1964). Vitiligo in a case of vaccinia virus-treated melanoma. *Cancer* **17**, 708–712.

Cameron, J. M., McDougall, I., Marsden, H. S., Preston, V. G., Ryan, D. M., and Subak-Sharpes, J. H. (1988). Ribonucleotide reductase encoded by herpes simplex virus is a determinant of the pathogenicity of the virus in mice and a valid antiviral target. *J. Gen. Virol.* **69**, 2607–2612.

Cassel, W. A., Murray, D. R., and Phillips, H. S. (1983). A phase II study on the postsurgical management of stage II malignant melanoma with a Newcastle disease virus oncolysate. *Cancer* **52**, 856–860.

Chatis, P. A., Miller, C. H., Schrager, L. E., and Crumpacker, L. S. (1989). Successful treatment with foscarnet of an acyclovir-resistant mucocutaneous infection with herpes simplex virus in a patient with acquired immunodeficiency syndrome. *N. Engl. J. Med.* **320**, 297–300.

Chen, S. H., Shine, H. D., Goodman, J. C., Grossman, R. G., and Woo, S. L. (1994). Gene therapy for brain tumors: Regression of experimental gliomas by adenovirus-mediated gene transfer *in vivo*. *Proc. Natl. Acad. Sci. USA* **91**(8), 3054–3057.

Chou, J., Kern, E. R., Whitley, R. J., and Roizman, B. (1990). Mapping of herpes simplex virus-I neurovirulence to gamma 34.5, a gene nonessential for growth in culture. *Science* **250**, 1262–1266.

Coen, D. M., Furman, P. A., Gelep, P. T., and Schaffer, P. A. (1982). Mutations in the herpes simplex virus DNA polymerase gene can confer resistance to 9-beta-D-arabinofuranosyladenine. *J. Virol.* **41**, 909–918.

Coen, D. M., Goldstein, D. J., and Weller, S. K. (1989). Herpes simplex virus ribonucleotide reductase mutants are hypersensitive to acyclovir. *Antimicrobiol. Agents Chemother.* **33**, 1395–1399.

Coen, D. M., Kosz-Vnenchak, M., Jacobson, J. G., Leib, D. A., Bogard, C. L., Schaffer, P. A., Tyler, K. L., and Knipe, D. M. (1989). Thymidine kinase-negative herpes simplex virus mutants establish latency in mouse trigeminal ganglia but do not reactivate. *Proc. Natl. Acad. Sci. USA* **86**, 4736–4740.

Corey, L., and Spear, P. G. (1986). Infections with herpes simplex viruses. *N. Engl. J. Med.* **314**, 749–757.

Culver, K. W., Ram, Z., Wallbridge, S., Ishii, H., Oldfield, E. H., and Blaese, R. M. (1992). *In vivo* gene transfer with retroviral vector-producer cells for treatment of experimental brain tumors. *Science* **256**, 1550–1552.

Daumas-Duport, C., Scheithauer, B., O'Fallon, J., and Kelly, P. (1988). Grading of astrocytomas. A simple and reproducible method. *Cancer* **62**, 2152–2165.

Davidson, B. L., Doran, S. E., Shewach, D. S., Latta, J. M., Hartman, J. W., and Roessler, B. J. (1994). Expression of *Escherichia coli* beta-galatosidase and rat HPRT in the CNS of Macaca mulatta following adenoviral mediated gene transfer. *Exp. Neurol.* **125**, 258–267.

DeArmond, B. (1991). Clinical trials of Ganciclovir. *Transplant. Proc.* **23**(Suppl. 3), 171–173.

Dolan, A., McKie, E., MacLean, A. R., and McGeoch, D. J. (1992). Status of the ICP34.5 gene in herpes simplex virus type 1 strain 17. *J. Gen. Virol.* **73**, 971–973.

Erlich, K. S., Jacobson, M. A., Koelher, J. E., Follansbee, S. E., Drennan, D. P., Gooze, L., Safrin, S., and Mills, J. (1989a). Foscarnet therapy for severe acyclovir-resistant herpes simplex virus type-2 infections in patients with the acquired immunodeficiency syndrome (AIDS). An uncontrolled trial. *Ann. Int. Med.* **110**, 710–713.

Erlich, K. S., Mills, J., Chantis, P., Mertz, G. J., Busch, D. F., Follansbee, S. E., Grant, R. M., and Crumpacker, L. S. (1989b). Acyclovir-resistant herpes simplex virus infections in patients with the acquired immunodeficiency syndrome. *N. Engl. J. Med.* **320**, 293–296.

Esiri, M. M. (1982). Herpes simplex encephalitis. An immunohistological study of the distribution of viral antigen with the brain. *J. Neurol. Sci.* **54**, 209–226.

Ezzeddine, Z. D., Martuza, R. L., Platika, D., Short, M. P., Malick, A., Choi, B., and Breakefield, X. O. (1991). Selective killing of glioma cells in culture and *in vivo* by retrovirus transfer of the herpes simplex virus thymidine kinase gene. *New Biol.* **3**, 608–614.

Field, H. J., and Coen, D. M. (1986). Pathogenicity of herpes simplex virus mutants containing drug resistance mutations in the viral DNA polymerase gene. *J. Virol.* **60,** 286–289.

Field, H. J., and Darby, G. (1980). Pathogenicity in mice of strains of herpes simplex virus which are resistant to acyclovir *in vitro* and *in vivo*. *Antimicrobiol. Agents Chemother.* **17,** 209–216.

Field, H. J., and Wildy, P. (1978). The pathogenicity of thymidine kinase-deficient mutants of herpes simplex virus in mice. *J. Hygiene* **81,** 267–277.

Fink, D. J., Sternberg, L. R., Weber, P. C., Mata, M., Goins, W. F., and Glorioso, J. C. (1992). *In vivo* expression of B-galactosidase in hippocampal neurons by HSV-mediated gene transfer. *Hum. Gene Ther.* **3,** 11–19.

Goldstein, D. J., and Weller, S. K. (1988a). Factor(s) present in herpes simplex virus type 1-infected cells can compensate for the loss of the large subunit of the viral ribonucleotide reductase: Characterization of an ICP6 deletion mutant. *Virology* **166,** 41–51.

Goldstein, D. J., and Weller, S. K. (1988b). Herpes simplex virus 1-induced ribonucleotide reductase activity is dispensible for virus growth and DNA synthesis: Isolation and characterization of an ICP6 lacZ insertion mutant. *J. Virol.* **62,** 196–205.

Hirsch, M. S., and Schooley, R. T. (1983a). Drug therapy. Treatment of herpesvirus infections. *N. Engl. J. Med.* **309,** 963–970.

Hirsch, M. S., and Schooley, R. T. (1983b). Drug therapy. Treatment of herpesvirus infections. *N. Engl. J. Med.* **309,** 1034–1039.

Jacobs, S. K., Wilson, D. J., Kornblith, P. L., and Grimm, E. A. (1986). Interleukin-2 autologous lymphokine-activated killer cell treatment of malignant glioma: Phase I trial. *Cancer Res.* **46,** 2101–2104.

Jacobson, J. G., Leib, D. A., Goldstein, D. J., Bogard, C. L., Schaffer, P. A., Weller, S. K., and Coen, D. M. (1989). A herpes simplex virus ribonucleotide reductase deletion mutant is defective for productive acute and reactivable latent infections of mice and for replication in mouse cells. *Virology* **173,** 276–283.

Jamieson, A. T., Gentry, G. A., and Subak-Sharpe, J. H. (1974). Induction of both thymidine and deoxycytidine kinase activity by herpes virus. *J. Gen. Virol.* **24,** 465–480.

Jia, W. W.-G., McDermott, M., Goldie, J., Cyander, M., Tan, J., and Tufaro, F. (1994). Selective destruction of gliomas in immunocompetent rats by thymidine kinase-defective Herpes Simplex Virus type I. *J. Natl. Cancer Inst.* **86,** 1209–1215.

Kaplitt, M. G., Tjuvajev, J. G., Leib, D. A., Berk, J., Pettigrew, K. D., Posner, J. B., Pfaff, D. W., Rabkin, S. D., and Blasberg, R. G. (1994). Mutant herpes simplex virus induced regression of tumors growing in immunocompetent rats. *J. Neuro-Oncol.* **19,** 137–147.

Kim, T. S., Halliday, A. L., Heldey-Whyte, E. T., and Convery, K. (1991). Correlates of survival and the Daumas–Duport grading system for astrocytomas. *J. Neurosurg.* **74,** 27–37.

Kuppner, M. C., Hamou, M.-F., Bodmer, S., Fontano, A., and DeTribolet, N. (1988). The glioblastoma-derived T-cell suppressor factor/transforming growth factor $beta_2$ inhibits the generation of lymphokine-activated killer(LAK) cells. *Int. J. Cancer* **42,** 562–567.

Le Gal La Salle, G., Robert, J. J., Berrard, S., Ridoux, V., Stratford-Perricaudet, L. D., Pericaudet, M., and Mallet, J. (1993). An adenovirus vector for gene transfer into neurons and glia in the brain. *Science* **259,** 988–990.

Lobayashi, H. (1979). Viral xenogenization of intact tumor cells. *Adv. Cancer Res.* **30,** 279–299.

MacLean, A. R., Ul-Fareed, M., Robertson, L., Harland, J., and Brown, S. M. (1991). Herpes simplex virus type 1 variants 1714 and 1716 pinpoint neurovirulence-related sequences in Glascow strain 17+ between immediate early gene and the 'a' sequence. *J. Gen. Virol.* **72,** 631–639.

Mahaley, M. S., Jr., Mettlin, C., Natarajan, N., Laws, E. R., Jr., and Peace, B. B. (1989). National survey of patterns of care for brain-tumor patients. *J. Neurosurg.* **71,** 826–836.

Market, J. M., Coen, D. M., Malick, A., Mineta, T., and Martuza, R. L. (1992). Expanded spectrum of viral therapy in the treatment of nervous system tumors. *J. Neurosurg.* **77,** 590–594.

Markert, J. M., Malick, A., Coen, D. M., and Martuza, R. L. (1993). Reduction and elimination of encephalitis in an experimental gliomas therapy model with attenuated herpes simplex mutants that retain susceptibility to acyclovir. *Neurosurgery* **32**, 597–603.

Martuza, R. L., Malick, A., Markert, J. M., Ruffner, K. L., and Coen, D. M. (1991). Experimental therapy of human glioma by means of a genetically engineered virus mutant. *Science* **252**, 854–856.

McGeoch, D. J., Dalrymple, M. A., Davison, A. J., Dolan, A., Frame, M. C., McNab, D., Perry, L. J., Scott, J. E., and Taylor, P. (1988). The complete DNA sequence of the long unique region in the genome of herpes simplex virus type I. *J. Gen. Virol.* **69**, 1531–1574.

Meignier, B., Longnecker, R., Mavromara-Nazos, P., Sears, A. E., and Roizman, B. (1988). Virulence of and establishment of latency by genetically engineered deletion mutants of herpes simplex virus. *Virology* **162**, 251–254.

Mineta, T., Rabkin, S. D., and Martuza, R. L. (1994). Treatment of malignant gliomas using ganciclovir-hypersensitive, ribonucleotide reductase-deficient herpes simplex viral mutant. *Cancer Res.* **54**, 3963–3966.

Moolten, F. (1986). Tumor chemosensitivity conferred by inserted thymidine kinase genes: Paradigm for a prospective cancer control strategy. *Cancer Res.* **46**, 5276–5281.

Moolten, F. L., and Wells, J. M. (1990). Curability of tumors bearing herpes thymidine kinase genes transferred by retroviral vectors. *J. Natl. Cancer Inst.* **82**, 297–300.

Moore, A. E. (1960). The oncolytic viruses. *Prog. Exp. Tumor Res.* **1**, 411–439.

Moyal, M., Berkowitz, C., Rosen-Wolff, A., Darai, G., and Becker, Y. (1992). Mutations in the UL53 gene of HSV-1 abolish virus neurovirulence to mice by the intracerebral route of infection. *Virus Res.* **26**, 99–112.

Nitta, T., Hishii, M., Sato, K., and Okumura, K. (1994). Selective expression of interleukin-10 gene within glioblastoma multiforme. *Brain Res.* **649**, 122–128.

Oberg, B. (1983). Antiviral effects of phosphonoformate (PFA, Foscarnet sodium). *Pharmacol. Ther.* **19**, 387–415.

Oldfield, E. H., Ram, Z., Culver, K. W., Blaese, R. M., Devroom, H. L., and Anderson, W. F. (1993). Gene therapy for the treatment of brain tumors using intra-tumoral transduction with the thymidine kinase gene and intravenous ganciclovir. *Hum. Gene Ther.* **4**, 39–69.

Onoyama, Y., Abe, M., Yabumoto, E., Sakamoto, T., and Nishidai, T. (1976). Radiation therapy in the treatment of glioblastoma. *Am. J. Roentgenol.* **126**, 481–492.

Pearson, H. E., and Lagerborg, D. L. (1956). Propagation of Thieler's virus in murine astrocytic tumors. *Proc. Soc. Exp. Biol. Med.* **92**, 551–553.

Preston, V. G., Darling, A. J., and McDonougall, I. M. (1988). The herpes simplex virus type I temperature sensitive mutant ts1222 has a single base pair deletion in the small subunit of ribonucleotide reductase. *Virology* **167**, 458–467.

Preston, V. G., Palfreyman, J. W., and Duita, B. M. (1984). Identification of a herpes simplex virus I polypeptide which is a component of the virus-induced ribonucleotide reductase. *J. Gen. Virol.* **65**, 1457–1466.

Pyles, R. B., Sawtell, N. M., and Thompson, R. L. (1992). Herpes simplex virus type 1 dUTPase mutants are attenuated for neurovirulence, neuroinvasiveness, and reactivation from latency. *J. Virol.* **66**, 607–671.

Rajcnai, J., Herget, U., and Kaerner, H. C. (1990). Spread of herpes simplex virus (HSV) strains SC16, ANG, ANGpath and its glyC minus and glyE minus mutants in DBA-2 mice. *Acta Virol.* **34**, 305–320.

Ram, Z., Culver, K. W., Walbridge, S., Blease, R. M., and Oldfield, E. H. (1993). *In situ* retroviral-mediated gene transfer for the treatment of brain tumors in rats. *Cancer Res.* **53**, 83–88.

Ridley, A., and Cavanagh, J. B. (1971). Lymphocytic infiltration in gliomas: Evidence of possible host resistance. *Brain* **94**, 117–124.

Roizman, B., and Jenkins, F. J. (1985). Genetic engineering of novel genomes of large DNA viruses. *Science* **229**, 1208–1214.

Roizman, B., and Sears, A. E. (1990). Herpes Simplex viruses and their replication. In "Fundamental Virology" (B. N. Fields and D. M. Knipe, Eds.), pp. 849–896. Raven Press, New York.

Rosenfeld, M. A., Yoshimura, K., Trapnell, B. C., Yoneyama, K., Rosenthal, E. R., Dalemans, W., Fukayama, M., Bargon, J., Stier, L. E., Stratford-Perricaudet, L., *et al.* (1992). *In vivo* transfer of the human cystic fibrosis transmembrane conductase regulator gene to the airway epithelium. *Cell* **68,** 143–155.

Roszman, T., Elliott, L., and Brooks, W. (1991). Modulation of T-cell function of gliomas. *Immunol. Today* **12,** 370–374.

Sacks, S. L., Wanklin, R. L., Reece, D. E., Hicks, K. A., Tyler, K. L., and Coen D. M. (1989). Progressive esophagitis from acyclovir-resistant herpes simplex. Clinical roles for DNA polymerase mutants and viral heterogeneity? *Ann. Int. Med.* **111,** 893–899.

Salazar, O. M., Rubin, P., Feldstein, M. L., and Pizzutiello, R. (1979). High dose radiation therapy in the treatment of malignant gliomas: Final report. *Int. J. Radiat. Oncol. Biol. Phys.* **5,** 1733–1749.

Saueter, C., Cavalli, F., Lindenmann, J., Gmur, J. P., Berchtold, W., Alberto, P., Obrecht, P., and Senn, H. J. (1978). Viral oncolysis: Its applications in maintenance treatment of acute myelogenous leukemia: Study analysis at 2.5 years. In "Current Chemotherapy: Proceedings of the Tenth International Congress of Chemotherapy" (W. Siegenthaler and R. Luthy, Eds.), pp. 1112–1114. American Society for Microbiology, Washington, DC.

Sears, A. E., Halliburton, I. W., Meignier, B., Silver, S., and Roizman, B. (1985). Herpes simplex virus 1 mutant deleted in the $\alpha 22$ gene: Growth and gene expression in permissive and restrictive cells and establishment of latency in mice. *J. Virol.* **55,** 338–346.

Siegal, M. M., Bernstein, A., and Pope, L. P. (1955). The interaction of menigopneumonitis virus and Kreabs-2 carcinoma. *J. Immunol.* **75,** 386.

Shapiro, W. R., Green, S. B., Burger, P. C., Mahaley, M. S., Jr., Selker, R. G., VanGilder, J. C., Robertson, J. T., Ransohoff, J., Mealey, J. Jr., Strike, T. A., and Pistemaa, D. A. (1989). Randomized trail of three chemotherapy regimens and two radiotherapy regimens in postoperative treatment of malignant glioma. Brain Tumor Cooperative Group Trial 8001. *J. Neurosurg.* **71,** 1–9.

Short, M. P., Choi, B. C., Lee, J. K., Malick, A., Breakefield, X. O., and Martuza, R. L. (1990). Gene delivery to glioma cells in rat brain by grafting of a retrovirus packaging cell line. *J. Neurosci. Res.* **27,** 427–439.

Southam, C. M. (1960). Present status of oncolytic virus studies. *Tran. N.Y. Acad. Sci.* **22,** 657–673.

Takamiya, Y., Short, M. P., Ezzeddine, Z. D., Moolten, F. L., Breakefield, X. O., and Martuza, R. L. (1992). Gene therapy of malignant brain tumors: A rat glioma line bearing the herpes simplex virus type-I thymidine kinase gene and wild type retrovirus kills other tumor cells. *J. Neurosci. Res.* **33,** 493–503.

Tenser, R. B., Miller, R. L., and Rapp, F. (1979). Trigeminal ganglion infection by thymidine kinase-negative mutants of herpes simplex virus. *Science* **205,** 915–917.

Tepper, R. I., and Mulé, J. J. (1994). Experimental and clinical studies of cytokines gene-modified tumor cells. *Hum. Gene Ther.* **5,** 153–164.

Tepper, R. I., Coffman, R. L., and Leder, P. (1992). An eosinophil-dependent mechanism for the antitumor effect of IL-4. *Science* **257,** 548–551.

Tepper, R. I., Pattengale, P. K., and Leder, P. (1989). Murine interleukin-4 displays potent antitumor activity *in vivo*. *Cell* **57,** 503–511.

Thompson, R. L., and Stevens, J. G. (1983). Biological characterization of a herpes simplex virus intertypic recombinant which is completely and specifically non-neurovirulent. *Virology* **131,** 171–179.

Thompson, R. L., Rogers, S. K., and Zerhusen, M. A. (1989). Herpes simplex virus neurovirulence and productive infection of neural cells is associated with a function which maps between 0.82 and 0.832 map units on the HSV genome. *Virology* **172,** 432–450.

Ugolini, G., Kuypers, H. G. J. M., and Strick, P. L. (1989). Transneuronal transfer of herpes virus from peripheral nerves to cortex and brainstem. *Science* **243**, 89–91.

Walker, M. D., Alexander, E., Jr., Hunt, W. E., MacCarty, C. S., Mahaley, M. S., Mealy, J., Norrell, H. A., Owens, G., Ransohoff, J., Nilson, C. B., Gehan, E. A., and Strike, T. A. (1978). Evaluation of BCNU and/or treatment of anaplastic gliomas: A cooperative clinical trial. *J. Neurosurg.* **49**, 333–343.

Walker, M. D., Green, S. B., Byar, D. P., Alexander, E., Jr., Batzdorf, U., Brooks, W. H., Hunt, W. E., MacCarty, C. S., Mahaley, M. S., Mealey, J., Jr., Owens, G., Ransohoff, J., Robertson, J. T., Shapiro, W. R., Smith, K. R., Wilson, C. B., and Strike, T. A. (1980). Randomized comparison of radiotherapy and nitroureas for the treatment of malignant glioma after surgery. *N. Engl. J. Med.* **303**, 1323–1329.

Whitley, R. J., Kern, E. R., Chatterjee, S., Chou, J., and Roizman, B. (1993). Replication, establishment of latency, and induced reactivation of herpes simplex virus gamma 34.5 deletion mutants in rodent models. *J. Clin. Invest.* **91**, 2837–2843.

Whitley, R. J., Middlebrooks, M., and Gnann, J. W. J. (1990). Acyclovir: The past ten years. In "Immunobiology and Prophylaxis of Human Herpesvirus Infections" (C. Lopez, R. Mori, B. Roizman, and R. J. Whitley Eds.), pp. 243–253. Plenum Press, New York.

Whitley, R. J., Soong, S., Dolin, R., Galasso, G. J., Ch-Ien, L., Alford, C., and the Collaborative Study Group. (1977). Adenine arabinoside therapy of biopsy-proved herpes simplex encephalitis: National Institute of Allergy and Infectious Diseases Collaborative Antiviral Study. *N. Engl. J. Med.* **297**, 289–294.

Yamada, Y., Kimura, H., Morishima, T., Daikoku, T., Maeno, K. M., and Nishiyama, Y. (1991). The pathogenicity of ribonucleotide reductase-null mutants of herpes simplex virus type 1 in mice. *J. Infect. Dis.* **164**, 1091–1097.

Yamada, M., Shimizu, K., Miyao, Y., Hayakawa, T., Ikenaka, K., Nakahira, K., Nakajima, K., Kagawa, T., and Mikoshiba, K. (1992). Retrovirus-mediated gene transfer targeted to malignant glioma cells in murine brain. *Jpn. J. Cancer Res.* **83**, 1244–1247.

Youle, M. M., Hawkins, D. A., Collins, P., Shanson, D. C., Evans, R., Oliver, N., and Lawrence, A. (1988). Acyclovir-resistant herpes in AIDS treated with foscarnet. *Lancet* **2**, 341–342.

Yu, J. S., Wei, M. X., Chiocca, A., Martuza, R. L., and Tepper, R. I. (1993). Treatment of Glioma by engineered interleukin 4-secreting cells. *Cancer Res.* **53**, 3125–3128.

Zhang, Y., Sirko, D. A., and McKnight, J. L. C. (1991). Role of herpes simplex virus type 1 UL46 and UL47 in a TIF-mediated transcriptional induction: Characterization of three viral deletion mutants. *J. Virol.* **65**, 829–841.

CHAPTER 16

Transfer and Expression of Antioncogenes and Paraneoplastic Genes in Normal and Neoplastic Cells in Vitro and in Vivo

Myrna R. Rosenfeld
Jan J. Verschuuren
Josep Dalmau
Department of Neurology and the Cotzias Laboratory of Neuro-Oncology
Memorial Sloan–Kettering Cancer Center and the Department of Neurology
Cornell University Medical College
New York, New York

I. Introduction

Developments in the field of viral vectors and new insights into the genetic basis of cancer have lead to the idea that gene therapy may have a role in the study and treatment of human cancer. We are particularly interested in two aspects of oncology: the treatment of primary brain tumors and the study of paraneoplastic neurologic diseases.

Primary tumors of the nervous system are the second most common cancer in infants and young children. In adolescents and young adults brain tumors range from the fifth to the eighth most frequent cancer. In adults, there are approximately 17,000 new cases of primary brain tumors diagnosed each year in the United States, almost as many as ovarian cancer (Laws and Thapar, 1993). Furthermore, in the elderly population the incidence of primary brain tumors is increasing (Greig *et al.*, 1990). The large majority of these tumors are malignant and to date remain incurable. Long-term survivals of patients with malignant primary brain tumors are rare (Salford *et al.*, 1988; Vertosick and Selker, 1992). For example, malignant gliomas, the most common malignant brain tumor in all age groups, have a median post-treatment survival of less than 1 year (Chandler *et al.*, 1993). Many of those patients who do survive are damaged not just by their tumor, but also by the therapy they receive since our current standard postsurgical therapies, including cranial irradiation and systemic or local chemotherapy, each have serious adverse side effects (DeLattre and Posner, 1989). Clearly, novel approaches to the treatment of these tumors are needed. Gene therapy offers a potential alternative to current treatment protocols.

In addition to primary brain tumors and metastases of systemic cancer to the brain, patients with cancer may develop neurologic symptoms as a result of nonmetastatic, indirect, or "remote" effects of cancer on nervous system function. The term "paraneoplastic neurologic syndromes" refers to neurologic disorders of unknown cause that occur at higher frequency in patients with cancer than in the general population. These disorders can involve any part of the central or peripheral nervous system, and for some of them, such as paraneoplastic cerebellar degeneration or the Lambert–Eaton myasthenic syndrome, there is a strong association with specific histological types of neoplasms (O'Neill *et al.*, 1988; Posner and Furneaux, 1990).

Although paraneoplastic neurologic syndromes are rare, occuring in less than 1% of all cancer patients, they are important for several reasons; (1) in almost 50% of patients the neurological symptoms precede the diagnosis of the tumor (Posner and Furneaux, 1990), (2) their identification directs the search for a neoplasm and may lead to early diagnosis and treatment which can result in a better tumor prognosis, (3) paraneoplastic tumors appear to have a more indolent course than tumors of the same histological type not associated with paraneoplastic symptoms (Dalmau *et al.*, 1990, 1992b), and (4) paraneoplastic neurologic syndromes may have important biological implications. Some of these disorders are associated with antibodies that react with antigens shared by the tumor and the nervous system. These antibodies are used as markers of the paraneoplastic disorder and the presence of specific types of tumors. Using these antibodies as probes, several neuronal-specific proteins which are also expressed by the tumor have been isolated. Some of these proteins appear to have a crucial role in neurogenesis and neuronal maintenance. Furthermore, it is likely (but as yet unproven) that the immunological response which results in the paraneoplastic syndrome is also responsible for the more "benign" course of the associated tumor. A better understanding of these immune mechanisms may result in new therapeutic strategies for treatment of both the tumor and the paraneoplastic disorder.

Recent efforts utilizing viral vectors for gene therapy have employed retroviral vectors (Oldfield *et al.*, 1993). Retroviruses can only insert and replicate in actively growing cells and therefore may be useful in tumors with high mitotic activity. However, the requirement of active cell division limits the transformation efficiency of retroviral vectors in most tumors which contain heterogeneously malignant cell populations. In addition, retroviruses must generate a DNA copy of their genome in order to express a foreign gene, which further limits their use in nongrowing but potentially reactivatable cells.

DNA viral vectors based upon defective herpes simplex viruses (HSV) (Spaete and Frenkel, 1982; Kwong and Frenkel, 1985) or recombinant adenoassociated viruses (AAV) (Samulski *et al.*, 1989) have several features, many of which have been discussed in earlier chapters, that make them particularly well suited to transfer genes to nervous tissue. Briefly, these viruses can enter the cell nucleus and express foreign genes in the absence of active cell division so that even nondividing tumor cells are potential targets of these vectors. For their use as therapeutic agents, DNA viral vectors do not need to specifically target neoplastic cells if the foreign gene has no effect on normal cells. However, if necessary, gene expression can be limited to tumor cells through the use of cell-specific promoters. Additionally, since DNA viral vectors will drive expression of foreign genes in terminally differentiated cells, such as neurons (Kaplitt *et al.*, 1991), they are also useful tools to study tumor oncogenesis and neurogenesis.

This chapter will review the work in our laboratory using defective HSV vectors and AAV vectors in the study of primary brain tumors and paraneoplastic neurologic syndromes. As described elsewhere in this volume, the AAV vector is similar to the defective HSV vector in that it contains only viral-recognition sequences but no viral genes. Adenovirus is used as a helper virus to provide replication proteins. The AAV vector is packaged into an AAV coat which is distinct from the helper adenovirus. Therefore, in contrast to defective HSV vectors, the helper adenovirus may be completely eliminated by taking advantage of the structural differences between the AAV vector and the helper adenovirus. The ability to generate a stock of AAV vector free of helper adenovirus makes the AAV vector particularly well suited to serve as a therapeutic agent.

II. Tumor-Suppressor Genes and Primary Brain Tumors

Tumor-suppressor genes encode for proteins that regulate normal cell functions including cell proliferation and differentiation (Sager, 1989; Weinberg, 1991). The loss of function of these proteins, either due to mutation or deletion, has been associated with tumor formation (Bishop, 1991).

The p53 gene is a tumor-suppressor gene that is often found mutated or deleted in human cancers including carcinomas of the colon, lung, breast, prostate, bladder, and osteosarcomas (Nigro *et al.*, 1989; Hollstein *et al.*, 1991; Levine *et al.*, 1991). Overexpression of mutant p53 protein is the most common genetic abnormality detected in human cancers to date (Levine *et al.*, 1991). The p53 gene product has a central role in normal cell proliferation. p53 functions as a negative growth regulator and as a transcriptional activator that suppresses transformation (Finlay *et al.*, 1982; Baker *et al.*, 1990; Mercer *et al.*, 1990). The wild-type gene also acts by controlling transit of the cell through the cell cycle (Kuerbitz *et al.*, 1992) and is involved in programmed cell death (apoptosis) (Yonish-Rouach *et al.*, 1991; Shaw *et al.*, 1992). Mutations of p53 are usually point mutations that result in loss of gene function and the production of mutant p53 protein (Finlay *et al.*, 1982; Eliyahu *et al.*, 1988).

Mutations of p53 are found in almost 50% of all low-grade astrocytomas suggesting a role in early tumorigenesis (Frankel *et al.*, 1992; Von Deimling *et al.*, 1992). The clonal expansion of cells that have acquired p53 mutations has been associated with brain tumor progression (Sidransky *et al.*, 1992). In addition to astrocytomas, mutations of p53 have also been reported in medulloblastomas, a common brain tumor of childhood (Saylors *et al.*, 1991) and oligodendrogliomas (Ohgaki *et al.*, 1991). In some studies, replacement of the wild-type p53 gene in human cancer cells has lead to growth arrest and/or apoptosis (Diller *et al.*, 1990; Shaw *et al.*, 1992; Ramqvist *et al.*, 1993).

We have used a defective HSV vector to replace wild-type p53 in primary brain tumor cell lines *in vitro* (Rosenfeld *et al.*, in press). For these experiments we used a medulloblastoma cell line, DAOY. This cell line overexpresses a mutant form of p53 due to the presence of a point mutation in exon 7 of the p53 gene (Saylors *et al.*, 1991). The presence of mutant p53 can be demonstrated by immunohistochemical techniques. Wild-type p53 has a short half-life, whereas mutant forms are more stable and have longer half-lives (Rogel *et al.*, 1985; Finlay *et al.*, 1988; Iggo *et al.*, 1990). Unless wild-type p53 is overexpressed, the short half-life of this protein makes it undetectable using standard immunohistochemical techniques, whereas mutant proteins are easily detected. Confirmation that the expressed protein is mutant is obtained from sequence analysis of cellular DNA.

Figure 1A shows DAOY cells immunoreacted with a monoclonal antibody specific for wild-type p53, PAb1620 (Ab-5, Oncogene Science, Uniondale, NY) (Harlow *et al.*, 1981; Ball *et al.*, 1984), demonstrating no reactivity, consistent with the lack of expression of any wild-type p53 protein by these cells. Figure 1B shows DAOY cells immunoreacted with the monoclonal antibody PAb1801 (Ab-2, Oncogene Science) that recognizes both mutant and wild-type p53 (Banks *et al.*, 1986). There is strong nuclear immunoreactivity typical of cells that overexpress mutant p53.

To perform gene-transfer experiments we constructed a HSV amplicon that contained a transcription unit consisting of the cytomegalovirus immediate-early promoter, wild-type p53 cDNA (obtained from Dr. Arnold Levine, Princeton University), a polyadenylation signal, and two HSV recognition signals, the origin of replication and HSV cleavage/packaging signal (Kaplitt *et al.*, 1991). In addition, the amplicon contained the bacterial sequence for ampicillin resistance and a bacterial origin of replication. The amplicon was transfected into rabbit skin cells using calcium chloride followed by glycerol shock. After 24 hr these cells were superinfected with a helper HSV virus. To offset any potential toxicity from intact helper virus in the viral stocks we used a temperature-sensitive helper virus (Graham and Van der Eb, 1973). Defective viral vector was propagated at 31°C, the permissive temperature for the helper virus, and a viral stock containing both defective and helper virus was obtained. As a control we propagated a defective viral vector from an amplicon containing the bacterial marker gene for β-galactosidase (lacZ) (Kaplitt *et al.,* 1991). Concentrations of helper virus in the viral stock were determined by standard viral plaque assay. Titers of defective p53 vector were obtained by immunocytochemical detection of novel p53 expression in a p53 null cell line (Saos-2) infected with serial dilutions of the viral stock.

Figure 1C shows DAOY cells 24 hr after infection with defective virus carrying wild-type p53 and reacted with PAb1620, the monoclonal antibody against wild-type p53. Compared with DAOY cells which did not express wild-type p53 before viral gene transfer (Fig. 1A), these cells now demonstrate intense nuclear immunoreactivity with PAb1620, indicating synthesis of wild-type p53. Cells infected with control virus showed no reactivity with this antibody (not shown). When reacted with antibody PAb1801, which recognizes both mutant and wild-type p53, we found loss of the expected nuclear immunostaining (Figure 1D). Since the mutation identified in DAOY cells results in the loss of a restriction enzyme digestion site, we were able to confirm that gene transfer resulted in novel expression of wild-type p53 by examining cDNA from cells before and after gene transfer.

One question raised by these studies is why does the immunostaining pattern of PAb1801 change after gene transfer? It is known that p53 is capable of undergoing conformational changes that could result in altered antibody recognition (Halazonetis *et al.,* 1993). Furthermore, it has been demonstrated that mutant and wild-type p53 complex to each other when they are co-translated but do not form complexes if mixed post-translationally (Milner *et al.,* 1991). Another possibility is that the decreased nuclear staining of mutant p53 represents decreased expression of the mutant transcript due to downregulation by wild-type p53. This is supported by several studies that have shown that wild-type p53 downregulates the activity of several promoters including an autoregulatory effect on its own promoter (Ginsberg *et al.,* 1991; Mercer *et al.,* 1991; Santhanam *et al.,* 1991; Shiio *et al.,* 1992). While our current studies do not specifically address these issues, the questions raised by our findings demonstrate the power of gene transfer in studying basic molecular mechanisms.

Having demonstrated that we could achieve expression of wild-type p53 through viral gene transfer we proceeded to examine if gene transfer resulted in any physiological effects, especially those that could be of potential therapeutic value. p53 is involved in regulation of cell transit through the cell cycle (Kastan *et al.,* 1992a). For example, when DNA of fibroblasts is damaged by ionizing radiation, the cells arrest in G1 and G2. Studies have shown that after DNA damage, these cells show increased levels of wild-type p53 protein, and that the ability to arrest in G1 is dependent upon wild-type p53 expression (Kastan *et al.*, 1992b). Cells that have lost wild-type p53 function will not arrest in G1. In other cells, such as thymocytes, when DNA is damaged by irradiation p53 plays a crucial role in the pathways that lead to apoptosis (Lowe *et al.*, 1993). Therefore, we were interested to examine if the overexpression of wild-type p53 after gene transfer would result in either growth arrest or apoptosis in brain tumor cells.

Progression of transduced cells through the cell cycle was studied using a panel of anti-cyclin antibodies. Figures 2A and 2B show the reactivity of DAOY cells with monoclonal antibodies against cyclin E and cyclin A prior to gene transfer. Cyclin E is a marker for cells in late G1 and early S phase, and cyclin A is a marker for cells in mid S phase and G2 (Reed *et al.*, 1992); the positive immunoreactivity indicates that under normal conditions DAOY cells are actively progressing through the cell cycle. After gene transfer with wild-type p53 there is loss of cyclin E and cyclin A reactivity (Figs. 2C and 2D) indicating that these cells are arrested, most likely at G1.

Apoptosis is a form of cell death that occurs during normal development and in some tissues throughout the life of the individual (Wyllie *et al.*, 1981; Brusch *et al.*, 1990). Unlike cells undergoing necrosis in which there is a generalized swelling of the cell and nuclear chromatin, apoptotic cells shrink in size and the chromatin condenses (Wyllie *et al.*, 1984). In this process there is activation of endonucleases which results in double-stranded breaks of the internucleosomal DNA and generation of a population of DNA fragments. This type of DNA fragmentation is relatively specific to cells undergoing apoptosis and is a useful biochemical marker (Arends and Wyllie, 1991). The fragmented DNA in apoptotic cells contains free 3′ hydroxyl groups not found in normal, nonfragmented DNA. To visualize apoptotic cells *in situ*, terminal deoxynucleotidyl transferase is used to end label the fragmented DNA with biotinylated dUTP which is then visualized by avidin–biotin peroxidase complexes reacted with diaminobenzidine (TUNEL method) (Gavrieli *et al.*, 1992). In cultured DAOY cells, only a few cells were identified undergoing apoptosis, but 24 hr after wild-type p53 gene transfer, the number of apoptotic cells increased to approximately 20%. Transfer of the β-galactosidase gene did not lead to apoptosis.

In the above studies viral gene transfer of wild-type p53 resulted in the expected physiological effects of cell cycle arrest and apoptosis. However, it is important to know through which pathways a transferred gene acts. One way to approach this question is to examine the effect of gene transfer on cellular components which under normal conditions are known to interact with the transduced gene. We approached this question by examining the effect of gene transfer on mdm-2 (murine double-minute) protein. Mdm-2 and wild-type p53 are involved in an autoregulatory feedback loop (Momand *et al.*, 1992; Wu *et al.*, 1993). Wild-type p53 protein upregulates the expression of mdm-2 at the level of transcription, while the mdm-2 protein inhibits the function of p53 by forming complexes with p53 protein. After gene transfer we found increased levels of mdm-2 expression using immunohistochemical techniques (Rosenfeld *et al.*, 1994). Similar studies examining interactions with other proteins such as Waf1 (a.k.a. Cip1, p21) (El-Deiry *et al.*, 1993; Harper *et al.*, 1993) have confirmed that transduced p53 is using the expected cellular pathways which under normal conditions drive to the same physiological effects.

III. Gene Transfer and the Study of Paraneoplastic Syndromes

DNA viral vectors are versatile tools that can be used not only as potential therapeutic agents in the treatment of tumors of the nervous system, but also in the study of other neurologic disorders. We are using these vectors to gain insights into the pathogenesis of some paraneoplastic neurologic syndromes. Patients with these disorders may develop dementia, cerebellar degeneration, brain stem dysfunction, spinal cord symptoms, and severe sensory deficits. Pathological findings include neuronal degeneration, gliosis, and inflammatory infiltrates involving one or several areas of the nervous system (Henson and Urich, 1989). There is increasing evidence that some of these disorders have an immune basis. However, an autoimmune pathogenesis has only been established for the Lambert–Eaton myasthenic syndrome (LEMS), a disorder in which the presence of antibodies against voltage-gated calcium channels of the presynaptic neuromuscular junction interferes with the quantal release of acetylcholine resulting in muscle weakness and fatigability (Lang *et al.*, 1981; Roberts *et al.*, 1985). Passive transfer of immunoglobulin fractions from patients with LEMS results in neuromuscular transmission defects in mice (Lang *et al.*, 1981).

In addition to LEMS, there are other paraneoplastic syndromes, including paraneoplastic encephalomyelitis (PEM) often combined with sensory neuronopathy (PSN), paraneoplastic cerebellar degeneration, and paraneoplastic opsoclonus–myoclonus, in which the presence of antibodies that react with antigens shared by the nervous system and tumor have been demonstrated (Furneaux *et al.*, 1990a; Luque *et al.*, 1991; Dalmau *et al.*, 1992a). These antibodies have provided the opportunity to isolate proteins (CDR34, CDR62, HuD, Nova) that under normal circumstances are specifically expressed in neurons, and that when expressed by tumors evoke a profound immune response (Dropcho *et al.*, 1987; Fathallah-Shaykh *et al.*, 1991; Szabo *et al.*, 1991; Buckanovich *et al.*, 1993). Identification of these antibodies allows physicians (1) to make the diagnosis of a disorder as paraneoplastic, (2) to anticipate

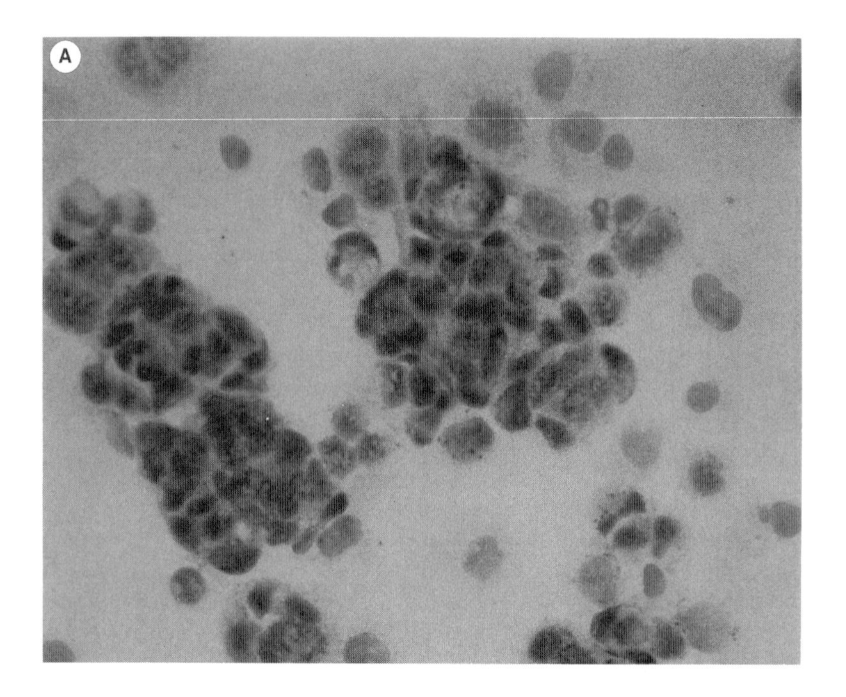

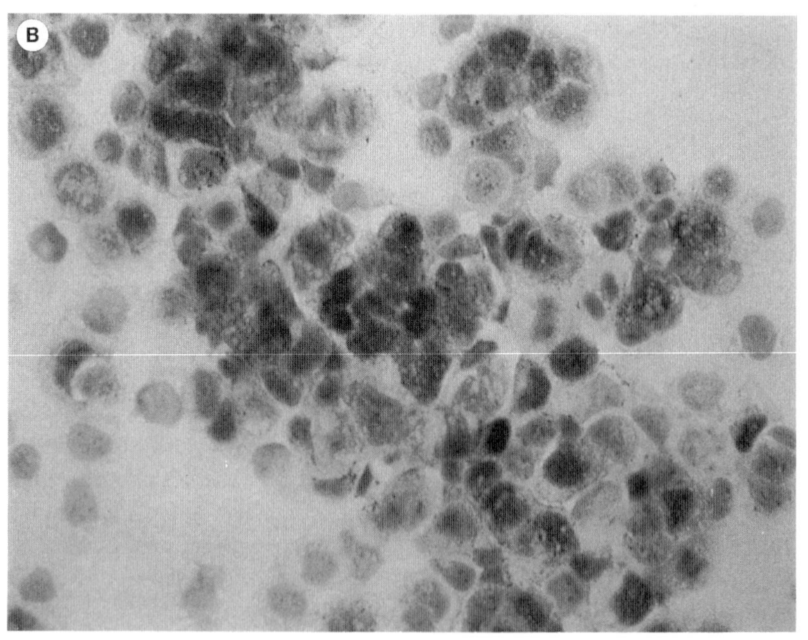

Figure 2 Effect of wild-type p53 on cell cycle progression. DAOY cells reacted with antibodies against cyclin E before (A) and after (C) gene transfer of wild-type p53, and DAOY cells reacted with antibodies against cyclin A before (B) and after (D) gene transfer. The loss of reactivity after gene transfer (C and D) indicates that there is cell cycle arrest.

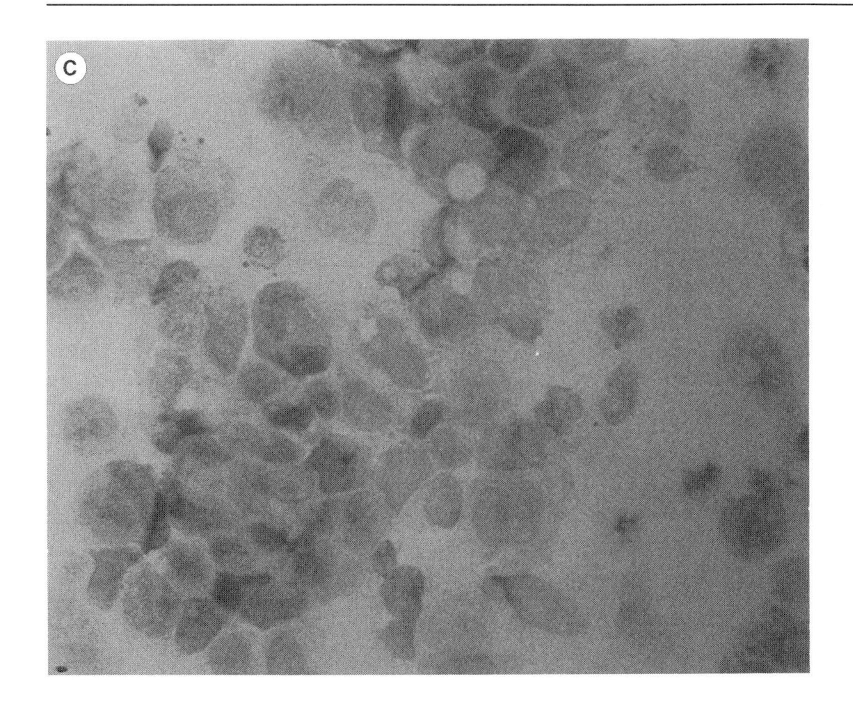

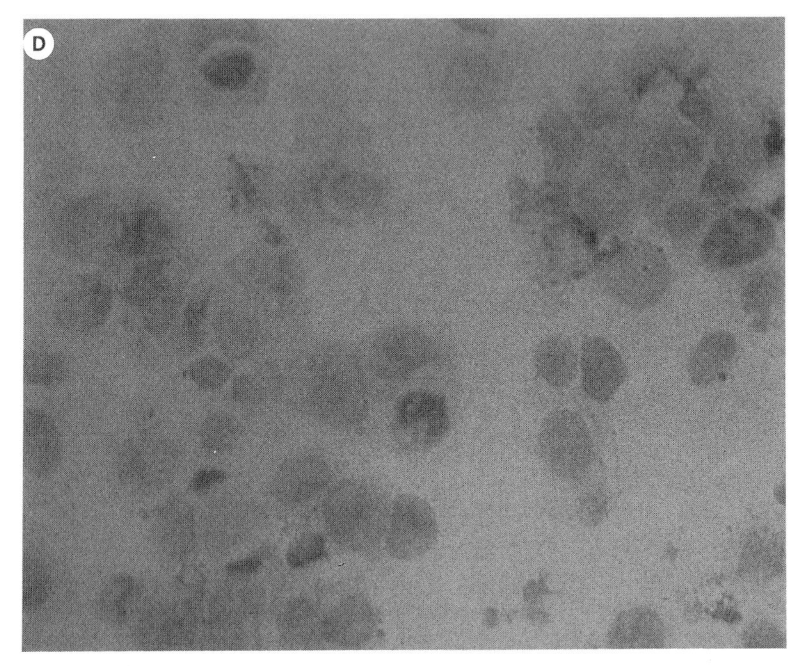

Figure 2 (Continued)

the presence of specific type(s) of cancer, and (3) to direct the search of the tumor to a few organs.

Of particular interest to us is anti-Hu-associated PEM and PSN, a paraneoplastic neurologic disorder characterized by multifocal inflammatory infiltrates and degeneration of the central and peripheral nervous system (Dalmau *et al.*, 1992b). Patients with this disorder develop antibodies (called anti-Hu) that react with 35- to 40-kDa proteins that are expressed both in neurons and in the tumor, usually small cell lung cancer (SCLC) (Graus *et al.*, 1986; Dalmau *et al.*, 1992a). Characteristically, the tumor of most of these patients remains localized and it is the neurological disorder that frequently causes death (Dalmau *et al.*, 1990; 1992b). Treatment of the tumor and reduction of the serum titer of antibodies have not been successful in controlling the disease. Development of an animal model is therefore crucial to better understand the pathogenesis of the disorder and develop strategies to treat these patients.

Using anti-Hu sera to screen a cDNA expression library, a gene encoding a neuronal RNA-binding protein, called HuD, was isolated (Szabo *et al.*, 1991). In normal tissues, the expression of HuD is restricted to neurons; HuD is also expressed in SCLC. HuD has high homology to the *Drosophila* proteins Elav (*E*mbryonic *l*ethal *a*bnormal *v*isual system), sex-lethal, and RBP9, sharing more than 65% identity with the RNA-binding regions of these proteins (Robinow and White, 1988; Robinow *et al.*, 1988). Expression of Elav is one of the earliest events in maturation of neurons in the *Drosophila* and mutation of this protein results in disruption of development of the nervous system and death (Campos *et al.*, 1985). Similarly, developmental studies in birds and mammals have shown that HuD-related proteins are expressed very early in neurogenesis (Graus and Ferrer, 1990; Marusich and Weston, 1992). These studies suggest that HuD plays a role in the maturation and maintenance of vertebrate neurons. However, the exact function of HuD in neurons and tumors is unknown.

The nervous system and tumor of patients with PEM/PSN contain deposits of anti-Hu IgG (Dalmau *et al.*, 1991). This finding, along with the demonstration of intrathecal synthesis of anti-Hu antibodies (Furneaux *et al.*, 1990b), has suggested that this antibody may have a role in the pathogenesis of the neurologic dysfunction. However, attempts to develop an animal model of the disease by passive transfer of anti-Hu IgG have not been successful (DeLattre and Posner, unpublished results). Furthermore, immunization of animals with purified recombinant protein (HuD) has not reproduced the disease despite the generation of high titers of antibodies (Sillevis-Smitt *et al.*, 1994). A possible explanation for this finding is that epitopes contained in native human HuD may be lost in the recombinant protein produced in bacteria, which is used to immunize the animal. Another possibility is that a cell-mediated cytotoxic mechanism, either alone or in association with the humoral immune response, may be involved in the pathogenesis of the disorder. This idea is supported by

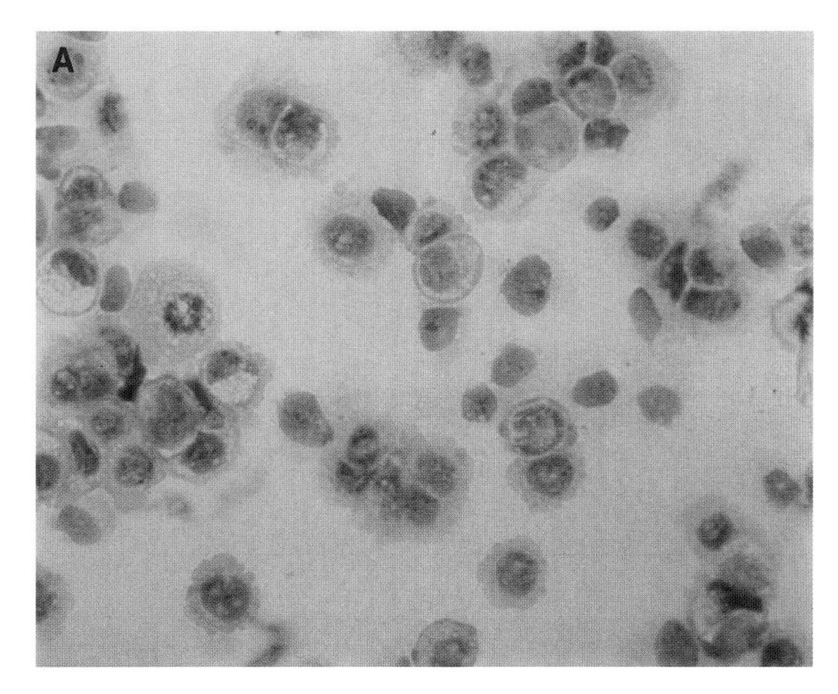

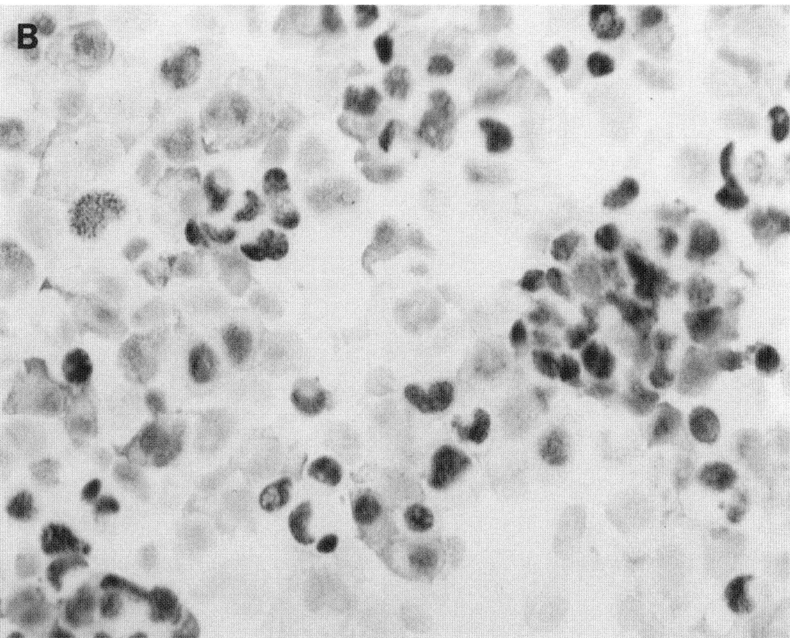

Figure 1 Immunocytochemical analysis of the expression of p53 in DAOY cells after gene transfer. (A) Cells immunoreacted with PAb1620, a monoclonal antibody that only reacts with wild-type p53. No reactivity is observed indicating the lack of expression of wild-type p53 by DAOY cells. (B) Cells immunoreacted with PAb1801, a monoclonal antibody that reacts with both wild-type and mutant p53. There is strong nuclear reactivity indicating

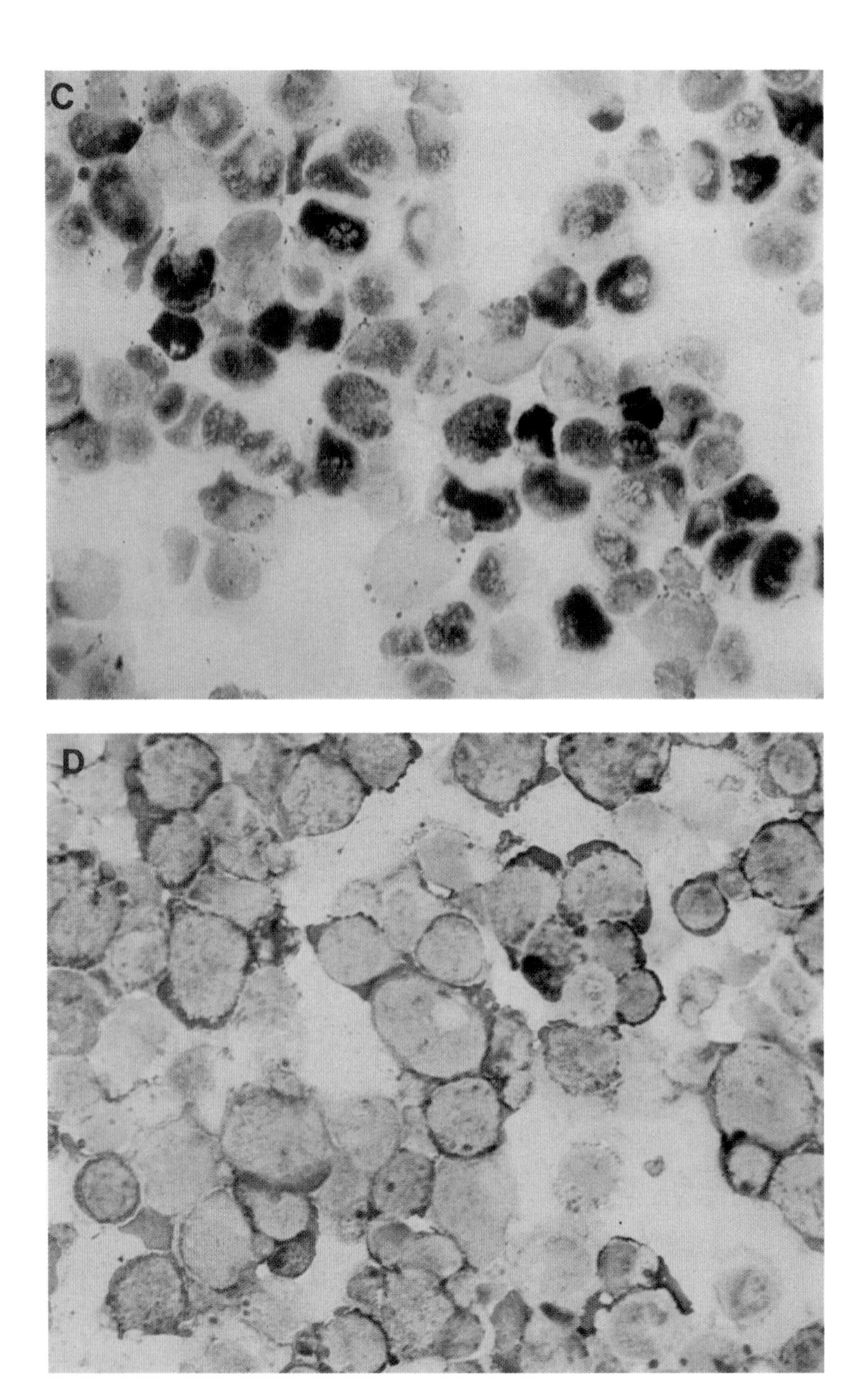

that DAOY cells overexpress mutant p53. (C) Twenty-four hours after gene transfer of wild-type p53 to DAOY cells there is intense nuclear reactivity with PAb1620 indicating novel synthesis of wild-type p53. (D) DAOY cells after gene transfer of wild-type p53 do not react with the PAb1801 (see text).

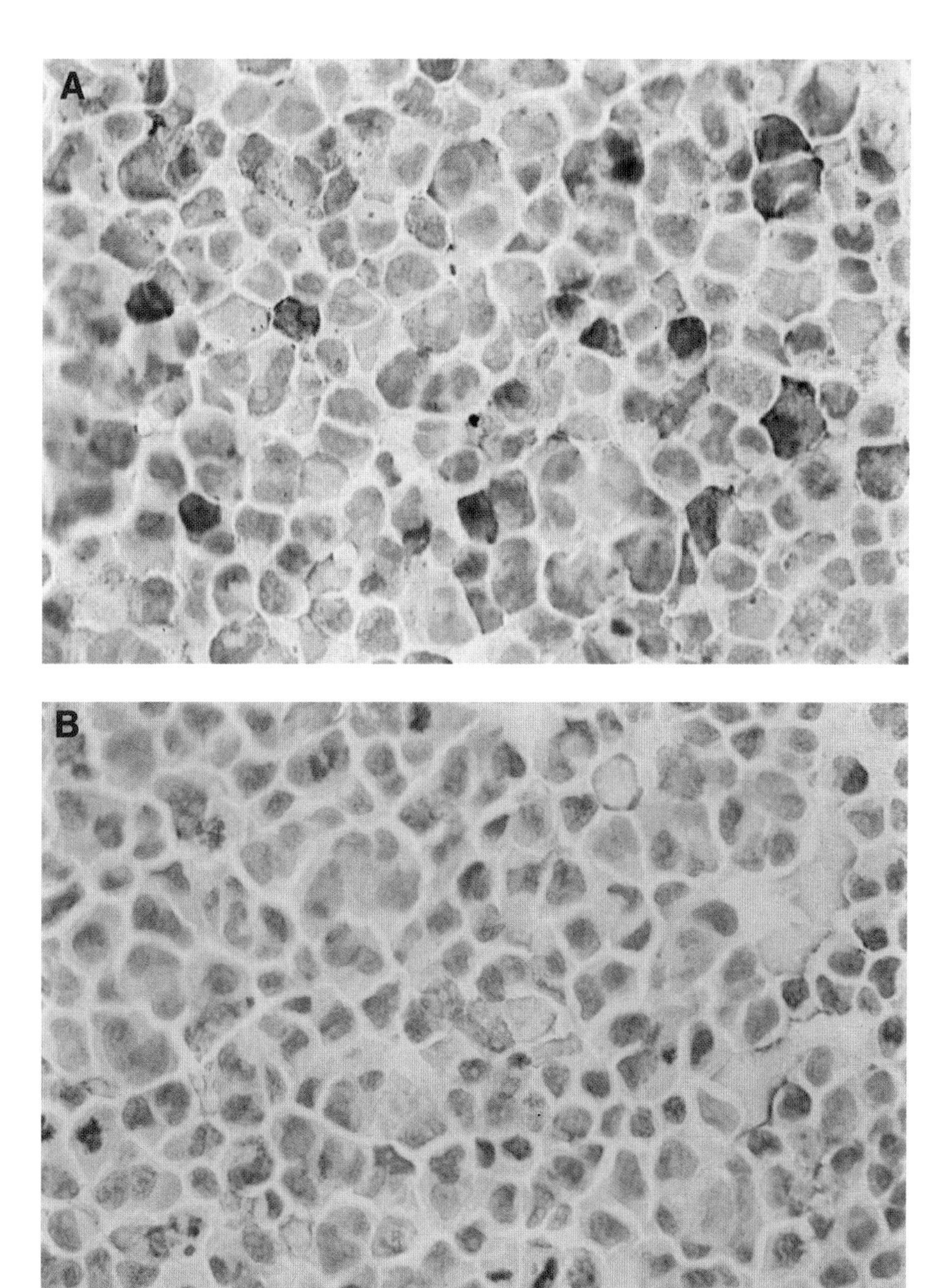

Figure 5 Immunocytochemical analysis of 293T cells after viral gene transfer of HuD. A and B correspond to 293T cells after viral gene transfer of HuD and C and D to 293T cells after viral gene transfer of the LacZ gene. Novel synthesis of HuD is demonstrated in A, which shows nuclear and cytoplasmic reactivity with human anti-Hu IgG. The same cells reacted with normal human IgG (B) do not demonstrate reactivity. C demonstrates that viral gene transfer of the control LacZ gene does not result in any reactivity with anti-Hu IgG; the presence of control virus is demonstrated by X-gal histochemisty (D).

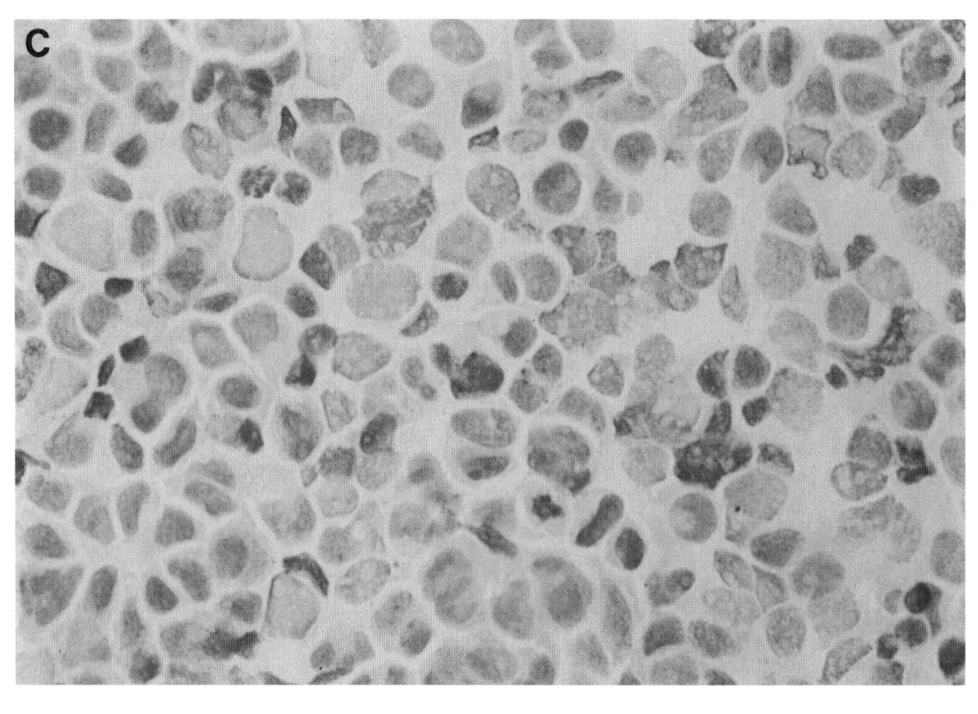

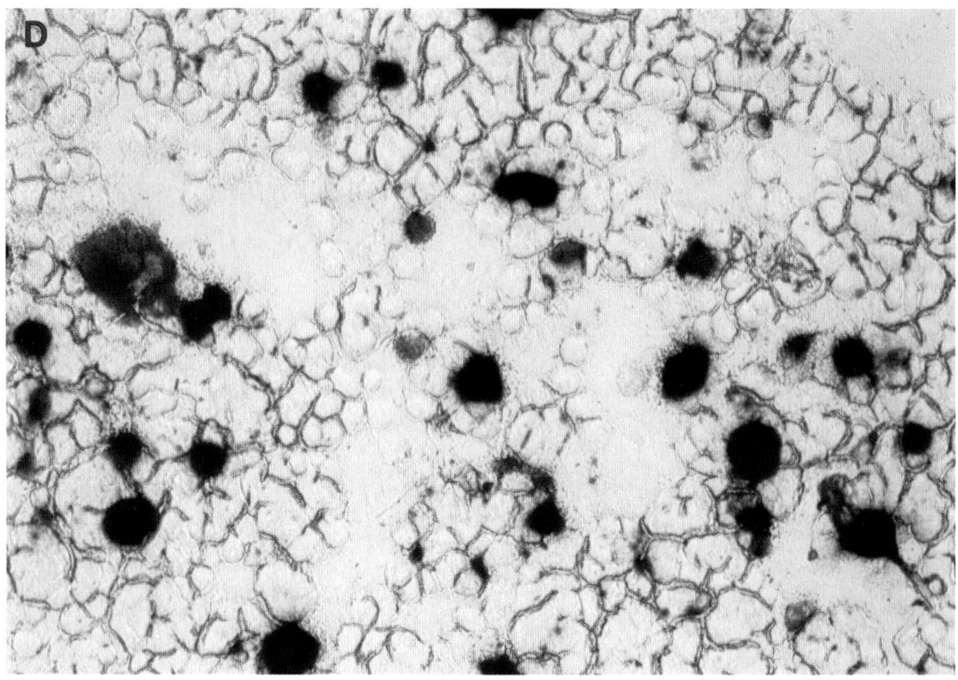

Figure 5 *(continued)*

the finding of conspicuous infiltrates of cytotoxic T cells in the nervous system of patients who die of PEM (Jean *et al.*, 1994).

One major limitation to develop an animal model of a disease mediated by a cytotoxic T cell immunoresponse is that the protein antigen has to be processed and presented to the immune system in conjunction with MHC class I proteins of the same (or syngeneic) animal (Royer and Reinherz, 1987; Townsend *et al.*, 1986). To circumvent this limitation, we use DNA viral vectors to transfer paraneoplastic genes to terminally differentiated mammalian tissues which have both MHC class I and class II pathways of antigen presentation.

For these experiments we generated both a defective HSV vector and an AAV vector carrying HuD cDNA. Similar defective viral vectors containing the bacterial marker gene for β-galactosidase were used as controls. Viral gene transfer of HuD to cell lines, including 293T cells (a transformed human kidney cell line), Saos-2 (a human osteosarcoma cell line), Vero cells (an African green monkey kidney cell line), and L929 cells (a mouse fibroblast cell line), resulted in the novel synthesis of immunoreactive HuD in all cell lines. Expression of HuD protein was demonstrated by immunocytochemistry (Fig. 3) and Western blot analysis (Fig. 4) using polyclonal anti-Hu antibodies. Immunocytochemical studies demonstrated expression of HuD in the nucleus and cytoplasm of cells; in some cells reactivity was more intense in the nucleus and in others, it was more intense in the cytoplasm. This pattern of immunostaining differs from the HuD immunoreactivity observed in neurons and SCLC, which always

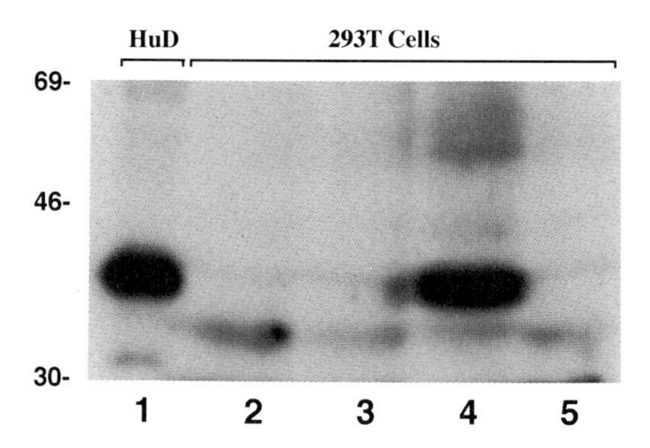

Figure 4 Western blot analysis of 293T cells after viral gene transfer of HuD. Lanes 2 and 5 correspond to protein extracts of 293T cells before gene transfer. Lane 3 corresponds to 293T cells after gene transfer of the control LacZ gene, and lane 4 corresponds to 293T cells after gene transfer of HuD. The same amount of protein has been used in all lanes. Reactivity with anti-Hu sera is only observed in protein extracts of cells that received the HuD gene (lane 4). Lane 1 corresponds to recombinant HuD protein and is used as a positive control.

predominates in the nuclei of the cells. However, a time course analysis of the expression of HuD in cytoplasmic and nuclear protein fractions obtained from cells after viral gene transfer demonstrated a rapid (1 hr) nuclear localization of HuD with a later (after 6 hr) predominant localization in the cytoplasm. Since HuD has no nuclear localization signal, these findings may represent saturation of the translocation system(s) that transport HuD from cytoplasm to the nucleus (Dalmau *et al.,* 1994a).

To study viral gene transfer of HuD *in vivo* we selected liver as an easily accessible target for viral injection. Previous studies using immunohistochemistry, Western blot analysis, and reverse transcriptase/polymerase chain reaction have demonstrated that liver does not contain HuD mRNA or express HuD protein (Szabo *et al.,* 1991; Dalmau *et al.,* 1992a). Eighteen hours after injection of the viral vector carrying HuD, the novel synthesis of HuD protein was demonstrated by immunohistochemical analysis of sections of tissue using a monoclonal antibody against HuD (Dalmau *et al.,* 1994b).

These studies demonstrated that immunoreactive HuD can be expressed in nonneuronal mammalian cells and as occurs in neurons, HuD localizes to the nuclei of cells. In order to develop an animal model of PEM we are currently using these defective viral vectors to transfer HuD to cells which have the ability to present antigens in association with MHC class I proteins, which may then be used to induce a cytotoxic T cell response. We selected the mouse fibroblast cell line (L929) which has been used to induce synthesis of other antibodies and is known to be a good target for cytotoxic T cell studies (Bennink and Yewdell, 1990). Preliminary experiments demonstrate that L929 cells do not express HuD and that novel expression of this protein is successfully achieved after viral gene transfer. L929 cells expressing HuD will be injected subcutaneously to syngeneic mice, and studies will be done in order to evaluate if animals develop symptoms of the disease, produce anti-Hu antibodies, or develop a cytotoxic T cell response.

Another, perhaps more direct, approach to achieve appropriate antigen presentation of HuD to the immunological system may be to transfer the HuD gene directly to antigen-presenting cells (APC) using defective viral vectors. To examine if this is feasable we harvested rat splenocytes and infected them with the defective control virus carrying the gene for β-galactosidase. Using an approximately equal number of viral particles as splenocytes we achieved expression of β-galactosidase in 5% of the cells (Fig. 5). Similar experiments using defective virus carrying the HuD gene are currently in progress. If expression of HuD is achieved, splenocytes will be returned to syngeneic animals for induction of an immune response. One advantage in using either L929 cells or APC cells is that these cell types are syngeneic to their host. Thus, the only nonself protein which will be expressed is the product of the transferred paraneoplastic gene; the appropriate presentation of this protein to the immune system may result in a T cell immune response.

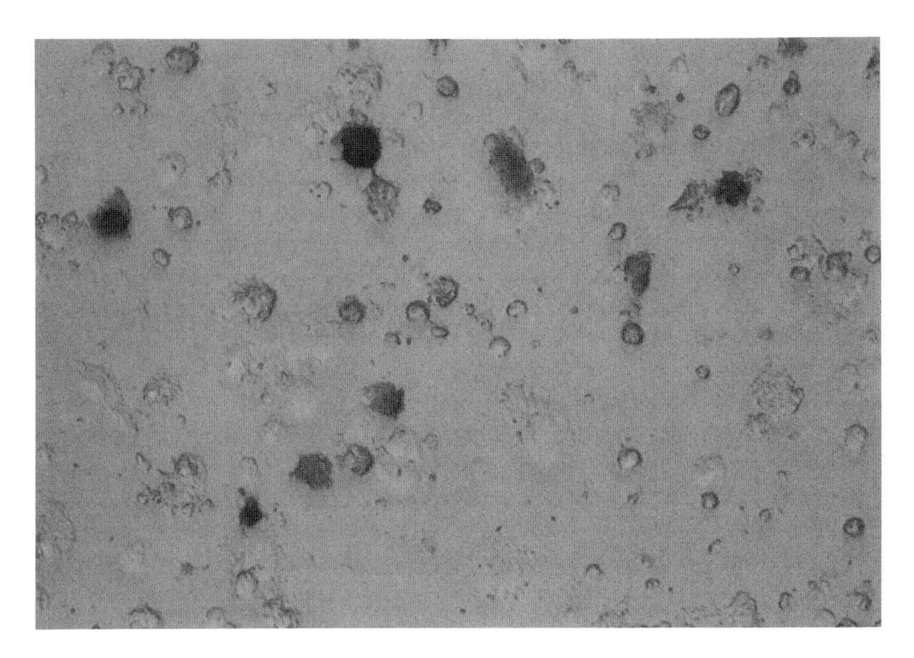

Figure 5 Rat splenocytes after transfer of the LacZ gene. The novel synthesis of β-galactosidase is demonstrated by X-gal histochemistry.

IV. Summary

The ability to transfer genes using DNA viral vectors is becoming a powerful tool for the neuroscience investigator and has a promising future for the treatment of cancer. DNA viral vectors provide the neuroscientist with the ability to manipulate gene expression in terminally differentiated cells of the nervous system, to express neuronal proteins in nonneuronal tissues, and to study immunological responses against neuronal proteins expressed in tumors, which may lead to the development of new strategies for the treatment of cancer and paraneoplastic syndromes.

As the pathways of oncogenesis are elucidated, the appropriate targets of gene therapy will be found. Since multiple genetic defects are involved in tumorigenesis, replacement of all defective genes may not be possible. However, replacement of critical genes may be sufficient to suppress cell growth or induce cell death, avoiding the toxic side effects of current therapies. This is supported by our studies showing that replacement of one tumor-suppressor gene, p53, leads to growth arrest and, in some cells, apoptosis. Through the use of inducible or cell-specific promoters, gene therapy may be directed to the target cell. The development of safe and efficient viral vectors is important. The use of the helper virus-free AAV vector is a step in that direction.

In addition to the potential therapeutic implications of our studies, we are using defective viral vectors to study paraneoplastic syndromes. It is believed that the ectopic expression of neuronal proteins by tumors that have the ability to present these proteins to the immunological system results in an immune response against the tumor that cross-reacts with the nervous system. Some of these disorders are mediated by a humoral immune response, but for others a cytotoxic T cell response appears to have a pathogenic role. Using defective viral vectors we have obtained expression of immunoreactive HuD (a paraneoplastic neuronal protein) in cells that have the ability to present antigens to the immunological system. An advantage of this approach is that the antigen-presenting cells that receive the paraneoplastic gene are syngeneic to their host. This may be relevant for developing an animal model of paraneoplastic neurologic diseases, since it allows us to overcome the restriction that cytotoxic T cell responses are only induced when antigens are presented to the immune system in conjunction with MHC class I proteins of the same animal.

Acknowledgments

We thank Dr. Jerome B. Posner for his support. M.R.R. is supported in part by Grant 1 KO8 NS 01626-01 from the National Institutes of Health, J.J.V. is supported by a fellowship from the Netherlands Organization for Scientific Research (NWO), and J.D. is supported in part by a grant from the American Cancer Society.

References

Arends, M. J., and Wyllie, A. H. (1991). Apoptosis: Mechanisms and roles in pathology. In "International Review of Pathology," Vol. 32, pp. 233–254. Academic Press, New York.

Baker, S. J., Markowitz, S., Fearon, E. R., Willson, J. K. V., and Vogelstein, B. (1990). Suppression of human colorectal carcinoma cell growth by wild-type p53. *Science* **249**, 217–221.

Ball, R. K., Siegl, B., Quellhorst, S., Brandner, G., and Braun, D. B. (1984). Monoclonal antibodies against simian virus 40 nuclear large T tumor antigen: Epitope mapping, papova virus cross-reaction and cell surface staining. *EMBO J.* **3**, 1485–1491.

Banks, L., Matlashewski, G., and Crawford, L. (1986). Isolation of human p53-specific monoclonal antibodies and their use in the studies of human p53-expression. *Eur. J. Biochem.* **159**, 529–534.

Bennink, J. R., and Yewdell, J. W. (1990). Recombinant vaccinia viruses as vectors for studying T lymphocyte specificity and function. *Curr. Topics Microbiol. Immunol.* **163**, 153–184.

Bishop, J. M. (1991). Molecular themes in oncogenesis. *Cell* **64**, 235–248.

Brusch, W., Kleine, L., and Tenniswood, M. (1990). The biochemistry of cell death by apoptosis. *Biochem. Cell. Biol.* **88**, 1071–1074.

Buckanovich, R. J., Posner, J. B., and Darnell, R. B. (1993). Nova, the paraneoplastic Ri antigen is homologous to an RNA-binding protein and is specifically expressed in the developing motor system. *Neuron* **11**, 657–672.

Campos, A. R., Grossman, D., and White, K. (1985). Mutant alleles at the locus Elav in Drosophila melangaster lead to nervous system defects. A developmental-genetic analysis. *J. Neurogenet.* **2**, 197–218.

Chandler, K. L., Prados, M. D., Malec, M., and Wilson, C. B. (1993). Long-term survival in patients with glioblastoma multiforme. *Neurosurgery* **32**, 716–720.

Dalmau, J., Furneaux, H. M., Gralla, R. J., Kris, M.G., and Posner, J. B. (1990). Detection of the anti-Hu antibody in the serum of patients with small cell lung cancer—A quantitative western blot analysis. *Ann. Neurol.* **27**, 544–552.

Dalmau, J., Furneaux, H. M., Rosenblum, M. K., Graus, F., and Posner, J. B. (1991). Detection of the anti-Hu antibody in specific regions of the nervous system and tumor from patients with paraneoplastic encephalomyelitis/sensory neuropathy. *Neurology* **41**, 1757–1764.

Dalmau, J., Furneaux, H. M., Cordon-Cardo, C., and Posner, J. B. (1992a). The expression of the Hu (paraneoplastic encephalomyelitis/sensory neuronopathy) antigen in human normal and tumor tissues. *Am. J. Pathol.* **141**, 881–886.

Dalmau, J., Graus, F., Rosenblum, M. K., and Posner, J. B. (1992b). Anti-Hu associated paraneoplastic encephalomyelitis/sensory neuronopathy: A clinical study of 71 patients. *Medicine* **71**, 59–72.

Dalmau, J., Kaplitt, M. G., Meneses, P., Posner, J. B., and Rosenfeld, M. R. (1994a). Novel expression of immunoreactive HuD (paraneoplastic encephalomyelitis) antigen in non-neuronal mammalian cells using a defective herpes viral (HSV) vector. *J. Neurol.* **241**(1), S146.

Dalmau, J. D., Kaplitt, M. G., Meneses, P., Rosenfeld, M. R., and Posner, J. B. (1994b). Novel expression of the HuD antigen in mammalian cells after gene transfer with a defective herpes viral (HSV) vector. *Neurology* **44**(suppl 2), A163.

DeLattre, J. Y., and Posner, J. B. (1989). Neurological complication of chemotherapy and radiation therapy. In "Neurology and General Medicine" (M. J. Aminoff, Ed.), pp. 365–387. Churchill–Livingston, New York.

Dropcho, E. J., Chen, Y-T., Posner, J. B., and Old, L. J. (1987). Cloning of a brain protein identified by autoantibodies from a patient with paraneoplastic cerebellar degeneration. *Proc. Natl. Acad. Sci. USA* **84**, 4552–4556.

El-Diery, W. S., Tokino, T., Velculescu, V. E., Levy, D. P., Parsons, R., Trent, J. M., Lin, D., Mercer, W. E., Kinzler, K. W., and Vogelstein, B. (1993). WAF1, a potential mediator of p53 tumor suppression. *Cell* **75**, 817–825.

Eliyahu, D., Goldfinger, N., Pinhasi-Kimhi, O., Shaulsky, Y., Skurnik, N., Arai, V., Rotter, V., and Oren, M. (1988). Meth A fibrosarcoma cells expresses two transforming mutant p53 species. *Oncogene* **3**, 313–321.

Fathallah-Shaykh, H., Wolf, S., Wong, E., Posner, J. B., and Furneaux, H. M. (1991). Cloning of a leucine zipper protein recognized by the sera of patients with antibody-associated paraneoplastic cerebellar degeneration. *Proc. Natl. Acad. Sci. USA* **88**, 3451–3454.

Finlay, C. A., Hinds, P. W., and Levine, A. J. (1982). The p53 proto-oncogene can act as a suppressor of transformation. *Cell* **57**, 1083–1093.

Finlay, C. A., Hinds, P. W., Tan, T-H., Eliyahu, D., Oren, M., and Levine, A. J. (1988). Activating mutations for transformation by p53 produce a gene product that forms an hsc70–p53 complex with an altered half-life. *Mol. Cell. Biol.* **8**, 531–539.

Frankel, R. H., Bayona, W., Koslow, M., and Newcomb, E. (1992). p53 mutations in human malignant gliomas: Comparison of loss of heterozygosity with mutation frequency. *Cancer Res.* **52**, 1427–1433.

Furneaux, H. M., Rosenblum, M. K., Dalmau, J., Wong, E., Woodruff, P., and Posner, J. B. (1990a). Selective expression of Purkinje cell antigens in tumor tissue from patients with paraneoplastic cerebellar degeneration. *N. Engl. J. Med.* **322**, 1844–1851.

Furneaux, H. M., Reich, L., and Posner, J. B. (1990b). Autoantibody synthesis in the central nervous system of patients with paraneoplastic syndromes. *Neurology* **40**, 1085–1091.

Gavrieli, Y., Sherman, Y., and Ben-Sasson, S. A. (1992). Identification of programmed cell death in situ via specific labeling of nuclear DNA fragmentation. *J. Cell. Biol.* **119,** 493–501.

Ginsberg, D., Mechta, F., Yaniv, M., and Oren, M. (1991). Wild-type p53 can down-modulate the activity of various promoters. *Proc. Natl. Acad. Sci. USA* **88,** 9979–9983.

Graham, F. L., and Van der Eb, A. J. (1973). A new technique for the assay of infectivity of human adenovirus 5 DNA. *Virology* **52,** 456–467.

Graus, F., Elkton, K. B., Cordon-Cardo, C., and Posner, J. B. (1986). Sensory neuronopathy and small cell lung cancer: Antineuronal antibody that also reacts with the tumor. *Am J. Med.* **80,** 45–52.

Graus, F., and Ferrer, I. (1990). Analysis of a neuronal antigen (Hu) expression in the developing rat brain detected by autoantibodies from patients with paraneoplastic encephalomyelitis. *Neurosci. Lett.* **112,** 14–18.

Greig, N. H., Ries, L. G., Yancik, R., Hellman, K., and Rapoport, S. I. (1990). Increasing annual incidence of primary malignant brain tumors in the elderly. *Proc. Am. Assoc. Cancer Res.* **31,** 229.

Halazonetis, T. D., Davis, L. J., and Kandil, A. N. (1993). Wild-type p53 adopts a "mutant"-like conformation when bound to DNA. *EMBO J.* **12,** 1021–1028.

Harlow, E., Crawford, L. V., Pim, D. C., and Williamson, N. M. (1981). Monoclonal antibodies specific for simian virus 40 tumor antigens. *J. Virol.* **39,** 861–869.

Harper, J. W., Adami, G. R., Wei, N., Keyomari, K., and Elledge, S. J. (1993). The p21 Cdk-interacting protein Cip1 is a potent inhibitor of G1 cyclin-dependent kinases. *Cell* **75,** 805–816.

Henson, R. A., and Urich, H. (1989). Encephalomyelitis with carcinoma. In "Cancer and the Nervous System" (R. A. Henson and H. Urich, Eds.), pp. 314–345. Blackwell Scientific, Oxford, England.

Hollstein, M., Sidransky, D., Vogelstein, B., and Harris, C. C. (1991). p53 mutations in human cancers. *Science* **253,** 49–53.

Iggo, R., Gatter, K., Bartek, J., Lane, D. P., and Harris, A. (1990). Increased expression of mutant forms of p53 oncogene in primary lung cancer. *Lancet* **335,** 675–679.

Jean, W. C., Dalmau, J., Ho, A., and Posner, J. B. (1994). Analysis of the IgG subclass distribution and inflammatory infiltrates in patients with anti-Hu associated paraneoplastic encephalomyelitis. *Neurology* **44,** 140–147.

Kaplitt, M. G., Pfaus, J. G., Kleopoulos, S. P., Hanlon, B. A., Rabkin, S. D., and Pfaff, D. W. (1991). Expression of a functional foreign gene in adult mammalian brain following *in vivo* transfer via a Herpes Simplex virus type 1 defective viral vector. *Mol. Cell. Neurosci.* **2,** 320–330.

Kastan, M. B., Zhan, Q., El-Diery, W. S., Carrier, F., Jacks, T., Walsh, W. V., Plunkett, B. S., Vogelstein, B., and Fornace, A. J., Jr. (1992a). A mammalian cell cycle checkpoint pathway utilizing p53 and GADD45 is defective in ataxia–telangiectasia *Cell* **71,** 587–597.

Kastan, M. B., Onyekwere, O., Sidransky, D., Vogelstein, B., and Craig, R. W. (1992b). Participation of p53 protein in the cellular response to DNA damage. *Cancer Res.* **51,** 6304–6311.

Kuerbitz, S. J., Plunkett, B. S., Walsh, W. V., and Kastan, M. (1992). Wild-type p53 is a cell cycle checkpoint determinant following irradiation. *Proc. Natl. Acad. Sci. USA* **89,** 7491–7495.

Kwong, A. D., and Frenkel, N. (1985). The herpes simplex virus amplicon, IV. Efficient expression of a chimeric chicken ovalbumin gene amplified within defective virus genomes. *Virology* **142,** 421–425.

Lang, B., Newsom-Davis, J., Wray, D., Vincent, A., and Murray, N. (1981). Autoimmune etiology for myasthenic (Lambert–Eaton) syndrome. *Lancet* **2,** 224–226.

Laws, E. R., and Thapar, K. (1993). Brain tumors. *CA Cancer J. Clin.* **43,** 263–271.

Levine, A. J., Momand, J., and Finlay, C. A. (1991). The p53 tumour suppressor gene. *Nature* **351,** 453–456.

Lowe, S. W., Schmitt, E. M., Smith, S. W., Osborne, B., and Jacks, T. (1993). p53 is required for radiation-induced apoptosis in mouse thymocytes. *Nature* **362,** 847–849.

Luque, F. A., Furneaux, H. M., Ferziger, R., Rosenblum, M. K., Wray, S. H., Schold, S. C., Glantz, M. J., Jaeckle, K. A., Biran, H., Lesser, M., Paulsen, W. A., River, M. E., and Posner, J. B. (1991). Anti-Ri: An antibody associated with paraneoplastic opsoclonus and breast cancer. *Ann. Neurol* **29,** 241–251.

Marusich, M. F., and Weston, J. A. (1992). Identification of early neurogenic cells in the neural crest lineage. *Dev. Biol.* **149,** 295–306.

Mercer, W. E., Shields, M. T., Amin, M., Sauve, G. J., Appella, E., Romano, J. W., and Ullrich, S. J. (1990). Negative growth regulation in a glioblastoma tumor cell line that conditionally expresses human wild-type p53. *Proc. Natl. Acad. Sci. USA* **87,** 6166–6170.

Mercer, W. E., Shields, M. T., Lin, D., Appella, E., and Ullrich, S. J. (1991). Growth suppression induced by wild-type p53 protein is accompanied by selective down-regulation of proliferating-cell nuclear antigen expression. *Proc. Natl. Acad. Sci. USA* **88,** 1958–1962.

Milner, J., Medcalf, E. A., and Cook, A. C. (1991). Tumor suppressor p53: Analysis of wild-type and mutant p53 complexes. *Mol. Cell Biol.* **11,** 12–19.

Momand, J., Zambetti, G. P., Olson, D. C., George, D. L., and Levine, A. J. (1992). The mdm-2 oncogene product forms a complex with the p53 protein and inhibits p53- mediated trans-activation. *Cell* **69,** 1237–1245.

Nigro, J. M., Baker, S. J., Preisinger, A. C., Jessup, J. M., Hostetter, R., Cleary, K., Bigner, S. H., Davidson, N., Baylin, S., Devilee, P., Glover, T., Collins, F. S., Weston, A., Modall, R., Harris, C. C., and Vogelstein, B. (1989). Mutations in the p53 gene occur in diverse tumour types. *Nature* **342,** 705–708.

Ohgaki, H., Eibl, R. H., Wiestler, O. D., Yasargil, M. G., Newcomb, E. W., and Kleihues, P. (1991). p53 mutations in nonastrocytic brain tumors. *Cancer Res.* **51,** 6202–6205.

Oldfield, E. H., Culver, K. W., Ram, Z., and Blaese, R. M. (1993). A clinical protocol: Gene therapy for the treatment of brain tumors using intra-tumoral transduction with the thymidine kinase gene and intravenous ganciclovir. *Hum. Gene Ther.* **4,** 39–69.

O'Neill, J. H., Murrary, N. M. F., and Newsom-Davis, J. (1988). The Lambert–Eaton myasthenic syndrome. A review of 50 cases. *Brain* **111,** 577–596.

Posner, J. B., and Furneaux, H. M. (1990). Paraneoplastic syndromes. In "Immunologic Mechanisms in Neurologic and Psychiatric Disease" (B. H. Waksman, Ed), pp. 187–219. Raven Press, New York.

Ramqvist, T., Magnusson, K. P., Wang, Y., Szekely, L., Klein, G., and Wiman, K. (1993). Wild-type p53 induces apoptosis in a Burkitt lymphoma (BL) line that carries mutant p53. *Oncogene* **8,** 1495–1500.

Reed, S. I., Wittenberg, C., Lew, D. J., Dulic, V., and Henze, M. (1992). G1 control in yeast and animal cells. In "The Cell Cycle" (D. Beach, B. Stillman, and J. D. Watson, Eds.), Vol. 56, pp. 61–67. Cold Spring Harbor Laboratory Press, Cold Spring Harbor, New York.

Roberts, A., Perera, S., Lang, B., Vincent, A., and Newsom-Davis, J. (1985). Paraneoplastic myasthenic syndrome IgG inhibit $45Ca2^+$ flux in human small cell carcinoma line. *Nature* **317,** 737–739.

Robinow, S., and White, K. (1988). The locus of Drosophila melangaster is expressed in neurons at all developmental stages. *Dev. Biol.* **126,** 294–303.

Robinow, S., Campos, A., Yao, K., and White, K. (1988). The elav gene product of Drosophila, required in neurons, has three RNP consensus motifs. *Science* **242,** 1570–1572.

Rogel, A., Popliker, M., Webb, C. G., and Oren, M. (1985). p53 cellular tumor antigen: Analysis of mRNA levels in normal adult tissues, embryos, and tumors. *Mol. Cell. Biol.* **5,** 2851–2855.

Rosenfeld, M. R., Meneses, P., Dalmau, J., Drobjnak, M., Cordon-Cardo, C., and Kaplitt, M. G. (1994). Gene transfer of wildtype p53 results in restoration of tumor suppressor function in a medulloblastoma cell line. *Neurology,* (in press).

Rosenfeld, M. R., Meneses, P., Kaplitt, M. G., Dalmau, J., Posner, J. B., and Cordon-Cardo, C. (1994). Expression of wild-type p53 after gene transfer with a defective herpes viral vector results in down-regulation of mutant p53 expression and up-regulation of mdm-2. *J Neurol.* **241**(1), S146.

Royer, H. D., and Reinherz, E. L. (1987). T-lymphocytes: Ontogeny, function, and relevance to clinical disorders. *N. Engl. J. Med.* **317**, 1136–1142.

Sager, R. (1989). Tumor suppressor genes: The puzzle and the promise. *Science* **246**, 1406–1412.

Salford, L., Brun, A., and Nirfalk, S. (1988). Ten-year survival among patients with supratentorial astrocytomas grade III and IV. *J. Neurosurg.* **69**, 506–509.

Samulski, R. J., Chang, L. S., and Shenk, T. S. (1989). Helper-free stocks of recombinant adeno-associated viruses: Normal integration does not require viral gene expression. *J. Virol.* **63**, 3822–3828.

Santhanam, U., Ray, A., and Sehgal, P. (1991). Repression of the interleukin 6 gene promoter by p53 and the retinoblastoma susceptibility gene product. *Proc. Natl. Acad. Sci. USA* **88**, 7605–7609.

Saylors, R. L., Sidransky, D., Friedman, H. S., Bigner, S. H., Bigner, D. D., Vogelstein, B., and Brodeur, G. M. (1991). Infrequent p53 gene mutations in medulloblastoma. *Cancer Res.* **51**, 4721–4723.

Shaw, P., Bovey, R., Tardy, S., Sahli, R., Sordat, B., and Costa, J. (1992). Induction of apoptosis by wild-type p53 in a human colon tumor-derived cell line. *Proc. Natl. Acad. Sci. USA* **89**, 4495–4499.

Shiio, Y., Yamamoto, T., and Yamaguchi, N. (1992). Negative regulation of Rb expression by the p53 gene product. *Proc. Natl. Acad. Sci. USA* **89**, 5206–5210.

Sidransky, D., Mikkelsen, T., Schwechheimer, K., Rosenblum, M. L., Cavanee, W., and Vogelstein, B. (1992). Clonal expansion of p53 mutant cells is associated with brain tumour progression. *Nature* **355**, 846–847.

Sillevis-Smitt, P., Manley, G., and Posner, J. B. (1994). High titer antibody but no disease in mice immunized with the paraneoplastic antigen HuD. *Neurology* **44**(2), A378.

Spaete, R. R., and Frenkel, N. (1982). The herpes simplex virus amplicon: A new eukaryotic defective-virus cloning-amplifying vector. *Cell* **30**, 295–304.

Szabo, A., Dalmau, J., Manley, G., Rosenfeld, M., Wong, E., Henson, J., Posner, J. B., and Furneaux, H. M. (1991). HuD, a paraneoplastic encephalomyelitis antigen, contains RNA-binding domains and is homologous to Elav and Sex-lethal. *Cell* **67**, 325–333.

Townsend, A. R. M., Rothbard, J., Gotch, F. M., Bahadur, G., Wraith, D., and McMichael, A. J. (1986). The epitopes of influenza nucleoprotein recognized by cytotoxic T lymphocytes can be defined with short synthetic peptides. *Cell* **44**, 959–968.

Vertosick, F., and Selker, R. (1992). Long-term survival after the diagnosis of malignant glioma: A series of 22 patients surviving more than 4 years after diagnosis. *Surg. Neurol.* **38**, 359–363.

Von Deimling, A., Eibl, R. H., Ohgaki, H., Louis, D. N., von Ammon, K., Petersen, I., Kleihues, P., Chung, R. Y., Wiestler, O. D., and Seizinger, B. R. (1992). p53 mutations are associated with 17p allelic loss in grade II and grade III astrocytoma. *Cancer Res.* **52**, 2987–2990.

Weinberg, R. A. (1991). Tumor suppressor genes. *Science* **254**, 1138–1146.

Wu, X., Bayle, H., Olson, D., and Levine, A. J. (1993). The p53-mdm-2 autoregulatory feedback loop. *Genes Dev.* **7**, 1126–1132.

Wyllie, A. H. (1981). Cell death: a new classification separating apoptosis from necrosis. In "Cell Death in Biology and Pathology" (I. D. Bowen and R. A. Lockshin, Eds.), pp. 9–34. Chapman and Hall, London.

Wyllie, A. H., Morris, R. G., Smith, A. L., and Dunlop, D. (1984). Chromatin cleavage in apoptosis: Association with condensed chromatin morphology and dependence on marcromolecular synthesis. *J. Pathol.* **142**, 67–77.

Yonish-Rouach, E., Resnitzky, D., Lotem, J., Sachs, L., Kimchi, A., and Oren, M. (1991). Wild type p53 induces apoptosis of myeloid leukaemic cells that is inhibited by interleukin-6. *Nature* **352**, 345–347.

CHAPTER 17

Transneuronal Tracing with Alpha-herpesviruses: A Review of the Methodology

Gabriella Ugolini
Laboratoire de Génétique des Virus, CNRS
Gif-Sur-Yvette, France

I. Introduction

Conventional tracing methods rely on the use of cellular markers that are transported in axons in the anterograde or retrograde direction from the neuronal cell body (Kuypers and Huisman, 1984; Mesulam, 1982). Minute quantities of these markers are injected into a specific peripheral or CNS site and then, after a sufficient time has elapsed for uptake and transport to occur, the animals are perfused with fixatives and the distribution of the marker is studied in tissue sections. Although considerable amounts of connectional data have been collected in this manner, these tracing methods have substantial limitations. For example, it is usually difficult to restrict an injection of tracer to a particular cell group without diffusion and uptake by nearby neurons or fibers of passage. Thus, false-positive labeling is a major problem associated with these methods. It is also very difficult to trace a chain of serially connected neurons. A solution to both of these problems has been provided by the development of transneuronal tracers, i.e., markers that are transferred between synaptically connected neurons. With these tracers, it is now possible to label a whole chain of connected neurons that control a given end target (e.g., a muscle or a specific CNS site).

For a marker to be effective as a transneuronal tracer, transfer should occur specifically between connected neurons. Moreover, transneuronal labeling should be easy to detect and should involve all groups of neurons that are part of a given neuronal circuit. There are two classes of transneuronal markers: conventional tracers, viz. the wheat germ agglutinin–horseradish peroxidase conjugate (WGA–HRP) and nontoxic fragments of tetanus toxin (Alstermark

VIRAL VECTORS

and Kummel, 1986, 1990a,b; Alstermark *et al.,* 1987; Harrison *et al.,* 1984; Horn and Büttner-Ennever, 1990; Jankowska, 1985), and some neurotropic viruses, viz. alpha-herpesviruses, such as herpes simplex virus type 1 (HSV-1) and pseudorabies virus (PrV, suid herpesvirus 1), and rhabdoviruses (reviewed in Kuypers and Ugolini, 1990; Strick and Card, 1992). Transneuronal transfer can occur both in the anterograde direction, i.e., from afferent fibers to postsynaptic neurons, and in the retrograde direction, i.e., from neuronal cell bodies and their dendrites to presynaptic terminals on their surface and then retrogradely from these terminals to their parent cell bodies (Fig. 1).

With conventional tracers, such as WGA–HRP, transneuronal transfer is inefficient, especially in the retrograde direction, because it occurs only if first-order neurons are filled with great quantities of tracer (Alstermark and Kummel, 1986, 1990a,b; Alstermark *et al.,* 1987; Harrison *et al.,* 1984; Jankowska, 1985). Even under these extreme conditions, only trace amounts of

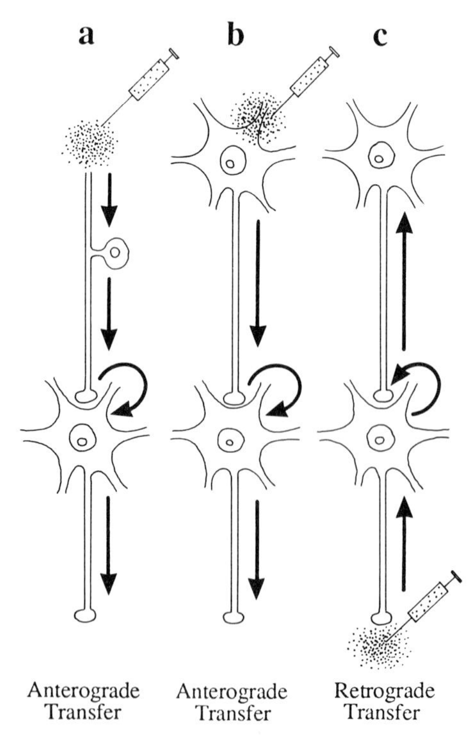

Figure 1 (A and B) Anterograde transneuronal transfer from bipolar peripheral sensory neuron (A) and from multipolar neuron (B) to postsynaptic neurons. (C) Retrograde transneuronal transfer, e.g., from motoneuron to presynaptic terminals of connected premotor interneuron. Reprinted with permission from Kuypers and Ugolini (1990).

these markers cross the synapse and label second-order neurons (Fig. 2). The second-order neuronal labeling is generally very weak and can be detected only in some groups of second-order neurons and the transfer is largely dependent on experimental neuronal activation (Alstermark and Kummel, 1986, 1990a,b; Kankowska, 1985). Moreover, the label disappears with time and third-order neurons cannot be detected (Alstermark *et al.*, 1987). Because of these pitfalls, use of these markers for transneuronal labeling has been restricted to only a few studies. In contrast, the viral transneuronal labeling method is more dependable and has already been used to study connections in a great variety of neural systems as will be discussed in this chapter.

A unique feature of viruses is their ability to replicate in recipient neurons after transneuronal transfer, thus functioning as a self-amplifying marker. As a result, second- and third-order neurons show the same intensity of labeling as first-order neurons (Figs. 2–4). In addition, viral transneuronal labeling does not require experimental activation of neuronal circuits and can be very extensive, involving all groups of second-order neurons, and even higher-order neurons (Kuypers and Ugolini, 1990) (see Section III,C).

The introduction of HSV-1 and PrV as transneuronal tracers has been a relatively recent advance. Although the anterograde transneuronal transfer

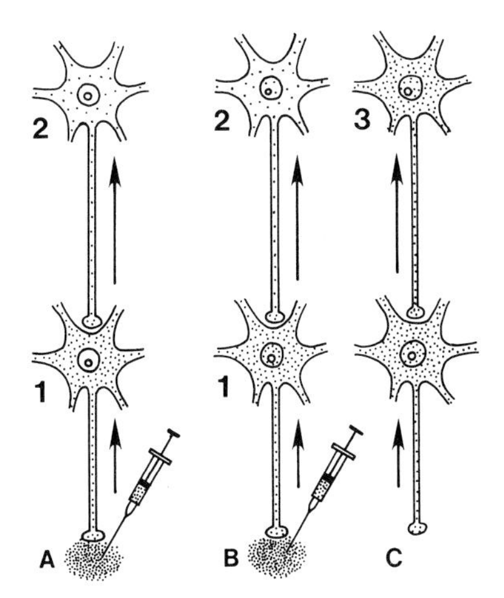

Figure 2 Comparison of the differences in staining intensity of the retrograde transneuronal labeling obtained with nonviral tracers (WGA–HRP, A) and alphaherpesviruses (HSV-1, B and C). (A) With WGA–HRP, only a small amount of tracer is transferred from first-order neurons (1) to second-order neurons, resulting in weak transneuronal labeling (2). (B and C) With alphaherpesviruses, transfer to second-order neurons (2) is followed by viral replication that results in intense transneuronal labeling (3). Adapted with permission from Kuypers and Ugolini (1990).

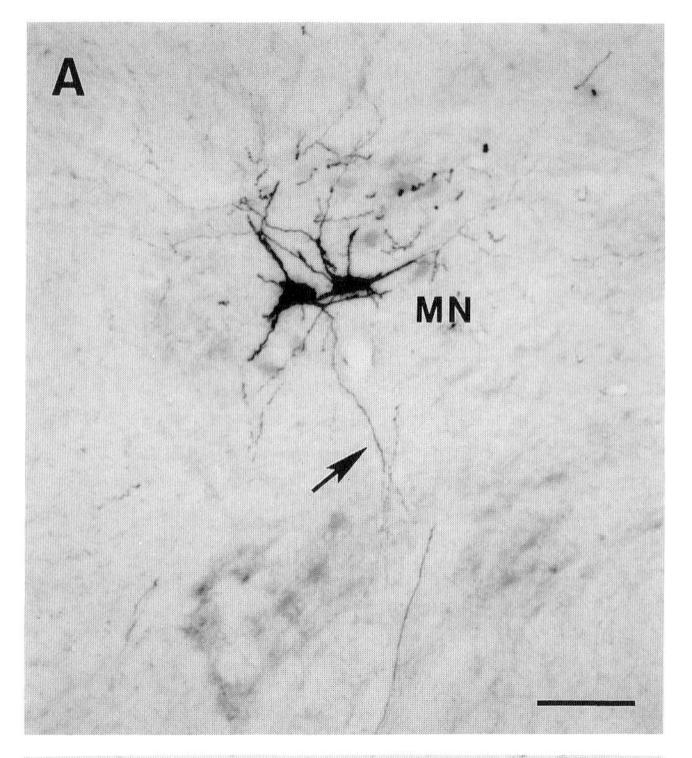
A
MN

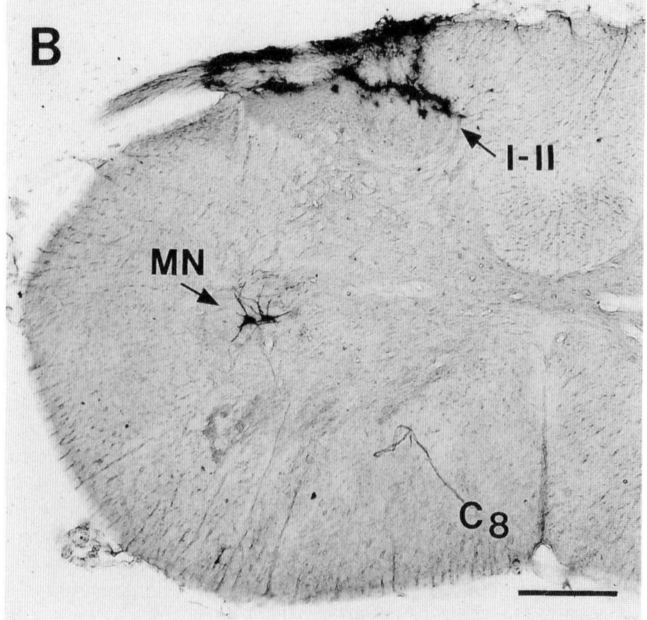
B
I-II
MN
C8

of HSV-1 through a chain of neurons was first demonstrated by Kristensson and collaborators in the 1970s (Kristensson *et al.*, 1974; 1982), the first evidence that HSV-1 could be used as an effective transneuronal tracer was provided by Ugolini and collaborators in 1987 (Ugolini *et al.*, 1987). Similarly, modern transneuronal labeling studies using PrV were initiated by Martin and Dolivo (1983), although much earlier reports had suggested that this was a possibility (Sabin, 1938). From the numerous studies that have been published in the past few years in which HSV-1 and PrV were used as transneuronal tracers, it has become clear that these two alphaherpesviruses have very similar properties as tracers. When using either virus for neural circuit analysis, optimal results require a correct choice of several experimental parameters (e.g., virus strain and dose) and knowledge of the kinetics of viral transfer. The aim of this chapter is to review the methodological aspects of this emerging technology.

II. Structure, Tropism, and Replication Cycle of HSV-1 and PrV

HSV-1 and PrV have a similar structure. The DNA of both viruses is enclosed by a nucleocapsid, which in turn is surrounded by an envelope (Roizman and Sears, 1990; Wildy, 1985). The envelope contains several surface glycoproteins that mediate binding and penetration of the viruses into host cells (reviewed in Mettenleiter, 1991, 1994). Since the surface glycoproteins of HSV-1 and PrV have several homologies, their nomenclature was unified in 1993, and now the original HSV-1 nomenclature is applied to both viruses. Both can infect all main categories of primary sensory neurons, motoneurons, sympathetic and parasympathetic neurons, as well as a great variety of CNS neurons (see Kuypers and Ugolini, 1990). In addition, both viruses can infect cells of epithelial origin as well as several classes of CNS glial cells, but peripheral glial cells do not support viral replication, at least in adult animals (Cook and Stevens, 1973; Dillard *et al.*, 1972).

Figure 3 HSV-1-positive elements stained immunohistochemically by the peroxidase–antiperoxidase method in the ipsilateral spinal segment C8 at short times (28–36 hr) after injection of HSV-1 into the ulnar and median nerves. (A) Retrograde labeling of forelimb motoneurons (MN). Note the complete labeling of these neurons including their axons (arrow). Glial cells are not labeled. Bar = 100 μm. (B) Low-power photomicrograph showing viral retrograde labeling of motoneurons (MN) and anterograde transneuronal labeling in laminae I and II of the dorsal horn that reflects transfer through small diameter sensory afferents. Also note labeling of sensory afferents and glial cells in cuneate fasciculus. Bar = 300 μm. Taken with permission from Ugolini (1992). (Copyright 1992 of Wiley–Liss, a division of Wiley & Sons, Inc.).

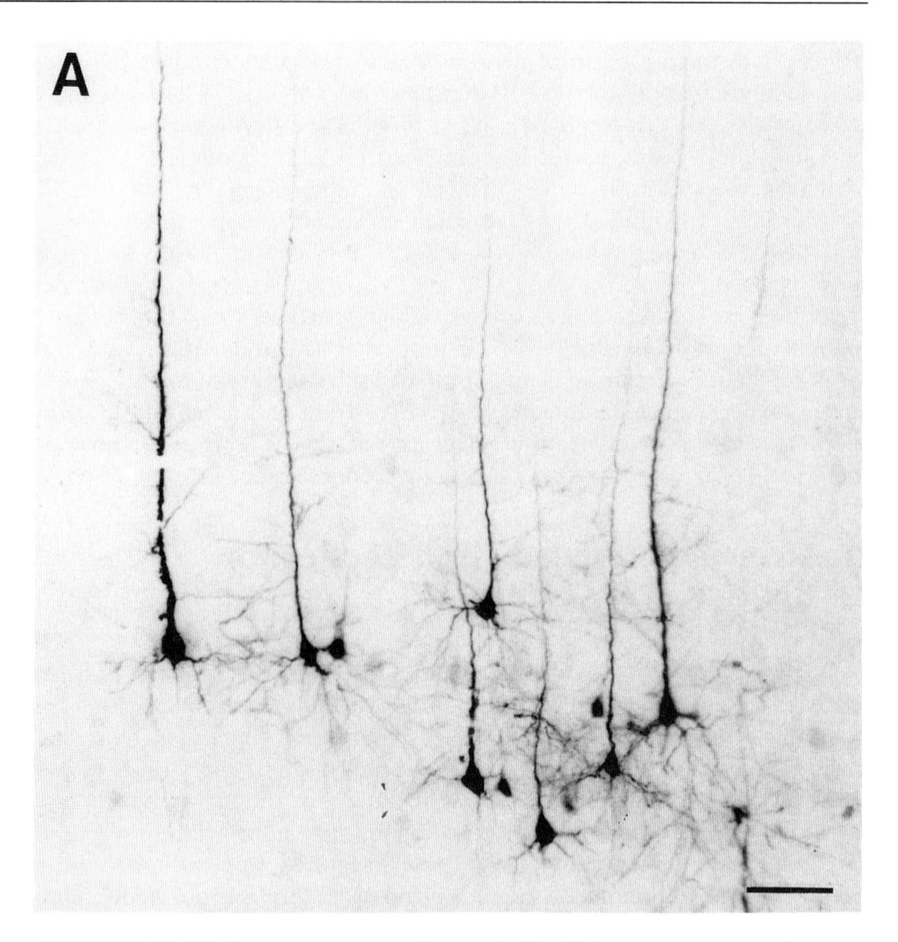
A

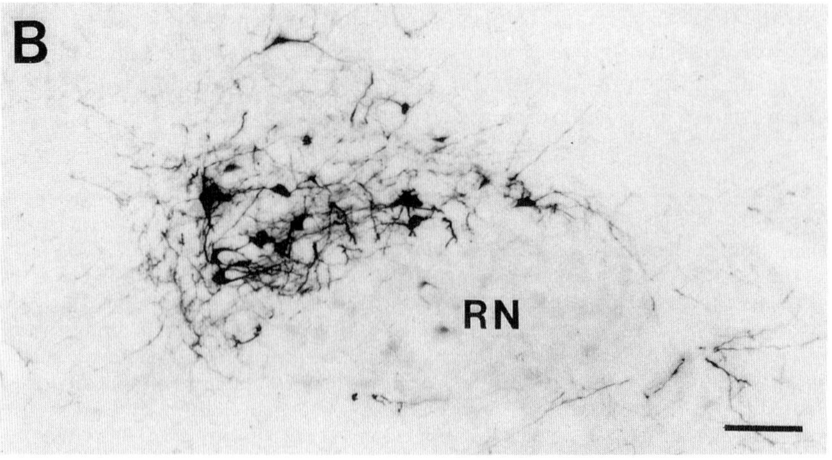
B
RN

Attachment of the viruses to neurons and glial cells is receptor mediated (Vahlne *et al.*, 1978). The precise nature of the cellular receptor(s) for HSV-1 and PrV is still unknown. Cell-surface glycosaminoglycans, especially heparan sulfate (HS), have a role in the primary step of binding that involves the viral envelope glycoprotein gC (previously designated gIII for PrV). However, secondary binding and penetration are also mediated by the interaction of other envelope glycoproteins with at least one non-HS cellular component (see Mettenleiter, 1994). The entry, replication, and release of HSV-1 and PrV in neurons has been well documented *in vitro* (Lycke *et al.*, 1984; Price *et al.*, 1982) and *in vivo* (see, e.g., Card *et al.*, 1993; Kristensson *et al.*, 1974). After attachment, the viruses are internalized by fusion of the envelope with the cell membrane followed by intracellular release of the nucleocapsid. This is then transported in the retrograde direction to the cell nucleus, where viral DNA enters to begin the replication process (Lycke *et al.*, 1984). After a cascade of replication steps, progeny nucleocapsids are assembled in the nucleus and acquire an envelope from the nuclear membrane during exit from the nucleus (Card *et al.*, 1993; Kuypers and Ugolini, 1990). The enveloped nucleocapsids enter the endoplasmic reticulum and are transported to the cell surface where they are released (Lycke *et al.*, 1984). Both the uptake and the release of the viruses can take place through dendrites or axons. Axonal transport of the viruses occurs at a fast rate (2–10 mm/hr) and often over long distances (Kristensson *et al.*, 1974; Kuypers and Ugolini, 1990; Price *et al.*, 1982). Any agent that interferes with axoplasmic transport, such as colchicine, vinblastine, and cytochalasin B, will block this process (Kristensson *et al.*, 1971; Lycke *et al.*, 1984). One cycle of viral replication, that is required for release and transneuronal transfer of the viruses, can take 8–16 hr. Labeling of first- and second-order neurons occurs sequentially, and the interval is due to the time required for axonal transport and replication. This interval may be highly variable, depending on the type of neuron and its axoplasmic transport properties, and is a factor that needs to be determined empirically for any given neuronal system (see Section V).

Figure 4 Retrograde transneuronal labeling of third-order neurons in forelimb-related regions of contralateral sensorimotor cortex and red nucleus after injection of HSV-1 into forelimb (ulnar and median) nerves (Ugolini *et al.*, 1989). HSV-1-positive neurons are stained immunohistochemically (peroxidase–antiperoxidase method). (A) Third-order corticospinal neurons in layer V in forelimb area of sensorimotor cortex. Intense transneuronal labeling of cell bodies, basal and long apical dendrites, and axons. Glial cells are not labeled. Bar = 180 μm. (B) Third-order rubrospinal neurons. Labeling is restricted to dorsal and medial (forelimb-related) sectors of red nucleus. Bar = 200 μm.

III. Parameters of Importance in Designing Viral Transneuronal Tracing Studies

The success of transneuronal labeling with HSV-1 and PrV is dependent upon several experimental parameters, some related to the virus (strain and dose) and others related to the host (species, strain, and age). This is because viral transneuronal transfer is the outcome of a struggle between the neuroinvasiveness of the virus and the defense of the host.

A. Host Variables: Species, Strain, and Age

In any susceptible host species, susceptibility to HSV-1 varies among individuals. An example of this variability is the clinical difference of recurrent HSV-1 infection in humans (i.e., the natural host). Some individuals have recurrent infections, while others are unaffected even though most of the population has been exposed to the virus since childhood (Wildy, 1986). Genetic differences in susceptibility to HSV-1 have been best documented in inbred strains of mice: on the basis of lethal dose 50 (LD_{50}) tests, it has been found that AKR and A/J strains are very susceptible, Balb/c and CBA are moderately susceptible, and C57BL mice are resistant to HSV-1 infection (Lopez, 1975, 1985). In addition, we have shown that these genetic differences are correlated with dramatic differences in the extent of transneuronal transfer of HSV-1. For example, in Balb/c mice (8 weeks of age), inoculation of high doses of virus into the hypoglossal (XII) nerve results in extensive infection of XII motoneurons and retrograde transneuronal transfer to all main groups of connected second-order neurons (Ugolini *et al.*, 1987) (Fig. 5). In contrast, similar inoculations in C57BL mice of the same age produce infection of only a few XII motoneurons and transneuronal transfer does not occur (Ugolini, unpublished observations). The genetic resistance to HSV-1 is transmitted as a dominant trait and is mediated by natural defense cellular mechanisms that are immediately active at the onset of infection (Lopez, 1985).

Within a given animal species and strain, the susceptibility to HSV-1 and PrV is age dependent, being greatest at early age (Nash and Wildy, 1983). It is well known that even in the natural host of HSV-1 (man) and PrV (pig), serious infections occur in newborns but only rarely in adults. Similarly, C57BL mice, that are resistant to HSV-1 as adults, can develop serious infection as newborns (Zawatzky *et al.*, 1982). In susceptible mice strains (Balb/c), the susceptibility to HSV-1 decreases with age: a first change occurs by the fourth week, an intermediate stage is reached between the fourth and eighth week, and great resistance is obtained by 15–20 weeks of age (Hirsch *et al.*, 1970;

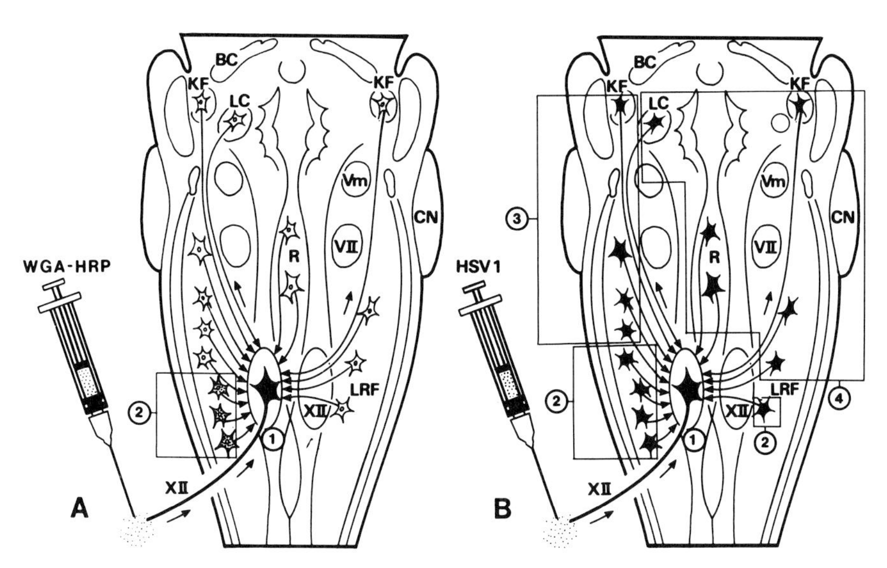

Figure 5 Differences in the extent of retrograde transneuronal transfer of WGA–HRP (A, left) and HSV-1 (B, right) from hypoglossal motoneurons (XII) (1) to second-order (premotor) interneurons in medullary and pontine lateral reticular formation (LRF), raphe nuclei (R), nucleus coeruleus and subcoeruleus (LC), and nucleus of Kölliker–Fuse (KF) (see Ugolini *et al.*, 1987). (A) Transneuronal transfer of WGA–HRP involves only a small subset of second-order neurons, i.e., in medullary LRF ipsilateral to XII nucleus. (B) With HSV-1, transneuronal labeling occurs sequentially in different groups of second-order neurons in stages 2–4. All groups of second-order neurons are labeled in stage 4. Note: the onset of transneuronal labeling with HSV-1 (B, 2) involves the only subset of neurons labeled by transfer of WGA–HRP (A, 2). Other abbreviations: BC, Brachium conjunctivum; CN, coclear nucleus; Vm, trigeminal motor nucleus; VII, facial nucleus). Adapted with permission from Kuypers and Ugolini (1990).

Nash and Wildy, 1983). This appears to reflect maturation of natural defense mechanisms.

Transneuronal transfer of HSV-1 is successful in rodents, i.e., Balb/c mice (LaVail *et al.*, 1990; Ugolini *et al.*, 1987), ICR mice (Margolis *et al.*, 1989), Wistar rats (Ugolini, 1992; Ugolini *et al.*, 1989), Sprague–Dawley and Long–Evans rats (McLean *et al.*, 1989), Fisher rats (Blessing *et al.*, 1991), and New Zealand White rabbits (Gieroba *et al.*, 1991; Li *et al.*, 1992a,b). Similarly, transneuronal transfer of PrV is very efficient in rodents, i.e., Swiss and Balb/c mice (Babic *et al.*, 1993; Heffner *et al.*, 1993; Peeters *et al.*, 1993) and Sprague–Dawley rats (Card *et al.*, 1990, 1992, 1993; Loewy *et al.*, 1991; Strack and Loewy, 1990; Strack *et al.*, 1989a, b). In contrast, PrV is not transferred transneuronally in primates (Strick and Card, 1992). With HSV-1, transneuronal transfer in experimental primates can be obtained after cortical injections in cebus monkey (Zemanick *et al.*, 1991), but is not successful after injection into peripheral nerves of macaca fascicularis, even in young animals (Ugolini, unpublished observations).

In rodents, PrV is more virulent than HSV-1, causing serious respiratory infection as well as "mad itch"—this is a dramatic phenomenon that is the result of PrV infection of the somatic sensory system and is presumably due to abnormal spontaneous activity of PrV-infected neurons (Dolivo, 1980; Dolivo *et al.*, 1978; Liao *et al.*, 1991). The extent and time of onset of these symptoms depend on the PrV strain that is used, being most pronounced with wild-type strains (see, e.g., Card *et al.*, 1991; Rinaman *et al.*, 1993; Strick and Card, 1992). In contrast, HSV-1 infection is accompanied by focal neurological deficits that reflect degeneration of specifically infected neurons (Ugolini, 1992; Ugolini *et al.*, 1987), but respiratory infection and mad itch are not seen.

B. Virus Strains

HSV-1 and PrV are classified as biohazardous agents and must be handled according to biosafety level II regulations (Strick and Card, 1992). Some institutions and countries may impose additional restrictions. For example, experimental use of PrV is not authorized in the United Kingdom, where PrV has been eradicated.

Numerous laboratory strains of HSV-1 and PrV are available, and these can differ considerably in neurovirulence (see, e.g., Dix *et al.*, 1983). Although only a few of these strains have been characterized with regard to their transneuronal transfer characteristics, striking differences in extent and even the direction of transfer have already been observed. For PrV, these differences are beginning to be mapped to specific viral genes (see below). Since the role of many viral genes in neuronal tropism and transneuronal transfer is still unknown, it is recommended to use a well-characterized virus strain because even the mutation of a single gene could affect transneuronal transfer (Babic *et al.*, 1993).

1. HSV-1

The wild-type strain SC16 of HSV-1 is transferred very effectively both in the anterograde and in the retrograde directions in Wistar rats and Balb/c mice (Ugolini, 1992; Ugolini *et al.*, 1987, 1989). After injection of high doses, transfer of the SC16 strain can involve all groups of second-order neurons and also third-order neurons (Ugolini *et al.*, 1989) (Fig. 4). Transneuronal labeling has also been obtained with other HSV-1 strains, i.e., the FMC-1 and FMC-134 strains in rats and rabbits (Blessing *et al.*, 1991; Li *et al.*, 1992a), the F strain in Balb/c mice and the McKrea strain in ICR mice (LaVail *et al.*, 1990; Margolis *et al.*, 1989), the McIntyre strain in Sprague–Dawley rats (McLean *et al.*, 1989), and the HFEM strain in Balb/c mice (Ugolini, unpublished results). In primates, two wild-type strains of HSV-1 have been shown to exhibit differences in the preferential direction of transfer: following intracortical inoculations, strain H129 is transferred mainly in the anterograde direction, whereas strain McIntyre-B is transferred almost exclusively in the retrograde direction (Hoover and Strick, 1993; Zemanick *et al.*, 1991). However, this may not hold in all species or systems. For example, the McIntyre strain produces bidirectional transneuronal labeling in the rat olfactory system (McLean *et al.*, 1989). Although the directionally selective transfer properties of these strains offer some interesting insights into this technology, further data in additional systems and species will be needed before these properties can be fully exploited.

2. PrV

Very extensive transneuronal transfer is obtained with wild-type strains of PrV, like the Becker strain, and some PrV mutants, like, for example, mutants deleted for the gene encoding the nonstructural glycoprotein gG (previously called gX). In mice and rats, the extent of transfer of either the wild-type Becker strain or the PrV gG-negative mutant in the XII model (Babic *et al.*, 1993; Card *et al.*, 1990) is similar to that of the SC16 strain of HSV-1 (Ugolini *et al.*, 1987). However, other PrV strains are able to infect only some of the neuronal pathways that are targeted by wild-type virus. Dramatic differences in the extent of anterograde transneuronal labeling of visual pathways have been reported for the wild-type Becker strain and the attenuated Bartha strain of PrV. After intraocular inoculation, Becker is transferred from retinal ganglion cells through all second-order visual relay centers, although not in synchrony, whereas Bartha–PrV labels only a small subset of the neurons targeted by Becker–PrV (Card *et al.*, 1991, 1992). In addition, preliminary findings suggest that transneuronal transfer of the Becker strain from the prefrontal cortex occurs both in the anterograde and in the retrograde directions, whereas Bartha is transferred only retrogradely (Enquist *et al.*, 1993). Differences between the Becker and Bartha strains have also been shown with regard to the extent of retrograde transneuronal transfer from the heart to the motor dorsal vagal nucleus that is readily infected with Becker but only rarely with Bartha (Standish *et al.*, 1994). This cannot reflect a general lack of tropism of Bartha for dorsal vagal neurons since the dorsal vagal nucleus is heavily infected by Bartha–PrV after inoculation into the stomach wall or the pancreas (Card *et al.*, 1990, 1993; Rinaman *et al.*, 1993). The Bartha strain is an attenuated PrV vaccine strain that carries several deletions and mutations which mainly affect the envelope glycoproteins gE, gI, and gC (that were previously designated gI, gp63, and gIII) (see, e.g., Card *et al.*, 1991, 1992). The selective impairment of transneuronal transfer of Bartha in visual and autonomic pathways described above has been reproduced by deletion of the PrV genes encoding the envelope glycoproteins gE and gI (previously designated gI and gp63) that are also affected in the Bartha strain (Card *et al.*, 1992; Standish *et al.*, 1994; Whealy *et al.*, 1993). An obvious implication of these findings is that the Bartha strain, as well as the gE-negative and gI-negative mutants, may fail to label some groups of neurons. Nonetheless, the Bartha strain has been successfully exploited to map autonomic pathways (Haxhiu *et al.*, 1993; Spencer *et al.*, 1990; Standish *et al.*, 1994; Strack *et al.*, 1989a, b). The advantage of Bartha relative to wild-type strains of PrV is that respiratory infection as well as other symptoms of generalized disease are delayed or less pronounced and the kinetics of transfer is slowed down (see, e.g., Card *et al.*, 1991; Rinaman *et al.*, 1993; Strack *et al.*, 1989b).

C. Virus Dose

Virus infectivity is influenced by the titer (concentration) of the virus, measured in Plaque-forming units per milliliter (PFU/ml). Conditions of virus storage are important: the virus stock should be aliquoted and stored at −70°C, as the virus titer is reduced by fluctuations in temperature and freezing/thawing of the virus stock. The success, extent, and specificity of transfer of HSV-1 and PrV are critically dependent upon the dose of virus that is inoculated.

1. HSV-1

With the wild-type strain SC16, successful infection and transneuronal transfer from peripheral nerves require use of high-titer virus stocks and injection of a high dose, viz. at least 10^4 PFU (as contained in 0.8 μl of stock with titer of 10^7 PFU/ml) (Ugolini *et al.*, 1987). Even when the titer of the virus stock is high (10^9 PFU/ml), small variations in the injected quantity (0.2–0.8 μl) can result in pronounced differences in the extent of transneuronal transfer (Ugolini *et al.*, 1987). Use of virus stocks with high titer (10^9 PFU/ml) and injection of a high dose (10^7 PFU) results in 100% infectivity (Ugolini, 1992). Under these conditions and if the number of PFU injected is kept constant, transneuronal transfer is strictly time dependent and all known groups of second-order neurons as well as third-order neurons will become labeled (Ugolini, 1992; Ugolini *et al.*, 1987, 1989).

2. PrV

In order to achieve 100% infectivity and extensive transneuronal transfer with PrV, high titers are required (e.g., 5×10^8 PFU/ml) and high doses per injection (10^6 PFU) (see e.g., Card *et al.*, 1991; Strick and Card, 1992). In contrast, use of virus stocks with lower titers (e.g., 2×10^5 PFU/ml) and injection of a small number of infectious units (e.g., 10^3 PFU) result in a low rate of infectivity (only 20%). Problems with using these low doses are that probably not all the first-order neurons that innervate a particular site will be labeled, and certainly there will be an underrepresentation of labeling in second-order neurons, and third-order neuronal labeling will not occur (Strack and Loewy, 1990; Strack *et al.*, 1989a,b). The advantage of this approach is that the infection may remain more controlled and the chance of spurious labeling is minimized (see Section V,A).

Thus, there is a trade-off between the two approaches. If the aim is to obtain 100% infectivity and to maximize the extent of transfer to second- and third-order neurons, one should use virus stocks with high titers (10^8 or 10^9 PFU/ml) and inoculate a high dose (10^7 or 10^8 PFU), but this increases the probability of spurious labeling due to local (nonspecific) viral spread from first-order neurons (see Section V,A). With this approach, specificity of the labeling can be maintained with a careful monitoring of the postinoculation survival period because local spread is time dependent and partially delayed relative to transneuronal transfer (Kuypers and Ugolini, 1990; Ugolini, 1992). Local spread can be minimized or abolished by injecting a small dose of virus (Loewy *et al.*, 1991; Strack and Loewy, 1990; Strack *et al.*, 1989a,b), but this also reduces the success rate of transneuronal transfer and it does not ensure labeling of all groups of second-order neurons or transfer to third-order neurons.

IV. Methods for Detection of HSV-1 and PrV and Combination with Other Methodologies

Excellent visualization of HSV-1 and PrV is provided by standard immunohistochemical methods, i.e., immunofluorescence and immunoperoxidase (peroxidase–antiperoxidase and avidin–biotin complex method) in frozen sections (see, e.g., Card *et al.*, 1990; Li *et al.*, 1992b; Strack and Loewy, 1990; Ugolini, 1992) (see Figs. 3 and 4). Immunoperoxidase staining can also be carried out in paraffin sections (Ugolini *et al.*, 1987) (Fig. 6). HSV-1 polyclonal primary antibodies are commercially available (e.g., Dako) and PrV polyclonal antibodies can be obtained from a number of laboratories. HSV-1 and PrV antigens are quite resistant to fixation with 4% paraformaldehyde that also inactivates the viruses, whereas fixatives containing glutaraldehyde produce much less satisfactory results. HSV-1 and PrV immunolabeling of first-, second-, and third-order neurons often has the appearance of Golgi preparations (Figs. 3 and 4). In addition, viral immunolabeling can be used in combination with other neuroanatomical methods. For example, infected first-order neurons can be double labeled with conventional tracers like WGA–HRP and DiI (Li *et al.*, 1992a; Strack and Loewy, 1990). However, the fluorochrome dye Fluorogold is not suitable because it impairs viral replication and can produce false-negative results (LaVail *et al.*, 1993). HSV-1 and PrV immunofluorescence labeling can also be combined with staining of several neurotransmitter-related antigens (Blessing *et al.*, 1991; Haxhiu *et al.*, 1993; Strack *et al.*, 1989a,b) although this may not necessarily apply to all neurotransmitters, especially when their turnover is rapid, because viral replication shuts off protein synthesis in infected cells.

Another method for detection of infected cells is based on systemic

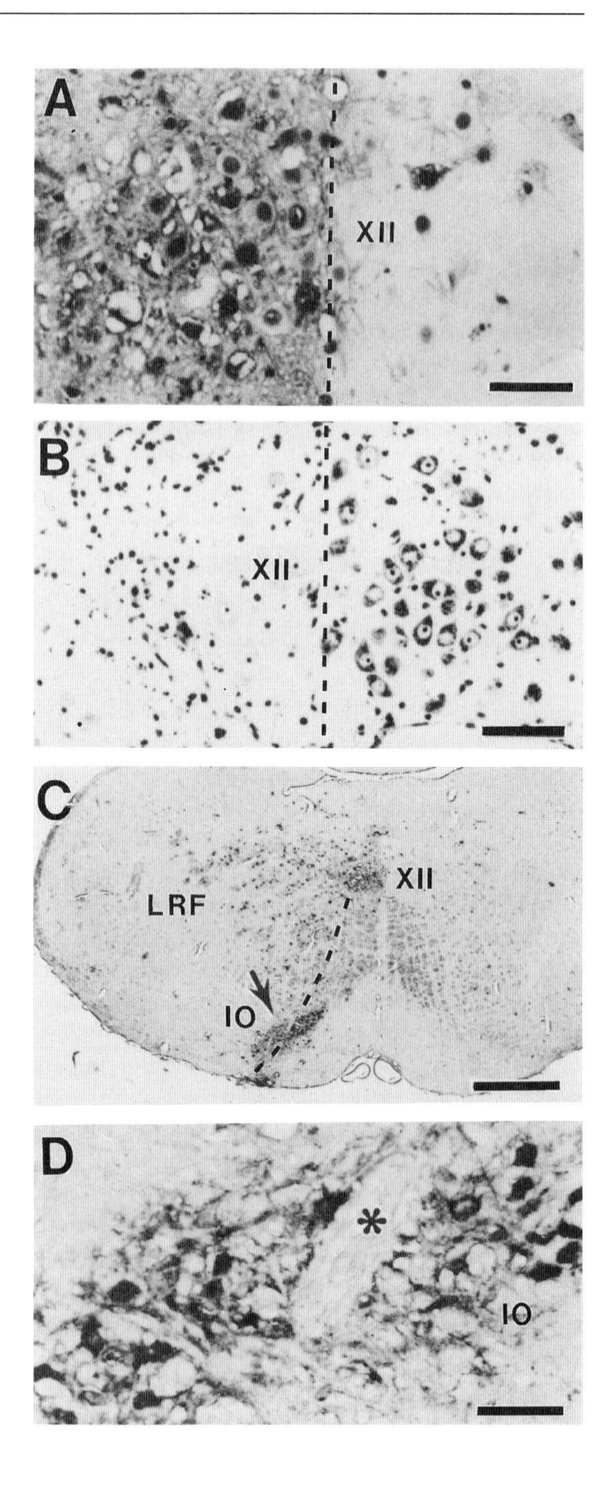
A
XII
B
XII
C
XII
LRF
IO
D
*
IO

administration of tritiated thymidine that is incorporated by infected neurons and glial cells during viral replication (Margolis *et al.*, 1987, 1989). However, this method is less satisfactory than immunohistochemistry for two reasons. First, it depends upon autoradiographic visualization which requires long periods of exposure (4 weeks). Second, the thymidine label is largely confined to cell nuclei, which precludes any possibility of studying the complete morphology of infected cells.

A more recent method for detection of replicating alphaherpesviruses is based on insertion into the viral genome of the marker LacZ gene that codes for the enzyme β-galactosidase. In the current PrV constructs, the LacZ gene has been inserted within the gene encoding the nonstructural, nonessential glycoprotein gG (previously designated gX) (Mettenleiter and Rauh, 1990) that does not play a role in transneuronal and local transfer of PrV (Babic *et al.*, 1993; Peeters *et al.*, 1993). After insertion in this region, β-gal activity is expressed during the last stages of viral replication. The marker can be detected with a simple histochemical reaction, providing intense labeling of all parts of the neuron (Babic *et al.*, 1993; Loewy *et al.*, 1991). This method is very convenient for a rapid screening of infected neurons. When the substrate Bluo-Gal (halogenated indoyly-β-D-galactoside, Gibco Bethesda Res. Lab., Gaithersburg, MD) is used, the cellular distribution of the virus can also be studied with electron microscopy because an insoluble reaction product is formed in virally infected neurons (Loewy *et al.*, 1991). However, the standard reaction product (X-gal) can diffuse from heavily infected cells and is somewhat less satisfactory (Babic *et al.*, 1993).

V. Interpretation of the Results: Kinetics of Transfer

Transneuronal transfer of HSV-1 and PrV can label chains of connected neurons, including first-, second-, third-, and even higher-order neurons (Fig.

Figure 6 Degeneration of hypoglossal (XII) motoneurons and local transfer from XII axons at 5 days after injection of HSV-1 into the XII nerve. (A, C, and D) Photomicrographs of HSV-1-positive neurons stained immunohistochemically in paraffine-embedded sections. (B) Cresyl violet (Nissl) staining of the XII nucleus. (A and B) Degeneration of XII motoneurons (first-order neurons) after long-standing infection. Labeled XII motoneurons (A, left) show loss of Nissl staining (B, left) and are swollen and vacuolized (compare A, left, with B, right). Labeling also involves glial cells. Bar = 80 μm. (C) Labeled first-order neurons in XII nucleus and second-order neurons in lateral reticular formation (LRF). Note the spurious (local) transfer of virus from XII axons (stippled line) to adjoining glial cells and neurons [e.g., in inferior olive (IO), arrow, see D]. Bar = 550 μm. (D) Labeling of neurons in inferior olive resulting from local transfer of HSV-1 from XII axons (asterisk). Bar = 50 μm. Taken with permission from Ugolini *et al.* (1987).

4) (see, e.g., Card *et al.*, 1991; McLean *et al.*, 1989; Rinaman *et al.*, 1993; Ugolini *et al.*, 1987, 1989). The possibility to identify hierarchical levels of neuronal systems is the most significant attraction of the viral method, but one that is not straightforward. In fact, differentiation between first-, second-, and third-order neurons cannot be made on the basis of their labeling intensity, which is pretty similar in all infected cells. To make this distinction, one should study the propagation of the viruses at different postinoculation times after injection of a constant dose. In this manner, first-, second-, and third-order neurons can be distinguished on the basis of their sequential labeling (Card *et al.*, 1990, 1991; Margolis *et al.*, 1989; Rinaman *et al.*, 1993; Ugolini, 1992; Ugolini *et al.*, 1987). In fact, the viruses progress from neuron to neuron in sequential steps, and the interval between labeling of each set of connected neurons is due to viral replication, which is necessary for transfer. The interpretation of the kinetics of viral transfer is often complicated at long postinoculation times because viral infection causes degeneration of first-order neurons, which is eventually accompanied by local (nonspecific) spread of the virus to nearby cells. Moreover, asynchronous labeling of different groups of second-order neurons may occur as well to complicate the analysis (see below).

A. Virus-Induced Neuronal Degeneration and Local Transfer

The kinetics of neuronal infection with HSV-1 and PrV comprises a characteristic sequence of events. Immunolabeling is initially restricted to the neuronal nucleus, cytoplasm, and the most proximal part of the dendritic tree, i.e., the sites of assembly and packaging of progeny virus. Later, it also extends to distal dendrites and axons, in parallel with centrifugal transport of progeny virus (see above). At these times, neuronal morphology is well preserved (Fig. 3) (Card *et al.*, 1990; Rinaman *et al.*, 1993; Ugolini, 1992; Ugolini *et al.*, 1987). At longer times, infected neurons start degenerating, showing a great increase in cell size (up to four times) and loss of Nissl staining (Figs 6A and 6B) (Babic *et al.*, 1993; Martin and Dolivo, 1983; Rinaman *et al.*, 1993; Ugolini, 1992; Ugolini *et al.*, 1987). Neuronal degeneration is probably unavoidable with all strains of HSV-1 and PrV that are used as tracers because it is induced by specific viral genes that are required for neurovirulence and that shut off protein synthesis in the host cell (Chou *et al.*, 1990; Roizman and Sears, 1990). The progressive degeneration of infected neurons explains why the anterograde axonal transport and transneuronal transfer of virus from these neurons is very efficient over short distances, e.g., from primary sensory neurons in dorsal root ganglia to connected spinal regions (Fig. 7), whereas it is largely absent over long distances, e.g., from dorsal root ganglion neurons to dorsal column nuclei (Ugolini, 1992; Ugolini *et al.*, 1989). In contrast, retrograde transneuronal transfer tcond-order and third-order neurons is very efficient even over long distances, e.g., from forelimb or hindlimb mixed

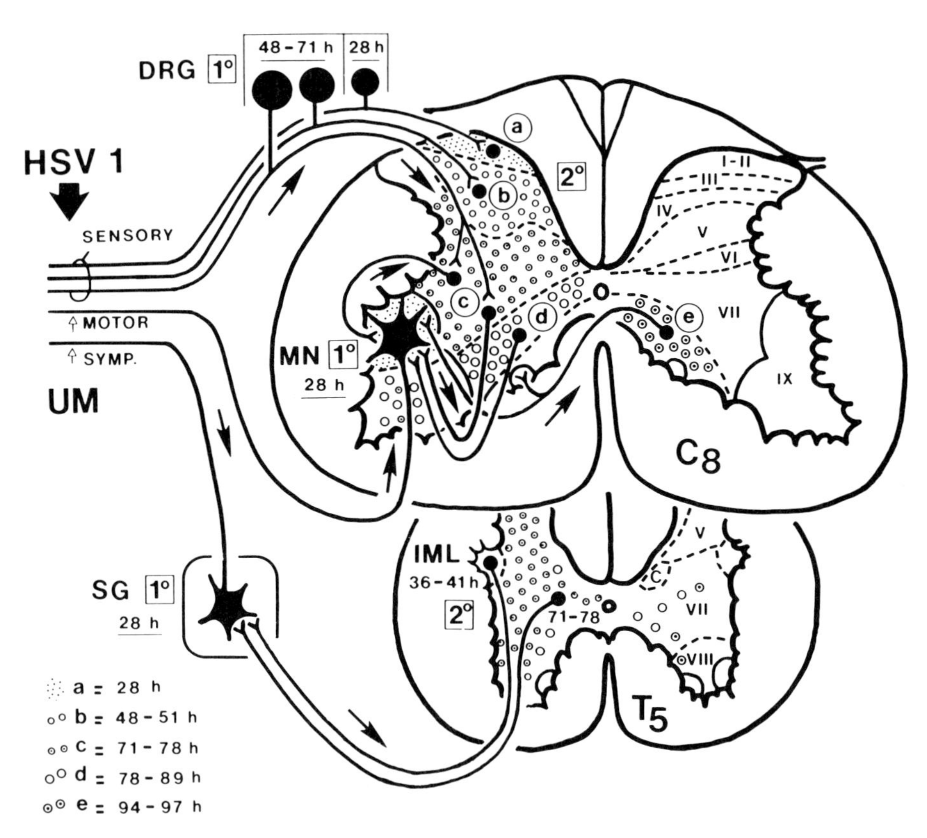

Figure 7 Asynchronous transneuronal transfer of HSV-1 from sensory, motor, and sympathetic axons of mixed forelimb nerves (ulnar + median, UM) to connected cell groups in spinal cord. Right: boundaries of laminae I–IX. Left: different symbols indicate the increase in spinal distribution at different postinoculation times (see a–e). All groups of first-order neurons (1°) are infected: motoneurons (MN), sympathetic neurons in stellate ganglion (SG), and primary sensory neurons in dorsal root ganglia (DRG). However, anterograde transneuronal transfer from small primary sensory neurons to ipsilateral laminae I and II occurs already at 28 hr (see also Fig. 3B), whereas transfer from sensory afferents of intermediate and large diameter to deeper laminae (III–VII) is obtained later (see b and c). Retrograde transneuronal transfer from sympathetic neurons in SG to second-order (2°) preganglionic neurons in intermediolateral cell group (IML) occurs already at 36–41 hr, whereas labeling consistent with retrograde transneuronal transfer from motoneurons is obtained much later (c–e). This sequence indicates that transneuronal transfer from small axons (small cutaneous afferents and sympathetic axons, largely unmyelinated) precedes transfer from large axons (myelinated sensory afferents and motor axons). In addition, nonsynchronous labeling of second-order neurons also occurs during transfer from the same group of second-order neurons (e.g., sympathetic): preganglionic cell groups (e.g., in IML at T5) that provide the strongest input to the SG are already labeled at 36 hr, whereas cell groups that provide only a few connections to SG (e.g., medial sympathetic preganglionic cell groups at T5) are labeled later. Adapted from Ugolini, (1992). (Copyright 1992 of Wiley–Liss, a division of Wiley & Sons, Inc.).

nerves to cortical and brainstem cell groups of descending pathways (Ugolini *et al.*, 1989) (Fig. 4). The explanation is simple: the neurons infected via retrograde transneuronal transfer are still healthy when axonal transport of the virus takes place because this transport precedes replication of the virus in their cell body (Ugolini *et al.*, 1989).

Virus-induced degeneration of specifically infected neurons is accompanied by local (nonsynaptic) transfer of progeny virus from their cell bodies and axons to adjoining glial cells (Babic *et al.*, 1993; Blessing *et al.*, 1991; Card *et al.*, 1990; Rinaman *et al.*, 1993; Li *et al.*, 1992a; Ugolini, 1992; Ugolini *et al.*, 1987). HSV-1 and PrV-infected glial cells have been identified with viral immunolabeling on the basis of their characteristic morphology (Card *et al.*, 1990; La Vail *et al.*, 1990; Ugolini, 1992; Ugolini *et al.*, 1987), as well as by specific glial markers (Blessing *et al.*, 1991; Rinaman *et al.*, 1993). Labeling of glial cells occurs as a "centrifugal wave" that begins at the cell bodies and progresses gradually down their axons (Ugolini, 1992). HSV-1 and PrV infection of glial cells around sensory, motor, sympathetic, and parasympathetic axons does not result in viral spread to passing fibers (Blessing *et al.*, 1991; Card *et al.*, 1990; Ugolini, 1992; Ugolini *et al.*, 1987, 1989). On the basis of these observations, it has been suggested that infected glial cells form a barrier to local spread of the viruses to nearby neuronal elements (see, e.g., Card *et al.*, 1990). This is probably due to the fact that viral replication in glial cells is not efficient, at least in early stages of infection (Card *et al.*, 1993). However, after long-standing infection, glial cells degenerate (Rinaman *et al.*, 1993; Ugolini, 1992). The barrier provided by glial cells against nonspecific viral spread can be only temporary, since spread of HSV-1 and PrV from specifically infected axons to glial cells can be followed by spurious infection of local neurons at long survival times. In the XII model, a clear example of this phenomenon is the spurious labeling of neurons that occurs in the inferior olivary nucleus near the site where infected XII axons exit the medulla oblongata. This occurs some time after alphaherpesviruses have infected glial cells in this region (Babic *et al.*, 1993; Ugolini *et al.*, 1987) (Figs. 6C and 6D). Neuronal degeneration and local spread can occur with all strains of PrV and HSV-1 that are transferred transneuronally, including the attenuated Bartha strain (Card *et al.*, 1993; Loewy *et al.*, 1991; Rinaman *et al.*, 1993). It is often difficult to judge whether the neuronal labeling obtained in the proximity of specifically infected neurons and their axons is the result of local spread or whether it may reflect transneuronal transfer to local interneurons.

Like transneuronal transfer, local spread is dependent on the dose of virus injected and the postinoculation period (see Section III,C). Even with the attenuated PrV–Bartha strain, local spread is minimal after administration of small doses in combination with short survival times (Jansen *et al.*, 1993; Strack and Loewy, 1990; Strack *et al.*, 1989a,b), but injection of high doses produces local spread (Loewy *et al.*, 1991). It may be impossible to construct

HSV-1 or PrV strains that are transferred transneuronally in the absence of local spread because both transneuronal transfer and local spread depend upon the presence of the same surface glycoprotein (gB, previously designated gII for PrV) on the viral envelope (Babic *et al.*, 1993). However, local spread is less efficient than transneuronal transfer, and it can be avoided or minimized by injecting a low dose of virus, although this lowers the infectivity considerably (to 20%) and it may also reduce the extent of specific transneuronal transfer to second-order neurons (Jansen *et al.*, 1993; Strack and Loewy, 1990; Strack *et al.*, 1989a,b). After injection of high doses of virus, which are required to obtain transfer to third-order neurons, local spread can be minimized with a careful control of the postinoculation period, which is therefore necessary to ensure specificity of the labeling (Kuypers and Ugolini, 1990; Ugolini, 1992; Ugolini *et al.*, 1987).

B. Asynchronous Labeling of Different Groups of Second-Order Neurons

Another potential difficulty of the viral transneuronal labeling method is that different groups of second-order neurons may be labeled at different times. Following intraocular injections of Becker–PrV, anterograde transneuronal transfer to different groups of second-order neurons in visual relay centers is not obtained in synchrony (Card *et al.*, 1991). Similarly, after inoculation of the SC16 strain of HSV-1 into forelimb and hindlimb mixed nerves (Ugolini, 1992), transneuronal transfer does not occur in synchrony via all classes of axons, regardless of whether this transfer is in the anterograde or the retrograde direction (Fig. 7). The speed of viral transneuronal transfer is inversely correlated with the fiber diameter of the transferring first-order neurons (Ugolini, 1992) (Fig. 7). Thus, retrograde transneuronal transfer from small (largely unmyelinated) sympathetic axons to sympathetic preganglionic neurons precedes transfer from large (myelinated) motor axons (Fig. 7). Similarly, anterograde transneuronal transfer of HSV-1 through small-diameter (unmyelinated or small myelinated) sensory afferents to the superficial laminae (I and II) of the dorsal horn precedes transfer via large (myelinated) sensory axons to deeper spinal laminae (Figs. 7 and 3B). The same sequence is obtained during anterograde transneuronal transfer of WGA–HRP through sensory afferents (Robertson and Grant, 1985). This phenomenon may reflect intrinsic differences of the neuronal systems involved (e.g., density of connections and/or level of activity), that may be unavoidable, regardless of which transneuronal tracer is used (Ugolini, 1992). Differences in number and/or type of neuronal receptors may also play a role (Card *et al.*, 1991).

The different speed of transfer of the viruses through the various classes of axons is probably also determined by the route of inoculation. For example, after inoculation into peripheral mixed nerves, the myelin sheet of Schwann cells may delay viral penetration into large (myelinated) axons (Ugolini, 1992) because peripheral Schwann cells do not support HSV-1 replication (Cook and Stevens, 1973; Dillard *et al.*, 1972). This can explain the earlier uptake of HSV-1 (SC16 strain) from small relative to large primary sensory neurons that is obtained after inoculation into forelimb and hindlimb mixed nerves (Ugolini, 1992). It may also explain why the uptake of HSV-1 (FMC-1 strain) from myelinated axons of dorsal vagal motoneurons is delayed relative to uptake and anterograde transfer from vagal sensory afferents to the visceral relay nucleus—the nucleus of the solitary tract—after inoculation into the vagus nerve (Blessing *et al.*, 1991). In contrast, uptake and retrograde transneuronal transfer from dorsal vagal motoneurons precedes anterograde transfer via vagal afferents when the virus is inoculated into the stomach wall (Card *et al.*, 1990), probably because the access of virus to vagal motor axons is facilitated by inoculation into their terminal regions.

Interestingly, asynchronous labeling of different groups of second-order neurons can also be obtained during retrograde transneuronal transfer of HSV-1 and PrV from the *same* group of first-order neurons, e.g., sympathetic neurons, parasympathetic neurons, or motoneurons (Rinaman *et al.*, 1993; Ugolini, 1992; Ugolini *et al.*, 1987) (Fig. 5). This also occurs during transfer of nonviral tracers (tetanus toxin fragments) from motoneurons (Horn and Büttner-Ennever, 1990). For HSV-1, the speed of transfer to the various groups of second-order neurons appears to be dependent upon the strength of their connections with the transferring (first-order) neurons (Ugolini, 1992). For example, retrograde transneuronal transfer of HSV-1 from the stellate ganglion (SG) first labels the sympathetic preganglionic cell groups that contain the greatest number of neurons that supply the SG; preganglionic cell groups that provide only a few connections to the SG are labeled later (Ugolini, 1992). Interestingly, the first group of second-order neurons labeled by retrograde transneuronal transfer of HSV-1 from XII motoneurons (i.e., in lateral reticular formation at the level of the XII nucleus) (Ugolini *et al.*, 1987) is also the *only* cell group labeled by the nonefficient transneuronal transfer of WGA–HRP in the XII model (Ugolini, unpublished results) (Fig. 5). Since the neuronal receptors for HSV-1 and WGA–HRP are likely to be different, this suggests once again that the efficiency of transneuronal transfer of viral and nonviral tracers is influenced by the organization of neuronal connections.

VI. Conclusions

When using HSV-1 and PrV as transneuronal tracers, it is important to realize that the extent and specificity of transfer depend upon many factors, including the virus strain and dose, the postinoculation time, and the species and age of the host. At this time, no perfect virus has been found that can produce transneuronal transfer without causing degeneration and local spread at long postinoculation times. While some researchers find it advantageous to inject small doses of virus to minimize local spread, this results in low rates of infectivity (20%) and the possibility that some groups of second-order neurons will not be labeled. In addition, third-order neurons may not be labeled at all under these conditions. In contrast, inoculation of high viral doses ensures 100% infectivity and robust labeling of second- and third-order neurons. Under these conditions, the potential pitfall is that spurious labeling will also occur at long times. Thus, experiments need to be designed to accommodate these differences. It is very important to study several different postinoculation periods in order to define the extent of transneuronal transfer in a given system and to distinguish first-, second-, and third-order neurons on the basis of their sequential labeling (Ugolini, 1992). Even with this approach, one should bear in mind that not all second-order neurons will necessarily be labeled in synchrony due to differences in synaptic density as well as additional factors. In practice, all neurons labeled at the *onset* of transneuronal transfer can be regarded as second-order, although the labeling at these times may comprise only some of the second-order cell groups. At longer time points, the transfer should involve all groups of second-order neurons, including those with weak connections. At long times, however, it may be difficult to distinguish the delayed labeling of second-order neurons from early labeling of third-order neurons as well as spurious labeling due to local spread from degenerating first-order neurons and their axons. Therefore, new data generated from experiments using the viral transneuronal labeling method will ultimately require careful cross-referencing with other methods (Ugolini, 1992).

Although the viral transneuronal labeling method has limitations, it is important to stress that this technique has highly significant advantages over conventional tracing methods: at least up to a certain time, the labeling occurs in a very specific manner via transfer of the viruses across synapses and the fibers-of-passage problem of conventional tracers is avoided. When used correctly, transneuronal tracing with HSV-1 and PrV is an extremely powerful technique that can serve to elucidate several aspects of neuronal connectivity that could never have been studied adequately with conventional tracing methods like, for example, the organization of autonomic networks (Blessing *et al.*, 1991; Card *et al.*, 1990; Gieroba *et al.*, 1991; Haxhiu *et al.*, 1993; Li *et al.*, 1992a,b; Spencer *et al.*, 1990; Standish *et al.*, 1994; Strack *et al.*, 1989a,b; Ugolini, 1992). Clearly, this method offers numerous possibilities that will be exploited in future years.

Acknowledgments

I thank Dr. P. Coulon for useful comments and S. Brasiles for assistance with photography. This work was supported by the CNRS (UPR 2431).

References

Alstermark, B., and Kümmel, H. (1986). Transneuronal labelling of neurones projecting to forelimb motoneurones in cats performing different movements. *Brain Res.* **376**, 387–391.

Alstermark, B., and Kümmel, H. (1990a). Transneuronal transport of wheat germ agglutinin conjugated horseradish peroxidase into last order spinal interneurones projecting to acromio- and spinodeltoideus motoneurones in the cat. 1. Location of labelled interneurones and influence of synaptic activity on the transneuronal transport. *Exp. Brain Res.* **80**, 83–95.

Alstermark, B., and Kümmel, H. (1990b). Transneuronal transport of wheat germ agglutinin conjugated horseradish peroxidase into last order spinal interneurones projecting to acromio- and spinodeltoideus motoneurones in the cat. 2. Differential labelling of interneurones depending on movement type. *Exp. Brain Res.* **80**, 96–103.

Alstermark, B., Kümmel, H., and Tantisira, B. (1987). Monosynaptic raphespinal and reticulospinal projection to forelimb motoneurones in cats. *Neurosci. Lett.* **74**, 286–290.

Babic, N., Mettenleiter, T. C., Flamand, A., and Ugolini, G. (1993). Role of essential glycoproteins gII and gp50 in transneuronal transfer of Pseudorabies virus from the hypoglossal nerve of mice. *J. Virol.* **67**, 4421–4426.

Blessing, W. W., Li, Y.-W., and Wesselingh, S. L. (1991). Transneuronal transport of herpes simplex virus from the cervical vagus to brain neurons with axonal input to central vagal sensory nuclei in the rat. *Neuroscience* **42**, 261–274.

Card, J. P., Rinaman, L., Lynn, R. B., Lee, B.-H., Meade, R. P., Miselis, R. R., and Enquist, L. W. (1993). Pseudorabies virus infection of the rat central nervous system: Ultrastructural characterization of viral replication, transport and pathogenesis. *J. Neurosci.* **13**, 2515–2539.

Card, J. P., Rinaman, L., Schwaber, J. S., Miselis, R. R., Whealy, M. E., Robbins, A. K., and Enquist, L. W. (1990). Neurotropic properties of pseudorabies virus: Uptake and transneuronal passage in the rat central nervous system. *J. Neurosci.* **10**, 1974–1994.

Card, J. P., Whealy, M. E., Robbins, A. K., and Enquist, L. W. (1992). Pseudorabies virus envelope glycoprotein gI influences both neurotropism and virulence during infection of the rat visual system. *J. Virol.* **66**, 3032–3041.

Card, J. P., Whealy, M. E., Robbins, A. K., Moore, R. Y., and Enquist, L. W. (1991). Two α-herpesvirus strains are transported differentially in the rodent visual system. *Neuron* **6**, 957–969.

Chou, J., Kern, E. R., Whitley, R. J., and Roizman, B. (1990). Mapping of herpes simplex virus-1 neurovirulence to $_{\gamma 1}34.5$, a gene nonessential for growth in culture. *Science* **250**, 1262–1266.

Cook, M. L., and Stevens, J. G. (1973). Pathogenesis of herpetic neuritis and ganglionitis in mice: Evidence for intra-axonal transport of infection. *Infect. Immun.* **7**, 272–288.

Dillard, S. H., Cheatham, W. J., and Moses, H. L. (1972). Electron microscopy of zosteriform herpes simplex infection in the mouse. *Lab. Invest.* **26**, 391–402.

Dix, R. D., McKendall, R. R., and Baringer, J. R. (1983). Comparative neurovirulence of herpes simplex virus type 1 strains after peripheral or intracerebral inoculation of Balb/c mice. *Infect. Immun.* **40**, 103–112.

Dolivo, M. (1980). A neurobiological approach to neurotropic viruses. *Trends Neurosci.* **3**, 149–152.

Dolivo, M., Beretta, E., Bonifas, V., and Foroglou, C. (1978). Ultrastructure and function in sympathetic ganglia isolated from rats infected with pseudorabies virus. *Brain Res.* **140**, 111–123.

Enquist, L. W., Levitt, P., and Card, J. P. (1993). Connections of the adult rat prefrontal cortex revealed by intracerebral injection of a swine alpha herpesvirus. *Soc. Neurosci. Abstr.* **19**(part II), 1442. [Abstract 589.11]

Gieroba, Z. J., Li, Y.-W., Wesselingh, S. L., and Blessing, W. W. (1991). Transneuronal labeling of neurons in rabbit brain after injection of herpes simplex virus type 1 into the aortic depressor nerve. *Brain Res.* **558**, 264–272.

Harrison, P. J., Hultborn, H., Jankowska, E., Katz, R., Storai, B., and Zytnicki, D. (1984). Labelling of interneurones by retrograde transsynaptic transport of horseradish peroxidase from motoneurones in rats and cats. *Neurosci. Lett.* **45**, 15–19.

Haxhiu, M. A., Jansen, A. S. P., Cherniack, N. S., and Loewy, A. D. (1993). CNS innervation of airway-related parasympathetic preganglionic neurons: A transneuronal labeling study using pseudorabies virus. *Brain Res.* **618**, 115–134.

Heffner, S., Kovács, F., Klupp, B. G., and Mettenleiter, T. C. (1993). Glycoprotein gp50-negative Pseudorabies virus: A novel approach towards a non-spreading live herpesvirus vaccine. *J. Virol.* **67**, 1529–1537.

Hirsch, M. S., Zisman, B., and Allison, A. C. (1970). Macrophage and age-dependent resistance to herpes simplex virus in mice. *J. Immunol.* **104**, 1160–1165.

Hoover, J. E., and Strick, P. L. (1993). Multiple output channels in the basal ganglia. *Science* **259**, 819–821.

Horn, A. K. E., and Büttner-Ennever, J. A. (1990). The time course of retrograde transsynaptic transport of tetanus toxin fragment C in the oculomotor system of the rabbit after injection into extraocular eye muscles. *Exp. Brain Res.* **81**, 353–362.

Jankowska, E. (1985). Further indications for enhancement of retrograde transneuronal transport of WGA-HRP by synaptic activity. *Brain Res.* **341**, 403–408.

Jansen, A. S. P., Farwell, D. G., and Loewy, A. D. (1993). Specificity of pseudorabies virus as a retrograde marker of sympathetic preganglionic neurons: Implications for transneuronal labeling studies. *Brain Res.* **617**, 103–112.

Kristensson, K., Ghetti, B., and Wiśniewski, H. M. (1974). Study on the propagation of Herpes Simplex virus (type 2) into the brain after intraocular injection. *Brain Res* **69**, 189–201.

Kristensson, K., Lycke, E., and Sjöstrand, J. (1971). Spread of herpes simplex virus in peripheral nerves. *Acta Neuropathol.* **17**, 44–53.

Kristensson, K., Nennesmo, I., Persson, L., and Lycke, E. (1982). Neuron to neuron transmission of herpes simplex virus. Transport of virus from skin to brainstem nuclei. *J. Neurol. Sci.* **54**, 149–156.

Kuypers, H. G. J. M., and Huisman, A. M. (1984). Fluorescent neuronal tracers. In "Advances in Cellular Neurobiology" (S. Fedoroff, Ed.), Vol. 5, pp. 307–340. Academic Press, Orlando.

Kuypers, H. G. J. M., and Ugolini, G. (1990). Viruses as transneuronal tracers. *Trends Neurosci.* **13**, 71–75.

La Vail, J. H., Carter, S. R., and Topp, K. S. (1993). The retrograde tracer Fluoro-Gold interferes with the infectivity of Herpes Simplex Virus. *Brain Res.* **625**, 57–62.

La Vail, J. H., Zhan, J., and Margolis, T. P. (1990). HSV (type 1) infection of the trigeminal complex. *Brain Res.* **514**, 181–188.

Li, Y.-W., Ding, Z.-Q., Wesselingh, S. L., and Blessing, W. W. (1992a). Renal and adrenal sympathetic preganglionic neurons in rabbit spinal cord: Tracing with Herpes simplex virus. *Brain Res.* **573**, 147–152.

Li, Y.-W., Wesselingh, S. L., and Blessing, W. W. (1992b). Projections from rabbit caudal medulla to C1 and A5 sympathetic premotor neurons, demonstrated with phaseolus agglutinin and herpes simplex virus. *J. Comp. Neurol.* **317**, 379–395.

Liao, G. S., Maillard, M., and Kiraly, M. (1991). Ion channels involved in the presynaptic hyperexcitability induced by herps virus suis in rat superior cervical ganglion. *Neuroscience* **41**, 797–807.

Loewy, A. D., Bridgman, P. C., and Mettenleiter, T. C. (1991). β-galactosidase expressing recombinant pseudorabies virus for light and electron microscopic study of transneuronally labeled CNS neurons. *Brain Res.* **555**, 346–352.

Lopez, C. (1975). Genetics of natural resistance to herpesvirus infections in mice. *Nature* **258**, 152–153.

Lopez, C. (1985). Natural resistance mechanisms in herpes simplex virus infection. In "The Herpesviruses" (B. Roizman, Ed.), Vol. 4, pp. 37–68. Plenum Press, New York.

Lycke, E., Kristensson, K., Svennerholm, B., Vahlne, A., and Ziegler, R. (1984). Uptake and transport of herpes simplex virus in neurites of rat dorsal root ganglia cells in culture. *J. Gen. Virol.* **65**, 55–64.

Margolis, T. P., La Vail, J. H., Setzer, P. Y., and Dawson, C. R. (1989). Selective spread of herpes simplex virus in the central nervous system after ocular inoculation. *J. Virol.* **63**, 4756–4761.

Margolis, T. P., Togni, B., La Vail, J. H., and Dawson, C. R. (1987). Identifying HSV infected neurons after ocular inoculation. *Curr. Eye Res.* **6**, 119–126.

Martin, X., and Dolivo, M. (1983). Neuronal and transneuronal tracing in the trigeminal system of the rat using the Herpes Virus Suis. *Brain Res.* **273**, 253–276.

McLean, J. H., Shipley, M. T., and Bernstein, D. I. (1989). Golgi-like, transneuronal retrograde labelling with CNS injections of Herpes Simplex Virus type 1. *Brain Res. Bull.* **22**, 867–881.

Mesulam, M.-M. (1982). "Tracing Neuronal Connections with Horseradish Peroxidase." IBRO Handbook Series: Methods in the Neurosciences, Wiley, New York.

Mettenleiter, T. C. (1991). Molecular biology of pseudorabies (Aujeszky's disease) virus. *Comp. Immunol. Microbiol. Infect. Dis.* **14**, 151–163.

Mettenleiter, T. C. (1994). Initiation and spread of α-herpesvirus infections. *Trends Microbiol.* **2**, 2–4.

Mettenleiter, T. C., and Rauh, I. (1990). A glycoprotein gX-β-galactosidase fusion gene as insertional marker for rapid identification of pseudorabies virus mutants. *J. Virol. Methods* **30**, 55–66.

Nash, A. A., and Wildy, P. (1983). Immunity in relation to the pathogenesis of herpes simplex virus. In "Human Immunity to Viruses" (F. Ennis, Ed.), pp. 179–192. Academic Press, New York.

Peeters, B., Pol, J., Gielkens, A., and Moormann, R. (1993). Envelope glycoprotein gp50 of Pseudorabies virus is essential for virus entry but it is not required for viral spread in mice. *J. Virol.* **67**, 170–177.

Price, R. W., Rubenstein, R., and Khan, A. (1982). Herpes simplex virus infection of isolated autonomic neurons in culture: Viral replication and spread in a neuronal network. *Arch. Virol.* **71**, 127–140.

Rinaman, L., Card, J. P., and Enquist, L. W. (1993). Spatiotemporal responses of astrocytes, ramified microglia, and brain macrophages to central neuronal infection with pseudorabies virus. *J. Neurosci.* **13**, 685–702.

Robertson, B., and Grant, G. (1985). A comparison between wheat germ agglutinin- and choleragenoid-horseradish peroxidase as anterogradely transported markers in central branches of primary sensory neurones in the rat with some observations in the cat. *Neuroscience* **14**, 895–905.

Roizman, B., and Sears, A. E. (1990). Herpes simplex viruses and their replication. In "Fields Virology, Second Edition" (B. N. Fields, and D. M. Knipe, Eds.), pp. 1795–1841. Raven Press, New York.

Sabin, A. B. (1938). Progression of different nasally instilled viruses along different nervous pathways in the same host. *Proc. Soc. Exp. Biol. Med.* **38**, 270–275.

Spencer, S. E., Sawyer, W. B., Wada, H., Platt, K. B., and Loewy, A. D. (1990). CNS projections to the pterygopalatine parasympathetic preganglionic neurons in the rat: A retrograde transneuronal viral cell body labeling study. *Brain Res.* **534**, 149–169.

Standish, A., Enquist, L. W., and Schwaber, J. S. (1994). Innervation of the heart and its central medullary origin defined by viral tracing. *Science* **263**, 232–234.

Strack, A. M., and Loewy, A. D. (1990). Pseudorabies virus: A highly specific transneuronal cell body marker in the sympathetic nervous system. *J. Neurosci.* **10**, 2139–2147.

Strack, A. M., Sawyer, W. B., Hughes, J. H., Platt, K. B., and Loewy, A. D. (1989a). A general pattern of CNS innervation of the sympathetic outflow demonstrated by transneuronal pseudorabies viral infections. *Brain Res.* **491**, 156–162.

Strack, A. M., Sawyer, W. B., Platt, K. B., and Loewy, A. D. (1989b). CNS cell groups regulating the sympathetic outflow to adrenal gland as revealed by transneuronal cell body labeling with pseudorabies virus. *Brain Res.* **491**, 274–296.

Strick, P. L., and Card, J. P. (1992). Transneuronal mapping of neural circuits with alpha herpesviruses. In "Experimental Neuroanatomy. A Practical Approach" (J. P. Bolam, Ed.), pp. 81–101. IRL Press.

Ugolini, G. (1992). Transneuronal transfer of Herpes Simplex Virus type 1 (HSV 1) from mixed limb nerves to the CNS. I. Sequence of transfer from sensory, motor and sympathetic nerve fibres to the spinal cord. *J. Comp. Neurol.* **326**, 527–548.

Ugolini, G., Kuypers, H. G. J. M., and Simmons, A. (1987). Retrograde transneuronal transfer of Herpes Simplex Virus type 1 (HSV 1) from motoneurones. *Brain Res.* **422**, 242–256.

Ugolini, G., Kuypers, H. G. J. M., and Strick, P. L. (1989). Transneuronal transfer of Herpes Virus from peripheral nerves to cortex and brainstem. *Science* **243**, 89–91.

Vahlne, A., Nyström, B., Sandberg, M., Hamberger, A., and Lycke, E. (1978). Attachment of herpes simplex virus to neurons and glial cells. *J. Gen. Virol.* **40**, 359–371.

Whealy, M. E., Card, J. P., Robbins, A. K., Dubin, J. R., Rziha, H.-J., and Enquist, L. W. (1993). Specific pseudorabies virus infection of the rat visual system requires both gI and gp63 glycoproteins. *J. Virol.* **67**, 3786–3797.

Wildy, P. (1985). Herpes viruses: A background. *Br. Med. Bull.* **41**, 339–344.

Wildy, P. (1986). Herpesvirus. In "Portraits of Viruses" (F. Rapp, Ed.), Intervirology Vol. 25, pp. 117–140. Krager AG, Basel.

Zawatzky, R., Engler, H., and Kirchner, H. (1982). Experimental infection of inbred mice with herpes simplex virus. III. Comparison between newborn and adult C57BL/6 mice. *J. Gen. Virol.* **60**, 25–29.

Zemanick, M. C., Strick, P. L., and Dix, R. D. (1991). Direction of transneuronal transport of herpes simplex virus 1 in the primate motor system is strain-dependent. *Proc. Natl. Acad. Sci. USA* **88**, 8048–8051.

CHAPTER 18

Pseudorabies Virus Replication and Assembly in the Rodent Central Nervous System

J. Patrick Card
Department of Neuroscience
University of Pittsburgh
Pittsburgh, Pennsylvania

I. Introduction

Alphaherpesviruses possess a well-documented capacity to invade the nervous system and replicate within synaptically linked populations of neurons. These properties were initially exploited for analysis of neuronal circuitry in classical studies conducted by Kristensson (Kristensson, 1970; Kristensson *et al.*, 1971, 1974, 1978, 1982) and Dolivo (Dolivo *et al.*, 1978, 1979; Dolivo, 1980; Martin and Dolivo, 1983) and, with the increased availability of well-defined attenuated strains of virus, there has been a dramatic increase in the application of this method for circuit analysis (see Kuypers and Ugolini, 1990; Strick and Card, 1992; Card and Enquist, 1995, for recent reviews). The tropism and invasive characteristics of herpes simplex virus (HSV) and a swine alphaherpesvirus known as pseudorabies (PRV) have also been employed for delivery of foreign genes to the nervous system as reporters for analysis of neural circuitry (Loewy *et al.*, 1991; Kovacs and Mettenleiter, 1991), therapeutic manipulation of neuronal function (Breakefield and Geller, 1987; Geller *et al.*, 1990; Kaplitt *et al.*, 1991; Chang *et al.*, 1991; Glorioso *et al.*, 1992; Federoff *et al.*, 1992), or treatment of malignant brain tumors (Martuza *et al.*, 1991; Takamiya *et al.*, 1992). Thus, it is clear that insight into the mechanisms underlying alphaherpesvirus replication and transport has important implications for improving our understanding of the functional organization of the nervous system, alteration of neuronal physiology, and treatment of central nervous system (CNS) disease.

Among the many factors that must be considered in evaluating the utility of using viruses for tract tracing or gene delivery are (1) the route of inoculation,

VIRAL VECTORS

(2) the invasive characteristics of the virus, (3) the neurovirulence and cytopathic consequences of infection, and (4) the cell-specific aspects of viral invasion and replication. Prior studies have demonstrated selective tropism of different alphaherpesviruses for specific classes of neurons (Card *et al.*, 1991; Norgren and Lehman, 1992), and it is clear from a number of studies that these viruses are also capable of replicating in glial cells (see Card and Enquist, 1994, for recent review). Furthermore, the direction of transport and intracellular distribution of different stains of virus can vary dramatically (Zemanick *et al.*, 1991; Enquist *et al.*, 1993). These findings have clear implications for evaluating both the specificity and the extent of viral transport through neural circuitry as well as the functional consequences of the expression of genes delivered with a viral vector. The goal of this chapter is to review these issues and consider their impact upon the use of alphaherpesviruses for gene delivery. The principal emphasis is placed upon the swine alphaherpesvirus known as pseudorabies, but general principals derived from studies of HSV and other alphaherpesviruses are included.

II. Virion Replication and Assembly in Neurons

A. Virion Structure

Alphaherpesvirus particles possess a variety of distinctive morphologies that reflect the differing stages of assembly as well as the infectious nature of the particle (Rixon, 1993). Extracellular infectious virus characteristically exhibits four morphologically distinct structural components. The central core of the particle consists of the *viral DNA* and this distinctive electron-dense mass of linear double-stranded DNA is packaged within a structural protein capsule known as the *capsid*. The capsid is surrounded by an amorphous *tegument* whose constituent proteins play important roles in the regulation of host cell metabolism and the initiation of the cascade of virally induced DNA transcription produced by infection. The capsid and tegument are surrounded by a *lipid envelope* derived from the parent cell that contains at least nine virally encoded glycoproteins in PRV (Mettenleiter, 1991; Card and Enquist, 1995). These envelope glycoproteins play essential roles in target cell recognition, attachment, and receptor-mediated invasion of a permissive cell (Wittmann and Rziha, 1989; Mettenleiter, 1991). Removal of the viral envelope or deletion of certain essential envelope glycoproteins eliminates the ability of virions to invade a permissive cell (see below). Similarly, failure of an infected cell to envelope newly assembled capsids prevents virion release from the parent cell.

B. DNA Replication and Packaging

The life cycle of PRV is illustrated in Fig. 1. As noted above, invasion of permissive cells in a glycoprotein-mediated event that culminates in the fusion of the viral envelope with the plasma membrane of the neuron (Spear, 1993). Liberation of the capsid and surrounding tegument within the cytoplasm of the neuron under these circumstances is the first step in a highly orchestrated cascade of intracellular events that ultimately leads to production and release of infectious progeny. Early in this process viral capsids are transported to the cell nucleus where they disassemble at the nuclear pore and release viral DNA

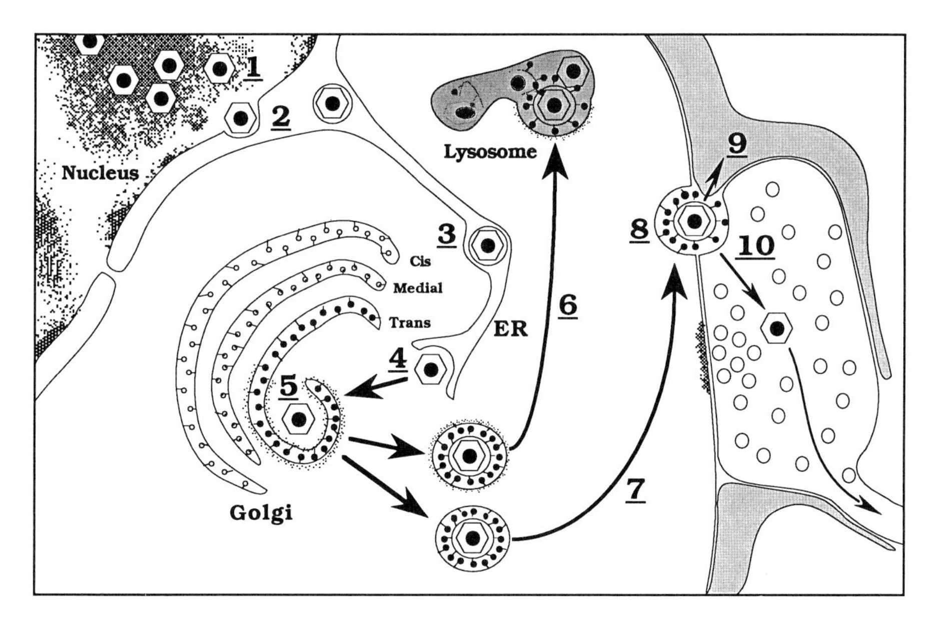

Figure 1 Illustrated is the model we have proposed for alphaherpesvirus assembly and intracelluar trafficking derived from studies of virulent and attenuated strains of pseudorabies virus in the rat CNS. (1) Replication and packaging of viral DNA forms nucleocapsids in the nucleus of the afflicted cell. (2) Nucleocapsids acquire a single membrane by budding through the inner leaf of the nuclear envelope to gain access to a compartment in the cell continuous with the ER. (3) Enveloped capsids traverse the ER and are released by a fusion event (4) that liberates naked capsids adjacent to the Golgi apparatus. (5) The trans cisterns of the Golgi complex wrap capsids with a bilaminar envelope endowed with mature virally encoded envelope glycoproteins. (6) Capsids that are enveloped early in the course of infection also are endowed with a spiked coat resembling clathrin which targets virions to primary lysosomes for degradation. (7) Capsids enveloped later in the course of infection generally do not exhibit a spike coat and escape the lysosomal pathway. These virions leave the cell (8) to either infect reactive astrocytes (9) or invade afferent terminals in synaptic contact with the parent cell (10).

into the cell nucleus. Although the specific functions of PRV tegument proteins have not been elucidated, it is likely that they serve roles similar to those in HSV by modulating the regulation of cellular macromolecular synthesis and the initiation of viral transcription early in the course of infection. The product of the single immediate-early gene of PRV (IE180) is necessary for the temporally organized expression of early and late genes, a subset of which is required for efficient replication of PRV DNA (Mettenleiter, 1991; Enquist, 1995). Collectively, this process ultimately leads to replication and packaging of viral DNA in newly assembled capsids in the nucleus of the afflicted cell. With advancing replication, these capsids become prevalent throughout the cell nucleus and often are found in large paracrystalline arrays adjacent to the nuclear envelope (Fig. 2).

C. Capsid Envelopment

Acquisition of a membrane envelope from the host cell is an essential component of the virus life cycle. The integral membrane proteins that contribute to this envelope are encoded by the virus and have been classified as either essential or nonessential based upon their requirement for viral growth in tissue culture. Following synthesis in the endoplasmic reticulum these proteins are transported to the Golgi complex and are modifed by glycosylation as they pass sequentially through the cisternae of this organelle. Mature envelope glycoproteins are enriched in the trans face of the Golgi cisterns and are also distributed to other membranes in the cell by vesicular transport. Although subcellular localization of these glycoproteins has been used to postulate potential sites of capsid envelopment, the diverse roles that these molecules play in virion assembly, intracellular trafficking, and viral release (see subsequent discussion) make it clear that the mere localization of a virally encoded glycoprotein in a cellular membrane cannot be taken as evidence that the membrane is a source of the virion envelope. Conclusions regarding the intracellular source of the viral envelope are further complicated by the differing conclusions drawn from studies of HSV, PRV, and varicella zoster virus (VZV) assembly. In studies of HSV replication, a number of investigators have concluded that *final* envelopment occurs at the nuclear membrane in response to capsids budding through the limiting membranes of this organelle to gain access to the cell cytoplasm (Johnson and Spear, 1982; Poliquin *et al.*, 1985). Johnson and Spear (1983) further postulate that glycosylation of viral glycoproteins occurs by physical migration of enveloped viral particles through the Golgi apparatus. A role for the nuclear envelope in envelopment has also been proposed for PRV by Pol and colleagues (Pol *et al.*, 1989, 1991a,b) in their studies of the assembly of this virus in porcine nasal mucosa, both *in vivo* and in explant cultures. On the basis of their analysis of the assembly of three strains of the virus, these investigators came to the conclusion that virulent PRV derived an envelope from the nuclear envelope, while attenuated strains were enveloped at the endoplasmic reticulum. They further suggested that the inability of the attenuated strains to acquire an envelope from the nuclear envelope was due to the absence of a gene responsible for encoding one of the viral envelope glycoproteins known as gE (nomenclature adopted at the 18th International Herpesvirus Workshop in 1993).

A number of other laboratories have suggested that acquisition of a membrane at the nuclear envelope is only the first event in a multistep cascade responsible for moving the capsid through the cell and ultimately endowing it with a membrane enriched with mature viral glycoproteins. In quantitative electron microscopic autoradiographic analyses incorporating [^{3}H]fucose labeling, Jones and Grose (1988) provided strong evidence for a cytoplasmic site of VZV envelopment and further postulated a central role for the Golgi apparatus in providing the mature viral glycoproteins that were incorporated within this envelope. These observations were consistent with earlier studies of HSV that proposed a cytoplasmic site of envelopment (Nii, 1971; Rodriquez and Dubois-Dalcq, 1978). More recently, we examined the assembly of PRV both *in vitro* and *in vivo* and came to a similar conclusion. Following infection of immortalized pig kidney fibroblasts with a virulent strain of PRV known as Becker (PRV–Becker; Becker, 1967) and serial transmission electron microscopic analysis we observed a multistep intracellular pathway of capsid envelopment that involved sequential acquisition and shedding of cellular membranes derived from the nuclear envelope and cisternae of the Golgi complex (Whealy *et al.*, 1991). The initial envelopment in this pathway occurred at the nucleus in response to budding of capsids through the inner leaf of the nuclear envelope. Enveloped capsids produced by this budding were surrounded by a single membrane and were afforded access to an intracellular pathway continuous with the endoplasmic reticulum (ER). Proliferation of the ER in response to viral infection produced a large trabecullar network that permitted these particles to travel throughout the soma of the cell, and subsequent fusion of the nuclear-derived membrane with the ER resulted in release of capsids into the cell cytoplasm in the vicinity of the Golgi complex. Final envelopment of the capsid occurred at the *trans* face of the Golgi cisternae (Figs. 1 and 3). Acquisition of this bilaminar envelope suggests a mechanism in which capsids were not only provided with the necessary membranes to excape the parent cell and gain access to the next permissive cell, but also receive the necessary complement of mature viral glycoproteins essential for this process. In testing this hypothesis we examined the effects of brefeldin A (BFA) upon PRV assembly in immortalized pig kidney fibroblasts (Whealy *et al.*, 1991). BFA is a potent inhibitor of transport between the ER and Golgi apparatus and ultimately results in dissolution of the Golgi apparatus (Lippincott-Schwartz *et al.*, 1989, 1990). When cells infected with PRV–Becker were exposed to BFA, analysis of single-step growth curves revealed an abrupt interruption in the production of infec-

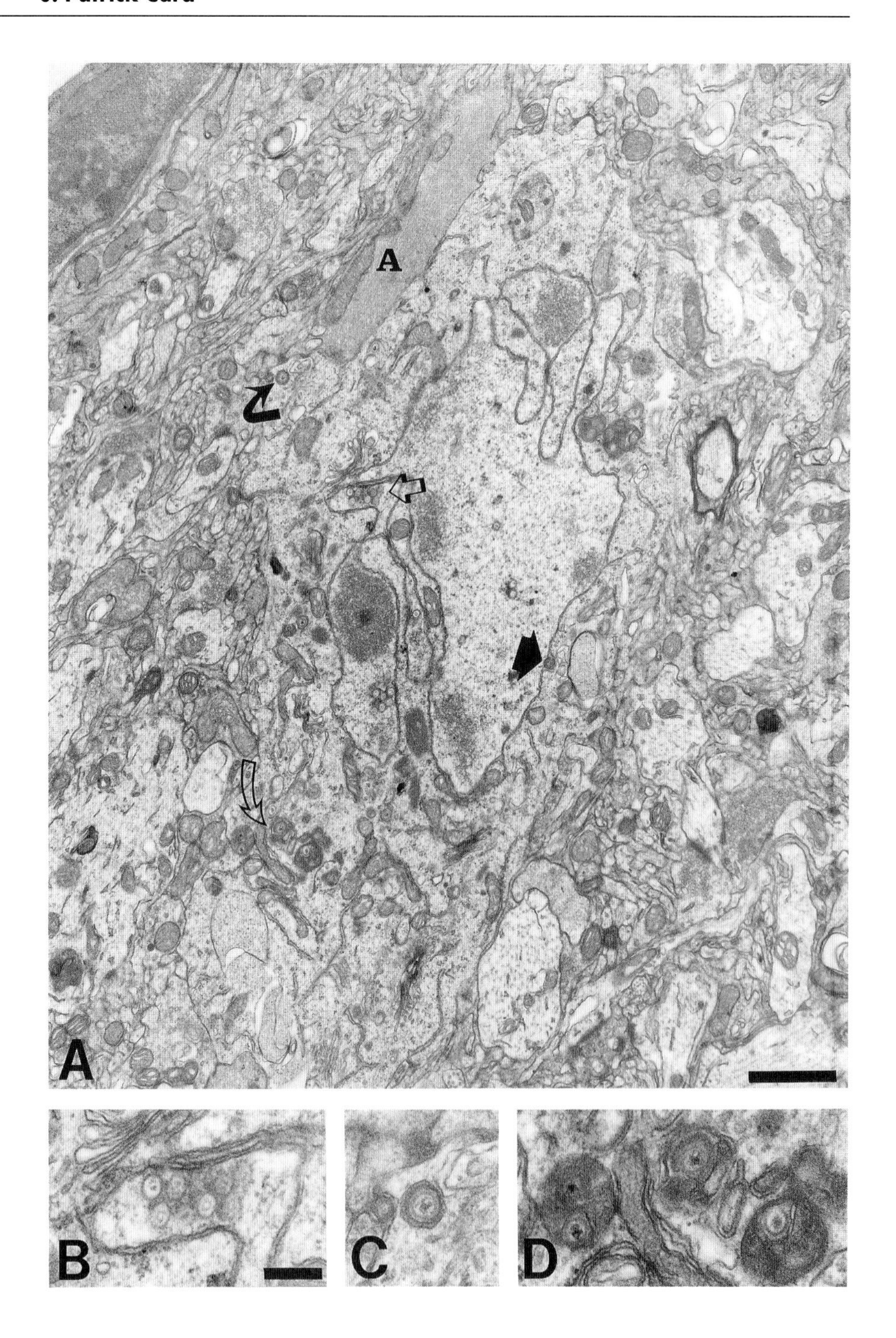
A
A
B
C
D

tious virus, an effect that could be reversed by removal of BFA. Ultrastructural analysis of cells treated with BFA demonstrated an accumulation of single-enveloped capsids between the leafs of the nuclear envelope and in the ER, as well as large numbers of naked capsids in the cell cytoplasm. These observations led us to propose that final envelopment of PRV occurs at the Golgi complex, and we subsequently confirmed that his mode of envelopment also occurs *in vivo* following infection of neurons with either virulent or attenuated strains of PRV (Card *et al.*, 1993).

Other recent studies of HSV envelopment also support the conclusion that the Golgi complex is the source of the virion envelope. Koyama and Uchida (1994) examined the effects of BFA on the envelopment of HSV-1 *in vitro* and reported effects essentially identical to those reported in our study of PRV (Whealy *et al.*, 1991). Additionally, Komuro and colleagues (1989) reported cytoplasmic envelopment of capsids by Golgi cisternae as well as the presence of acid phosphatase (an enzymatic marker of *trans*Golgi membrane) in the envelope newly replicated virions. Their analysis further demonstrated that an enzymatic marker of the nuclear envelope and ER (glucose-6-phosphatase) is not present in virion envelopes. More recently, van Genderen and colleagues (1994) have analyzed the phospholipid composition of the extracellular herpes simplex virions and find that it differs from that exhibited by the host cell nucleus, but is similar to that of the Golgi apparatus. Collectively, these data are consistent with the conclusion that the Golgi complex plays a central role in virion envelopment in infected neurons. Nevertheless, differences in the mode of viral assembly in different cell types (Card *et al.*, 1993; see below) as well as reported differences in the envelopment of different strains of virus (Pol *et al.*, 1989, 1991a,b) indicate that other routes of virion envelopment must be considered in evaluating virion assembly and transport in the CNS.

D. Intracellular Transport of Virions

Although there is now ample evidence for specific passage of alphaherpes-viruses through synaptically linked neurons (Card *et al.*, 1990, 1993; Strack

Figure 2 (A) The morphology of a brain stem neuron infected with PRV is illustrated. Virus injected into the stomach invaded the peripherally projecting axon of this cell and capsids were then retrogradely transported to the cell soma to initiate viral replication. The early hallmarks of infection are found in the cell nucleus which becomes highly invaginated and contain accumulations of nucleocapsids (large open arrow and B). Capsids escape the cell nucleus by budding through the inner leaf of the nuclear envelope (solid arrow) and traversing the endoplasmic reticulum. Final envelopment occurs at the Golgi complex producing mature virions enclosed within a bilaminar envelope. These virions are either targeted to lysosomes (open curved arrow shown at higher magnification in D) or escape the lysosomal pathway and are transported to the periphery of the soma (solid curved arrow shown at higher magnification C) and into the dendritic tree. Marker bars: A = 1 μm; B–D = 200 nm.

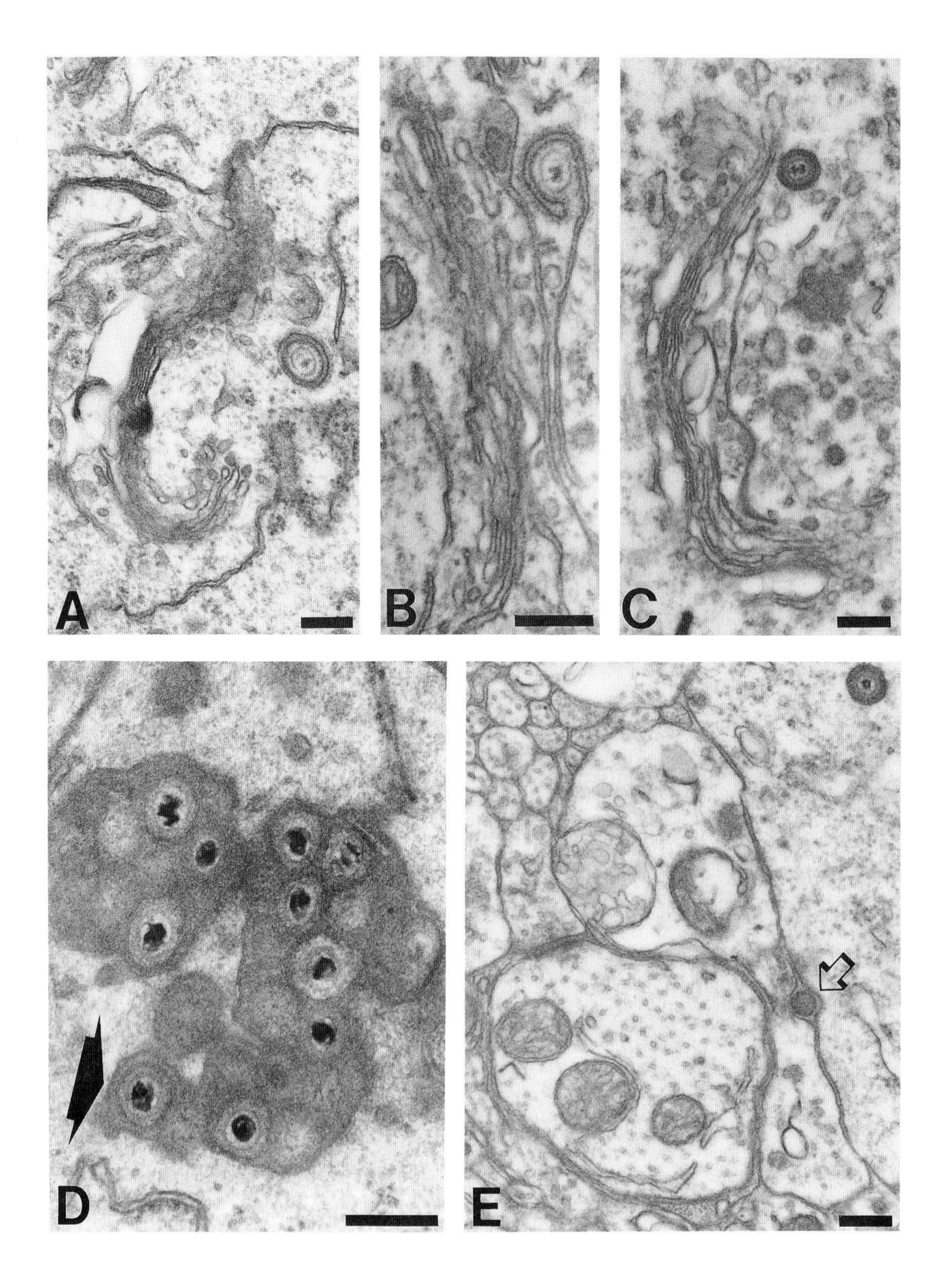
A
B
C
D
E

and Loewy, 1990; Rinaman *et al.*, 1993), the precise mechanisms that lead to targeted intracellular delivery of virions to sites of afferent synaptic contact remain to be determined. Available data suggest that the intracellular transit of virions occurs in a highly ordered manner that contributes to specific passage of virions through a neural circuit and also implicates viral glycoproteins as important determinants of the intracellular sorting and release of newly replicated virions. As noted above, a number of studies support the conclusion that virions acquire an envelope from the membrane of the *trans* face of the Golgi complex (Komuro *et al.*, 1989; Card *et al.*, 1993; Koyama and Uchida, 1994; van Genderen *et al.*, 1994). This means of envelopment takes advantage of the normal functional properties of the Golgi complex (Mollenhauer and Morre, 1991; Moore, 1991) to glycosylate virally encoded glycoproteins and insert them in membranes that will ultimately be applied to the capsid. This process also appears to be part of an intracellular targeting pathway responsible for delivery of vesicles to lysosomes. Indeed, we have demonstrated that virions accumulate in primary lysosomes and are observed in various stages of degradation (Card *et al.*, 1993; Figs. 2D and 3D). The appearance of virions in lysosomes occurs early in the course of infection, but the number of lysosomes containing viral particles does not appear to increase progressively with advancing viral replication. Thus, we have postulated that targeting of virions to lysosomes is an early protective response of the cell to infection that is ultimately overwhelmed as the viral infection alters the metabolism of infected cells. Evidence in support of this conclusion is apparent in the morphology of enveloped virions at different stages of infection. At early stages postinfection, the Golgi complex endows capsids with a bilaminar envelope that contains a spiked coat (Figs. 3A and 3B). This coat is similar to the clathrin coat that is responsible for delivery of Golgi-derived vesicles to lysosomes (see Mellman and Simons, 1992, for a recent review). In contrast, virions produced later in the course of infection often lack this distinctive spiked coat (Fig. 3C) and are similar in morphology to virions throughout the dendritic tree and those observed in relation to sites of afferent synaptic contact (Figs. 3E and 4). Although it is clear that further studies examining the effects of viral infection on the expression of

Figure 3 A, B, and C illustrate envelopment of capsids at the trans face of the Golgi cisternae. A and B illustrate the spiked coat characteristically found on the outer surface of virions produced early in the course of infection. Note in B that this spiked coat is only present on the portion of the *trans* cistern that will ultimately become the viral envelope. Virions produced later in the course of infection generally lack the spiked coat (C). Two lysosomal inclusions filled with virions in various stages of degradation are illustrated in D. The arrow indicates a virion with a bilaminar envelope approaching the larger lysosome. E illustrates a mature virion in the peripheral aspects of the soma of an infected neuron (note the absence of the spiked coat) and a virion (open arrow) in the process of passing transneuronally into an afferent terminal. Marker bars for all figures = 200 nm.

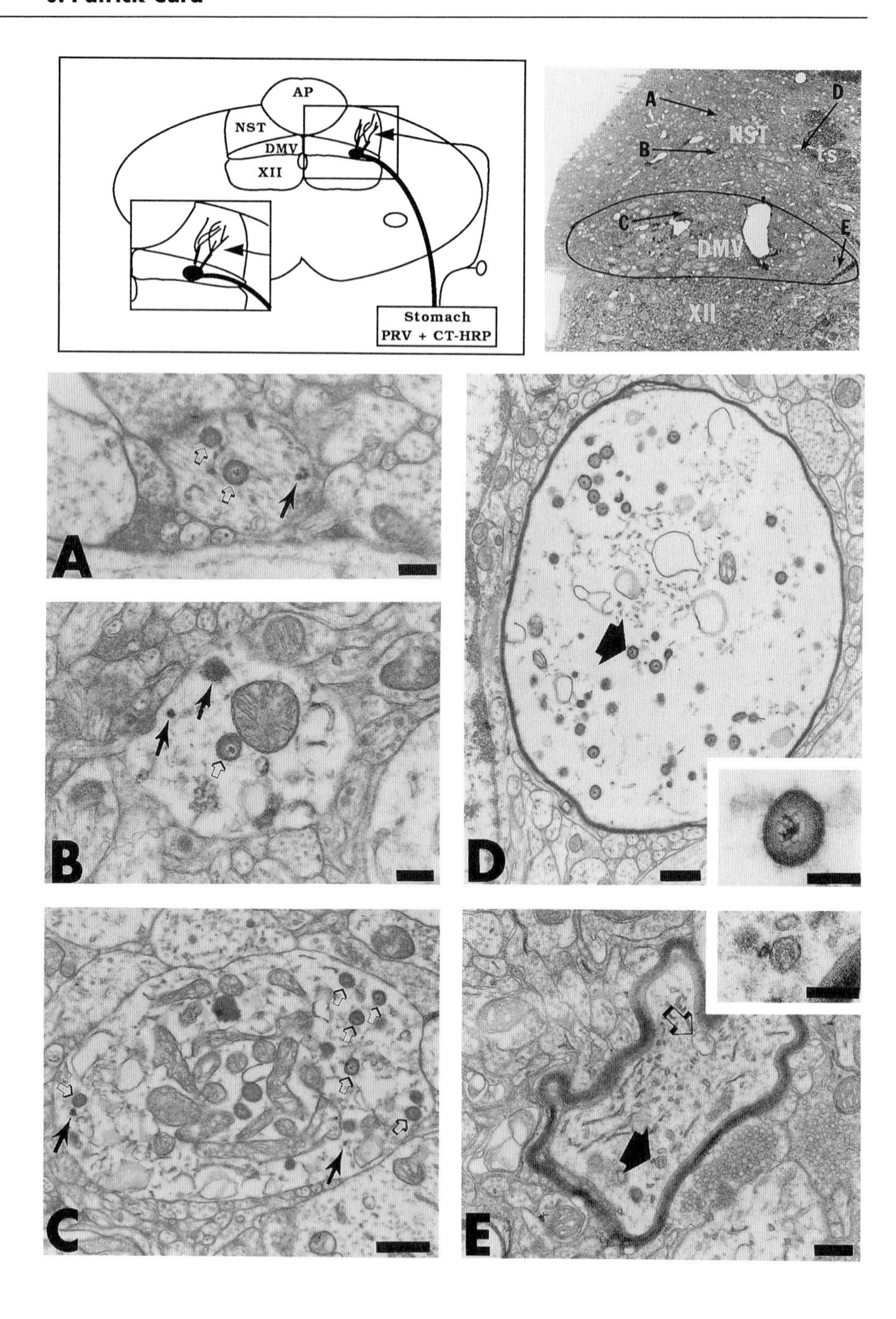
AP
NST
DMV
XII
Stomach
PRV + CT-HRP
A
B
C
D
E
NST
TS
DMV
XII

clathrin and other host proteins involved in intracellular trafficking are necessary, these morphological observations suggest that the virus is effectively parasitizing the well-known functional capacities of the Gogli complex (Geuze and Morre, 1991) for assembly and intracellular distribution of virions.

Evidence also supports the conclusion that virally encoded glycoproteins subserve important roles in intracellular trafficking and release of virions. Differential sorting of alphaherpesvirus glycoproteins in polarized epithelial cells has been demonstrated (Spear, 1984) and Dotti and collaborators (Dotti and Simons, 1990; Dotti *et al.*, 1993) have shown that glycoproteins of vesicular stomatitus virus are preferentially sorted into the somata and dendrites of cultured hippocampal neurons. Our *in vivo* studies using monospecific antisera generated against PRV envelope glycoproteins have shown that these proteins are differentially concentrated in the perikarya and dendrites of these cells (Rinaman *et al.*, 1993). Thus, it is possible that the cellular processes that account for the intracellular sorting of Golgi-derived viral glycoproteins could similarly influence the sorting of the viral particles that contain these same glycoproteins. In particular, considered with the documented fusogenic functions of some viral glycoproteins (Campadelli-Fiume *et al.*, 1988a,b, 1990), the differential localization of viral glycoproteins could influence the ultimate site of virion release. Further analysis of the subcellular distribution of these glycoproteins in infected neurons is necessary to determine the validity of this hypothesis.

Preferential sorting of virions into the somatodendritic compartment is also consistent with the majority of studies that have reported retrograde, transneuronal passage of virus through the CNS (Ugolini *et al.*, 1987, 1989; Strack *et al.*, 1989a,b; McLean *et al.*, 1989; Strack and Loewy, 1990; Blessing *et al.*, 1991; Jansen *et al.*, 1992; Rotto-Percelay *et al.*, 1992; Ugolini, 1992;

Figure 4 This series of micrographs illustrates the morphology and intacellular distribution of virions in neurons of the dorsal motor vagal complex (DMV) infected by injection of a "cocktail" of virus and a cholera toxin–HRP (CT–HRP) conjugate into the wall of the stomach. The experimental paradigm is illustrated in the schematic diagram. The electron-dense CT–HRP histochemical reaction product (small arrows in A–C) permitted unambiguous identification of components of the vagal circuitry. The thick section at the right of the schematic diagram is from the region immediately adjacent to the thin sections used for the ultrastructural analysis. The labeled arrows in the photomicrograph of the thick section indicates the region from which the electron micrographs in A through E were taken. Newly replicated virions with bilaminar envelopes (small open arrows in A–C) are present through the full extent of the DMV dendritic tree, including the apical dendrites. They are also found in the centrally projecting process of sensory neurons that became infected with the virus (D). In contrast, axons of DMV neurons only contain naked capsids. The morphology and location of these particles support the proposed model of assembly illustrated in Fig. 1. Taken with permission from Card *et al.* (1993). Marker bars for A, B, and E = 200 nm and 500 nm for C and D. The inset marker bars for D and E represent 100 nm.

Barnett *et al.*, 1993). However, it is clear from a number of observations that the intracellular trafficking of virions can vary in different populations of neurons. This is particularly apparent in comparing the direction of virion transport through sensory neurons, retinal ganglion cells, and neurons in the CNS. Numerous studies have shown that sensory neurons have the capacity for anterograde transport of newly replicated virions (Kristensson *et al.*, 1978, 1982; Lycke *et al.*, 1984, 1988; Martin and Dolivo, 1983; Margolis *et al.*, 1987, 1989, 1992; Carter *et al.*, 1992; LaVail *et al.*, 1993). Similarly, our studies with PRV (Card *et al.*, 1991, 1992; Whealy *et al.*, 1993), and those of Kristensson (1974) and Norgren and Lehman (1992) with HSV demonstrate that infection of retinal ganglion cells results in anterograde transneuronal infection of retinorecipient neurons in the CNS. In contrast, once CNS neurons become infected, subsequent transneuronal transport of virus is usually consistent with what one would expect if virions were only passing in the retrograde direction through a multisynaptic circuit. That is to say, evidence for anterograde transport of newly replicated virus into the axons of infected CNS neurons has only been reported in a few investigations. For example, studies from Strick's laboratory (Zemanick *et al.*, 1991) have shown that two strains of HSV produce different patterns of infection following intracerebral injection into the primate motor cortex. One strain produces a pattern of infection that one would expect if the virus were only being transported retrogradely through a multisynaptic circuit, whereas the other strain infects neurons known to be the synaptic targets of the motor cortex in a pattern consistent with anterograde transport of virus from the injection site. The mechanisms underlying this differential transport of two closely related strains of virus have not been characterized, but the findings have obvious implications for interpreting data derived from viral tract-tracing studies as well as the transport of viral vectors injected directly into the CNS. For that reason, we initiated a comparative analysis of the transport of virulent and attenuated strains of PRV designed to examine the role of viral envelope glycoproteins in this process (Enquist *et al.*, 1993). The virulent wild-type strain of PRV (PRV–Becker; Becker, 1967) contains a full complement of envelope glycoproteins, whereas the attenuated strain (PRV–Bartha; Bartha, 1961) contains well-characterized deletions or mutations in the genes encoding three of the known envelope glycoproteins. We reasoned that the differing patterns of infection noted by Strick and coworkers might be due to differences in the glycoprotein content of the viral envelope since envelope glycoproteins are known to play important roles in virion attachment and invasiveness (Spear, 1993). Our analysis demonstrated strain-specific patterns of neuronal infection, but there was one important difference between our findings and those of Strick's group. Intracerebral injection of PRV–Becker produced patterns of infection consistent with *both* anterograde and retrograde transport from the site of injection. In contrast, injection of PRV–Bartha only infected a subset of this circuitry that would have been infected if the virus were being selectively taken up by terminals afferent to this region (e.g., selective retrograde infection). The loss of the ability of attenuated virus to infect a subset of the neurons that are susceptible to infection by wild-type virus suggests that different virally encoded gene products (i.e., envelope glycoproteins) are necessary for aspects of the viral life cycle necessary for anterograde transneuronal infection. For example, deletion of virally encoded glycoproteins necessary for invasion of neuronal perikarya would preclude the ability of the virus to gain access to cortical neurons giving rise to efferent projection systems. Loss of a virally encoded protein necessary for anterograde transport or release would similarly compromise the ability of the virus to infect the synaptic targets of the cortical neurons. Further analysis of isogenic strains of virus with one or more of the same mutations exhibited by PRV–Bartha is necessary to answer these questions.

The extent of viral transport through the CNS is also dependent upon the distribution of newly replicated virions within an infected cell. If virus only gains access to somata and proximal dendrites, then afferents terminating on the more distal portions of the dendritic arbor would be excluded from transneuronal infection, a finding that would have significant import for defining the extent of viral dissemination through the neuraxis. We have addressed that issue in electron microscopic studies of preganglionic parasympathetic neurons in the caudal brain stem that innervate the viscera (Card *et al.,* 1993). The dendritic tree of these neurons is highly polarized and in a previous study we devised a method of systematically identifying the proximal, intermediate, and distal dendrites of these cells at the ultrastructural level (Rinaman *et al.,* 1989). This approach takes advantage of a tracer [cholera toxin (CT) conjugated to HRP] that is retrogradely transported from the terminal axons of the cells in sufficient quantities to fill the neuron in a "Golgi-like" fashion (Shapiro and Miselis, 1985). We injected the stomach with a cocktail of the CT–HRP conjugate and PRV and then examined the distribution of newly replicated virions in the dendrites of the infected cells (Fig. 4). These virions were easily recognized by virtue of the double-membrane Golgi-derived envelope, and the histochemical localization of CT–HRP provided an electron-dense marker of dendrites of the preganglionic neurons. This analysis demonstrated that virions reached even the most distal dendrites of the afflicted cells and implied that all afferents were equally predisposed to transneuronal infection. However, our findings do not resolve the temporal dynamics of intracellular transport (i.e., do proximal afferents become infected first?) and it remains possible that some afferents are resistant to infection.

E. The Role of Viral Envelope Glycoproteins in Viral Assembly and Transport

A number of recent studies of PRV transport in rodent CNS have provided striking evidence that virally encoded envelope glycoproteins subserve important roles in the assembly and transneuronal transport of virus. In an effort to establish a model system that would permit us to examine the role of viral envelope glycoproteins in viral neurotropism and anterograde transport we initiated a systematic evaluation of the ability of genetically defined strains of PRV to invade central visual projection systems. We began our analysis by comparing the patterns of central infection produced by intravitreal injection of PRV–Becker and the attenuated PRV–Bartha strain and discovered a remarkable difference in the ability of these two strains of virus to invade the central visual projection pathways (Card *et al.*, 1991). PRV–Becker, which possesses a full complement of envelope glycoproteins, infected all retinorecipient nuclei, but did so in two temporally separated waves that targeted functionally distinct subdivisions of the visual system. The first wave of infection became apparent approximately 2 days postinoculation with the appearance of infected neurons in the dorsal geniculate nucleus (DGN) and superior colliculus of the tectum (Figs. 5 and 6). Infected neurons in these two regions were coextensive with the distribution of retinal afferents and their density also reflected the heavier crossed projection characteristic of the rat visual system. During this time frame no infected neurons were present in the suprachiasmatic nuclei of the hypothalamus (SCN) or the intergeniculate leaflet of the thalamus (IGL), two retinorecipient areas involved in the control of circadian function.

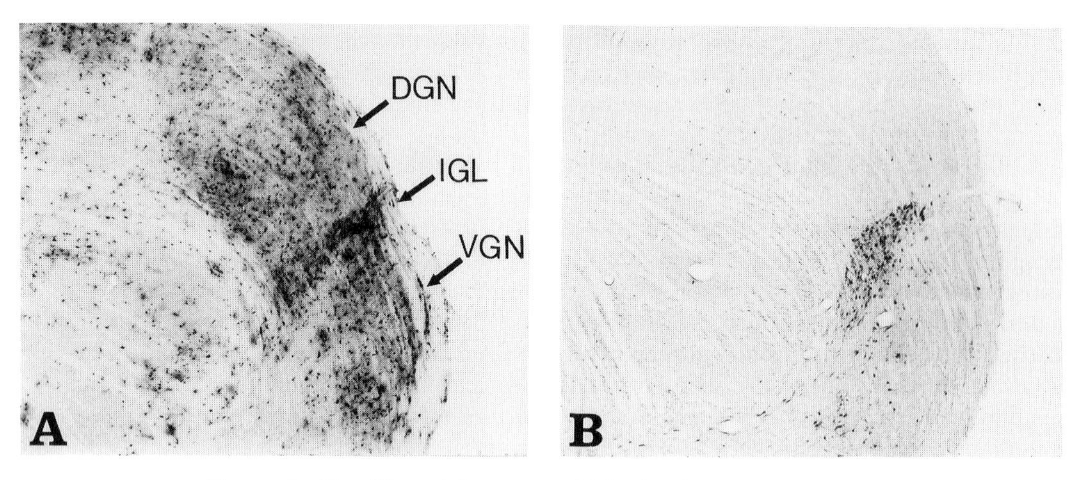

Figure 5 The differential patterns of infection in retinorecipient neurons of the lateral geniculate complex following injection of PRV–Becker or PRV–Bartha into the vitreous body of the eye are illustrated. PRV–Becker produces transneuronal infection of all three subdivisions of the geniculate (A). In contrast, PRV–Bartha produces selective productive infection of neurons in the intergeniculate leaflet (IGL) and, to a lesser extent, in the ventral geniculate nucleus (VGN). Neurons in the dorsal geniculate never exhibit a productive infection following injection of PRV–Bartha (B).

The second wave of infection began approximately 3 days postinoculation with the appearance of infected neurons in the ventrolateral retinorecipient subdivision of the SCN and throughout the IGL. The pattern of central infection produced by identical administration of PRV–Bartha was dramatically different. This strian never infected neurons in the DGN or tectum at postinoculation intervals extending to 5 days, but retained the ability to infect neurons in the SCN and IGL with temporal parameters essentially identical to those exhibited by PRV–Becker (Figs. 5 and 6).

In an effort to explain the restricted neurotropism exhibited by PRV–Bartha we focused upon the mutations of the viral genome present in this strain. PRV–Bartha is characterized by a large deletion in the unique short region of the viral genome that eliminates the genes for the gE and gI envelope glycoproteins (Mettenleiter, 1991). It also possesses a signal sequence mutation in the gC gene of the unique long segment that reduces the concentration of this glycoprotein in the viral envelope (Robbins *et al.*, 1989). We hypothesized that one or more of these mutations were responsible for the restricted phenotype characteristic of the Bartha strain and prepared a set of isogenic strains of PRV–Becker that contained one or more of the glycoprotein gene deletions characteristic of PRV–Bartha (Card *et al.*, 1992). Analysis of the infectivity of these null mutants demonstrated that the deletion in the unique short segment of the genome produced the restricted phenotype. For example, deletion of the gC gene in the unique long segment produced patterns of central infection that were identical to those produced by the wild-type virus, while deletion of the unique short gE gene produced a phenotypic pattern of infection identical to that produced by PRV–Bartha. In a subsequent investigation we demonstrated that deletion mutants lacking the gene that encodes the gI glycoprotien also exhibited the restricted phenotype and provided evidence that gE and gI function as a heterodimer (Whealy *et al.*, 1993). Consequently, deletion of either of these unique short gene products removes the capacity for transneuronal infection of retinorecipient neurons in the DGN or tectum, but does not compromise the ability of these mutants to infect neurons involved in the regulation of circadian function.

Upon identifying the viral gene products responsible for the selective tropism of PRV mutants we sought to determine if the inability of these strains to invade functionally distinct components of the visual neuraxis was a property of the virus or the host neurons (Enquist *et al.*, 1994). We entertained three hypotheses. The first attributed the restricted neurotropism to a differential distribution of viral receptors on projection-specific classes of retinal ganglion cells. In this scenario, the gE/gI dimer was necessary for the virus to recognize a receptor differentially concentrated on ganglion cells projecting to the DGN and tectum. Consequently, deletion of one or both of the genes encoding these glycoproteins would remove the capacity of the virus to infect this functionally distinct class of ganglion cells and thereby prevent anterograde transneuronal

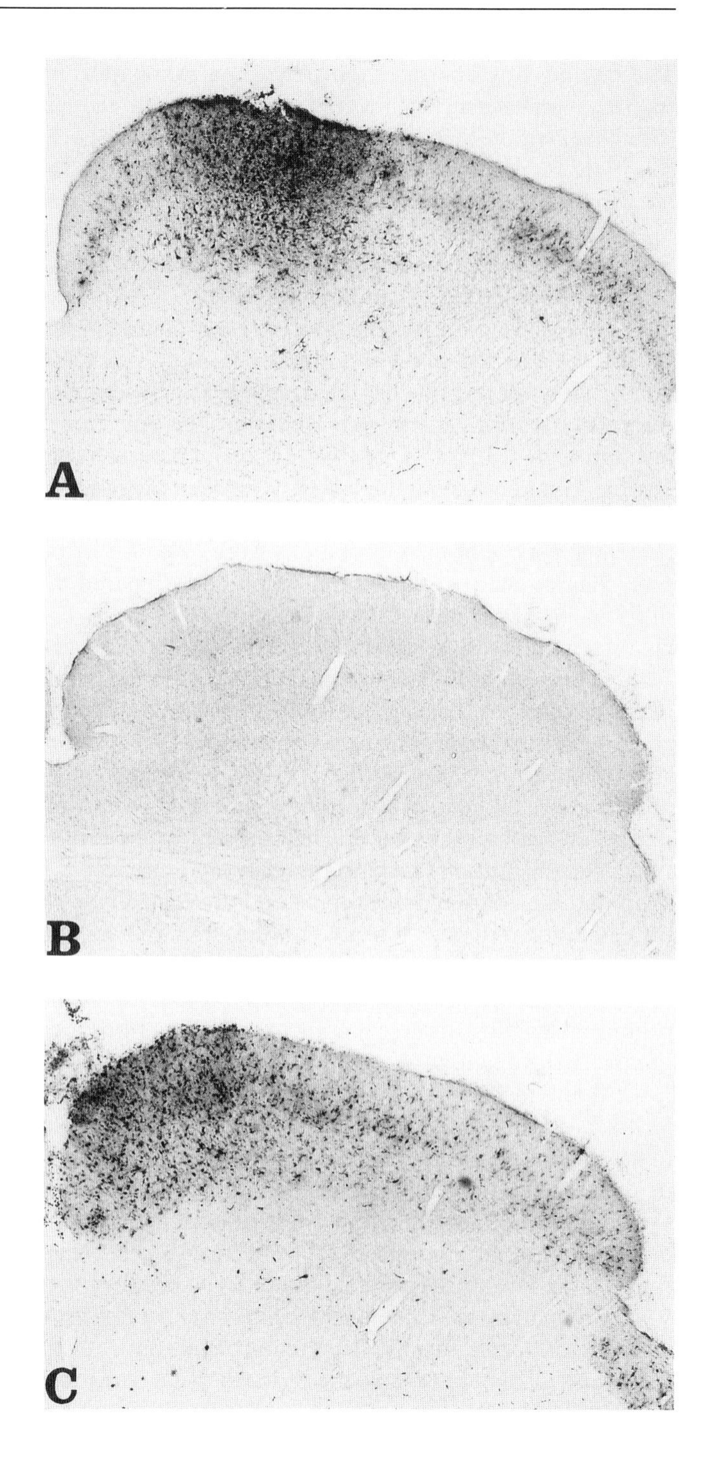
A
B
C

infection of their synaptic targets in the CNS. The second hypothesis assumed that the mutant virus was able to invade all ganglion cells and attributed the restricted phenotype to a defect in replication, intracellular transport, or release of virions from ganglion cells projecting to the DGN and tectum. The third hypothesis postulated that the inability to infect the DGN or tectum is a property of the target neurons in these regions; i.e., virus is released from retinal terminals and either cannot invade DGN and tectal neurons or establishes a latent rather than a productive infection. To test these hypotheses we injected one series of animals with the gI deletion mutant, a second group with the gI deletion mutant, and a third set with a cocktail containing both mutant viruses. As expected, the two groups injected with either gE or gI mutants exhibited the restricted phenotype characteristic of PRV–Bartha. However, simultaneous injection of both mutants restored the wild-type phenotype with respect to both the distribution of infected neurons and the temporal order of their appearance (Fig. 6). This apparent genetic complementation of the two viruses strongly suggests that the mutants gain access to all retinal ganglion cells, but are incapable of producing a productive infection in cells projecting to the dorsal geniculate or tectum. Further analysis is necessary to determine if this defect is a property of retinal ganglion cells or their synaptic targets in the CNS. However, these findings and other recent reports of differential patterns of infection of different viruses in the same circuitry (Barnett *et al.*, 1993; Standish *et al.*, 1994) illustrate the potential differences in susceptibility to viral infection that should be considered in the use of these viruses as vectors for gene delivery.

Further evidence of the importance of viral glycoproteins in the transneuronal spread of virus is evident in several recent studies of PRV. Numerous *in vitro* analyses have demonstrated that the virally encoded envelope glycoproteins play integral roles in the invasiveness and intercellular transport of virions (Ben-Porat and Kaplan, 1985; Wittmann and Rziha, 1989; Mettenleiter, 1991). This is particularly true of glycoproteins gB, gD, and gH. Although it has long been known that these glycoproteins are essential for growth of virus in culture, recent work has begun to provide insight into their roles in the assembly

Figure 6 A and B illustrate the differential patterns of productive infection in retinorecipient neurons of the tectum produced by injection of PRV–Becker or PRV–Bartha into the vitreous body of the eye. PRV–Becker produces productive infection of retinorecipient neurons in the superior colliculus (A), while identical injection of PRV–Bartha fails to produce a productive infection in these cells (B). C demonstrates the restoration of the wild-type pattern of infection in animals receiving an injection of a mixture of two deletion mutants isogenic with PRV–Becker. These mutants, which are lacking the genes that encode either the gE or the gI envelope glycoproteins, produce the restricted phenotype of PRV–Bartha when injected alone, but restore the wild-type phenotype when coadministered. See text or Enquist *et al.* (1994) for further details.

and transneuronal passage of virus in intact animals. Using null mutants and complementing cell lines to examine the functions of these envelope glycoproteins it has been shown that gD is necessary for invasion of permissive cells, but is not required for the subsequent spread of virus to other cells either *in vitro* or *in vivo* (Rauh and Mettenleiter, 1991; Rauh *et al.*, 1991; Peeters *et al.*, 1992a,b, 1993; Heffner *et al.*, 1993; Babic *et al.*, 1993). In contrast, the same experimental approach has demonstrated that gB and gH are required for *both* penetration and transneuronal passage of PRV in the CNS (Rauh and Mettenleiter, 1991; Rauh *et al.*, 1991; Babic *et al.*, 1993). When these data are considered with the findings of the previously discussed studies it is apparent that individual envelope glycoproteins subserve a number of important but specialized roles in virion invasiveness and assembly and therely exert a dramatic influence upon dissemination of virus through the CNS. These data also suggest that further insight into the specific functions of envelope glycoproteins may provide a mechanism for targeted delivery of viral vectors to specific populations of neurons.

III. Role of Glia in Dissemination of Viral Infection

Reactive gliosis is a fundamental response of the CNS to neuronal injury and it is clear from numerous studies that this response is ultimately evoked in response to infection of neurons with HSV or PRV (Weinstein *et al.*, 1990; Rinaman *et al.*, 1993). The extent and timing of this response varies with the virulence of the infecting strain of virus and it is apparent from a number of investigations that at least some classes of glia are susceptible to infection. Furthermore, recent studies conducted by Johnson and colleagues (1992) have shown that even replication-deficient HSV vectors can produce cytotoxic changes in cultured neurons and astrocytes. Thus, the neuropathological consequences of alphaherpesvirus infection and the resulting reactive gliosis are important considerations in evaluating the spread of virus through the nervous system as well as the utility of using these and other viruses for gene delivery.

Determining the susceptibility of different classes of glia to viral infection is an important prerequisite to defining the role of these cells in the dissemination of virus in the CNS. Early *in vitro* analyses conducted by Vahlne and colleagues (1978, 1980) revealed that HSV has a high affinity for astrocytic membranes and there have been many subsequent demonstrations of alphaherpesvirus replication in these cells *in vivo* (Ugolini and Kuypers, 1987; Card *et al.*, 1990, 1993; Rinaman *et al.*, 1993). However, other classes of glia have been shown to be resistant to infection by both HSV and PRV. For example, *in vivo* studies indicate that microglia are resistant to infection by HSV (Weinstein *et al.*, 1990) and PRV (Rinaman *et al.*, 1993; Card *et al.*, 1993) and Kastrukoff and colleagues (1986) have reported differential resistance of oligodendrocytes isolated from inbred strains of mice to infection by HSV type 1. Kastrukoff and co-workers (1987) further demonstrated that the degree of oligodendrocyte resistance *in vitro* correlated with the extent of CNS pathology and mortality in the strains of mice from which the cells had been isolated. In extending these studies, Thomas and colleagues (1991) demonstrated that resistance of cultured oligodendrocytes to HSV is influenced by the initial multiplicity of infection and speculated that resistance to infection by more resistant cells could be overcome by an increase in infectious dose. Additionally, Sarmiento (1988) has shown that mouse macrophages isolated from different organs exhibit variable resistance to HSV infection, an observation that is particularly relevant to CNS infections in light of recent demonstrations of recruitment of these cells into the CNS in response to infection of neurons with HSV or PRV (Weinstein *et al.*, 1990; Rinaman *et al.*, 1993; Card *et al.*, 1993). Collectively, these findings demonstrate the complex nature of alphaherpesvirus invasiveness in the CNS and the diversity of cell types that can serve as a host for viral replication. The data also suggest that extensive replication of virus by permissive cells in the CNS can alter the susceptibility of cells that are resistant to infection by lower doses of virus. These observations have clear implications for considering factors that may influence the targeting and expression of genes delivered to the CNS with alphaherpesvirus vectors.

Determining the consequences of alphaherpesvirus infection of glia is further complicated by data suggesting that some permissive cells may harbor defects that prevent the production of infectious progeny. In an early electron microscopic study of the pathogenesis produced in the dorsal root ganglia following peripheral inoculation of mice with PRV, Field and Hill (1974) noted that glia "either appeared uninfected or showed signs of abortive infection." The abortive infection that they described was characterized by the presence of naked capsids in the cytoplasm and the absence of any sign of capsid envelopment. Bak and colleagues (1977) made similar observations in an ultrastructural analysis of neurons in the substantia nigra infected by retrograde transport of HSV injected into the striatum. This analysis demonstrated robust replication and envelopment of capsids in infected neurons and, at 5 days postinoculation, capsids were observed in glial cells. However, enveloped capsids were not apparent in infected glia at any postinoculation interval. We have recently confirmed and extended these observations in our analysis of PRV-induced pathogenesis in the rat brain stem (Card *et al.*, 1993). In our model, preganglionic parasympathetic neurons in the dorsal motor nucleus are infected by injection of virus into their target organs in the abdomen (e.g., the stomach). Among the many advantages of this model (see Card and Enquist, 1994, for a recent review) is the ability to infect neurons in the CNS by peripheral injection, thereby eliminating the pathological consequences of direct injection of virus into the CNS. Using this approach we have characterized the temporal aspects of viral replication and transport through this circuitry and also conducted a systematic spatiotemporal characterization of the response of glia to infection of neurons with virulent and attenuated strains of PRV (Rinaman *et al.*, 1993; Card *et al.*, 1993). These studies have shown that astrocytes anticipate cytopathic changes in infected neurons by increasing in size and number at the site of infection (Rinaman *et al.*, 1993). This virus-induced hypertrophy and hyperplasia isolates infected neurons and effectively limits the indiscriminate dissemination of virus by restricting the diffusion of virions from lytically infected neurons (Card *et al.*, 1993). However, it also predisposes astrocytes to infection. Ultrastructural analysis confirmed that reactive astrocytes became infected with PRV and that the prevalence of infected astrocytes was strictly correlated with the virulence of the infecting strain of virus as well as the extent of pathological changes in adjacent neurons (Card *et al.*, 1993). Serial analysis of ultrathin sections through the same astrocytes also revealed robust replication and packaging of viral DNA and subsequent transport of these capsids to the cell cytoplasm. However, rather than acquiring an envelope from the Golgi complex that would permit the spread of virions to other cells, naked capsids accumulated in the cytoplasm of these cells (Fig. 7).

The mechanisms underlying the assembly defect harbored by astrocytes remain to be defined. Further analysis of the expression of virally encoded transcripts in these cells as well as the host cell response to viral invasion may provide some insight into this process. Nevertheless, this fascinating cellular adaptation to infection has clear implications for restricting the local dissemination of virus in the CNS, especially when considered with demonstrated affinity of astrocytic membranes for HSV (Vahlne *et al.*, 1978, 1980). In these circumstances, the high affinity of virus for astrocytes combined with the inability of these cells to produce infectious progeny provides an effective barrier to diffusion of virions released from chronically infected neurons. The long-term consequences of viral replication and intracytoplasmic accumulation of capsids in astrocytes also remain to be defined. Our light microscopic immunohistochemical studies demonstrated that there is a dramatic increase in glial fibrillar acidic protein (GFAP) immunoreactivity in reactive astrocytes early in the course of infection. However, at longer postinoculation intervals there was a paradoxical decrease in levels of GFAP immunoreactivity in the same areas, despite the increase in cytopathic changes of resident neurons. Ultrastructural analysis of the DMV at comparable stages of infection demonstrated that astrocytes remain prevalent throughout the region of infection, suggesting that the loss of GFAP immunoreactivity is not indicative of cell death. However, it remains formally possible that decrease in GFAP antigenicity represents an early change in the conformation of cellular proteins that is indicative of pathological changes in the cell. Unfortunately, the decrease in GFAP immunoreactivity occurs late in the course of infection and this has been a difficult problem to approach *in vivo*. As a result, *in vitro* studies of PRV replication in astrocytes may be the best approach to resolving the long-term consequences of viral replication in these cells.

Although astrocytes provide an effective barrier to local dissemination of virus, they do not prevent transneuronal passage of virus through neuronal circuits. Specific passage of virus through synaptically linked populations of neurons has been demonstrated even following inoculation with very virulent strains of virus that produce pronounced reactive gliosis early in the course of infection (Norgren and Lehman, 1989; Card *et al.*, 1990; Blessing *et al.*, 1991; Norgren *et al.*, 1992; Yu-Wen *et al.*, 1992; Rinaman *et al.*, 1993; Standish *et al.*, 1994). However, there are indications that reactive microglia may influence the extent of transneuronal passage from infected neurons. Our temporal studies of PRV-induced pathogenesis have shown that microglia become reactive shortly after the induction of reactive astrogliosis (Rinaman *et al.*, 1993). The reactive response exhibited by these cells is characterized by an increase in the size and number of cells found in relation to neurons exhibiting pathological response to infection. Furthermore, the extent of the response correlates with the virulence of the infecting strain of virus and reactive microglia are intimately associated with the most lytically infected neurons. Nevertheless, we have never observed productive infection of microglia, irrespective of the degree of cytopathic changes displayed by infected neurons (Rinaman *et al.*, 1993; Card *et al.*, 1993). In addition to their phagocytic function our ultrastructural studies indicate that reactive microglia participate in two additional processes that contribute to the restriction of viral dissemination. The high density of these cells in relation to the most chronically infected neurons and the associated increase in the number and size of the processes emanating from these cells appear to contribute to a glia barrier that surrounds the most severely infected neurons. In addition, reactive microglia are also involved in a process defined as "synaptic stripping" by Blinzinger and Kreutzberg (1968). In this response, thin processes of microglia cells surround afferents terminating on injured cells, ultimately invading the synaptic cleft and forming a physical barrier that separates the terminal from its synaptic target. While it is clear that this response does not prevent transynaptic passage of virus, it may influence the extent of transneuronal infection and should be considered in evaluating the targeting of viral vectors whose effectiveness is dependent upon transneuronal passage.

Recruitment of cells from the vasculature in response to viral infection of the CNS is also an important consideration in the use of viral vectors for gene delivery. Monocytes and other vascular cells invade the CNS in response to injury (Werkele *et al.*, 1986; Ling and Leong, 1988; Coffey *et al.*, 1990; Sloan *et al.*, 1992) and our own studies with PRV (Rinaman *et al.*, 1993; Card *et al.*, 1993) and those of Weinstein and colleagues with HSV (1990) have shown that viral infections of neurons induce a massive recruitment of these cells late in the course of infection. Stroop and colleagues (1990) have also

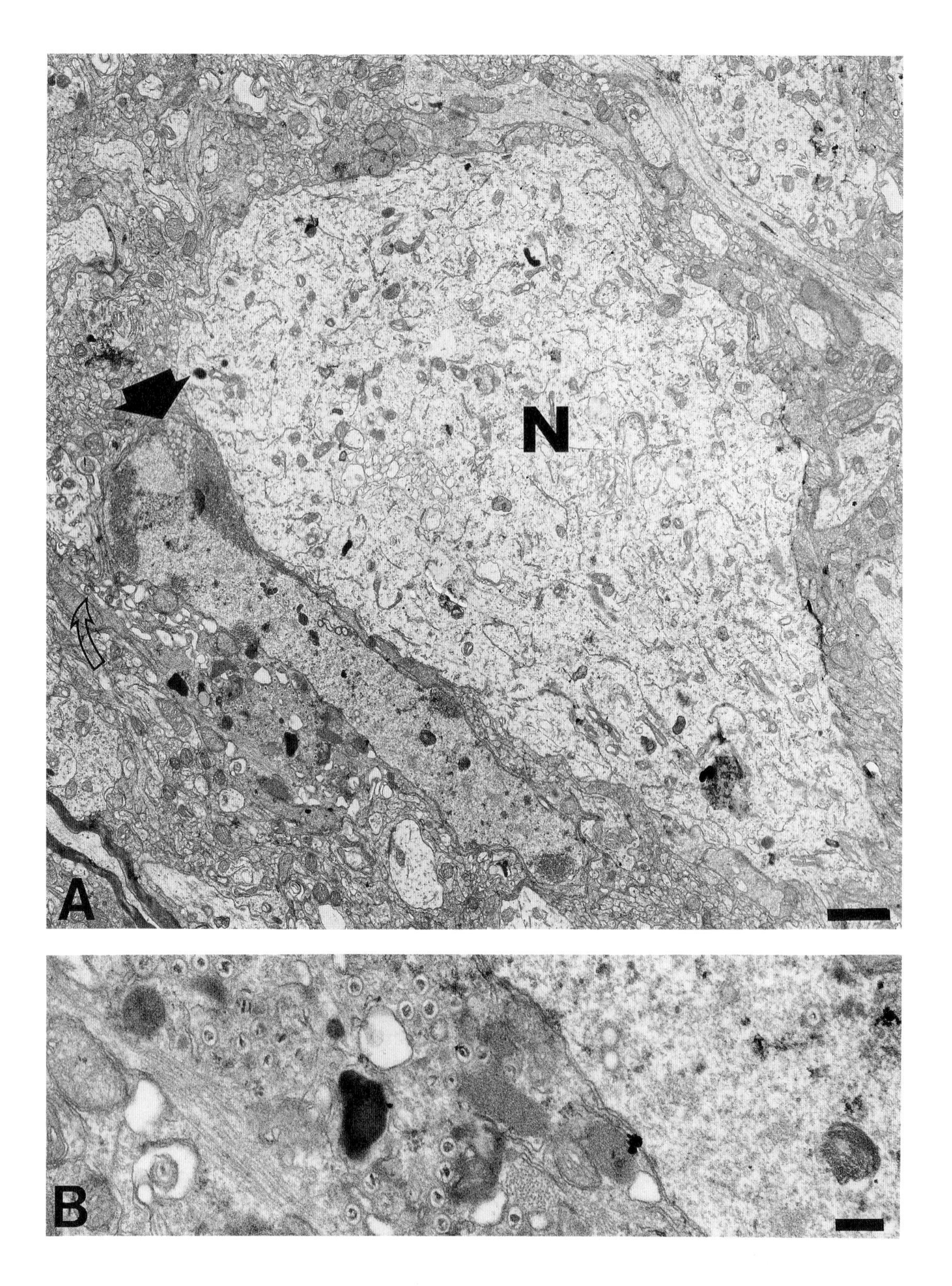
N
A
B

shown that drug-induced reactivation of HSV in the olfactory and temporal cortex produces focal breakdowns of the blood–brain barrier. The monocytes that invade the CNS in response to PRV infection differentiate into macrophages that participate in the phagocytosis of compromised neurons (Card *et al.*, 1993). It is also clear that macrophages are susceptible to infection by both PRV and HSV (Daniels *et al.*, 1978; Morahan, 1985; Sarmiento, 1988; Sarmiento and Kleinerman, 1990; Rinaman *et al.*, 1993). The fate of macrophages that enter the CNS in response to the virus-induced extravasation of monocytes from the vasculature has not been established with certainty, but it remains possible that macrophages infected in the CNS could reenter the circulation and spread the viral infection to peripheral organs. This is obviously an important consideration in evaluating the consequences of the expression of foreign genes delivered with viral vectors.

IV. Summary

Replication, assembly, and transport of alphaherpesviruses in the CNS is a complex multifaceted process. The same attributes of the viral life cycle (e.g., neurotropism, neuroinvasiveness, transneuronal passage) that have been exploited in using these viruses for definition of multisynaptic circuits have also made them attractive as potential vehicles for delivery of foreign genes to the CNS. However, selective tropism of different strains of virus for subpopulations of neurons, the fact that glia and other cells are permissive to infection by these viruses, and the cytopathic consequences of infection indicate that effective use of these viruses as vectors for gene delivery is dependent upon a thorough understanding of the mechansims that contribute to these processes.

Acknowledgments

I gratefully acknowledge the valuable contributions of Dr. Lynn Enquist, my principal collaborator on studies of PrV assembly and neuropathogenesis. I am also pleased to acknowledge the many contributions of Dr. Linda Rinaman, Mary Whealy, Raymond Meade, Alan Robbins, and Joan Dubin.

Figure 7 Abortive PRV replication in astrocytes of the rodent CNS is illustrated. The reactive astrocyte intimately associated with the chronically infected DMV neuron (N) is replicating the virus. Nucleocapsids are visible in the astrocyte nucleus (block arrow) and capsids surrounded by a single envelope are present in the swollen endoplasmic reticulum (open curved arrow). However, the higher magnification view of the cell shown in B illustrates the accumulation of naked capsids in the cytoplasm that results from the inability of these glial cells to provide a viral envelope. Marker bars for A = 1 μm and B = 200 nm.

References

Babic, N., Mettenleiter, T. C., Flamand, A., and Ugolini, G. (1993). Role of essential glycoproteins gII and gp50 in transneuronal transfer of pseudorabies virus from the hypoglossal nerves of mice. *J. Virol.* **67**, 4421–4426.

Bak, I. J., Markham, C. H., Cook, M. L., and Stevens, J. G. (1977). Intra-axonal transport of herpes simplex virus in the rat central nervous system. *Brain Res.* **136**, 415–429.

Barnett, E. M., Cassell, M. D., and Perlman, S. (1993). Two neurotropic viruses, herpes simplex virus type 1 and mouse hepatitis virus, spread along different neural pathways from the mail olfactory bulb. *Neuroscience* **57**, 1007–1025.

Bartha, A. (1961). Experimental reduction of virulence of Aujeszky's disease virus. *Magy. Allatorv. Lapja* **16**, 42–45.

Becker, C. H. (1967). Zur primaren Schadingung vegetativer Ganglien nach Infektion mit deme Herpes suis Virus bei verschiedenen Tierarten. *Experentia* **23**, 209–217.

Ben-Porat, T., and Kaplan, A. S. (1985). Molecular biology of pseudorabies virus. In "The Herpesviruses" (B. Roizman, Ed.), Vol. 3, pp. 105–173. Plenum Press, New York.

Blessing, W. W., Li, Y.-W., and Wesselingh, S. L. (1991). Transneuronal transport of herpes simplex virus from the cervical vagus to brain neurons with axonal inputs to central vagal sensory nuclei in the rat. *Neuroscience* **42**, 261–274.

Blinzinger, K., and Kreutzberg, G. W. (1968). Displacement of synaptic terminals from regenerating motoneurons by microglial cells. *Z. Zellforsch, Mikrosk. Anat.* **85**, 145–157.

Breakefield, X. O., and Geller, A. I. (1987). Gene transfer into the nervous system. *Mol. Neurobiol.* **1**, 339–371.

Campadelli-Fiume, G., Arsenakis, M., Farabegoli, F., and Roizman, B. (1988a). Entry of herpes simplex virus 1 in BJ cells that constitutively express glycoprotein D is by nedocytosis and results in degradation of the virus. *J. Virol.* **62**, 159–167.

Campadelli-Fiume, G., Avitabile, E., Fini, S., Arsenakis, M., and Roizman, B. (1988b). Herpes simplex virus glycoprotein D is sufficient to induce spontaneous pH-independent fusion in a cell line that constitutively expresses the glycoprotein. *Virology* **166**, 589–602.

Campadelli-Fiume, G., Qi, S., Avitabile, E., Foa-Tomasi, L., Brandimarti, R., and Roizman, B. (1990). Glycoprotein D of herpes simplex virus encodes a domain which precludes penetration of cells expressing the glycoprotein by superinfecting herpes simplex virus. *J. Virol.* **64**, 6070–6079.

Card, J. P., and Enquist, L. W. (1995). The neurovirulence of pseudorabies virus. *Crit. Rev. Neurobiol.* (in press).

Card, J. P., Rinaman, L., Schwaber, J. S., Miselis, R. R., Whealy, M. E., Robbins, A. K., and Enquist, L. W. (1990). Neurotropic properties of pseudorabies virus: Uptake and transneuronal passage in the rat central nervous system. *J. Neurosci.* **10**, 1974–1994.

Card, J. P., Whealy, M. E., Robbins, A. K., Moore, R. Y., and Enquist, L. W. (1991). Two α-herpesvirus strains are transported differentially in the rodent visual system. *Neuron* **6**, 957–969.

Card, J. P., Whealy, M. E., Robbins, A. K., and Enquist, L. W. (1992). Pseudorabies virus envelope glycoprotein gI influences both neurotropism and virulence during infection of the rat visual system. *J. Virol.* **66**, 3032–3041.

Card, J. P., Rinaman, L., Lynn, R. B., Lee, B.-H., Meade, R. P., Miselis, R. R., and Enquist, L. W. (1993). Pseudorabies virus infection of the rat central nervous system: Ultrastructural characterization of viral replication, transport, and pathogenesis. *J. Neurosci.* **13**, 2515–2539.

Carter, S. R., Pereira, L., Pax, P., and LaVail, J. H. (1992). A quantitative assay of retrograde transported HSV in the trigeminal ganglion. *Invest. Ophthal. Vis. Sci.* **33**, 1934–1939.

Chang, J. Y., Johnson, E. M., and Olivo, P. D. (1991). A gene delivery/recall system for neurons which utilizes reductase-negative herpes simplex viruses. *Virology* **185**, 437–440.

Coffey, P. J., Perry, V. H., and Rawlins, J. N. P. (1990). An investigation into the early stages of inflammatory response following ibotenic acid-induced neuronal degeneration. *Neuroscience* **35**, 121–132.

Daniels, C. A., Kleinerman, E. S., and Snyderman, R. (1978). Abortive and productive infections of human mononuclear phagocytes by types 1 herpes simplex virus. *Am. J. Pathol.* **91**, 119–129.

Dolivo, M. (1980). A neurobiological approach to neurotropic viruses. *TINS* **3**, 149–152.

Dolivo, M., Beretta, E., Bonifas, V., and Foroglou, C. (1978). Ultrastructure and function in sympathetic ganglia isolated from rats infected with pseudorabies virus. *Brain Res.* **140**, 111–123.

Dotti, C. G., and Simons, K. (1990). Polarized sorting of viral glycoproteins to the axon and dendrites of hippocampal neurons in culture. *Cell* **62**, 63–72.

Dotti, C. G., Kartenbeck, J., and Simons, K. (1993). Polarized distribution of the viral glycoproteins of vesicular stomatitis, fowl plaque and Semliki Forest viruses in hippocampal neurons in culture: A light and electron microscopy study. *Brain Res.* **610**, 141–147.

Dunn, W. A., Jr. (1990). Studies on the mechanisms of autophagy: Formation of the autophagic vacuole. *J. Cell Biol.* **110**, 1923–1933.

Enquist, L. W. (1994). Infection of the mammalian nervous system by pseudorabies virus (PRV). *Semin. Virol.* **5**, 221–231.

Enquist, L. W., Levitt, P., and Card, J. P. (1993). Connections of the adult rat prefrontal cortex revealed by intracerebral injection of a swine alpha herpesvirus. *Soc. Neurosci. Abstr.* 589.11.

Enquist, L. W., Dubin, J., Whealy, M. E., and Card, J. P. (1994). Complementation analysis of pseudorabies virus gE and gI mutants in retinal ganglion cell neurotropism. *J. Virol.* **68**, 5275–5279.

Federoff, H. J., Geschwind, M. D., Geller, A. I., and Kessler, J. A. (1992). Expression of nerve growth factor in vivo from a defective herpes simplex virus 1 vector prevents effects of axotomy on sympathetic ganglia. *Proc. Natl. Acad. Sci.* **89**, 1636–1640.

Field, H. J., and Hill, T. J. (1974). The pathogenesis of pseudorabies in mice following peripheral inoculation. *J. Gen. Virol.* **23**, 145–157.

Geller, A. I., Keyomarsi, K., Bryan, J., and Pardee, A. B. (1990). An efficient deletion mutant packaging system for defective herpes simplex virus vectors: Potential applications to human gene therapy and neuronal physiology *Proc. Natl. Acad. Sci. USA* **87**, 8950–8954.

Geuze, H. J., and Morre, D. J. (1991). Trans-Golgi reticulum. *J. Elec. Micros. Tech.* **17**, 24–34.

Glorioso, J. C., Goins, W. F., and Fink, D. J. (1992). Herpes simplex virus-based vectors. *Virology* **3**, 265–276.

Gustafson, D. P. (1975). Pseudorabies. In "Diseases of Swine" (H. W. Dunne and A. D. Leman, Eds.), 4th ed., pp. 391–410. Iowa State Univ. Press, Ames, IA.

Heffner, S., Kovacs, F., Klupp, B. G., and Mettenleiter, T. C. (1993). Glycoprotein gp50-negative pseudorabies virus: A novel approach toward a nonspreading live herpesvirus vaccine. *J. Virol.* **67**, 1529–1537.

Inglesias, G., Pijoan, C., and Molitor, T. (1989). Interactions of pseudorabies virus with swine alveolar macrophages. I. Virus replication. *Arch. Virol.* **104**, 107–115.

Johnson, D. C., and Spear, P. G. (1982). Monensin inhibits the processing of herpes simplex virus glycoproteins, their transport to the cell surface, and the egress of virions from infected cells. *J. Virol.* **43**, 1102–1112.

Johnson, D. C., and Spear, P. G. (1983). O-linked oligosaccharides are acquired by herpes simplex virus glycoproteins in the Golgi apparatus. *Cell* **32**, 987–997.

Johnson, P. A., Yoshida, K., Gage, F. H., and Friedmann, T. (1992). Effects of gene transfer into cultured CNS neurons with a replication-defective herpes simplex virus type 1 vector. *Mol. Brain Res.* **12**, 95–102.

Jones, F., and Grose, C. (1988). Role of cytoplasmic vacuoles in varicella-zoster virus glycoprotein trafficking and virion envelopment. *J. Virol.* **62**, 2701–2711.

Kaplitt, M. G., Pfaus, J. G., Kleopoulos, S. P., Hanlon, B. A., Rabkin, S. D., and Pfaff, D. W. (1991). Expression of a functional foreign gene in adult mammalian brain following in vivo transfer via a herpes simplex virus type 1 defective viral vector. *Mol. Cell. Neurosci.* **2**, 320–330.

Kastrukoff, L. F., Lau, A. S., and Puterman, M. L. (1986). Genetics of natural resistance to herpes simplex virus type 1 latent infection of the peripheral nervous system in mice. *J. Gen. Virol.* **67**, 613–631.

Kastrukoff, L. F., Lau, A. S., and Kim, S. U. (1987). Multifocal CNS demyelination following peripheral inoculation with herpes simplex virus type 1. *Ann. Neurol.* **22**, 52–59.

Kimman, T. G., De Wind, N., Oei-Lie, N., Pol, J. M. A., Berns, A. J. M., and Gielkens, A. J. J. (1992). Contribution of single genes within the unique short region of Aujeszky's disease virus (suid herpesvirus type 1) to virulence, pathogenesis and immunogenicity, *J. Gen. Virol.* **73**, 243–251.

Komuro, M., Tajima, M., and Kato, K. (1989). Transformation of Golgi membrane into the envelope of herpes simplex virus in rat anterior pituitary cells. *Eur. J. Cell Biol.* **50**, 398–406.

Kovacs, F., and Mettenleiter, T. C. (1991). Firefly luciferase as a marker for herpesvirus (pseudorabies virus) replication in vitro and in vivo. *J. Gen. Virol.* **72**, 2999–3008.

Koyama, A. H., and Uchida, T. (1994). Inhibition by Brefeldin A of the envelopment of nucleocapsids in herpes simplex virus type 1 infected cells. *Arch. Virol.* **135**, 305–317.

Kristensson, K. (1970). Morphological studies of the neural spread of herpes simplex virus to the central nervous system. *Acta Neuropathol.* **16**, 54–63.

Kristensson, K., Lycke, E., and Sjostrand, J. (1971). Spread of herpes simplex virus in peripheral nerves. *Acta Neuropathol.* **17**, 44–53.

Kristensson, K., Ghetti, G., and Wisniewski, H. M. (1974). Study on the propagation of herpes simplex virus (type 2) into the brain after intraocular injection. *Brain Res.* **69**, 189–201.

Kristensson, K., Vahlne, A., Persson, L. A., and Lycke, E. (1978). Neural spread of herpes simplex virus types 1 and 2 in mice after corneal or subcutaneous (footpad) inoculation. *J. Neurol. Sci.* **35**, 331–340.

Kristensson, K., Nennesmo, I., Persson, L., and Lycke, E. (1982). Neuron to neuron transmission of herpes simplex virus—Transport of virus from skin to brainstem nuclei. *J. Neurol. Sci.* **54**, 149–156.

Kuypers, H. G. J. M., and Ugolini, G. (1990). Viruses as transneuronal tracers. *TINS* **13**, 71–75.

LaVail, J. H., Johnson, W. E., and Spenser, L. C. (1993). Immunohistochemical identification of trigeminal ganglion neurons that innervate the mouse cornea: Relevance to intracellular spread of herpes simplex virus. *J. Comp. Neurol.* **327**, 133–140.

Ling, E. A., and Leong, S. K. (1988). Infiltration of carbon-labelled monocytes into the dorsal motor nucleus following an intraneural injection of ricinus communis agglutinin-60 into the vagus nerve in rats. *J. Anat.* **159**, 207–218.

Lippencott-Schwartz, J., Yuan, L. C., Bonifacino, J. S., and Klausner, R. D. (1989). Rapid redistribution of Golgi proteins into the ER in cells treated with brefeldin A: Evidence for membrane cycling from Golgi to ER. *Cell* **56**, 801–813.

Lippencott-Schwartz, J., Donaldson, J. G., Schweizer, A., Berger, E. G., Hauri, H., Yuan, L. C., and Klausner, R. D. (1990). Microtubule dependent retrograde transport of proteins into the ER in the presence of brefeldin A suggests an ER recycling pathway. *Cell* **60**, 821–836.

Loewy, A. D., Bridgman, P. C., and Mettenleiter, T. C. (1991). β-galactosidase expressing recombinant pseudorabies virus for light and electron microscopic study of transneuronally labeled CNS neurons. *Brain Res.* **555**, 346–352.

Lycke, E., Kristensson, K., Svennerholm, B., Vahlne, A., and Ziegler, R. (1984). Uptake and tansport of herpes simplex virus in neurites of dorsal root ganglia cells in culture. *J. Gen. Virol.* **65**, 55–64.

Lycke, E., Hamark, B., Johansson, M., Krotochwil, A., Lycke, J., and Svennerholm, B. (1988). Herpes simplex virus infection of the human sensory neuron. An electron microscopic study. *Arch. Virol.* **101**, 87–104.

Marchand, C. F., and Schwab, M. E. (1987). Binding, uptake and retrograde axonal transport of herpes virus suis in sympathetic neurons. *Brain Res.* **383**, 262–270.

Margolis, T., Togni, B., LaVail, J., and Dawson, C. R. (1987). Identifying HSV infected neurons after ocular inoculation. *Curr. Eye Res.* **6**, 119–126.

Margolis, T. P., LaVail, J. H., Setzser, P. Y., and Dawson, C. R. (1989). Selective spread of HSV in the central nervous system after ocular inoculation. *J. Virol.* **63**, 4756–4761.

Margolis, T. P., Sedarati, F., Dobson, A. T., Feldman, L. T., and Stevens, J. G. (1992). Pathways of viral gene expression during acute neuronal infection with HSV-1. *Virology* **189**, 150–160.

Martin, X., and Dolivo, M. (1983). Neuronal and transneuronal tracing in the trigeminal system of the rat using herpes virus suis. *Brain Res.* **273**, 253–276.

Martuza, R. L., Malick, A., Markert, J. M., Ruffner, K. L., and Coen, D. M. (1991). Experimental therapy of human glioma by means of a genetically engineered virus mutant. *Science* **252**, 854–856.

Matthews, R. E. F. (1982). Classification and nomenclature of viruses. *Intervirology* **17**, 1–200.

McLean, J. H., Shipley, M. T., and Bernstein, D. I. (1989). Golgi-like, transneuronal retrograde labeling with CNS injections of herpes simplex virus type 1. *Brain Res. Bull.* **22**, 867–881.

Mellman, I., and Simons, K. (1992). The Golgi complex: In vitro veritas? *Cell* **68**, 839–840.

Mettenleiter, T. C. (1991). Molecular biology of pseudorabies (Aujeszky's disease) virus. *Comp. Immunol. Microbiol. Infect. Dis.* **14**, 151–163.

Mettenleiter, T. C., Schreurs, C., Zuckermann, F., and Ben-Porat, T. (1987). Role of pseudorabies virus glycoprotein gI in virus release from infected cells. *J. Virol.* **61**, 2764–2769.

Morahan, P. S., Conner, J. R., and Leary, K. R. (1985). Viruses and the versatile macrophage. *Br. Med. Bull.* **41**, 15–21.

Nii, S. (1971). Electron microscopic observations on FL cells infected with herpes simplex virus. II. Envelopment. *Biken J.* **14**, 325–348.

Norgren, R. B., Jr., and Lehman, M. N. (1989). Retrograde transneuronal transport of herpes simplex virus in the retina after injection in the superior colliculus, hypothalamus and optic chiasm. *Brain Res.* **479**, 374–378.

Norgren, R. B., Jr., McLean, J. H., Bubel, H. C., Wander, A., Bernstein, D. I., and Lehman, M. N. (1992). Anterograde transport of HSV-1 and HSV-2 in the visual system. *Brain Res. Bull.* **28**, 393–399.

Peeters, B., Pol, J., Gielkens, A., and Moorman, R. (1993). Envelope glycoprotein gp50 of pseudorabies virus is essential for virus entry but is not required for viral spread in mice. *J. Virol.* **67**, 170–177.

Peeters, B., de Wind, N., Hooisma, M., Wagenaar, F., Gielkens, A., and Moormann, R. (1992a). Pseudorabies virus envelope glycoproteins gp50 and gII are essential for virus penetration, but only gII is involved in membrane fusion. *J. Virol.* **66**, 894–905.

Peeters, B., de Wind, N., Broer, R., Gielkens, A., and Moorman, R. (1992b). Glycoprotein H of pseudorabies virus is essential for entry and cell-to-cell spread of the virus. *J. Virol.* **66**, 3888–3892.

Pol, J. M. A., Gielkens, A. L. J., and van Oirschot, J. T. (1989). Comparative pathogenesis of three strains of pseudorabies virus in pigs. *Microbiol. Pathol.* **7**, 361–371.

Pol, J. M. A., Quint, W. G. V., Kok, J. L., and Broekhuysen-Davies, J. M. (1991a). Pseudorabies virus infections in explants of porcine nasal mucosa. *Res. Vet. Sci.* **50**, 45–53.

Pol, J. M. A., Wagenaar, F., and Gielkens, A. (1991b). Morphogenesis of three pseudorabies virus strains in porcine nasal mucosa. *Intervirology* **32**, 327–337.

Poliquin, L., Levine, G., and Shore, G. C. (1985). Involvement of Golgi apparatus and a restructured nuclear envelope during gliogenesis and transport of herpes simplex virus glycoproteins. *J. Histochem. Cytochem.* **33**, 875–883.

Rauh, I., and Mettenleiter, T. C. (1991). Pseudorabies virus glycoproteins gII and gp50 are essential for virus penetration. *J. Virol.* **65**, 5348–5356.

Rauh, I., Weiland, F., Fehler, F., Keil, G. M., and Mettenleiter, T. C. (1991). Pseudorabies virus glycoproteins gII and gp50 are essential for virus penetration. *J. Virol.* **65**, 5348–5356.

Rinaman, L., Card, J. P., and Enquist, L. W. (1993). Spatiotemporal responses of astrocytes, ramified microglia, and brain macrophages to central neuronal infection with pseudorabies virus. *J. Neurosci.* **13**, 685–702.

Rinaman, L., Card, J. P., Schwaber, J. S., and Miselis, R. R. (1989). Ultrastructural demonstration of a gastric monosynaptic vagal circuit in the nucleus of the solitary tract in rat. *J. Neurosci.* **9**, 1985–1996.

Rixon, F. J. (1993). Structure and assembly of herpesviruses. *Semin. Virol.* **4**, 135–144.

Robbins, A. K., Ryan, J. P., Whealy, M. E., and Enquist, L. W. (1989). The gene encoding the gIII envelope protein of pseudorabies virus vaccine strain Bartha contains a mutation affection protein localization. *J. Virol.* **63**, 250–258.

Rodriguez, M., and Dubois-Dalcq, M. (1978). Intramembrane changes occurring during maturation of herpes simplex virus type 1: Freeze-fracture study. *J. Virol.* **26**, 435–447.

Sarmiento, M. (1988). Intrinsic resistance to viral infection. Mouse macrophage restriction of herpes simplex virus replication. *J. Immunol.* **141**, 2740–2748.

Sarmiento, M., and Kleinerman, E. S. (1990). Innate resistance to herpes simplex virus infection. Human lymphocyte and monocyte inhibition of viral replication. *J. Immunol.* **144**, 1942–1953.

Shapiro, R. E., and Miselis, R. R. (1985). The central organization of the vagus nerve innervating the stomach of the rat. *J. Comp. Neurol.* **238**, 473–488.

Sloan, D. J., Wood, M. J., and Charlton, H. M. (1992). Leucocyte recruitment and infiltration in the CNS. *TINS* **15**, 276–278.

Spear, P. G. (1984). Glycoproteins specified by herpes simplex viruses. In "The Herpesviruses" (B. Roizman, Ed.), Vol. 3, pp. 315–356. Plenum, New York.

Spear, P. G. (1993). Entry of alphaherpesviruses into cells. *Semin. Virol.* **4**, 167–180.

Stackpole, C. W. (1969). Herpes-type virus of the frog renal adenocarcinoma. I. Virus development in tumor transplants maintained at low temperature. *J. Virol.* **4**, 75–93.

Standish, A., Enquist, L. W., and Schwaber, J. S. (1994). Innervation of the heart and its medullary origin defined by viral tracing. *Science* **263**, 232–234.

Strack, A. M., and Loewy, A. D. (1990). CNS cell groups regulating the sympathetic outflow of the adrenal gland as revealed by transneuronal cell body labeling with pseudorabies virus. *Brain Res.* **491**, 274–296.

Stroop, W. G., McKendall, R. R., Battles, E.-J. M. M., Schaefer, D. C., and Jones, B. G. (1990). Spread of herpes simplex virus type 1 in the central nervous system during experimentally reactivated encephalitis. *Microbial. Pathogen.* **8**, 119–134.

Takamiya, Y., Short, M. P., Ezzeddine, Z. D., Moolten, F. L., Breakefield, X. O., and Martuza, R. L. (1992). Gene therapy of malignant brain tumors: A rat glioma line bearing the herpes simplex virus type 1-thymidine kinase gene and wild type retrovirus kills other tumor cells. *J. Neurosci. Res.* **33**, 493–503.

Thomas, E. E., Lau, A. S., Kim, S. U., Osborne, D., and Kastrukoff, L. F. (1991). Variation in resistance to herpes simplex virus type 1 of oligodendrocytes derived from inbred strains of mice. *J. Gen. Virol.* **72**, 2051–2057.

Ugolini, G. (1992). Transneuronal transfer of herpes simplex virus type 1 (HSV-1) from mixed limb nerves to the CNS. I. Sequence of transfer from sensory, motor, and sympathetic nerve fibres to the spinal cord. *J. Comp. Neurol.* **326**, 527–548.

Ugolini, G., Kuypers, H. G. J. M., and Simons, A. (1987). Retrograde transneuronal transfer of herpes simplex virus type 1 from motoneurons. *Brain Res.* **422**, 242–256.

Ugolini, G., Kuypers, H. G. J. M., and Strick, P. L. (1989). Transneuronal transfer of herpes virus from peripheral nerves to cortex and brainstem. *Science* **243**, 89–91.

Vahlne, A., Nystrom, B., Sandberg, M., Hamberger, A., and Lycke, E. (1978). Attachment of herpes simplex virus to neurons and glial cells. *J. Gen. Virol.* **40**, 359–371.

Vahlne, A., Sennerholm, B., Sandberg, M., Hamberger, A., and Lycke, E. (1980). Differences in attachment between herpes simplex type 1 and type 2 viruses to neurons and glial cells. *Infect. Immunol.* **28**, 675–680.

Van Genderen, I. L., Brandimarti R., Torrisi, M. R., Campadelli, G., and Van Meer, G. (1994). The phospholipid composition of the extracellular herpes simplex virions differs from that of host cell nuclei. *Virology* **200**, 831–836.

Weinstein, D. L., Walker, D. G., Akiyama, H., and McGeer, P. L. (1990). Herpes simplex virus type 1 infection of the CNS induces major histocompatibility complex antigen expression on rat microglia. *J. Neurosci. Res.* **26**, 55–65.

Wekerle, H., Linington, C., Lassmann, H., and Myermann, R. (1986). Cellular immune reactivity within the CNS. *TINS,* 271–277.

Whealy, M. E., Card, J. P., Meade, R. P., Robbins, A. K., and Enquist, L. W. (1991). Effect of Brefeldin A on alpha herpesvirus membrane protein glycosylation and virus egress. *J. Virol.* **65**, 1066–1081.

Whealy, M. E., Card, J. P., Robbins, A. K., Dubin, J. R., Rziha, H.-J., and Enquist, L. W. (1993). Specific pseudorabies virus infection of the rat visual system requires both gI and gp63 glycoproteins. *J. Virol.* **67**, 3786–3797.

Wittmann, G., and Rziha, H. J. (1989). Aujeszky's disease (pseudorabies) in pigs. In "Herpesvirus Diseases of Cattle, Horses and Pigs" (G. Wittmann, Ed.), pp. 230–325. Kluwer, Boston.

Yu-Wen, L., Wesselingh, S. L., and Blessing, W. W. (1992). Projections from rabbit caudal medulla to C1 and A5 sympathetic premotor neurons, demonstrated with phaseolus leucoagglutinin and herpes simplex virus. *J. Comp. Neurol.* **317**, 379–395.

Zemanick, M. C., Strick, P. L., and Dix, R. D. (1991). Direction of transneuronal transport of herpes simplex virus 1 in the primate motor system is strain-dependent. *Proc. Natl. Acad. Sci. USA* **88**, 8048–8051.

Zsak, L., Mettenleiter, T. C., Sugg, N., and Ben-Porat, T. (1989). Release of pseudorabies virus from infected cells is controlled by several viral functions and is modulated by cellular components. *J. Virol.* **63**, 5475–5477.

CHAPTER 19

Pseudorabies Virus: A Transneuronal Tracer for Neuroanatomical Studies

Arthur D. Loewy
Department of Anatomy and Neurobiology
Washington University School of Medicine
St. Louis, Missouri

I. Introduction

The ultimate goal of neurobiology is to understand how the human brain works and one of the research strategies being used to address this enormously complex problem is the study of the brains of laboratory animals in order to formulate some general principles regarding brain organization. To reach this level of understanding, technological advances will be needed, particularly in the area of neuroanatomy.

The objective of neuroanatomical research is to obtain information on the chemical nature of central neuronal networks and, subsequently, relate these data to the physiology of each neural system. Progress in this field has been swift with the development of a number of highly sensitive tracing methods that have exploited basic cellular processes. For example, Gerfen and Sawchenko (1984) discovered that the plant lectin *Phaseolus vulgaris* leucoagglutinin could be used as an exquisitely sensitive anterograde axonal tracer. Minute amounts of this lectin are injected into discrete central nervous system (CNS) areas of laboratory animals and after several days this tracer is taken up by the local neurons and transported centrifugally by all of its neuronal processes. A complete labeling of individual neurons is achieved, including the axon and all of its terminals which permits researchers to trace neural connections throughout the CNS. In addition to this method, the retrograde cell body labeling method, which originally used horseradish peroxidase as a tracer, has now been superceded by a second generation of more sensitive markers. Fluorochrome dyes (Fluoro-Gold and Fast Blue) as well as protein markers, such as wheat germ agglutin–horseradish peroxidase and cholera toxin subunit β, are now used. These are now easily combined with immunohistochemical methods so it is now a routine technique to identify the chemical nature of retrogradely labeled neurons. Even though each of these methods represents significant advancements, these techniques are limited because they only label sets of first-order neurons. Functionally connected chains of neurons cannot be visualized by these methods. To obviate this problem, two new methods have been developed: the C-*fos* labeling technique and the viral transneuronal labeling method.

The immediate-early gene c-*fos* is rapidly and transiently expressed in neurons by a variety of stimuli, producing a protein product, Fos, that appears to be a universal marker of activated neurons (see Dragunow and Faull, 1989; Morgan and Curran, 1989, 1991; Sheng and Greenberg, 1990, for reviews). While this is not the only transcription factor expressed by neurons, it is the one that most investigators have used to date in neuroanatomical studies. For example, it has been relatively straightforward to use Fos immunohistochemical method to identify the central cell groups activated by stress (Ceccatelli *et al.*, 1989; Chan *et al.*, 1993; Ericsson *et al.*, 1994) or pain (Bullitt, 1990; Herdegen *et al.*, 1991a,b; Lantéri-Minet *et al.*, 1994). Even though this method represents an important advancement, it is limited because of a major shortcoming—it only labels neurons that are excited during a particular response. Neurons that are inhibited remain unstained. Thus, the Fos immunohistochemical method provides only a limited view of a neural circuit.

The second major development has been the viral transneuronal labeling method—a technique that depends on the ability of certain viruses to sequentially infect functionally related chains of neurons. The infection spreads in a hierarchical fashion along synaptically connected groups of neurons regardless of whether they are excitatory or inhibitory (Card and Enquist, 1994; Kuypers and Ugolini, 1990; Strick and Card, 1992; Ugolini, 1995). This method depends on using relatively weak neurotropic viruses to produce highly selective, nonlytic infections that spread via a transsynpatic mechanism. Under ideal conditions, it has been possible to label first-, second-, and sometimes third-order neurons of CNS circuits and use immunohistochemical procedures to visualize both viral products and putative neurotransmitters or related enzymes within the same neurons (Haxhiu *et al.*, 1993; Li *et al.*, 1992; Spencer *et al.*, 1990; Strack *et al.*, 1989a).

II. Historical Development of the Viral Transneuronal Labeling Method

Sabin (1938) developed a bioassay that permitted him to examine the specific neural routes taken by various neurotropic viruses as they enter the brain after they were injected into the nasal cavity of mice. Two major ideas emerged from these studies: viral specificity and transneuronal labeling. First, Sabin's studies demonstrated that viruses have affinities for particular types of neurons, causing unique patterns of infection in the brain. For example, pseudorabies virus (PrV) entered the CNS via both components of the autonomic nervous system as well as the trigeminal system, but not by the olfactory nerves. In contrast, other viruses, such as vesicular stomatitis virus and equine encephalomyelitis virus, entered the CNS only via the olfactory system. Second, viruses were capable of producing transneuronal infections of functionally related chains of neurons. For example, PrV produced transneuronal infections in the second-order autonomic cell groups (both parasympathetic and sympathetic systems) that regulate the nasal blood vessels and mucous glands.

The idea that viruses could be used as a neuroanatomical tracer was first presented by Martin and Dolivo (1983). This was the first time a neurotropic virus was used in a systematic analysis of CNS circuits. In separate experiments, three different cranial targets were injected with PrV—eye, nose, and masseter muscle—producing specific patterns of transneuronal labeling in brain for each of these systems. Unfortunately, the details regarding their method were not presented. So another hiatus occurred until Ugolini and co-workers (1987) demonstrated the usefulness of HSV-1 as a transneuronal tracer. Others soon followed (Blessing *et al.*, 1991; Li *et al.*, 1992a,b; McLean *et al.*, 1989).

In 1986, we began our studies with PrV. Our decision to use PrV, as opposed to HSV-1 or rabies virus, was based on the fact that this virus does not infect humans. While attenuated strains of these other neurotropic viruses exist and may be equally useful for CNS mapping studies, we decided to concentrate on PrV. Three different strains were studied: wild type, BUK, and Bartha (Strack *et al.*, 1989b). The wild-type strain proved to be useless because it produced nonspecific infections and killed rats within 72 hr after inoculation. Also, the BUK strain was not useful because it produced diffuse, nonspecific CNS infections. Serendipitously, we discovered that one of these viruses, Bartha–PrV, was a highly specific transneuronal tracer (Jansen *et al.*, 1993; Strack and Loewy, 1990; Strack *et al.*, 1989a,b). This virus has now been used extensively for the analysis of central autonomic pathways (Haxhiu *et al.*, 1993; Loewy and Haxhiu, 1992; Loewy *et al.*, 1994; Marson *et al.*, 1993; Rotto-Percelay *et al.*, 1992; Schramm *et al.*, 1993; Spencer *et al.*, 1990; Strack *et al.*, 1989a,b; Ter Horst *et al.*, 1993). It also has been used for ultrastructural studies (Loewy *et al.*, 1991).

III. Viral Transneuronal Tracers

Transneuronal labeling of a neural circuit depends on using a weak neurotropic virus that causes an infection in a chain of functionally related neurons without killing them so they can be subsequently visualized by histochemical methods. To date, the alphaherpesviruses, HSV-1 and PrV, are the two most commonly used viruses for transneuronal labeling studies. However, other neurotropic viruses, such as rabies (Coulon and Flamand, 1995; Lafay *et al.*, 1991), vesicular stomatitis (Lundh *et al.*, 1987, 1988), and mouse hepatitis virus (Barnett *et al.*, 1993), offer potential alternatives, but the majority of research in this field has focused on herpes viruses. Specific strains have been discovered that are transported preferentially in the retrograde (i.e., toward the cell body) or anterograde direction (i.e., away from the cell body), producing unique patterns of infection in neural systems (Fig. 1). For example, strain 129 of HSV-1 produces only anterograde transneuronal infections (Zemanick *et al.*, 1991), while the Bartha strain of PrV seems to produce mainly retrograde transneuronal infections (Rotto-Percelay *et al.*, 1992). However, some viruses, such as the McIntyre strain of HSV-1, produce infections that spread in a bidirectional fashion and thus produce data that are difficult to interpret (McLean *et al.*, 1989).

The two properties that make viruses useful transneuronal tracers are: (1) they are self-ampflying markers and (2) they are transferred from neuron to neuron via synapses. Finding particular strains of neurotropic viruses with these properties is an empirical matter. Most of the basic information regarding viral tracers has been gathered in several well-defined neural systems such as the sympathetic nervous system (Strack and Loewy, 1990; Strack *et al.*, 1989a,b) and visual system (Card and Enquist, 1995a,b; Card *et al.*, 1990, 1991, 1992). While this method can be applied in any neural network, it should be stressed that the optimal conditions (viz., viral strain, titer, injection volume, postoperative survival, and species) need to be determined empirically on a case by case basis.

From a practical standpoint, the retrograde transneuronal method works in the following manner. An attenuated form of a herpes virus (although this is not the only type of virus that will produce transneuronal labeling) is injected into a particular target such as an autonomic ganglion, visceral organ, or CNS site. The viruses infect the first-order neurons by a series of steps (Fig. 2; also see Card, 1995 for further discussion). First, they bind preferentially to synapses as demonstrated by Valhne and co-workers (1980) (see Fig. 3). After attaching to the presynaptic membrane, they penetrate the cell membrane, discarding their envelope. Once in the axoplasm, nucleocapsids are retrogradely transported to the cell body where they enter the nucleus to begin a cascade of replicative steps resulting in the production of progeny virions. Then, mature virions are distributed centrifugally throughout the cytoplasm, eventually budding from all neuronal surfaces. Since synaptic regions have the greater affinity for binding herpes viruses, the transfer to second-order neurons occurs with greatest efficiency and specificity at these sites (Fig. 4). If first-order neurons lyse and release large numbers of viruses into the neuropil, there will be a high probability that

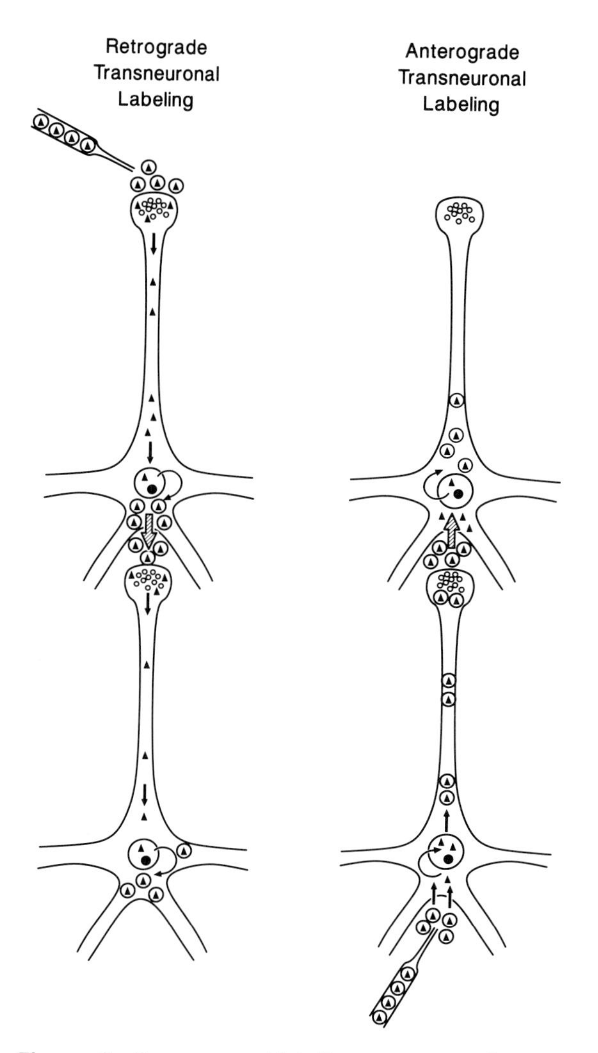

Figure 1 Transneuronal labeling can occur either in an anterograde or in a retrograde direction relative to the direction in which neural information is transferred.

nearby unrelated neurons will become infected. This type of nonspecific spread has been termed "local spread" as opposed to the specific transsynaptic transfer mechanism (Fig. 4). Assuming the transfer occurs via the transsynaptic mechanism, the viruses attach and penetrate the second-order neurons at these synaptic sites, and the infectious process is then repeated. By selecting the appropriate virus and postinoculation period, this method can be used to infect networks of functionally related CNS neurons.

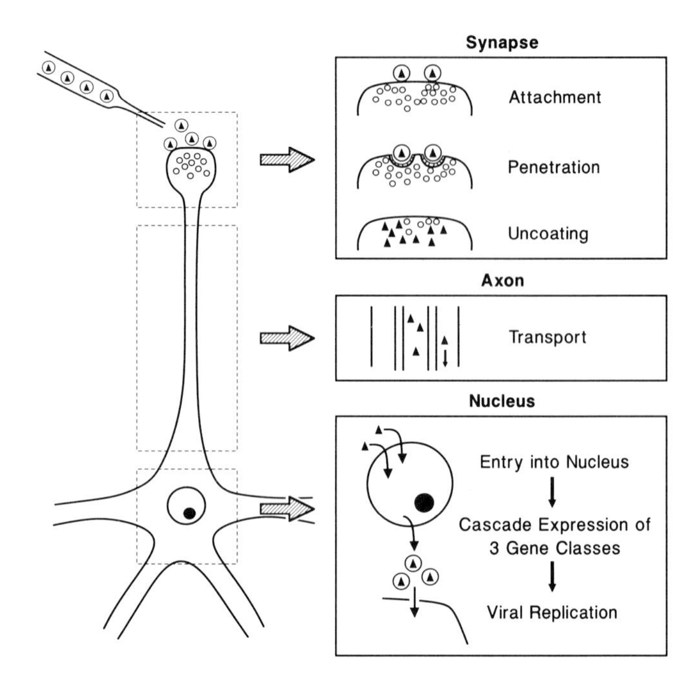

Figure 2 Schematic summary of a herpes virus infection of a neuron. Viruses bind preferentially to the synaptic region, then they penetrate the cell membrane, discarding their envelope as the nucleocapsids enter the axoplasm. Next, the nucleocapsids are transported via neurotubules to the nucleus where replication is directed. After progeny virions are produced, they exit the nucleus and are transported centrifugally in all neuronal processes. As they reach the surface of the neuron, viral budding occurs. Since there is a higher propensity that viral binding occurs at synapses, as opposed to other neuronal membrane surfaces, second-order neurons will become infected at these sites either by a direct transsynaptic mechanism or by preferentially binding to the freely dispersed viruses found within the neuropil.

A fine balance exists between an uncontrolled CNS viral infection and highly specific transneuronal labeling. The outcome depends upon a number of different parameters including the intrinsic properties of the virus itself, its binding and penetration characteristics, replication cycle, virulence, as well as the number of virions injected, density of innervation, survival period, and species. This means that while the viral transneuronal labeling method is very powerful, as previously discussed, it should be considered to be a highly empirical method; but in our view, the power of the method greatly overshadows its shortcomings.

Finally, even though the weakened herpes virus strains are useful tract tracing tools, the severity of the infection varies as a result of specific immune mechanisms of the host (Nash *et al.*, 1987; Oakes, 1975; Simmons and Nash, 1987). Moreover, individual strains of rodents react differently to specific viruses (Ugolini, 1995) and variations in infectivity occur in outbred strains,

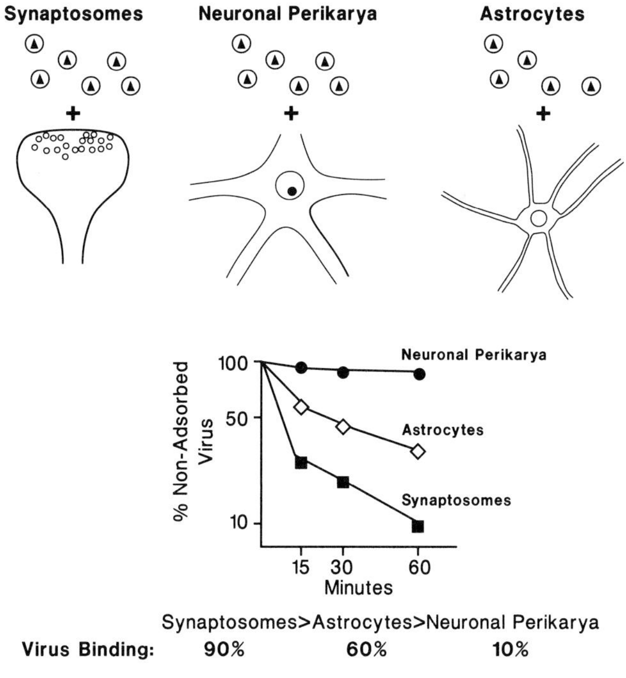

Figure 3 Herpes viruses bind preferentially to synaptosomes compared to astrocytes or neuronal perikarya. The data presented in this figure came from a report by Valhne *et al.* (1980).

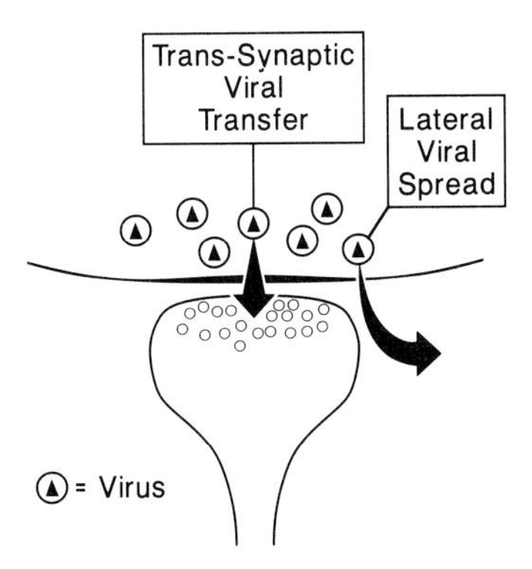

Figure 4 When a first-order neuron is infected with virus, second-order neurons subsequently become infected either by a highly specific transsynaptic transfer mechanism or the infection may become dispersed locally in the neuropil by lateral spread. This second type of transmission may result in nonspecific labeling patterns of neural pathways. (From Strack and Loewy, 1990, with permission).

such as the Sprague–Dawley rat, which are undoubtedly related to variations in the immune response of an individual rat.

IV. PrV as a Transneuronal Marker

Bartha–PrV is a herpes virus used in the pig industry as a vaccine to prevent a lethal disease called the "mad itch" or Aujesky's disease. It also can be used in rats as a specific retrograde transneuronal marker of CNS neurons (Jansen *et al.*, 1993; Strack and Loewy, 1990; Strack *et al.*, 1989a,b). This naturally occurring attenuated strain has been weakened by the deletion of two genes in its unique short segment coding for nonessential envelope glycoproteins (gI = gE of HSV-1 and gp63 = gD of HSV-1), an abnormal gene coding for a nonessential glycoprotein in its unique long segment ($gIII^B$ = gC of HSV-1), and a mutation of the gene in the unique long segment coding for UL21 protein that is a nonessential capsid protein (see Mettenleiter, 1991; Mettenleiter, 1995—this volume for review).

Bartha–PrV can be used as a specific retrograde transneuronal marker (Strack and Loewy, 1990). The evidence supporting this claim comes from a series of labeling experiments in which Bartha–PrV produced transneuronal patterns of labeling in the sympathetic nervous system that correlated remarkably well with physiological data. For example, after PrV was injected into the anterior chamber of the eye, transneuronally labeled sympathetic preganglionic neurons were found concentrated in the T1–T3 spinal segments, the exact segments established in earlier physiological studies to provide the majority of sympathetic preganglionic outflow regulating the eye (Fig. 5). Similarly, PrV injections into the skin of the ear resulted in transneuronal labeling in the T4 and T5 spinal segments, also the precise spinal levels that regulate the blood flow to this tissue. In addition to these observations, Jansen *et al.* (1993) found that Bartha–PrV infections of one functional class of sympathetic preganglionic neurons remain confined to these neurons and do not spread to nearby sympathetic preganglionic neurons, even those lying in the same cell cluster only a couple hundred micrometers away. One of the factors aiding in the transsynaptic transport of PrV seems to be the glial partitioning that surrounds infected neurons (Rinaman *et al.*, 1993).

Bartha–PrV has an additional property that makes it extremely useful for neuronanatomical studies: it is only mildly cytopathetic. This factor is important because by 4 to 6 days after a PRV injection into a peripheral target, transneuronally infected neurons are still intact and can be histologically identified with methods that permit the immunohistochemical demonstration of both neuropeptides or transmitter enzymes (or other neurochemicals) and PrV in the infected neurons (Fig. 6). Thus, the chemical codes used in a particular neural circuit can be analyzed, as has been done for the CNS autonomic

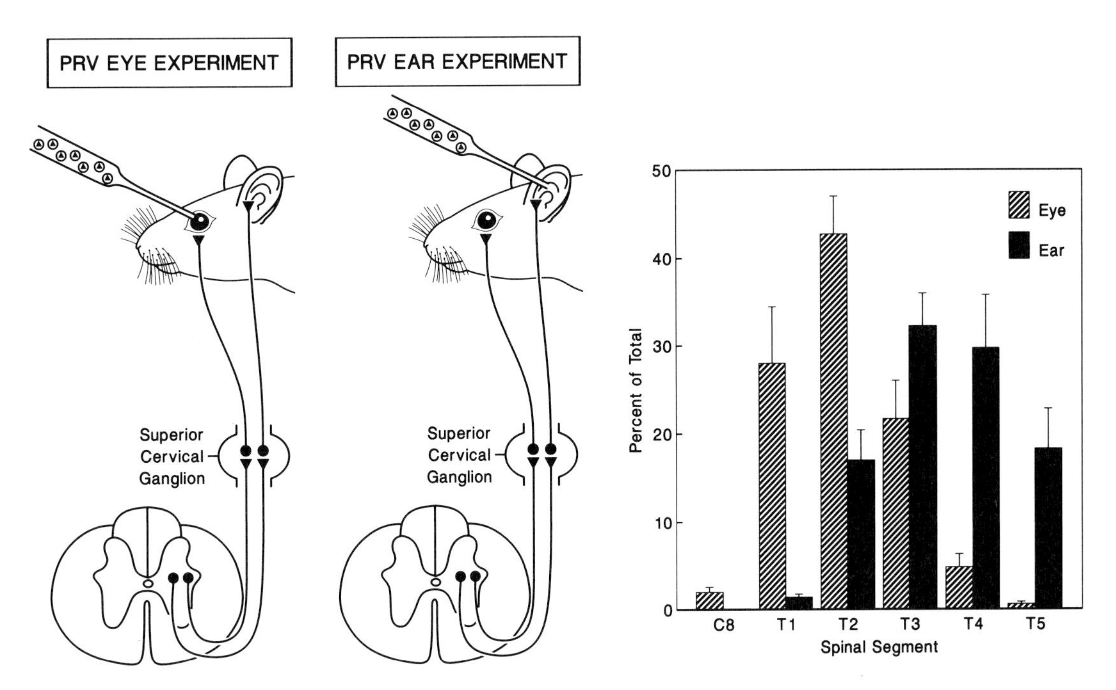

Figure 5 Bartha–PrV produces specific patterns of transneuronal labeling. For example, PrV injections into the anterior chamber of eye label mainly the sympathetic preganglionic neurons of the T1–T3 spinal segments which correlates with physiological evidence. (From Strack and Loewy, 1990, with permission).

groups that innervate the sympathoadrenal and PrV in the infected preganglionic neurons (Fig. 7; Strack *et al.*, 1989b).

V. Limitations of Viral Transneuronal Labeling Method

Although Bartha–PrV is an excellent transneuronal marker, it cannot be considered to be a universal transneuronal marker because certain types of neurons are refractory to infection (Martin and Dolivo, 1983; Sabin, 1983). Thus, when PrV (or any other virus) is used as a tracer, the absence of labeling does not necessarily indicate that a particular cell group does not project to a particular target. There may be a variety of reasons why some neurons do not become infected. For example, some neurons may lack PrV receptors and thus, the virus will not bind to them. In addition, PrV may not be able to penetrate all neurons with equal efficiency. Furthermore, there may be differences in the rate at which infections occur within different neurons that are related to

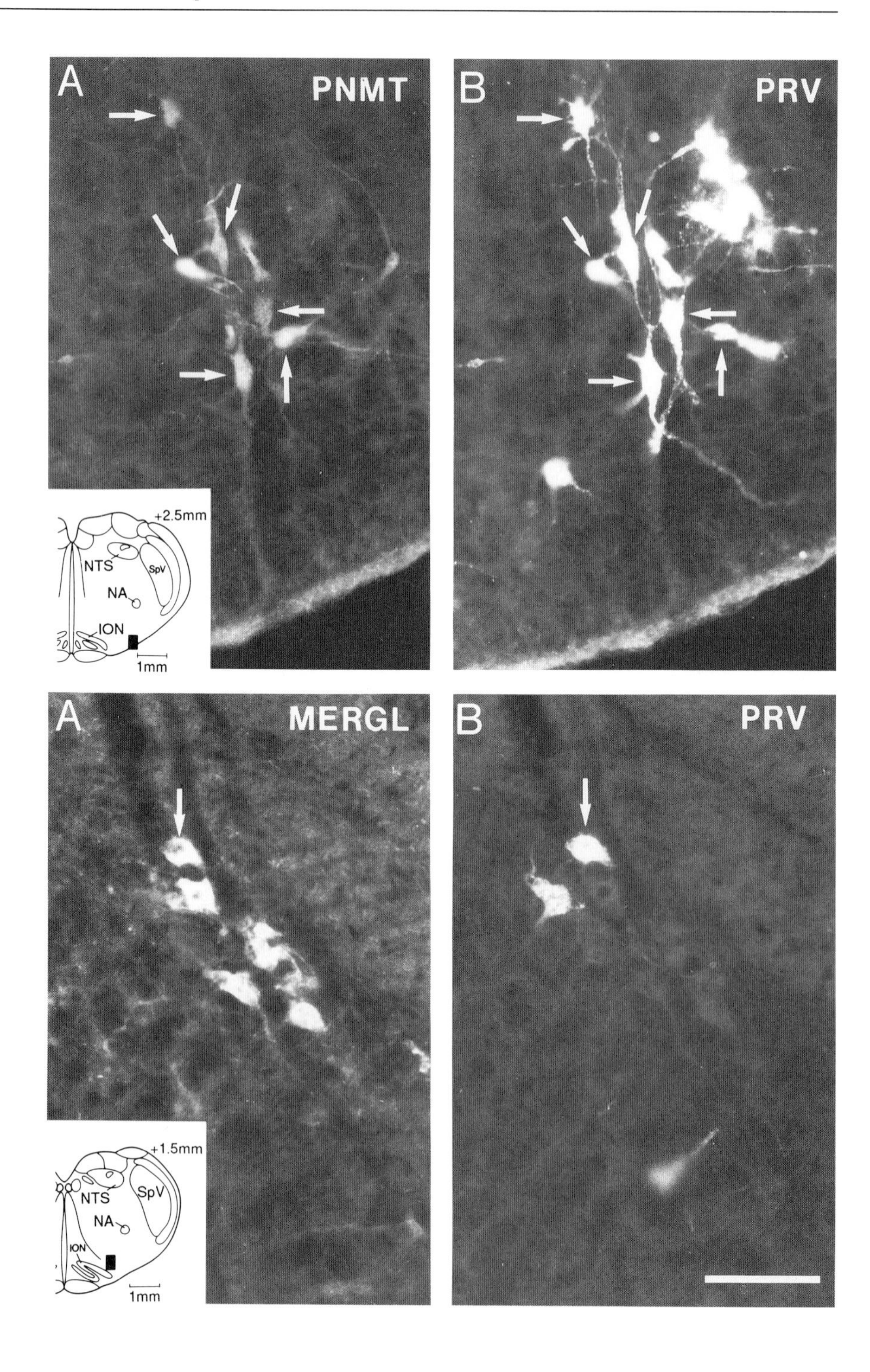
A
PNMT
B
PRV
+2.5mm
NTS
SpV
NA
ION
1mm
A
MERGL
B
PRV
+1.5mm
NTS
SpV
NA
ION
1mm

axoplasmic transport rates. Moreover, the failure to label some systems may reflect differences in innervation density. If a particular class of neurons supplies only few synaptic inputs to a particular site, then fewer viruses will bind to this type of neuron, resulting in a lower probability that an infection will occur.

These differences have been well documented by examining the patterns of labeling that occur after Bartha–PrV was injected into skeletal muscle (Rotto-Percelay *et al.*, 1992). In this experiment, only the sympathetic and somatic motor systems were labeled. The absence of labeling in the sensory system was a surprise since dorsal root ganglion cells innervate skeletal muscle. PrV could not be visualized with histochemical procedures in the sensory ganglia, but viral DNA was present as demonstrated by PCR analysis. Several interpretations can be offered, but the one we favor is that PrV virions reached the ganglion cell body but did not multiply in sufficient number to be visualized by immunohistochemical methods at 4 days postinoculation. Possibly, PrV was transported faster in sympathetic and somatic neurons than in sensory neurons. Alternatively, differential rates of replication may occur in different types of neurons and under the experimental conditions of this study, perhaps Bartha–PrV did not optimally reproduce in dorsal root ganglia to be visualized with histochemical methods. This implies that when working with weakened strains of neurotropic virus, optimal conditions need to be determined.

When a neuron is infected with a herpes virus, particular viral genes are activated that mediate cytopathic effects and one of these is the virion host shut-off gene which inhibits or prevents the expression of certain host enzymes (Kwong and Frenkel, 1989). Activation of this gene or similar genes may disrupt the synthesis of various neuropeptides or neurotransmitter enzymes but this has not interfered with our ability to visualize catecholamine-synthesizing enzymes (Loewy *et al.*, 1994; Strack *et al.*, 1989b) or some neuropeptides (Strack *et al.* 1989b) in PrV-infected neurons.

In our studies, the overall success rate of PrV experiments has been low ($\approx$ 10–20%) because we have tended to use very low doses of PrV and have been highly conservative in our selection of cases. We used this approach to avoid material with false-positive labeling. Typically, we use titers in the range of 4×10^7 pfu/ml and inject volumes on the order of 100 nl. In other words, a single injection contains only about 4000 pfu per autonomic ganglion or multiples of this for particular organs. Others have used higher concentrations,

Figure 6 Transneuronally labeled neurons can be dually immunostained for both PrV and neurotransmitter enzymes or neuropeptides. Top, C1 adrenergic neurons (medulla oblongata) were transneuronally labeled 4 days after a PrV injection in the rat celiac ganglion (PNMT, phenylethanolamine-*N*-methyltransferase). Lower bottom, Met–Arg–Leu–Enkephalin (MERGL) containing neurons from the medial part of the medulla oblongata were transneuronally labeled 4 days following a PrV injection into rat adrenal gland. Scale bar = 100 μm.

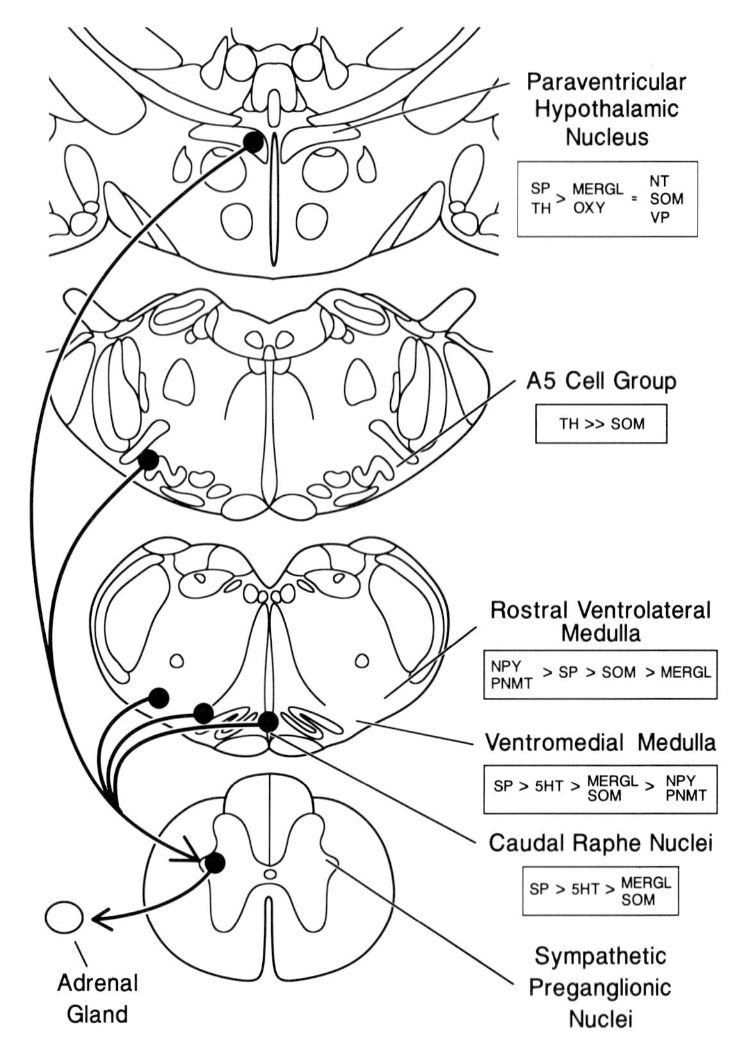

Figure 7 Different classes of chemically coded central neurons that project to sympathoadrenal preganglionic neurons (From Strack *et al.,* 1989b, with permission). Abbreviations: MERGL, Met–Arg–Leu–enkephalin; NPY, neuropeptide Y; NT, neurotensin; OXY, oxytocin; PNMT, phenylethanolamine-*N*-methyltransferase; Som, somatostatin; SP, substance P; 5-HT, serotonin; TH, tyrosine hydroxylase; VP, vasopressin

viz., 10^8 pfu/ml. Clearly, this concentration yields a higher success rate, but it also increases the probability that nonspecific infections will occur. To avoid this type of problem, we have erred on the side of having a low success rate, but high confidence that the results were specific. It may turn out that this has been a conservative approach, but as more data are accumulated on the properties of Bartha–PrV and other types of weakened neurotropic viruses, further refinements of this method will surely be made.

Another useful method for evaluating the quality of a given transneuronal labeling experiment is to compare the pattern of the PrV labeling in the first-order neurons with the cell body labeling obtained using conventional retrograde markers such as Fluoro-Gold or cholera toxin ß-subunit. The cytoarchitectonic patterns should be identical. Ideally, quantitative data, although difficult to obtain and to interpret because of the possibility that local interneurons will be also labeled, are extremely useful and have been obtained in some studies, such as the innervation of skeletal muscles where the two methods showed similar results, although the PrV labeling of α-motoneurons was somewhat less than that obtained with conventional tracers (Rotto-Percelay *et al.*, 1992).

A further problem associated with using viruses as a transneuronal marker relates to their virulence (Babic *et al.*, 1993; Card and Enquist, 1995a; Card *et al.*, 1990, 1991, 1992)—a factor that is dependent on the virus's ability to adsorb and penetrate neurons. This is greatly influenced by the glycoproteins present on the surface of the viral envelope (Mettenleiter, 1991, 1995). By using a variety of PrV mutants that were deficient in specific glycoproteins, both Babic and co-workers (1993) and Card *et al.*, (1992) have studied the properties of different PrV mutants as transneuronal markers. The gI glycoprotein, which is homologous to gE of HSV-1, seems to greatly affect this process. This particular glycoprotein is absent from Bartha–PrV and may be one of the reasons why this particular virus is weak and works well as a transneuronal marker (see Chapter 20 by Mettenleiter for further discussion).

Some viruses are not particularly useful for the analysis of neural circuits because they travel in a bidirectional fashion, acting as both an anterograde and retrograde transneuronal marker. For example, both the McIntyre strain of HSV-1 (McLean *et al.*, 1989) and the CVS and Av01 strains of rabies virus produce bidirectional transneuronal labeling (Lafay *et al.*, 1991). In contrast, Bartha–PrV acts preferentially as a retrograde transneuronal marker (Rotto-Percelay *et al.*, 1992), except when large amounts of virus are used (Card *et al.*, 1990, 1991) which will also produce anterograde transneuronal labeling. Whether this latter labeling is specific has not yet been addressed because experiments demonstrating a point-to-point retinofugal projection have not yet been presented.

In summary, Bartha–PrV can be used to produce a highly specific transneuronal marker of hierarchical chains of CNS cell groups. The infections of first-, second-, and in some cases, third-order neurons proceed in a retrograde direction that appears to be highly specific. By comparing the pattern of first-order labeling obtained with Bartha–PrV to that obtained with a standard retrograde cell marker, such as cholera toxin ß-subunit, the severity and specificity of the infection can be assessed. If the first-order neurons have been lysed,

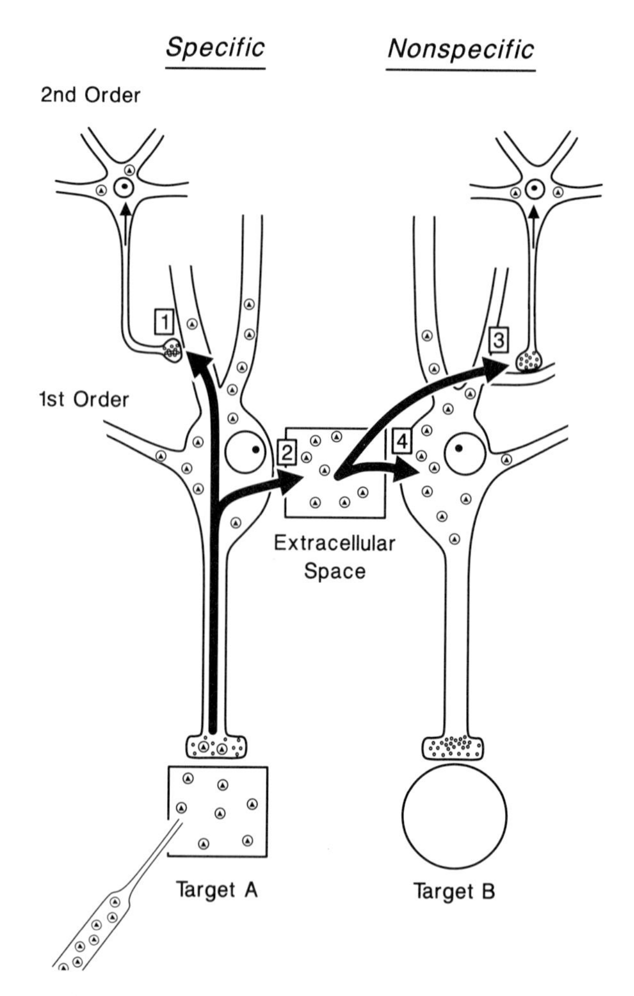

Figure 8 Potential different specific and nonspecific patterns of retrograde transneuronal labeling may occur. When PrV is injected into a target (target A), the virus will be retrogradely transported in the neurons innervating this structure and then transsynaptically transferred in a specific manner to second-order neurons synapsing on the infected neuron (1). Alternatively, if the viruses become dispersed in the neuropil (2), they may bind to synapses of nonrelated neurons (3) or infect local neurons (4). Conditions 3 and 4 would result in nonspecific labeling of central neural pathways. Experiments by Jansen *et al.* (1993) rule out condition 4 for Bartha–PrV. Condition 3 has not yet been tested. (From Jansen *et al.,* 1993, with permission).

this presents a significant problem because virus particles can potentially be dispersed throughout the neuropil, possibly causing false-positive labeling (Fig. 8). Usually this type of material should be discarded. Even with these shortcomings, neurotropic viruses can be used in combination with immunohistochemical methods to analyze the chemical nature of CNS neural circuits and this method offers a significant technological advancement for neuroanatomical investigations.

References

Babic, N., Mettenleiter. T. C., Flamand, A., and Ugolini, G. (1993). Role of essential glycoproteins gII and gp50 in transneuronal transfer of pseudorabies virus from the hypoglossal nerves of mice. *J. Virol.* **67**, 4421–4426.

Barnett, E. M., Cassell, M. D., and Perlman, S. (1993). Two neurotropic viruses, herpes simplex virus type 1 and mouse hepatitis virus, spread along different neural pathways from the main olfactory bulb. *Neuroscience* **57**, 1007–1025.

Blessing, W. W., Li, Y. W., and Wesselingh, S. L. (1991). Transneuronal transport of herpes simplex virus from the central vagus to brain neurons with axonal inputs to central vagal sensory nuclei in the rat. *Neuroscience* **42**, 261–274.

Bullitt, E. (1990). Expression of c-fos-like protein as a marker for neuronal activity following noxious stimulation in the rat. *J. Comp. Neurol.* **296**, 517–530.

Card, J. P. (1995). Pseudorabies virus replication and assembly in the rodent central nervous system. In "Viral Vectors: Tools for the Analysis and Genetic Manipulation of the Nervous System" (M. G. Kaplitt and A. D. Loewy, Eds.), pp. 319–347. Academic Press, Orlando.

Card, J. P., and Enquist, L. W. (1994). The use of pseudorabies virus for definition of synaptically linked populations of neurons. *Methods Mol. Genet.* **4**, 363–382.

Card, J. P., and Enquist, L. W. (1995). The neurovirulence of pseudorabies virus. *Crit. Rev. Neurobiol.* (in press).

Card, J. P., Rinaman, L., Schwaber, J. S., Miselis, R. R., Whealy, M. E., Robbins, A. K., and Enquist, L. W. (1990). Neurotropic properties of pseudorabies virus: Uptake and transneuronal passage in the rat central nervous system. *J. Neurosci.* **10**, 1974–1994.

Card, J. P., Whealy, M. E., Robbins, A. K., Moore, R. Y., and Enquist, L. W. (1991). Two α-herpesvirus strains are transported differentially in the rodent visual system. *Neuron* **6**, 957–969.

Card, J. P., Whealy, M. E., Robbins, A. K., and Enquist, L. W. (1992). Pseudorabies virus envelope glycoprotein gI influences both neurotropism and virulence during infection of the rat visual system. *J. Virol.* **66**, 3032–3041.

Ceccatelli, S., Villar, M. J., Goldstein, M., and Hökfelt, T. (1989). Expression of c-fos immunoreactivity in transmitter-characterized neurons after stress. *Proc. Natl. Acad. Sci. USA* **86**, 9569–9573.

Chan, R. K. W., Brown, E. R., Ericsson, A., Kovacs, K. J., and Sawchenko, P. E. (1993). A comparison of two immediate-early genes, c-fos and NGFI-B, as markers for functional activation in stress-related neuroendocrine circuitry. *J. Neurosci.* **13**, 5126–5138.

Coulon, P., and Flamand, A. (1995). Molecular analysis of rabies and pseudorabies neurotropism. In "Viral Vectors: Tools for the Analysis and Genetic Manipulation of the Nervous System" (M. G. Kaplitt and A. D. Loewy, Eds.), pp. 395–409. Academic Press, Orlando.

Dragunow, M., and Faull, R. (1989). The use of c-fos as a metabolic marker in neuronal pathway tracing. *J. Neurosci. Methods* **29**, 261–265.

Ericsson, A., Kovacs, K. J., and Sawchenko, P. E. (1994). A functional anatomical analysis of central pathways subserving the effects of interleukin-1 on stress-related neuroendocrine neurons. *J. Neurosci.* **14,** 897–913.

Gerfen, C. R., and Sawchenko, P. E. (1984). An anterograde neuroanatomical tracing method that shows the detailed morphology of neurons, their axons and terminals: Immunohistochemical localization of an axonally transported plant lectin, Phaseolus vulgaris leucoaggultinin (PHA-L). *Brain Res.* **290,** 219–238.

Haxhiu, M. A., Jansen, A. S. P., Cherniack, N. S., and Loewy, A. D. (1993). CNS innervation of airway-related parasympathetic preganglionic neurons: A transneuronal labeling study using pseudorabies virus. *Brain Res.* **618,** 115–134.

Herdegen, T., Kovary, K., Leah, J., and Bravo, R. (1991a). Specific temporal and spatial distribution of Jun, Fos, and Krox-24 proteins in spinal neurons following noxious transsynaptic stimulation. *J. Comp. Neurol.* **313,** 178–191.

Herdegen, T., Tölle, T. R., Bravo, R., Zieglgänsberger, W., and Zimmermann, M., (1991b). Sequential expression of JUN B, JUN D and FOS B proteins in rat spinal neurons: Cascade of transcriptional operations during nociception. *Neurosci. Lett.* **129,** 221–224.

Jansen, A. S. P., Farwell, D. G., and Loewy, A. D. (1993). Specificity of pseudorabies virus as a retrograde marker of sympathetic preganglionic neurons: Implications for transneuronal labeling studies. *Brain Res.* **617,** 103–112.

Kuypers, H. G. J. M., and Ugolini, G. (1990). Viruses as transneuronal tracers. *Trends Neurosci.* **13,** 71–75.

Kwong, A. D., and Frenkel, N. (1989). The herpes simplex virus virion host shutoff function. *J. Virol.* **63,** 4834–4839.

Lafay, F., Coulon, P., Astic, L., Saucier, D., Riche, D., Holley, A., and Flamand, A. (1991). Spread of the CVS strain of rabies virus and of the avirulent mutant Av01 along the olfactory pathways of the mouse after intranasal inoculation. *Virology* **183,** 320–330.

Lantéri-Minet, M., Weil-Fugazza, J., De Pommery, J., and Menétrey, D. (1994). Hindbrain structures involved in pair processing as revealed by the expression of c-Fos and other immediate early gene proteins. *Neurosciences* **58,** 287–298.

Li, Y. W, Ding, Z. Q., Wesselingh, S. L., and Blessing, W. W. (1992a). Renal and adrenal sympathetic preganglionic neurons in rabbit spinal cord: Tracing with herpes simplex virus. *Brain Res.* **573,** 147–152.

Li, Y. W., Wesselingh, S. L., and Blessing, W. W. (1992b). Projections from rabbit caudal medulla to C1 and A5 sympathetic premotor neurons, demonstrated with phaseolus leucoagglutinin and herpes simplex virus. *J. Comp. Neurol.* **317,** 379–395.

Loewy, A. D., and Haxhiu, M. A. (1992). CNS cell groups projecting to pancreatic parasympathetic preganglionic neurons. *Brain Res.* **620,** 323–330.

Loewy, A. D., Bridgman, P. C., and Mettenleiter, T. C. (1991). ß-galactosidase expressing recombinant pseudorabies virus for light and electron microscopic study of transneuronally labeled CNS neurons. *Brain Res.* **555,** 346–352.

Loewy, A. D., Franklin, M. F., and Haxhiu, M. A. (1994). CNS monoamine cell groups projecting to pancreatic vagal motoneurons: A transneuronal labeling study using pseudorabies virus. *Brain Res.* **638,** 248–260.

Lundh, B., Kristensson, K., and Norrby, E. (1987). Selective infections of olfactory and respiratory epithelium by vesicular stomatitis and sendai viruses. *Neuropathol. Appl. Neurobiol.* **13,** 111–112.

Lundh, B., Love, A., Kristensson, K., and Norrby, E. (1988). Nonlethal infection of aminergic reticular core neurons: Age-dependent spread of ts mutant vesicular stomatitis virus from the nose. *J. Neuropathol. Exp. Neurol.* **47,** 497–506.

Marson, L., Platt, K. B., and McKenna, K. E. (1993). Central nervous system innervation of the penis as revealed by the transneuronal transport of pseudorabies virus. *Neuroscience* **55,** 263–280.

Martin, X., and Dolivo, M. (1983). Neuronal and transneuronal tracing in the trigeminal system of the rat using herpes virus suis. *Brain Res.* **555**, 253–276.

McLean, J. H., Shipley, M. T., and Bernstein, D. I. (1989). Golgi-like, transneuronal retrograde labelling with CNS injections of herpes simplex virus type-1. *Brain Res. Bull.* **22**, 867–881.

Mettenleiter, T. C. (1991). Molecular biology of pseudorabies (Aujesky's disease) virus. *Comp. Immunol. Microbiol. Infect. Dis.* **14**, 151–163.

Mettenleiter, T. C. (1995). Molecular properties of alphaherpesviruses used in transneuronal pathway tracing. In "Viral Vectors: Tools for the Analysis and Genetic Manipulation of the Nervous System" (M. G. Kaplitt, and A. D. Loewy, Eds.), pp. 367–393. Academic Press, Orlando.

Morgan, J. I., and Curran, T. (1989). Stimulus-transcription coupling in neurons: Role of cellular primary response genes. *Trends Neurosci.* **12**, 459–462.

Morgan, J. I., and Curran, T. (1991). Stimulus-transcription coupling in the nervous system: Involvement of the inducible proto-oncogenes fos and jun. *Annu. Rev. Neurosci.* **14**, 421–451.

Nash, A. A., Jayasuria, A., Pheran, J., Cobbold, S. P., Waldmanner, H., and Prospero, T. (1987). Different roles for $L3T4^{+}$ and Lyt 2^{+} cell subsets in the control of an acute herpes simplex virus infection of the skin and nervous system. *J. Gen. Virol.* **68**, 825–833.

Oakes, J. E. (1975). Invasion of the central nervous system by herpes simplex virus type 1 after subcutaneous inoculation of immunosuppressed mice. *J. Infect. Dis.* **131**, 51–57.

Rinaman, L., Card, J. P., and Enquist, L. W. (1993). Spatio-temporal responses of astrocytes, ramified microglia and brain macrophages to central neuronal infection with pseudorabies virus. *J. Neurosci.* **13**, 685–702.

Rotto-Percelay, D. M., Wheeler, J. G., Osorio, F. A., Platt, K. B., and Loewy, A. D. (1992). Transneuronal labeling of spinal interneurons and sympathetic preganglionic neurons after pseudorabies virus injections in the rat medial gastrocnemius muscle. *Brain Res.* **574**, 291–306.

Sabin, A. B. (1938). Progression of different nasally instilled viruses along different nervous pathways in the same host. *Proc. Soc. Exp. Biol. Med.* **38**, 270–275.

Schramm, L. P., Strack, A. M., Platt, K. B., and Loewy, A. D. (1993). Peripheral and central pathways regulating the kidney: A study using pseudorabies virus. *Brain Res.* 251–262.

Sheng, M., and Greenberg, M. E. (1990). The regulation and function of c-fos and other immediate early genes in the nervous system. *Neuron* **4**, 477–485.

Simmons, A., and Nash, A. A. (1987). Effect of B cell suppression on primary infection and reinfection of mice with herpes simplex virus. *J. Infect. Dis.* **155**, 649–654.

Spencer, S. E., Sawyer, W. B., Wada, H., Platt, K. B., and Loewy, A. D. (1990). CNS projections to the pterygopalatine parasympathetic preganglionic neurons in the rat: A retrograde transneuronal viral cell body labeling study. *Brain Res.* **534**, 149–169.

Strack, A. M., and Loewy, A. D. (1990). Pseudorabies virus: A highly specific transneuronal cell body marker in the sympathetic nervous system. *J. Neurosci.* **10**, 2139–2147.

Strack, A. M., Sawyer, W. B., Hughes, J. H., Platt, K. B., and Loewy, A. D. (1989a). A general pattern of CNS innervation of the sympathetic outflow demonstrated by transneuronal pseudorabies viral infections. *Brain Res.* **491**, 156–162.

Strack, A. M., Sawyer, W. B., Platt, K. B., and Loewy, A. D. (1989b). CNS cell groups regulating the sympathetic outflow to the adrenal gland as revealed by transneuronal cell body labeling with pseudorabies virus. *Brain Res.* **491**, 274–296.

Strick, P. L., and Card, J. P. (1992). Transneuronal mapping of neural circuits with alpha herpesviruses. In "Experimental Neuroanatomy. A Practical Approach" (J. P. Bolam, Ed.), pp. 81–101. IRL Press at Oxford Univ. Press.

Ter Horst, G. J., Van den Brink, A., Homminga, S. A., Hautvast, R. W. M., Rakhorst, G., Mettenleiter, T. C., De Jongste, M. J. L., Lie, K. I., and Korf, J. (1993). Transneuronal viral labelling of rat heart left ventricle controlling pathways. *Neuroreport* **4**, 1307–1310.

Ugolini, G. (1995). Transneuronal tracing with alpha-herpesviruses: A review of the methodology. In "Viral Vectors: Tools for the Analysis and Genetic Manipulation of the Nervous System" (M. G. Kaplitt, and A. D. Loewy, Eds.), pp. 293–317. Academic Press, Orlando.

Ugolini, G., Kuypers, H. G. J. M., and Simmons, A. (1987). Retrograde transneuronal transfer of herpes simplex virus type 1 (HSV-1) from motorneurons. *Brain Res.* **422,** 242–256.

Valhne, A., Svennerholm, B., Sandberg, M., Hamberger, A., and Lycke, E. (1980). Differences in attachment between herpes simplex type 1 and 2 viruses to neurons and glial cells. *Infect. Immunol.* **28,** 675–680.

Zemanick, M. C., Strick, P. L., and Dix, R. D. (1991). Direction of transneuronal transport of herpes simplex virus 1 in the primate motor system is strain dependent. *Proc. Natl. Acad. Sci. USA* **88,** 8048–8051.

CHAPTER 20

Molecular Properties of Alphaherpesviruses Used in Transneuronal Pathway Tracing

Thomas C. Mettenleiter
Federal Research Centre for Virus Diseases of Animals
Institute of Molecular and Cellular Virology
Insel Riems, Germany

I. Introduction

Several groups of viruses exhibit a distinct tropism for the central nervous system of their host species. Among the most important are the picorna viruses, e.g., poliovirus, the rhabdoviruses, e.g., rabies virus, and the subfamily *Alphaherpesvirinae* of the herpes viruses (Johnson, 1982; Tyler, 1987a,b; Griffin, 1990). After peripheral infection these viruses enter neurons and spread within the central nervous system until they reach terminal target cells whose destruction ultimately leads to severe disease or death. Besides their clinical importance, neurotropic viruses are increasingly being used as tracers to follow neuronal networks. Transneuronal tracing has first been attempted using soluble compounds like tetanus toxin and wheat germ agglutinin. However, the transneuronal signal was usually weak and difficult to detect (Evinger and Erichsen, 1986; Spatz, 1989). This can be explained by dilution of the compound within the labeled neuron, and inefficient transsynaptic transfer from first- to second-order neurons. Higher-order connections remained inaccessible by this method.

The use of neurotropic viruses in neuronal tracing solves some of these problems. Viruses replicate within the infected cell producing a large number of progeny virions. Transneuronal infection by a single infectious virion is sufficient to initiate the processes leading to virus multiplication. A dilution effect as observed with nonreplicating tracers does not occur. Neurotropic viruses, therefore, appear to be the method of choice for following complex, hierarchically ordered neuronal pathways (reviewed in Kuypers and Ugolini, 1990). One prerequisite for their use in neuronal pathway tracing is that virus transfer from infected to uninfected neurons should occur preferentially or exclusively across synapses. However, neurotropic viruses often produce widespread infections in the nervous system with no exclusive transsynaptic transfer. Therefore, in several cases mutated viruses with reduced capability for lateral spread, i.e., infection of neighboring cells other than across synapses, have been used.

Four viruses, belonging to two different virus families, are most prominent in transneuronal pathway tracing. These are the rhabdoviruses vesicular stomatitis virus (VSV) and rabies virus (Lundh *et al.*, 1987; Astic *et al.*, 1993), and the alphaherpesviruses herpes simplex virus type 1 and suid herpes virus 1, also called pseudorabies virus (Kristensson *et al.*, 1982; Ugolini *et al.*, 1989; Martin and Dolivo, 1983; Strack *et al.*, 1989; Card *et al.*, 1990). This review focuses on the molecular properties of the alphaherpesviruses in initiation and spread of infection in cell culture and the role viral glycoproteins play in these processes. The importance of the results derived from cultured continuous cell lines will be discussed with regard to the situation in infection of the nervous system in the animal host. In a final section, the role of other viral gene products in neurotropism of alphaherpesviruses *in vivo* will also be discussed.

II. The Alphaherpesviruses

The subfamily *Alphaherpesvirinae* of the family *Herpesviridae* comprises mostly neurotropic viruses which cause disease in humans and animals (Roizman *et al.*, 1992; Table 1). This grouping was initially based on biological

Table I
Family *Herpesviridae*

	Subfamily	
Alphaherpesvirinae	*Betaherpesvirinae*	*Gammaherpesvirinae*
Herpes simplex virus 1[a]	Human cytomegalovirus[a]	Epstein–Barr virus[a]
Herpes simplex virus 2	Human herpes virus 6	Herpes virus saimiri[a]
Varicella zoster virus[a]	Human herpes virus 7	Bovine herpes virus 4
Pseudorabies virus	Murine cytomegalovirus	Alcelaphine herpes virus 1
Equine herpes virus 1[a]		
Bovine herpes virus 1		
Feline herpes virus 1		
Infect. laryngotracheitis virus		
Marek's disease virus (?)[b]		

[a] Complete genomic sequence is available.
[b] Sequence data show a close relationship of Marek's disease virus to alphaherpesviruses. List of viruses is not complete.

parameters, such as short replication cycles, rapid lytic growth, and wide host range in cell culture, as well as establishment of latency in neurons. Recently, complete or partial genomic sequence information of a number of herpes viruses has become available (Table 1). Comparison between these sequences revealed a high degree of relationship between the alphaherpesviruses at the molecular level. In general, most of the genes found in alphahesperviruses are conserved within the members of the subfamily which means that homologous genes are found in other members of this family in a mostly collinear gene arrangement. This conservation is indicative of a high conservation of basic mechanisms for multiplication of these viruses. Indeed, although differences do exist, there is a remarkable similarity at the molecular level in the replication of the different viruses.

The *Alphaherpesvirinae* have been subdivided into two genera (Roizman *et al.,* 1992). Genus *Varicellovirus* encompasses, e.g., the human pathogen varicella-zoster virus (VZV), as well as the animal pathogens equine herpes virus 1, bovine herpes virus 1, and pseudorabies virus (PrV). PrV is the causative agent of Aujeszky's disease, an illness that is of major importance in pigs and causes significant losses in the pig industry. Although the virus can infect and cause high mortality in nearly all mammalian species, it is regarded as nonpathogenic for higher primates including humans (Pensaert and Kluge, 1989). The herpes simplex viruses type 1 and 2 (HSV-1, HSV-2), which cause labial and genital herpes in humans, belong to the genus *Simplexvirus* (Roizman *et al.,* 1992). Based on sequence data, the tumorigenic chicken herpes virus Marek's disease virus is also being considered closely related to alphaherpesviruses, despite its apparent lack of neurotropism (Buckmaster *et al.,* 1988).

Herpes virions are approx 100–200 nm in diameter and consist of four electron microscopically distinguishable components. The double-stranded linear DNA genome of between 120,000 and 230,000 base pairs is tightly associated with proteins forming the nucleoprotein core. This structure is enclosed in a virus capsid consisting of 162 capsomers in an icosahedral symmetry. An amorphous proteinaceous structure termed the tegument encloses the herpes virus capsid and is in turn surrounded by a lipid envelope derived from cellular membranes during virus maturation. In the envelope, virus-encoded glycoproteins are embedded. These glycoproteins represent major antigens recognized by the immune system of the infected host. They have also been found to be of prominent importance for infectivity of free virions, as well as for the capability of herpes viruses to directly spread between infected and noninfected cells. So far, 11 glycoproteins have been described in HSV-1, designated as gB, gC, gD, gE, gG, gH, gI, gJ, gK, gL (reviewed in Spear, 1993a,b), and gM (Baines and Roizman, 1993). In most cases, homologous proteins or genes have also been found in the other members of the *Alphaherpesvirinae* analyzed. Exceptions are VZV which lacks a glycoprotein D homolog (Davison and Scott, 1986), and a gJ-encoding gene has so far only been found in HSV

(McGeoch, 1988; Spear, 1993a,b). Only four glycoproteins, gB, gH, gL, and gM, are conserved throughout the herpes virus family (reviewed in Mettenleiter, 1994b). From the sheer number of glycoproteins it is evident that herpes virus envelope structure has to be highly complex. (At the 18th International Herpes virus Workshop a common nomenclature for alphaherpesviral glycoproteins has been agreed upon based on the designations of homologous herpes simplex virus glycoproteins. This novel nomenclature is used throughout the review. For comparison of old and new designations for PrV glycoproteins see Table 2.)

With the advent of specific mutagenesis techniques these glycoproteins have become more amenable for analysis of their functions during the infectious cycle. Herpes viral gene products have operationally been divided into nonessential and essential proteins. Nonessential genes and proteins can be deleted from the virus without dramatically affecting virus replication in cell culture. In contrast, deletion of essential proteins abolishes infectivity. Whereas mutagenesis of nonessential glycoproteins is, therefore, easy to perform, mutations in essential glycoproteins that eliminate or severely impair their function are

Table 2
Properties of HSV-1 and PrV Glycoproteins

Designation[a]		Essential[b]		Virion component		Attachment[c]		Penetration		Cell–cell spread	
HSV-1	PrV	HSV-1	PrV	HSV-1	PrV	HSV-1	PrV	HSV-1	PrV	HSV-1	PrV
gB	(gII)	+	+	+	+	< + >	–	+	+	+	+
gC	(gIII)	–	–	+	+	< + >	< + >	–	–[d]	–	–
gD	(gp50)	+	+	+	+	< + >	< + >	+	+	+	< + >
gE	(gI)	–	–	+	+	–	–	–	–	< + >	< + >
gG	(gX)	–	–	+	–	–	–	–	–	–	–
gH	gH	+	+	+	+	?[e]	–	+	+	+	+
gI	(gp63)	–	–	+	+	–	–	–	–	< + >	< + >
gJ	[f]	–		?		?		–		–	
gK	gK	?	?	?	?	?	?	?	?	?	?
gL	gL	+	+	+	+	?	?	+	+	+	+
gM	gM	–	?	+	+	–	?	–	?	< + >	?

[a] A common nomenclature based on HSV-1 glycoprotein designations was agreed upon for homologous alphaherpesviral glycoproteins. Designations in parentheses show the original designations given to PrV glycoproteins.
[b] +, Essential; –, nonessential.
[c] + Indicates an essential function, < + > indicates a modulating function, and – indicates no involvement demonstrated.
[d] gC mutants of PrV and HSV-1 exhibit a delay in penetration. This is thought to be related to their attachment deficiency.
[e] ?, No information available.
[f] No entry: glycoprotein or gene has not been found.

lethal for the virus. The construction of transgenic cell lines that stably carry and express a wild-type form of the mutated protein which can *in trans* complement the lethal defect of the virus mutant greatly facilitated functional studies on essential herpes viral glycoproteins. Virus mutants grown on these transgenic cell lines are phenotypically complemented, i.e., they carry the wild-type glycoprotein in the virus envelope. However, upon passage in noncomplementing cells, e.g., after infection of animals, the glycoprotein is lost and the respective defect of the virus mutant becomes apparent (Fig. 1). Glycoproteins gB, gD, gH, and gL are considered essential proteins (Table 2).

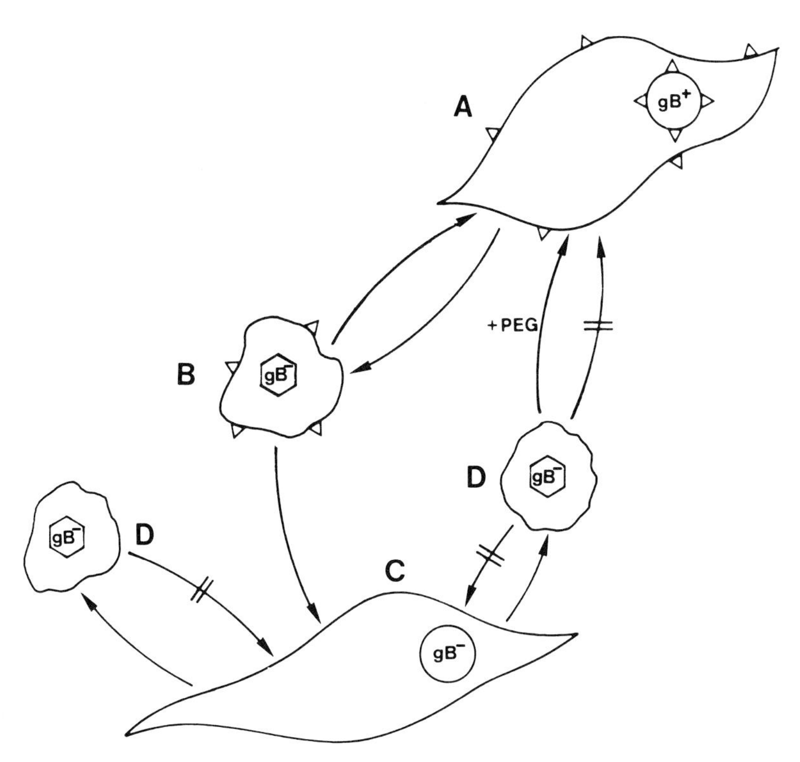

Figure 1 Phenotypic transcomplementation of virus mutants in essential glycoproteins. Viruses with lethal mutations in essential glycoproteins, in this case gB, can be propagated on transgenic cell lines expressing gB (A). Virus progeny (B) is phenotypically complemented, i.e., despite the genetic defect (indicated by the designation gB^-) carries a wild-type envelope containing all glycoproteins including gB obtained from the cell line (indicated by "spikes"). After infection of noncomplementing cells (C) by these phenotypically complemented virus mutants noninfectious virus particles lacking gB are produced (D). Infectivity can partially be restored when attached virus mutants lacking gB are treated with polyethylene glycol (PEG) which induces fusion of the virion envelope with the cellular cytoplasmic membrane. This scheme is valid for PrV and HSV-1 mutants lacking gB, gD, or gH/gL.

III. Replicative Cycle of Herpes Viruses

As enveloped viruses, herpes viruses attach to target cells via interaction of envelope glycoproteins with cellular surface components acting as receptors (Marsh and Helenius, 1989; Spear, 1993a; Haywood, 1994; Mettenleiter, 1994a). After this adsorption step, fusion between virion envelope and cellular cytoplasmic membrane occurs leading to release of the nucleocapsid into the cytoplasm (penetration). The capsid is translocated to nuclear pores, the genome is released into the nucleus, and transcription of the viral DNA begins. Herpes virus transcription is coordinately regulated in a cascade-like fashion. Initially, three distinct stages of viral gene expression have been differentiated (for review see Ben-Porat and Kaplan, 1985; Roizman and Batterson, 1985). Immediate-early (IE) or alpha (α) genes are the first genes to be expressed even when cellular protein synthesis is blocked by cycloheximide. They encode potent transcriptional regulatory proteins. In HSV-1, five immediate-early proteins have been described. In the PrV genome, only one IE-regulated gene has so far been identified (Ben-Porat and Kaplan, 1985). Early (E) or beta (ß) genes are characterized by their dependence on IE proteins for transactivation and their transcription starts before the onset of viral DNA replication. Experimentally, they can be identified by inhibition of viral DNA replication with phosphonoacetic acid. Several enzyme-encoding genes have been assigned to this group. Finally, late (L) or gamma (γ) genes, mostly encoding viral structural components, are expressed. However, in recent years it has become clear that not all genes can be divided into these classes and that a more complex regulatory cascade is indeed present.

After circularization of the incoming linear DNA molecule, replication proceeds via a rolling circle mechanism. Seven virus-encoded proteins including a viral DNA polymerase have been found to be necessary to initiate daughter strand synthesis at conserved sequences acting as origins of replication and to complete synthesis of full-length daughter strands (Wu *et al.*, 1988). Concatemeric head-to-tail fused linear genomes accumulate in the nucleus in high molecular-weight structures. A cleavage/encapsidation signal is formed by the fused genome ends which directs correct cleavage into linear unit-sized molecules and their incorporation into newly formed capsids (Roizman and Batterson, 1985; Ben-Porat and Kaplan, 1985). These capsids acquire a first envelope at the inner nuclear membrane. After traversing the endoplasmic reticulum deenvelopment and reenvelopment may occur. Alternatively, enveloped virions may reach the cytoplasmic membrane within membrane-bound structures (Campadelli-Fiume *et al.*, 1991). Finally, the mature virion is released from the cell.

In neural tissues, alphaherpesviruses can establish latency which is characterized by the persistence of the viral genome without production of virus progeny. Different stimuli can lead to reactivation and concomitant virus production. Although neuronal latency is one of the most interesting interactions between herpes viruses and the nervous system, this cannot be discussed in this chapter. Readers are referred to recent excellent reviews (Enquist, 1994; Rock, 1993).

IV. Initiation of Infection by Alphaherpesviruses

Herpes viruses depend on two consecutive processes for infection of target cells. First, free virions have to come into close physical contact with target cells. This process is called attachment or adsorption. In a second step, the virion envelope has to fuse with the adjacent cellular cytoplasmic membrane leading to release of the capsid into the cytoplasm. This step has been designated as entry or penetration. Virion envelope glycoproteins play important roles in both processes. Within the past several years, some of the steps leading to initiation of infection by alphaherpesviruses have been elucidated, although a detailed knowledge of molecular events during attachment and penetration is still mostly lacking.

A. Attachment

Viruses first contact their target cells by interaction of one or more virion proteins with at least one cellular surface molecule acting as virus receptor. Whereas specific receptors have been identified for a number of viruses including the betaherpesviruses human cytomegalovirus (HCMV; Söderberg *et al.*, 1993) and human herpes virus 7 (Lusso *et al.*, 1994) as well as the gammaherpesvirus Epstein–Barr Virus (reviewed in Nemerow and Cooper, 1992), the knowledge of cellular receptors for alphaherpesviruses is still rudimentary. Studies in cell culture revealed that the human pathogen HSV-1, as well as PrV which is not pathogenic for humans but infects a wide range of other mammals, make first contact with their target cell by interaction of a virion surface glycoprotein with cell-membrane-associated proteoglycan(s) carrying highly sulfated glycosaminoglycan side chains (WuDunn and Spear, 1989; Mettenleiter *et al.*, 1990; Sawitzky *et al.*, 1990). This primary interaction is mediated by the gC glycoproteins (Mettenleiter *et al.*, 1990; Herold *et al.*, 1991). Both HSV-1 and PrV primarily attach to heparan sulfate proteoglycans (HSPG). HSPG are found on a wide variety of cells which could explain the broad host range found for both viruses *in vitro*. They are also present on neurons (Stipp *et al.*, 1994 and references therein), indicating that infection *in vivo* may also be initiated by interaction of virion-bound gC with cellular HSPG.

Surprisingly, neither HSV-1 nor PrV gC are essential for infectivity of free virions. gC-negative HSV-1 and PrV mutants exhibit an approx 10- to 100-fold lower specific infectivity, correlating with lower virus titers obtained in cell culture (Mettenleiter *et al.*, 1990; Herold *et al.*, 1991). However, gC-negative mutants of both viruses are still highly infectious. The question, therefore, arises whether the HSPG-dependent attachment pathway is required for infection of cells by HSV and PrV. Analysis of gC-deletion mutants of both viruses revealed a significant difference. Whereas gC-negative HSV-1 is still able to bind to heparan sulfate (Herold *et al.*, 1994), this is not the case for gC-negative PrV (Mettenleiter *et al.*, 1990). It is clear, therefore, that binding to HSPG is not necessary for infection of cells by PrV, whereas gC^- HSV-1 virions still interact with heparan sulfate. In HSV-1, gB also binds HSPG, and it is the interaction between HSV-1 gB and heparan sulfate that substitutes for the missing gC-attachment function in gC^- HSV-1 (Herold *et al.*, 1994). In conclusion, a heparan sulfate-independent (and probably more generally proteoglycan-independent) pathway of attachment has to function in PrV infection.

Whereas attachment of wild-type HSV-1 and PrV to target cells is initially sensitive to competition by exogenous heparin, this heparin-sensitive attachment converts into a heparin-resistant binding (Karger and Mettenleiter, 1993; McClain and Fuller, 1994). This secondary binding step appears to be specifically mediated by the interaction of PrV and HSV-1 glycoprotein gD with a second cellular component. Absence of gD eliminates or dramatically decreases heparin-resistant attachment. In addition, a limited number of HSV-1 gD-specific receptor molecules has been found on cultured cells (Johnson *et al.*, 1990). Studies using gD-negative PrV and HSV-1 mutants have indicated that both viruses probably interact with common or overlapping second receptor(s) (Lee and Fuller, 1993). It is interesting in this context that a soluble form of HSV-1 gD prevented intraocular infection of rats by HSV-1 (Martin *et al.*, 1992).

Taken together, HSV-1 and PrV exhibit striking similarities in their mode of attachment. A primary interaction between virus and target cell is mediated by binding of virion envelope glycoprotein gC to heparan sulfate proteoglycans on the cell surface. This initial contact is followed by a secondary interaction which probably involves glycoprotein gD and a second cellular receptor. Besides HSV-1 gC, HSV-1 gB also interacts with heparan sulfate and probably mediates attachment of gC-negative HSV-1 mutants. In contrast, gC^- PrV does not bind heparan sulfate and, therefore, has to attach by a HSPG-independent pathway (reviewed in Mettenleiter, 1994a).

It is currently unclear whether the gC–HSPG interaction plays a prominent role in the neurotropism of HSV-1 and PrV. In mice, gC-deleted PrV mutants were as neurovirulent as wild-type virus (Zsak *et al.*, 1992). Invasion of the nervous system was also similar between wild-type and gC^- PrV after intranasal infection of PrVs natural host, swine (Kritas *et al.*, 1994a,b). Similar results have been obtained after infection of mice with gC$^-$ HSV-1 (Sunstrum *et al.*, 1988). In contrast, it has been shown that in PrV mutants lacking glycoprotein gE, the additional deletion of gC abolishes or dramatically decreases neurovirulence in pigs, mice, and chicken (Mettenleiter *et al.*, 1988; Zsak *et al.*, 1992). gE has been implicated in direct cell-to-cell transmission of virus (see Section V). It is conceivable that after impairment of direct cell-to-cell spread the role of gC in mediating attachment of free virions becomes more important.

Using polarized cultured canine cells it has been observed that, whereas wild-type HSV-1 was able to infect these cells equally well by the apical or basolateral route, gC$^-$ HSV-1 infection occurred nearly exclusively via the basolateral surface (Sears *et al.*, 1991). This led to the hypothesis that a gC-dependent interaction may be functional at both levels, whereas the basolateral membrane specifically contains receptors that interact with viral envelope proteins other than gC. Whether these findings truly reflect differences in the receptor repertoire of the respective cellular surfaces, and whether similar phenomena may be observed using other polarized cells, such as neurons, remains to be analyzed.

B. Penetration

Fusion between virus envelope and cellular cytoplasmic membrane is a prerequisite for infectious entry of the nucleocapsid into target cells. After attachment, herpes virus envelopes fuse at neutral pH directly with the cytoplasmic membrane. All viral glycoproteins that have been found to be involved in this process, i.e., gB, gD, gH, and gL, are essential for infectivity of free virions (reviewed in Spear, 1993b). gB consists of a homodimeric complex that may or may not be proteolytically processed by a *trans*-Golgi protease into disulfide-linked subunits. gH and gL form a noncovalently linked, probably heterodimeric complex that represents the functional entity. gB-, gH-, and gL-homologous glycoproteins have been found in members of all herpes virus subfamilies indicating a common mechanism of membrane fusion for all herpes viruses. In the absence of any one of these glycoproteins, virions are unable to initiate infection. Experimentally, infectivity of virions lacking gB, gD, or gH could be restored when membrane fusion was induced by polyethylene glycol. This indicates that, whereas attachment of free virions devoid of gB, gD, or gH/gL can occur, most likely via the gC–HSPG interaction, subsequent steps leading to infectious entry of the capsid into target cells are blocked. It was hypothesized mainly on the basis of electron microscopic studies that each of these glycoproteins is involved in a distinct step of the fusion process in either cascade-like sequential or simultaneous interactions with different cellular receptor proteins (Fuller and Lee, 1992). It has previously been described that gD is important for stable attachment of PrV and HSV-1 (Karger and

Mettenleiter, 1993; McClain and Fuller, 1994). In addition, gD is essential for penetration (Ligas and Johnson, 1988; Rauh and Mettenleiter, 1991; Peeters *et al.*, 1992). Although it is not clear whether the functions of gD in attachment and penetration are related, it has been suggested that gD is operational in triggring the fusion process. In HCMV, the gH-homologous protein may interact with a cellular membrane protein acting as receptor (Keay *et al.*, 1989). Given the considerable conservation between herpes viral gH homologs (see Klupp *et al.*, 1991), a similar interaction may also occur in the other herpes viruses, including HSV-1 and PrV.

The role of the most highly conserved herpes viral glycoprotein, gB, in fusion is completely unclear at present. Since mutations in the carboxyterminal part of HSV-1 gB have been shown to deregulate fusion and lead to so-called syncytial phenotypes characterized by extensive cell–cell fusion leading to polykaryocyte formation (Gage *et al.*, 1993), the function of gB could be in the regulated expansion of an already formed fusion pore.

The lethal penetration defect associated with lack of gB in PrV and HSV-1 can be complemented by heterologous gB polypeptides. BHV-1 gB was able to complement gB^- PrV (Rauh *et al.*, 1991; Kopp and Mettenleiter, 1992), and PrV gB functionally complemented gB^- HSV-1 (Mettenleiter and Spear, 1994). This capacity for heterologous transcomplementation probably reflects a highly conserved functional role for the gB homologs. However, the observation of unidirectional complementation between the gB proteins of PrV and HSV-1, as well as between gB proteins of PrV and BHV-1, also shows that these proteins are functionally related but not identical (Mettenleiter and Spear, 1994; Kopp and Mettenleiter, unpublished results). Heterologous transcomplementation has so far only been described for gB homologs. Experiments using the gD proteins of HSV-1, PrV, and BHV-1 did not show heterologous complementation of respective viral mutants (Peeters, personal communication; Mettenleiter, unpublished results).

Many viral fusion proteins are processed by a proteolytic cleavage event which leads to structural alterations within the protein and activation of the fusion function (see Marsh and Helenius, 1989). Since most herpes viral gB proteins, with the most prominent exception being HSV gB, are post-translationally processed by a protease located in the *trans*-Golgi (Whealy *et al.*, 1990), it was hypothesized that this cleavage event may also be required for the function of gB in fusion. However, recent studies with recombinant PrV and BHV-1 viruses expressing a mutated noncleavable form of BHV-1 gB showed that this proteolytic cleavage is not necessary for membrane fusion to occur (Kopp *et al.*, 1994). The reason for conservation of the proteolytic cleavage site in most herpes viral gB proteins is, therefore, unclear.

In summary, penetration, i.e., fusion between virion envelope and cellular cytoplasmic membrane, is thought to require a number of herpes virus glycoprotein–cell surface receptor interactions with several virion glycoproteins acting simultaneously or in a cascade-like fashion. Virion components involved include gB, gD, and gH/gL. Respective cellular reaction partners have not yet been identified.

V. Viral Spread by Direct Cell-to-Cell Transmission

Spread of infection within cell culture and the infected host can occur by two mechanisms: (i) release of free infectious virions from infected cells which initiate a novel round of infection from outside, and (ii) direct virus transfer from primary infected to adjacent noninfected cells (Fig. 2). The latter process is mainly responsible for the formation of plaques in cell culture. In HSV-1, all glycoproteins essential for penetration have also been found to be necessary for direct cell-to-cell spread, i.e., HSV-1 mutants lacking gB, gD, or gH/gL are unable to form plaques in noncomplementing cells (reviewed in Spear, 1993b). A similar situation exists for PrV mutants lacking gB or gH/

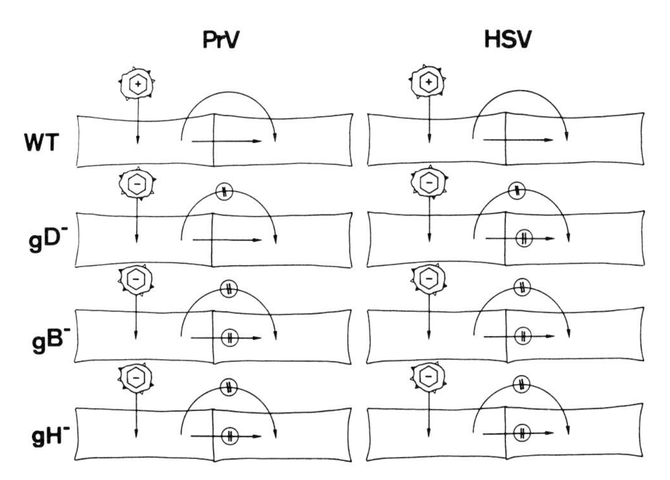

Figure 2 Cell-to-cell spread capabilities of different PrV and HSV-1 glycoprotein mutants. After phenotypic complementation on transgenic cell lines, all virus mutants are able to infect noncomplementing cells as does wild-type (WT) virus. Since gB and gH in the case of PrV, or gD, gB, and gH in the case of HSV, are necessary for infectivity of free virions and direct cell-to-cell spread, further spread by either means is blocked after infection of noncomplementing cells by phenotypically complemented mutants in any one of these glycoproteins. In contrast, after primary infection by phenotypically complemented virus, gD-negative PrV mutants are able to exclusively spread by direct cell-to-cell transfer. (+) indicates the presence of wild-type genome, (−) indicates mutated genome with deletion in the respective glycoprotein gene. Open and closed "spikes" indicate the presence of wild-type envelopes in wild-type and transcomplemented viruses.

gL. In striking contrast, gD^- PrV mutants are capable of forming plaques by direct cell-to-cell spread (reviewed in Mettenleiter, 1994b). Here, the interesting situation arises that virions released from gD^- PrV-infected noncomplementing cells are noninfectious due to their lack of gD in the envelope. However, virus is able to directly spread to neighboring cells leading to complete degeneration of the cell monolayer. Whereas the initial results on gB^- or gD^- HSV mutants led to the assumption that membrane fusion processes during infectious entry and cell-to-cell spread are more or less identical, the finding that gD is necessary for penetration but dispensable for cell-to-cell spread of PrV (Rauh and Mettenleiter, 1991; Peeters *et al.*, 1992) clearly showed that penetration and direct cell-to-cell spread, although related, are distinct and separable (Fig. 2). Using monoclonal antibodies it has also recently been suggested that different functional domains of gB are involved in penetration and cell-to-cell spread of HSV-1 (Navarro *et al.*, 1992). Although nonessential for direct cell-to-cell spread, the presence of PrV gD appears to augment this process since plaques formed by gD^- PrV in noncomplementing cells are smaller than those of wild-type PrV (Mettenleiter *et al.*, 1994).

Besides the gB homodimer and gH/gL heterodimer, a third glycoprotein complex is found in the envelope of alphaherpesviruses. It consists of glycoproteins gE and gI which form a noncovalently linked complex that is most likely stabilized by hydrophobic interactions (Johnson *et al.*, 1988; Zuckermann *et al.*, 1988). Both glycoproteins are nonessential for *in vitro* replication of HSV-1 and PrV. The HSV-1 gE/gI complex has been found to interact with Fc domains of immunoglobulin G (Johnson *et al.*, 1988). No such activity has so far been demonstrated for the PrV gE/gI complex. PrV gE/gI is involved in the release of mature virions from infected cells in a cell type-specific manner (Zsak *et al.*, 1989). In addition, both PrV and HSV-1 gE/gI have been shown to modulate direct cell-to-cell spread. Respective mutants of PrV and HSV-1 have an impaired ability for direct viral spread in cell culture leading to small plaque phenotypes (Jacobs *et al.*, 1993; Balan *et al.*, 1994; Dingwell *et al.*, 1994) and show a significant decrease in neurovirulence (Kritas *et al.*, 1994a,b; Kimman *et al.*, 1992).

In conclusion, the molecular mechanisms leading to direct cell-to-cell spread of herpes viruses are still unclear. Fusion between virion envelope and cellular cytoplasmic membrane during infectious entry and spread of infectivity by direct cell-to-cell transmission appear to be related but distinct processes that are probably regulated differently.

VI. Infection of Neurons

Generally, alphaherpesvirus infection of and productive replication in neurons appear to be similar in many aspects to the processes found in the commonly used immortalized cultured cells from different origins. Ultrastructural analyses, for example, indicated that the maturation steps leading to intracellular virion formation in infected neurons *in vivo* are similar to those seen in cultured fibroblasts (Card *et al.*, 1993). However, it is also clear that significant differences do exist (Stevens, 1991). Earlier studies indicated the presence of specific receptors for both HSV-1 and PrV on neurons and already proposed that viral glycoproteins might be important in this interaction (Marchand and Schwab, 1986; Vahlne *et al.*, 1978). In these experiments, radiolabeled virions bound especially to the synaptosomal fraction prepared from brain tissue of rabbits, rats, and mice after Ficoll gradient centrifugation (Vahlne *et al.*, 1978) and to axon termini in a two-chamber system (Marchand and Schwab, 1986). However, the nature of these receptors on neurons is not known. HSPG, which are present on neurons (Stripp *et al.*, 1994), might be involved in the attachment process of HSV-1 and PrV to nerve endings. However, as outlined above, at least in the case of PrV this mode of attachment does not appear to play a prominent role for invasion of the nervous system *in vivo*. Whether there are different alphaherpesvirus receptor molecules on bona fide neurons, compared to immortalized cultured cells which are commonly used for receptor studies, is unknown. There is indirect evidence that penetration of free infectious virions into neurons is governed by similar mechanisms as infection of cultured fibroblasts. PrV virions lacking gB or gD are noninfectious when inoculated intranasally into highly susceptible mice (Babic *et al.*, 1993; Heffner and Mettenleiter, unpublished results). Therefore, an "unspecific" uptake leading to productive infection does not appear to occur, at least not in the periphery.

VII. Transneuronal Transfer

The mechanisms that lead to circuit-specific transneuronal transfer of virus across synapses, the basic prerequisite for neuronal pathway tracing, are only beginning to be clarified. Recent evidence suggests that local immune reactions on one hand and an only limited permissivity for virus replication of nonneuronal cells within neural tissues on the other hand may lead to a preferentially transneuronal infection (Card *et al.*, 1993; Rinaman *et al.*, 1993). However, wild-type strains of both HSV-1 and PrV are also able to spread laterally between unconnected neurons which is especially evident using highly virulent virus strains and large inocula. Lateral nonspecific spread may also depend on the functioning of the immune system of the host. This nonspecific transfer is a major obstacle for unambiguous transneuronal circuit-specific tracing. The successful detection of specific transneuronal labeling, therefore, heavily depends on the virus inoculum, the site of infection, as well as the virus strain used. To overcome this problem virus mutants with only limited capacity for lateral spread have been used to follow neuronal connections (Strack and Loewy, 1990).

How virions cross the synaptic cleft is still a mystery. Infectious virions may be released at the postsynaptic membrane, cross the synaptic cleft, and initiate infection in the presynaptic neuron (Card *et al.*, 1993). This pathway would be similar to infection of target cells from the outside and would, therefore, follow the infectious process outlined above. However, recent evidence suggests that transneuronal transfer *in vivo* may indeed be related to direct cell-to-cell spread as observed in cell culture. This is mainly based on the analysis of PrV mutants deficient in the expression of glycoprotein gD. In contrast to HSV-1, where gD is necessary for penetration of free infectious virions as well as direct cell-to-cell spread, PrV gD is only required for infectivity of free virions. Direct cell-to-cell spread does occur even in the absence of gD. Therefore, in this situation both processes are clearly separable. To analyze behavior of these mutants *in vivo*, mice and pigs were infected with PrV mutants whose gD gene has been deleted but which have been transcomplemented after propagation on cell lines expressing gD. These phenotypically complemented virions can initiate a first round of infection. After a first cycle in complementing cells, however, virus spread can only occur by direct cell-to-cell transfer since virions released from infected noncomplementing cells lack gD and, therefore, are not infectious. After intranasal infection of mice, these transcomplemented mutants were as virulent as wild-type PrV indicating that direct cell-to-cell transmission is the pathogenetically important way of viral spread within these animals (Heffner *et al.*, 1993; Peeters *et al.*, 1993). After infection of pigs, a clear reduction in virulence was observed compared to wild-type PrV which might reflect the need for free infectious virions to disseminate infectivity efficiently in this larger animal species (Heffner *et al.*, 1993). Following injection into the hypoglossal nerve of mice (Babic *et al.*, 1993), specific transneuronal transfer of gD^- PrV mutants was observed. In contrast, infection by phenotypically complemented gB^- mutants led to infection of primary target cells with a lack of further viral spread, identical to the situation observed in cell culture (Babic *et al.*, 1993; Heffner *et al.*, 1993). How gD-negative PrV is able to initiate infection in the presynaptic neuron is unclear. It might be that the putative gD receptor is not necessary for transneuronal infection initiation by free virions, which would indicate a different pathway of infection at the presynaptic membrane compared to the cultured cells studied to date perhaps due to the confinement of the synaptic cleft. A "nonspecific" mechanism of virion uptake followed by infectious entry appears, however, unlikely since virions lacking gB are unable to be passaged transneuronally, i.e., are incapable of initiating the infectious cycle in the presynaptic neuron (Fig. 3). Virus mutants that are deficient in the expression of glycoproteins essential for cell-to-cell spread are therefore ideal tools for the identification of primary target cells within the infected organism. After propagation in complementing cell lines

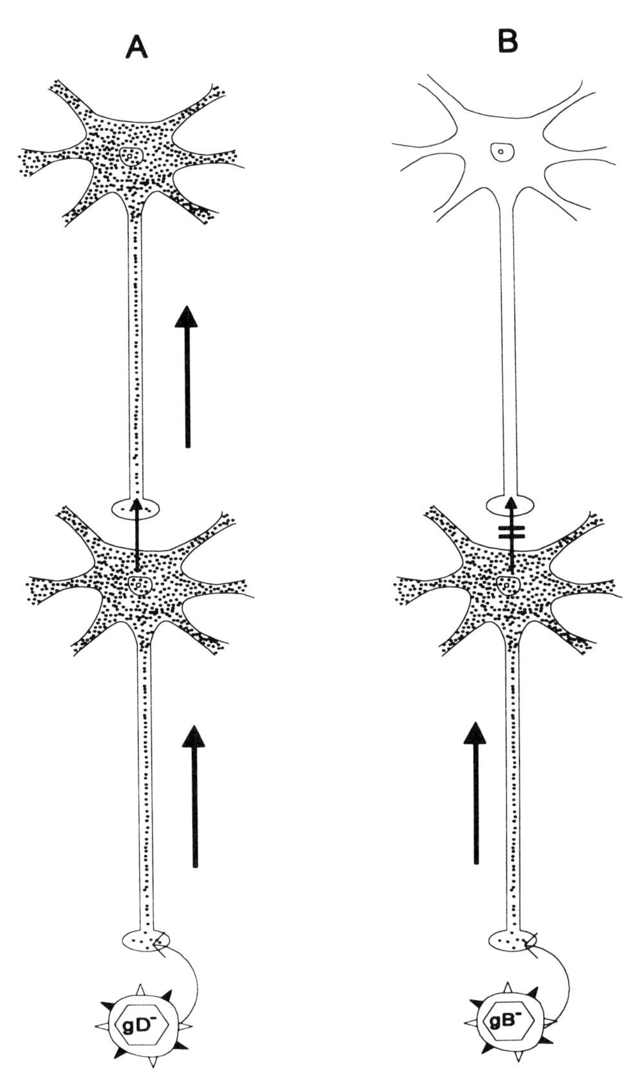

Figure 3 Transneuronal transfer of gD- or gB-negative PrV mutants. Phenotypically complemented gD- (A) or gB-negative PrV mutants (B) infect primary target cells, including neurons. However, whereas gD-negative PrV is able to transneuronally infect secondary order neurons, in the absence of gB transneuronal infection does not occur. Therefore, gB-negative PrV labels primary target cells, whereas gD-negative PrV is able to spread within the nervous system to higher-order neurons.

the phenotypically complemented viruses are able to infect primary target cells. However, only one replicative cycle can be completed by these viruses since progeny virions due to the lack of the respective glycoprotein are noninfectious, and cell-to-cell spread in noncomplementing cells does not occur (Fig. 3).

An influence of mutations in glycoproteins gB and gD of HSV-1 on viral neuroinvasiveness that do not impair function of these glycoproteins *in vitro* has also been observed. Replacement of the gB gene of a nonneuroinvasive HSV-1 strain by a "wild-type" form of the gene led to an increase in neuroinvasiveness (Yuhasz and Stevens, 1993). Along similar lines, a mutation in the glycoprotein gD gene was also correlated with a lack of neuroinvasiveness, although gD function in cell culture was not impaired by this mutation (Izumi and Stevens, 1990). It appears, therefore, that small changes in these glycoproteins can significantly affect neuroinvasion without detectably altering their function during infection of cultured cells.

Molecular analysis of attenuated PrV strains used as live vaccines against Aujeszky's disease had first demonstrated a striking absence of glycoprotein gE (Mettenleiter *et al.*, 1985), and sometimes gI (Petrovskis *et al.*, 1986), in these virus mutants. One of them is the vaccine strain Bartha (see Section VIII). The absence of gE correlated with the attenuated phenotype, i.e., decreased neurovirulence. Specific deletion of the gE gene from wild-type PrV genomes led to a more or less dramatic reduction in neurovirulence, depending on the virus strain used (Mettenleiter *et al.*, 1987; Kimman *et al.*, 1992). Deletion mutants of HSV-1 lacking gE or gI also exhibit a dramatic decrease in neurovirulence (Balan *et al.*, 1994; Dingwell *et al.*, 1994). Infection of the rat visual system has been shown to be heavily influenced by the absence of gE in that gE-deleted PrV strains were not able to infect certain subsets of retinorecipient neurons compared to wild-type PrV (Card *et al.*, 1992). This phenotype is also seen in gI-deleted mutants (Whealy *et al.*, 1993) indicating, as has been observed in cultured cells before, that the gE/gI complex represents the biologically functional entity. A similar situation has been found in a subset of brain stem neurons infected after injection of gE^- PrV into the ventricles of the heart (Standish *et al.*, 1993). After intranasal infection of pigs, gE- and gI-deleted PrV strains were able to efficiently replicate in the nasal mucosa and to infect first-order neurons in both the trigeminal and the olfactory pathway. However, viral spread to higher-order neuronal connections was severely impaired or abolished (Kritas *et al.*, 1994a,b). Interestingly, gE mutants exhibited a more severe defect than gI mutants, indicating that gE might exert functions in mediating neuronal infection in addition to those provided by the gE/gI complex. The reason for the decreased neuroinvasiveness of gE and gI mutants is unclear at present. It has been hypothesized that gE might interact with a specific receptor molecule on susceptible neurons and that gE-deleted virus mutants might be unable to enter specific target cells due to the abolishment of this putative virion gE–cellular gE-receptor interaction (Card *et al.*, 1992).

However, it also seems possible that the decrease in cell-to-cell spread, which has been shown to represent the major way of virus spread in PrV-infected animals, leads to a restriction in neuroinvasion.

Interestingly, whereas deletion of gE or gI alone did decrease but not abolish virulence of some PrV strains, concomitant inactivation of gE (and/or gI) and gC completely abolished neurovirulence (Mettenleiter *et al.*, 1988). Further studies indicated that, although direct cell-to-cell spread proved to be the most important mode of viral spread within the animal, gC-mediated attachment of free virions does also play a role. In the absence of gE, spread of PrV mainly occurs by dissemination of free infectious virions, whereas in wild-type and gC$^-$ PrV spread is mainly effected by direct cell-to-cell transfer (Zsak *et al.*, 1992). Therefore, inactivation of gE or gI leads to a significant decrease in neurovirulence by inhibiting direct transfer. Residual infectivity is mainly due to gC-mediated attachment of released virions. If this residual mode of infection is also impaired by deletion of gC, neurovirulence is completely abolished (Zsak *et al.*, 1992).

In summary, glycoproteins that are involved in cell-to-cell spread also function in neuroinvasion, whereas glycoprotein gC, which is prominent in mediating virion attachment in cultured cells, appears to play only a minor role in the neuroinvasiveness of HSV-1 and PrV. Glycoproteins essential for cell-to-cell spread in culture also appear essential for transneuronal spread *in vivo*. Nonessential glycoproteins gE and gI modulate cell-to-cell spread in culture and heavily influence neural spread *in vivo*.

VIII. Molecular Analysis of PrV Strain Bartha

Wild-type PrV has a strong capability for lateral nonspecific spread within the infected tissue which renders it difficult for use in transneuronal pathway tracing. Recently, a PrV strain that has been used for decades as an attenuated live vaccine against Aujeszky's disease in pigs was shown to be better suited for pathway tracing studies (Strack and Loewy, 1990). This PrV strain which was originally isolated by A. Bartha (1961), and is therefore called PrV–Bartha, was analyzed at the molecular level to identify viral functions that lead to its decreased neural spread. At least three independent defects have been localized by marker rescue and subsequent DNA sequence analysis. A deletion within the unique short portion of the viral genome abolishes expression of the gE and gI glycoproteins (reviewed in Mettenleiter *et al.*, 1989). A mutation within the glycoprotein C locus impairs signal sequence function leading to inefficient incorporation of gC into the virion envelope (Robbins *et al.*, 1989). In addition, monoclonal antibodies that differentiate wild-type PrV gC from PrV–Bartha gC have been isolated indicating further alterations in PrV–Bartha–gC (Ben-

Porat *et al.*, 1986). Other mutations have been mapped to a gene encoding a homolog of the UL21 protein of HSV (Klupp *et al.*, unpublished results). The PrV–UL21 product is a nonessential capsid protein which modulates capsid formation (de Wind *et al.*, 1992). Restoration of gE/gI expression to PrV–Bartha increased its neuronal spreading capability in chicken, rats, and mice, though not in pigs (Lomniczi *et al.*, 1984). Since gE/gI has been implicated in direct cell-to-cell spread *in vitro*,, this might reflect an increase in the potential to directly spread from neuron to neuron *in vivo*. Full virulence for pigs is only restored after correction of all three defects, gE/gI, gC, and UL21 (Lomniczi *et al.*, 1987).

IX. Other Proteins Influencing Neurotropism and Neurovirulence of Alphaherpesviruses

Besides viral envelope glycoproteins, a number of other viral functions have been reported to influence neurotropism, i.e., infection of neural target cells, and/or neurovirulence, i.e., extent of injury to these cells, in animals infected by HSV-1 and PrV. Prominent among them are genes encoding enzyme proteins involved in nucleotide metabolism. Most herpes viruses specify a thymidine kinase (TK) which plays an important role in the salvage pathway of nucleotide biosynthesis by its ability to directly monophosphorylate thymidine which is then in turn converted to its triphosphorylated form by cellular enzymes. Inactivation of TK in HSV-1 and PrV leads to a dramatic decrease in neurovirulence of both viruses (Field and Wildy, 1978; Kit *et al.*, 1985). In fact, a novel generation of genetically engineered live vaccines against PrV infection in pigs is based on TK-deleted virus mutants. TK has been shown to be nonessential *in vitro*, and viral mutants deficient in TK expression do not exhibit a growth disadvantage in cell culture compared to isogenic wild-type strains. Recent studies have shown that an oppositely oriented open reading frame, designated UL24, partially overlaps the 5′ end of the TK gene (McGeoch *et al.*, 1988) and that mutations thought to impair only TK expression may also affect the UL24 gene. Therefore, the *in vivo* phenotypes observed by various groups using "TK-negative" virus mutants have to be carefully reanalyzed for possible involvement of the UL24 gene.

Ribonucleotide reductase and dUTPase are other alphaherpesvirus-encoded nonessential enzymes whose inactivation leads to an attenuated phenotype in infected animals (Cameron *et al.*, 1988; de Wind *et al.*, 1993; Pyles *et al.*, 1992). As is true in most cases, the exact molecular basis for this effect on neurotropism/neurovirulence of the virus is unclear. However, it has been hypothesized that these enzymes, including TK, may be required for replication in nondividing cells, such as neurons, specifying only limited nucleotide metabolism compared to actively dividing cells such as immortalized cultured cells. Other mutations leading to avirulent phenotypes were mapped to the UL5 (helicase; Bloom and Stevens, 1994) and UL30 (DNA–polymerase; Day *et al.*, 1988) genes.

Recently, a novel neurovirulence-determining protein has been described in HSV-1. Its gene resides within the repeat regions bracketing the U_L portion of the viral genome and is, therefore, present in two copies. Regulated as a late (γ) gene it has been designated γ34.5. HSV-1 mutants deleted of this gene exhibited a dramatically reduced ability to replicate in the CNS of mice (Chou and Roizman, 1990; McLean *et al.*, 1991). The γ 34.5 protein has been shown to inhibit apoptosis, i.e., programmed cell death, in HSV-1 infected neurons but not in nonneuronal cells (Chou and Roizman, 1992). In the absence of γ34.5 protein, infected neuroblastoma cells undergo programmed cell death by a shut off of cellular protein synthesis before the viral replicative cycle can be completed. This leads to a concomitant cessation of the synthesis of viral proteins and an abortive infection.

X. Reporter Genes

Originally, transneuronal pathway tracing studies used immunological reagents to detect virus-infected cells in thin sections of organ tissues. Although in general satisfactory, the success of these studies was mainly dependent on the specificity and sensitivity of the antibodies used. In addition, the procedure proved to be relatively cumbersome. To overcome these problems, genes encoding enzyme proteins with activities, that are easily detectable using appropriate substrates, have been introduced into herpes viral genomes to produce stable virus recombinants. These include the *lacZ* gene from *Escherichia coli* encoding β-galactosidase (Ho and Mocarski, 1988) as well as the *luc* gene from *Photinus pyralis* encoding firefly luciferase (Kovács and Mettenleiter, 1991). Detection of β-gal activity in cells infected by recombinant virus is easily achieved by a single-step reaction with the substrate 5-bromo-4-chloro-3-indolyl-β-D-galactopyranosid (X-gal). By the action of β-gal an insoluble precipitate is formed which can be seen in light microscopy as a blue staining (Fig. 4) and in the electron microscope as an accumulation of electron-dense material (Loewy *et al.*, 1991). Since the *E. coli* enzyme has a different pH optimum than endogenous mammalian galactosidases, altering the pH in the substrate solution can, if necessary, suppress endogenous β-gal activity. β-gal staining can be performed on organ pieces prior to thin sectioning, thus allowing an easy and quick overview on the areas of virus infection which can be followed by specific histological examination of appropriate thin sections.

Another reporter gene which was used to study viral spread within infected animals is derived from the firefly *Photinus pyralis*. It encodes the enzyme

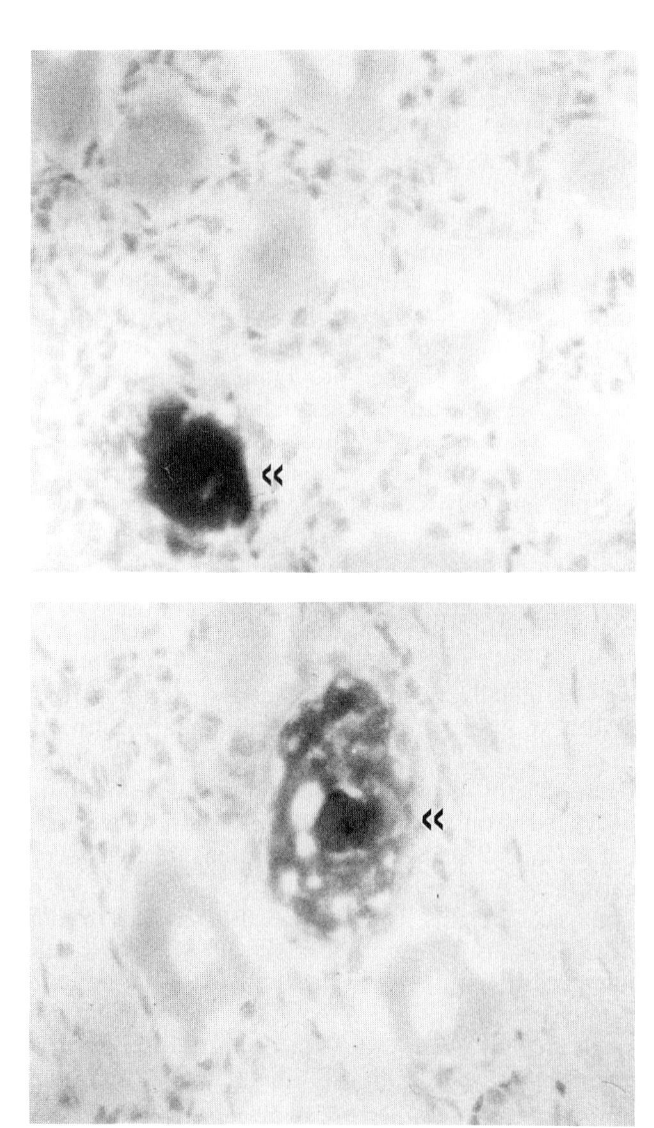

Figure 4 X-gal staining of PrV-infected porcine neurons. Five days after intranasal infection of pigs with a β-galactosidase-expressing PrV recombinant the trigeminal ganglia were cryosectioned, stained with X-gal, and counterstained with hematoxylin–eosin. Arrows mark blue-staining neurons. Magnification $\times 675$.

luciferase which, after addition of luciferin substrate, catalyzes a photon-emitting reaction. This can be fast and easily quantitated in scintillation counters or specialized luminometers. Detection of luciferase activity has the advantage of being essentially free of background (mammals do not have a luciferase gene). So far, luciferase activity was mainly measured in organ extracts allowing only a relatively gross estimation of viral spread (Kovács and Mettenleiter, 1991). However, recent advances in enhanced light microscopy might also allow the detection of luciferase activity in histological sections at the cellular level (Mettenleiter and Gräwe, unpublished results).

Recently, a luminometric assay has also been developed for β-gal. It proves to be significantly more sensitive than colorimetric tests and is easy to perform (Jain and Magrath, 1991). Since substrate requirements are different for the photon-emitting reactions by β-gal and luciferase, both enzymes can be assayed in parallel samples in a luminometer. This allows a fast and easy quantitation of two reporter proteins within the same organ extract.

XI. Conclusion

Studies performed in cultured cells using specifically engineered mutants of HSV-1 and PrV lead to an increased understanding of the molecular interactions required for infection initiation and spread of alphaherpesviruses. Testing of these viral mutants in several animal systems, including the natural virus–host system, PrV–swine, allows the correlation of *in vitro* findings with *in vivo* phenotypes. Based on these comparative studies, knowledge on the molecular mechanisms governing herpes viral neuroinvasion and neural spread is accumulating. However, we are still far from fully understanding at the molecular level the complex virus–cell interactions that are necessary for successful herpes virus infection *in vitro* and *in vivo*.

Acknowledgments

I thank A. Flamand, A. Karger, B. Klupp, and A. Loewy for critical reading of the manuscript, E. Mundt for help with the artwork, and I. Jakobi for secretarial assistance. Work in my laboratory was supported by the Deutsche Forschungsgemeinschaft, the European Union, and Intervet Intl.

References

Astic, L., Saucier, D., Coulon, P., Lafay, F., and Flamand, A. (1993). The CVS strain of rabies virus as transneuronal tracer in the olfactory system of mice. *Brain Res.* 619, 146–156.

Babic, N., Mettenleiter, Th. C., Flamand, A., and Ugolini, G. (1993). Role of essential glycoproteins gII and gp50 in transneuronal transfer of pseudorabies virus from the hypoglossal nerves of mice. *J. Virol.* **67**, 4421–4426.

Baines, J., and Roizman, B. (1993). The UL10 gene of herpes simplex virus 1 encodes a novel viral glycoprotein, gM, which is present in the virion and in the plasma membrane of infected cells. *J. Virol.* **67**, 1441–1452.

Balan, P., Davis-Poynter, N., Bell, S., Atkinson, H., Browne, H., and Minson, T. (1994). An analysis of the *in vitro* and *in vivo* phenotypes of mutants of herpes simplex virus type 1 lacking glycoproteins gG, gE, gI or the putative gJ. *J. Gen. Virol.* **75**, 1245–1258.

Bartha, A. (1961). Experimental reduction of virulence of Aujeszky's disease virus. *Magy. Allatory. Lapja* **16**, 42–45. [In Hungarian]

Ben-Porat, T., and Kaplan, A. S. (1985). Molecular biology of pseudorabies virus. In "The Herpesviruses" (B. Roizman, Ed.), Vol. III, pp. 105–173. Plenum Press, New York.

Ben-Porat, T., DeMarchi, J. M., Lomniczi, B., and Kaplan, A. S. (1986). Role of glycoproteins of pseudorabies virus in eliciting neutralizing antibodies. *Virology* **154**, 325–334.

Bloom, D. C., and Stevens, J. G. (1994). Neuron-specific restriction of a herpes simplex virus recombinant maps to the UL5 gene. *J. Virol.* **68**, 3761–3772.

Buckmaster, A. E., Scott, S., Sanderson, M., Boursnell, M., Ross, N., and Binns, M. (1988). Gene sequence and mapping data from Marek's disease virus and herpes virus of turkeys: Implications for herpes virus classification. *J. Gen. Virol.* **69**, 2033–2042.

Cameron, J., McDougall, I., Marsden, H., Preston, V., Ryan, D., and Subak-Sharpe, J. (1988). Ribonucleotide reductase encoded by herpes simplex virus is a determinant of the pathogenicity of the virus in mice and a valid antiviral target. *J. Gen. Virol.* **69**, 2607–2612.

Campadelli-Fiume, G., Farabegoli, F., di Gaeta, S., and Roizman, B. (1991). Origin of unenveloped capsids in the cytoplasm of cells infected with herpes simplex virus 1. *J. Virol.* **65**, 1589–1595.

Card, J. P., Rinaman, L., Schwaber, J. S., Miselis, R. R., Whealy, M. E., Robbins, A. K., and Enquist, L. W. (1990). Neurotropic properties of pseudorabies virus: Uptake and transneuronal passage in the rat central nervous system. *J. Neurosci.* **10**, 1974–1994.

Card, J. P., Whealy, M. E., Robbins, A. K., and Enquist, L. W. (1992). Pseudorabies virus envelope glycoprotein gI influences both neurotropism and virulence during infection of the rat visual system. *J. Virol.* **66**, 3022–3041.

Card, J. P., Rinaman, L., Lynn, R. B., Lee, B.-H., Meade, R. P., Miselis, R. R., and Enquist, L. W. (1993). Pseudorabies virus infection of the rat central nervous system: Ultrastructural characterization of viral replication, transport, and pathogenesis. *J. Neurosci.* **13**, 2515–2539.

Chou, J., and Roizman, B. (1992). The γ 34.5 gene of herpes simplex virus 1 precludes neuroblastoma cells from triggering total shutoff of protein synthesis characteristic of programmed cell death in neuronal cells. *Proc. Natl. Acad. Sci. USA* **89**, 3266–3270.

Chou, J., Kern, E., Whitley, R., and Roizman, B. (1990). Mapping of herpes simplex virus 1 neurovirulence to γ 34.5, a gene nonessential for growth in culture. *Science* **250**, 1262–1266.

Davison, A. J., and Scott, J. E. (1986). The complete DNA sequence of varicella-zoster virus. *J. Gen. Virol.* **67**, 1759–1816.

Day, S., Lausch, R., and Oakes, J. (1988). Evidence that the gene for herpes simplex virus type 1 DNA polymerase accounts for the capacity of an intertypic recombinant to spread from eye to central nervous system. *Virology* **163**, 166–173.

de Wind, N., Berns, A., Gielkens, A., and Kimman, T. (1993). Ribonucleotide reductase-deficient mutants of pseudorabies virus are avirulent for pigs and induce partial protective immunity. *J. Gen. Virol.* **74**, 351–359.

de Wind, N., Wagenaar, F., Pol, J., Kimman, T., and Berns, A. (1992). The pseudorabies virus homolog of the herpes simplex virus UL21 gene product is a capsid protein which is involved in capsid maturation. *J. Virol.* **66**, 7096–7103.

Dingwell, K. S., Brunetti, C. R., Hendricks, R. L., Tang, Q., Tang, M., Rainbow, A. J., and Johnson, D. C. (1994). Herpes simplex virus glycoproteins E and I facilitate cell-to-cell spread in vivo and across junctions of cultured cells. *J. Virol.* **68**, 834–845.

Evinger, C., and Erichsen, J. T. (1986). Transsynaptic retrograde transport of fragment C of tetanus toxin demonstrated by immunohistochemical localization. *Brain Res.* **380**, 383–388.

Enquist, L. W. (1994). Infection of the mammalian nervous system by pseudorabies virus (PRV). *Semin. Virol.* **5**, 221–231.

Field, H. J., and Wildy, P. (1978). The pathogenicity of thymidine kinase-deficient mutants of herpes simplex virus in mice. *J. Hyg.* **81**, 267–277.

Fuller, A. O., and Lee, W.-C. (1992). Herpes simplex virus type 1 entry through a cascade of virus-cell interactions requires different roles of gD and gH in penetration. *J. Virol.* **66**, 5002–5012.

Gage, P., Levine, M., and Glorioso, J. (1993). Syncytium-inducing mutations localize to two discrete regions within the cytoplasmic domain of herpes simplex virus type 1 glycoprotein B. *J. Virol.* **67**, 2191–2201.

Griffin, D. E. (1990). Viral infections of the central nervous system. In "Antiviral Agents and Viral Diseases of Man" (G. Galasso, R. Whitley, and T. Merigan, Eds.) 3rd ed., pp. 461–495. New York.

Haywood, A. M. (1994). Virus receptors: Binding, adhesion strengthening, and changes in viral structure. *J. Virol.* **68**, 1–5.

Heffner, S., Kovács, F., Klupp, B., and Mettenleiter, Th. C. (1993). Glycoprotein gp50-negative pseudorabies virus: A novel approach toward a nonspreading live herpes virus vaccine. *J. Virol.* **67**, 1529–1537.

Herold, B., Visalli, R., Susmarski, N., Brandt, C., and Spear, P. G. (1994). Glycoprotein C-independent binding of herpes simplex virus to cells requires cell surface heparan sulphate and glycoprotein B. *J. Gen. Virol.* **75**, 1211–1222.

Herold, B., WuDunn, D., Soltys, N., and Spear, P. G. (1991). Glycoprotein C of herpes simplex virus type 1 plays a principal role in the adsorption of virus to cells and in infectivity. *J. Virol.* **65**, 1090–1098.

Ho, D. Y., and Mocarski, E. (1988). β-galactosidase as a marker in the peripheral and neural tissues of the herpes simplex virus-infected mouse. *Virology* **167**, 279–283.

Izumi, K. M., and J. G. Stevens. (1990). Molecular and biological characterization of a herpes simplex virus type 1 (HSV-1) neuroinvasiveness gene. *J. Exp. Med.* **172**, 487–496.

Jacobs, L., Rziha, H.-J., Kimman, T., Gielkens, A., and van Oirschot, J. (1993). Deleting valine-125 and cysteine -126 in glycoprotein gI of pseudorabies virus strain NIA-3 decreases plaque size and reduces virulence in mice. *Arch. Virol.* **131**, 251–264.

Jain, V. K., and Magrath, I. T. (1991). A chemiluminescent assay for quantitation of β-galactosidase in the femtogram range: Application to quantitation of β-galactosidase in *lacZ*-transfected cells. *Anal. Biochem.* 119–124.

Johnson, D. C., Burke, R. L., and Gregory, T. (1990). Soluble forms of herpes simplex virus glycoprotein D bind to a limited number of cell surface receptors and inhibit virus entry into cells. *J. Virol.* **64**, 2569–2576.

Johnson, D. C., Frame, M. C., Ligas, M. W., Cross, A. M., and Stow, N. D. (1988). Herpes simplex virus immunoglobulin G Fc receptor activity depends on a complex of two viral glycoproteins, gE and gI. *J. Virol.* **62**, 1347–1354.

Johnson, R. T. (1982). "Viral Infections of the Nervous System." Raven Press, New York.

Karger, A., and Mettenleiter, Th. C. (1993). Glycoproteins gIII and gp50 play dominant roles in the biphasic attachment of pseudorabies virus. *Virology* **194**, 654–664.

Keay, S., and Baldwin, B. (1991). Anti-idiotype antibodies that mimic gp86 of human cytomegalovirus inhibit viral fusion but not attachment. *J. Virol.* **65**, 5124–5128.

Kimman, T., de Wind, N., Oei-Lie, N., Pol, J., Berns, A., and Gielkens, A. (1992). Contribution of single genes within the unique short region of Aujeszky's disease virus (suid herpes virus type 1) to virulence, pathogenesis and immunogenicity. *J. Gen. Virol.* **73,** 243–251.

Kit, S., Kit, M., and Pirtle, C. C. (1985). Attenuated properties of thymidine kinase-negative deletion mutant of pseudorabies virus. *Am. J. Vet. Res.* **46,** 1359–1367.

Klupp, B., and Mettenleiter, Th. C. (1991). Sequence and expression of the glycoprotein gH gene of pseudorabies virus. *Virology* **182,** 732–741.

Kopp, A., and Mettenleiter, Th. C. (1992). Stable rescue of a glycoprotein gII deletion mutant of pseudorabies virus by glycoprotein gI of bovine herpes virus 1. *J. Virol.* **66,** 2754–2762.

Kopp, A., Blewett, E., Misra, V., and Mettenleiter, Th. C. (1994). Proteolytic cleavage of bovine herpes virus 1 (BHV-1) glycoprotein gB is not necessary for its function in BHV-1 or pseudorabies virus. *J. Virol.* **68,** 1667–1675.

Kovács, F., and Mettenleiter, Th. C. (1992). Firefly luciferase as a marker for herpes virus (pseudorabies virus) replication of vitro and in vivo. *J. Gen. Virol.* **72,** 2999–3008.

Kristensson, K., Nennesmo, I., Persson, L., and Lycke, E. (1982). Neuron to neuron transmission of herpes simplex virus: Transport of virus from skin to brainstem nuclei. *J. Neurol. Sci.* **54,** 158–162.

Kritas, S. K., Pensaert, M. B., and Mettenleiter, Th. C. (1994a). Invasion and spread of single glycoprotein deleted mutants of Aujeszky's disease virus (ADV) in the trigeminal nervous pathway of pigs after intranasal inoculation. *Vet. Microbiol.* **40,** 323–334.

Kritas, S. K., Pensaert, M. B., and Mettenleiter, Th. C. (1994b). Role of gI, gp63, and gIII in the invasion and spread of Aujeszky's disease virus (ADV) in the olfactory nervous pathway of the pig. *J. Gen. Virol.* **75,** 2319–2327.

Kuypers, H. G. J. M., and Ugolini, G. (1990). Viruses as transneuronal tracers. *TINS* **13,** 71–75.

Lee, W.-C., and Fuller, A. O. (1993). Herpes simplex virus type 1 and pseudorabies virus bind to a common saturable receptor on Vero cells that is not heparan sulfate. *J. Virol.* **67,** 5088–5097.

Ligas, M. W., and Johnson, D. C. (1988). A herpes simplex virus mutant in which glycoprotein D sequences are replaced by β-galactosidase sequences binds to but is unable to penetrate into cells. *J. Virol.* **62,** 1486–1494.

Loewy, A. D., Bridgman, P. C., and Mettenleiter, Th. C. (1991). β-galactosidase expressing recombinant pseudorabies virus for light and electron microscopic study of transneuronally labeled CNS neurons. *Brain Res.* **555,** 346–352.

Lomniczi, B., Watanabe, S., Ben-Porat, T., and Kaplan, A. S. (1984). Genetic basis of the neurovirulence of pseudorabies virus. *J. Virol.* **52,** 198–205.

Lomniczi, B., Watanabe, S., Ben-Porat, T., and Kaplan, A. S. (1987). Genome location and identification of functions defective in the Bartha vaccine strain of pseudorabies virus. *J. Virol.* **61,** 796–801.

Lundh, B., Kristensson, K., and Norrby, E. (1987). Selective infections of olfactory and respiratory epithelium by vesicular stomatitis and Sendai viruses. *Neuropathol. Appl. Neurobiol.* **13,** 111–122.

Lusso, P., Secchiero, P., Crowely, R., Garzino-Demo, A., Bernemann, Z., and Gallo, R. C. (1994). CD4 is a critical component of the receptor for human herpes virus 7: Interference with human immunodeficiency virus. *Proc. Natl. Acad. Sci. USA* **91,** 2872–3876.

Marchand, C. F., and Schwab, M. E. (1986). Binding, uptake and retrograde axonal transport of herpes virus suis in sympathetic neurons. *Brain Res.* **383,** 262–270.

Marsh, M., and Helenius, A. (1989). Virus entry into animal cells. *Adv. Virus Res.* **36,** 107–151.

Martin, L. B., Montgomery, P. C., and Holland, T. C. (1992). Soluble glycoprotein D blocks herpes simplex virus type 1 infection of rat eyes. *J. Virol.* **66,** 5183–5189.

Martin, X., and Dolivo, M. (1983). Neuronal and transneuronal tracing in the trigeminal system of the rat using the herpes virus suis. *Brain Res.* **273,** 253–276.

McClain, D. S., and Fuller, O. A. (1994). Cell-specific kinetics and efficiency of herpes simplex virus type 1 entry are determined by two distinct phases of attachment. *Virology* **198**, 690–702.

McGeoch, D. J., Dalrymple, M. A., Davison, A. J., Dolan, A., Frame, M. C., McNab, D., Perry, L. J., Scott, J. E., and Taylor, P. (1988). The complete DNA sequence of the long unique region in the genome of herpes simplex virus type 1. *J. Gen. Virol.* **69**, 1531–1574.

McLean, A. R., Ul-Fareed, M., Robertson, L., Harland, J., and Brow, S. M. (1991). Herpes simplex virus type 1 deletion variants 1714 and 1716 pinpoint neurovirulence-related sequences in Glasgow strain 17^+ between immediate early gene 1 and the "a" sequence. *J. Gen. Virol.* **72**, 631–639.

Mettenleiter, Th. C., Lukács, N., and Rziha, H.-J. (1985). Pseudorabies virus avirulent strains fail to express a major glycoprotein. *J. Virol.* **56**, 307–311.

Mettenleiter, Th. C., Zsak, L., Kaplan, A., Ben-Porat, T., and Lomniczi, B. (1987). Role of a structural glycoprotein of pseudorabies in virus virulence. *J. Virol.* **61**, 4030–4032.

Mettenleiter, Th. C., Schreurs, C., Zuckermann, F., Ben-Porat, T., and Kaplan, A. S. (1988). Role of glycoprotein gIII of pseudorabies virus in virulence. *J. Virol.* **62**, 2712–2717.

Mettenleiter, Th. C., Lomniczi, B., Zsak, L., Medveczky, I., Ben-Porat, T., and Kaplan, A. S. (1989). Analysis of the factors that affect virulence of pseudorabies virus. *Curr. Topics Vet. Med. Animal Sci.* **49**, 3–11.

Mettenleiter, Th. C., Zsak, L., Zuckermann, F., Sugg, N., Kern, H., and Ben-Porat, T. (1990). Interaction of glycoprotein gIII with a cellular heparinlike substance mediates adsorption of pseudorabies virus. *J. Virol.* **64**, 278–286.

Mettenleiter, Th. C., Klupp, B. G., Weiland, F., and Visser, N. (1994). Characterization of a quadruple glycoprotein-deleted pseudorabies virus mutant for use as a biologically safe live virus vaccine. *J. Gen. Virol.* **75**, 1723–1733.

Mettenleiter, Th. C. (1994a). Initiation and spread of α-herpesvirus infections. *Trends Microbiol.* **2**, 2–4.

Mettenleiter, Th. C. (1994b). Pseudorabies (Aujeszky's disease) virus: State of the art. *Acta Vet. Hung.* **42**, 153–177.

Mettenleiter, Th. C., and Spear, P. G. (1994). Glycoprotein gB (gII) of pseudorabies virus can functionally substitute for glycoprotein gB in herpes simplex virus type 1. *J. Virol.* **68**, 500–504.

Navarro, D., Paz, P., and Pereira, L. (1992). Domains of herpes simplex virus 1 glycoprotein B that function in virus penetration, cell-to-cell spread, and cell fusion. *Virology* **186**, 99–112.

Nemerow, G. R., and Cooper, N. R. (1992). CR2 (CD21) mediated infection of B lymphocytes by Epstein–Barr virus. *Semin. Virol.* **3**, 117–124.

Peeters, B., de Wind, N., Hooisma, M., Wagenaar, F., Gielkens, A., and Moormann, R. (1992). Pseudorabies virus envelope glycoproteins gp50 and gII are essential for virus penetration, but only gII is involved in membrane fusion. *J. Virol.* **66**, 894–905.

Peeters, B., Pol, J., Gielkens, A., and Moormann, R. (1993). Envelope glycoprotein gp50 of pseudorabies virus is essential for virus entry but is not required for viral spread in mice. *J. Virol.* **67**, 170–177.

Pensaert, M., and Kluge, J. (1989). Pseudorabies Virus (Aujeszky's disease). In "Virus Infections of Porcines" (M. Pensaert, Ed.), pp. 39–64. Elsevier, Amsterdam.

Petrovskis, E., Timmins, J. G., Giermann, T. M., and Post, L. E. (1986). Deletions in vaccine strains of pseudorabies virus and their effect on synthesis of glycoprotein gp63. *J. Virol.* **60**, 1166–1169.

Pyles, R., Sawtell, N., and Thompson, R. (1992). Herpes simplex virus type 1 dUTPase mutants are attenuated for neurovirulence, neuroinvasiveness, and reactivation from latency. *J. Virol.* **66**, 6706–6713.

Rauh, I., and Mettenleiter, Th. C. (1991). Pseudorabies virus glycoproteins gII and gp50 are essential for virus penetration. *J. Virol.* **65**, 5348–5356.

Rauh, I., Weiland, F., Fehler, F., Keil, G., and Mettenleiter, Th. C. (1991). Pseudorabies virus mutants lacking the essential glycoprotein gII can be complemented by glycoprotein I of bovine herpes virus 1. *J. Virol.* **65**, 621–631.

Rinaman, L., Card, J. P., and Enquist, L. W. (1993). Spatiotemporal responses of astrocytes, ramified microglia, and brain macrophages to central neuronal infection with pseudorabies virus. *J. Neurosci.* **13**, 684–702.

Robbins, A. K., Ryan, J. P., Whealy, M. E., and Enquist, L. W. (1989). The gene encoding the gIII envelope protein of pseudorabies virus vaccine strain Bartha contains a mutation affecting protein localization. *J. Virol.* **63**, 250–258.

Rock, D. L. (1993). The molecular basis of latent infections by alphaherpesviruses. *Semin. Virol.* **4**, 157–165.

Roizman, B., and Batterson, W. (1985). Herpes viruses and their replication. In "Virology" (B. Fields *et al.*, Eds.), pp. 497–526. Raven Press, New York.

Roizman, B., Desrosiers, R., Fleckenstein, B., Lopez, C., Minson, A. C., and Studdert, M. J. (1992). The family Herpesviridae: An update. *Arch. Virol.* **123**, 425–449.

Sawitzky, D., Hampl, H., and Habermehl, K.-O. (1990). Comparison of heparin-sensitive attachment of pseudorabies virus (PRV) and herpes simplex virus type 1 and identification of heparin-binding PRV glycoproteins. *J. Gen. Virol.* **71**, 1221–1225.

Sears, A. E., McGwire, B. S., and Roizman, B. (1991). Infection of polarized MDCK cells with herpes simplex virus 1: Two asymmetrically distributed cell receptors interact with different viral proteins. *Proc. Natl. Acad. Sci. USA* **88**, 5087–5091.

Söderberg, C., Giugni, T. D., Zaia, J. A., Larsson, S., Wahlberg, J. M., and Möller, E. (1993). CD13 (human aminopeptidase N) mediates human cytomegalovirus infection. *J. Virol.* **67**, 6476–6585.

Spatz, W. B. (1989). Differences in transneuronal transport of horseradish peroxidase conjugated wheat germ agglutinin in the visual system: Marmoset monkey and guinea pig compared. *J. Hirnforsch.* **30**, 375–384.

Spear, P. G. (1993a). Entry of alphaherpesviruses into cells. *Semin. Virol.* **4**, 167–180.

Spear, P. G. (1993b). Membrane fusion induced by herpes simplex virus. In "Viral Fusion Mechanisms" (J. Bentz, Ed.), pp. 201–232. CRC Press, Boca Raton, FL.

Standish, A., Enquist, L. W., and Schwaber, J. S. (1994). Innervation of the heart and its central medullary origin defined by viral tracing. *Science* **263**, 232–234.

Stevens, J. G. (1991). Herpes simplex virus: Neuroinvasiveness, neurovirulence and latency. *Semin. Neurosci.* **3**, 141–147.

Stipp, C., Litwach, E., and Lander, A. D. (1994). Cerebroglycan: An integral membrane heparan sulfate proteoglycan that is unique to the developing nervous system and expressed specifically during neuronal differentiation. *J. Cell Biol.* **124**, 149–160.

Strack, A. M., and Loewy, A. D. (1990). Pseudorabies virus: A highly specific transneuronal cell body marker in the sympathetic nervous system. *J. Neurosci.* **10**, 2139–2147.

Strack, A. M., Sawyer, W. B., Hughes, J. H., Platt, K. B., and Loewy, A. D. (1989). A general pattern of CNS innervation of the sympathetic outflow demonstrated by transneuronal pseudorabies viral infections. *Brain Res.* **491**, 156–162.

Sunstrum, J. C., Chrisp, C. E., Levine, M., and Glorioso, J. (1988). Pathogenicity of glycoprotein C negative mutants of herpes simplex virus type 1 for the mouse central nervous system. *Virus Res.* **11**, 17–32.

Tyler, K. L. (1987a). Host and viral factors that influence viral neurotropism. I. Viral cell attachment proteins and target cell receptors. *TINS* **10**, 455–460.

Tyler, K. L. (1987b). Host and viral factors that influence viral neurotropism. II. Viral genes, host genes, site of entry and route of spread of virus. *TINS* **10**, 492–497.

Ugolini, G., Kuypers, H. G. J. M., and Strick, P. L. (1989). Transneuronal transfer of herpes virus from peripheral nerves to cortex and brainstem. *Science* **243**, 89–91.

Vahlne, A., Svennerholm, B., Sandberg, M., Hamberger, A., and Lycke, E. (1978). Attachment of herpes simplex virus to neurons and glial cells. *J. Gen. Virol.* **40**, 359–371.

Whealy, M. E., Card, J. P., Robbins, A. K., Dubin, J. R., Rziha, H.-J., and Enquist, L. W. (1993). Specific pseudorabies virus infection of the rat visual system requires both gI and gp63 glycoproteins. *J. Virol.* **67**, 3786–3797.

Whealy, M. E., Robbins, A. K., and Enquist, L. W. (1990). The export pathway of the pseudorabies virus gB homolog gII involves oligomer formation in the endoplasmic reticulum and protease processing in the golgi apparatus. *J. Virol.* **64**, 1946–1955.

Wu, C., Nelson, N., McGeoch, D., and Challberg, M. (1988). Identification of herpes simplex virus type 1 genes required for origin-dependent DNA synthesis. *J. Virol.* **62**, 435–443.

WuDunn, D., and Spear, P. G. (1989). Initial interaction of herpes simplex virus with cells is binding to heparan sulfate. *J. Virol.* **63**, 52–58.

Yuhasz, S. A., and Stevens, J. G. (1993). Glycoprotein B is a specific determinant of herpes simplex virus type 1 neuroinvasiveness. *J. Virol.* **67**, 5948–5954.

Zsak, L., Mettenleiter, Th. C., Sugg, N., and Ben-Porat, T. (1989). Release of pseudorabies virus from infected cells is controlled by several viral functions and is modulated by cellular components. *J. Virol.* **63**, 5475–5477.

Zsak, L., Zuckermann, F., Sugg, N., and Ben-Porat, T. (1992). Glycoprotein gI of pseudorabies virus promotes cell fusion and virus spread via direct cell-to-cell transmission. *J. Virol.* **66**, 2316–2325.

Zuckermann, F., Mettenleiter, Th. C., Schreurs, C., Sugg, N., and Ben-Porat, T. (1988). Complex between glycoprotein gI and gp63 of pseudorabies virus: Its effect on virus replication. *J. Virol.* **62**, 4622–4626.

CHAPTER 21

Molecular Analysis of Rabies and Pseudorabies Neurotropism

Patrice Coulon
Anne Flamand
Laboratoire de Génétique des Virus
CNRS, Gif sur Yvette Cedex, France

I. Introduction

Rabies and pseudorabies viruses are neurotropic viruses that belong respectively to the rhabdovirus or the herpes virus family. They induce sufficiently similar symptoms to justify their closely related common names. In this chapter, we will focus on how these viruses attach, enter, and propagate themselves in neurons. We will also examine mutants of these viruses with modified neurotropism. Finally, we will discuss the use of these two types of viruses as tracers for neuronal network or as vectors for gene therapy in the central nervous system (CNS).

II. General Features of the Molecular Biology of Rabies and Pseudorabies Viruses

Rabies is a negative single-stranded, unsegmented RNA virus with an enveloped bullet-shape virion composed of five different proteins (see Wagner, 1990, for a review). A single external protein that is called glycoprotein G is anchored by its C-terminus in a lipid envelope. It is organized as a trimer and responsible for any interaction that rabies virus makes with host cells during the first steps of the viral cycle. After binding to receptors on the cell membrane, rabies virus enters cells by endocytosis and, after acidification in endosomal

VIRAL VECTORS

vesicles, its envelope fuses with endosomal membranes liberating the nucleocapsid in the cytoplasm. The P and L proteins, which are constituents of the nucleocapsid, then start to transcribe the genome which is wrapped in the N protein. When a sufficient amount of the five-messenger RNAs has been translated, the N protein attaches to the nascent 5′ copy of the genome and replication starts. The viral cycle is entirely cytoplasmic and maturation occurs by budding of newly synthesized nucleocapsids through internal or external cellular membranes containing the viral glycoprotein. In cell culture, the cycle lasts around 24 hr.

Rabies virus infects a variety of mammals, but rodents have become the standard laboratory animals that are used to study the properties of this virus. Rabies is a rare example of a virus which infects almost exclusively neurons. It enters nerve terminals at the point of inoculation and produces transneuronal infections via viral transfer at synapses (Coulon *et al.*, 1989; Kucera *et al.*, 1985; Lafay *et al.*, 1991; Murphy, 1977; Murphy *et al.*, 1973; Ugolini, 1994). Local transfer to nonconnected, adjacent neurons or glial cells does not occur as the virus is propagating an infection in second- and third-order neurons of a neural circuit (Ugolini, 1994). Whether it could later infect nonneuronal cells is currently unknown. Since nonnervous tissues are not (or poorly) permissive to rabies infection, rabies virus is totally dependent on its ability to penetrate nerve terminals at the point of inoculation. In this aspect, rabies virus differs from other neurotropic viruses, such as pseudorabies virus, which also multiply in nonnervous tissues.

Rabies probably has a similar replication cycle in neurons as has been found in cells maintained in culture, but the morphology of the neuron with its extended axonal and dendritic processes imposes some differences. After internalization by endocytosis at nerve terminals, virions are retrogradely transported to the cell body, either as free nucleocapsids or in vesicles, and then undergo replication involving both viral RNA and protein synthesis. Several factors affect the duration of this cycle. For example, the length of axon and rate of axoplasmic transport varies for different classes of neurons, so the replication process does not begin in a temporally precise fashion. When viral synthesis is completed, viral budding occurs in the endoplasmic reticulum and mature virions are subsequently transported centrifugally in both dendrites and the axon inside the vesicles (see Matsumoto, 1975, for review). Virus budding may also occur at synapses as shown in Fig. 1. In this striking electromicrograph, the virion is still attached to the presynaptic membrane and, at the same time, appears to be in the process of being engulfed by the contiguous postsynaptic membrane. This process leading to transneuronal infection of a second-order neuron takes about 20 and 36 hr in the mouse. In contrast, a generalized fatal encephalitis requires about 7 days.

Pseudorabies virus (PrV) is a large double-stranded DNA virus that belongs to the alphaherpesvirus family and codes for more than 70 proteins.

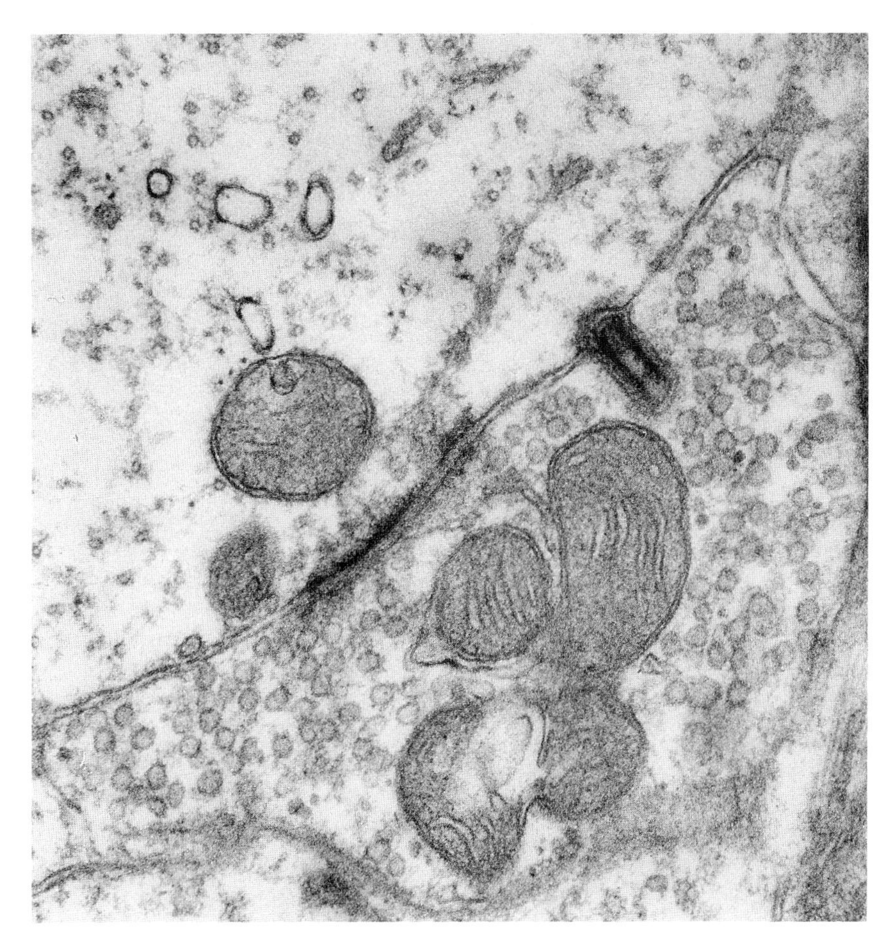

Figure 1 Transsynaptic passage of a rabies virus. The rabies virus is engulfed in a coated pit at a nerve terminal. Reprinted from Charlton and Casey (1979) with permission. Magnification ×60,000.

There are at least eight glycoproteins and several additional transmembrane proteins which may be involved in binding with target cells (Roizman and Sears, 1990). Penetration of this virus probably involves fusion of the viral envelope with the cell membrane and entry of the capsid into the cytoplasm. The capsid is then transported to the nucleus where its delivers the viral genome. Transcription of early and late genes as well as replication and maturation of capsids take place in the cell nucleus. Maturation of enveloped particles occurs first at the internal nuclear membrane after which the virus moves inside vesicles that are directed to the external cell membrane. Electromicroscopic data suggest that virions can be liberated from rough endoplasmic vesicles and wrapped in the Golgi membranes. In fibroblasts maintained in culture, the duration of the viral replication cycle is 8 to 12 hr. In neurons, this cycle is longer because axoplasmic transport of the virions to the nucleus is required before the replication process can begin.

PrV infects many tissues and animals die very rapidly: for example, Swiss mice die within 48–52 hr after inoculation and Balb/c mice survive a little longer. Forty-hour survival gives only time for two, or at most three, replication cycles to occur in the nervous system. Because of this relatively limited neuronal invasion, it is unlikely that animals die from infection of the CNS, although very pronounced symptoms of nervous origin, like the "mad itch," a generalized pruritus, occur shortly before death.

III. Direct Penetration of Rabies and PrV into Peripheral Neurons

Rabies and PrV can enter neurons directly without prior replication in local nonneural tissues (Babic *et al.*, 1994; Coulon *et al.*, 1989). For instance, the CVS strain of rabies does not infect myotubes or other nonneuronal tissues, but seems to penetrate directly into most categories of neurons innervating the site of inoculation (Table 1). Other strains of rabies are capable of limited multiplication at the site of inoculation. In the case of PrV, which infects many nonneuronal tissues, evidence that it is capable of direct penetration into neurons comes from experiments using a deletion mutant form of PrV lacking the gene for essential glycoprotein gB. The gB^- PrV mutants were capable of multiplication when grown in a complementing cell line. Complemented virus, that has the gB glycoprotein properly inserted into its viral membrane but lacks the gB gene, is phenotypically equivalent to wild-type PrV and can infect any cell type that is normally infected by this virus (Rauh *et al.*, 1991). However, the progeny virus, being devoid of the essential gB glycoprotein, is not infectious and cannot propagate by local or transneuronal transfer (Babic *et al.*, 1993). When this mutant was inoculated intranasally, a few neurons in the trigeminal ganglion were infected, indicating that their nerve endings were directly accessible to the virus in the nasal cavity (Babic *et al.*, 1994). Injection of this complemented mutant in the hypoglossal nerve also showed that it could penetrate directly into the axons of motoneurons (Babic *et al.*, 1993). Even though direct penetration can occur, most neurons become infected after local multiplication of PrV occurs at the site of inoculation (Babic *et al.*, 1994).

In order to examine the different types of first-order neurons capable of being infected with rabies or PrV, a series of different tissues was injected with these viruses. These included the anterior chamber of the eye, skeletal muscle, and nasal cavity (Table 1). Each of these has served as a useful model system.

Table 1
Permissivity of Various Nervous Structures to Rabies and Pseudorabies Virus

Route of inoculation	First-order neurons[a] (second-order neurons)	Permissivity for rabies		Permissivity for PrV
		CVS	AvO1	
Intramuscular	**Motoneurons**	+	+	+
	Sensory neurons	+	+	+
	(Interneurons)	+	–	+
Intraocular	**Ciliary ganglion**[b]	+	–	+
	(Edinger–Westphal nucleus)[b]	+	–	+
	Retinopetal fibers	+	–	–
	Trigeminal ganglion	+	+	+
	(Spinal trigeminal nucleus)	–	–	+
	Superior cervical ganglion[c]	–	–	+
	(Intermediolateral nucleus)[c]	–	–	+
Intranasal	**Olfactory receptors**	+	+	±
	(Periglomerular or tufted cells)	+	+	?[e]
	(Mitral cells)	+	–	?
	(Internal plexiform layer)	+	–	?
	(Horizontal diagonal band)	+	+	?
	Pterygopalatine ganglion[b]	ND[d]	ND	+
	(Superior salivatory nucleus)[b]	ND	ND	+
	Superior cervical ganglion[c]	ND	ND	+
	(Intermediolateral nucleus)[c]	ND	ND	+
	Trigeminal ganglion	+	+	+
	(Spinal trigeminal nucleus)	–	–	+

[a] Neurons primarily infectable are listed; corresponding second-order neurons are listed just below in parentheses.
[b] Parasympathetic system.
[c] Sympathetic system.
[d] ND, not done.
[e] Uninfected second-order neurons because of the nonpermissivity of the first-order neurons.

When rabies virus was injected into the anterior chamber of the eye, three types of neurons were infected (parasympathetic neurons, retinal ganglion cells, and trigeminal sensory neurons) but sympathetic neurons were not (Fig. 2) (Kucera *et al.*, 1985; Tsiang *et al.*, 1983). Following intramuscular injection, rabies virus produced infections in both motor and sensory neurons (Coulon *et al.*, 1989). After intranasal instillation, rabies virus infected the olfactory receptor cells and trigeminal sensory system, but since the autonomic nervous system was not examined, it is uncertain whether infections would have been present in the sympathetic and parasympathetic neurons that innervate the nasal cavity (Lafay *et al.*, 1991).

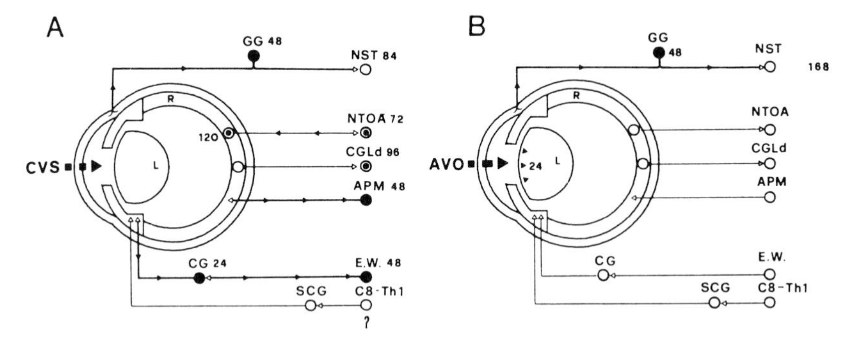

Figure 2 Propagation of the parental (A, abortive dissemination) and mutant (B, generalized encephalitis) rabies strains through the trigeminal (top), visual (center), and autonomic (bottom) interconnections between the eye and brain. Symbols: open arrows, direction of synaptic transmission; solid arrows, direction of propagation of the virus; circles, peripheral and central neuronal somata infected primarily (solid), secondarily (dots), and not infected (open) at each interval of time, indicated in hours after inoculation. Abbreviations: CG, ciliary ganglion; CGLd, lateral geniculate body (dorsal part); C8-Th1, spinal preganglionic sympathetic neurons; GG, trigeminal ganglion of Gasser; L, eye lens; NTOA, terminal nuclei of the accessory optic system; NST, terminal trigeminal sensory nucleus; R, retina; SCG, superior cervical sympathetic ganglion. Reprinted from Kucera *et al.* (1985) with permission.

The pattern of infection produced by PrV has also been studied at the same three sites (Babic *et al.*, 1994; Martin and Dolivo, 1983; Strack and Loewy, 1990). After intraocular inoculation, PrV infected sympathetic neurons of the superior cervical ganglion, parasympathetic neurons of the ciliary ganglion, as well as sensory neurons of the trigeminal ganglion (Table 1) (Martin and Dolivo, 1983). After intranasal injection, PrV infected the nasal mucosa, thus increasing its ability to infect the neurons that innervate this tissue (Babic *et al.*, 1994), including trigeminal sensory neurons. After intramasseter inoculation, all nerve endings were not equally accessible to PrV: the virus can enter into motor end plates and the Golgi tendon organ but not into neuromuscular spindles (Martin and Dolivo, 1983).

Viruses penetrate nerve endings in an extremely inefficient manner. For example, the number of neurons that were infected by rabies or PrV never exceeded several dozen when an inoculum of up to 4×10^7 pfu was used. Thus, a one-to-one correspondence between the number of viruses injected and the number of neurons infected does not occur. At least for PrV, the efficiency of penetration into neurons depends upon viral replication in nonneuronal tissues at the site of inoculation that increases the number of viruses in the proximity of nerve endings.

Infection of first-order neurons can be detected in sensory neurons and motoneurons as early as 18 hr after rabies is injected into skeletal muscle and in trigeminal ganglion cells within 9 hr after PrV is instilled in the nasal cavity (Babic *et al.*, 1994; Coulon *et al.*, 1989). This time takes into account the axonal transport of the viruses from nerve terminals to cell bodies where viral synthesis occurs, as well as the duration of the viral cycle itself, which is shorter in the case of PrV. Finally, it is clear that both viruses can penetrate into axons and dendrites and travel to and from cell bodies in both categories of neuronal processes.

IV. Propagation into the Nervous System

Generally speaking, the interval separating two waves of infection is of the order of 24 hr in the case of rabies and shorter in the case of PrV (around 18 hr). Of course, these are mean values with considerable variations due, for instance, to delayed penetration into nerve terminals or to the duration of the travel to, and from, the cell body. Within the limits of this imprecision, it is clear that second-order neurons become infected several hours after first-order neurons, and this for both viruses eliminates the possibility of passive transfer through neurons without replication. One can clearly observe two or three successive waves of infection affecting anatomically connected chains of neurons. Beyond the third cycle, the picture can be difficult to interpret due to the complexity of neuronal networks.

Viral infection spreads in a transneuronal fashion to infect most categories of neurons that synapse on first-order infected neurons (Table 1) (Ugolini, 1994). However, in the case of rabies virus, second-order neurons in trigeminal sensory nuclei are refractory to infection, at least after intraocular inoculation (Kucera *et al.*, 1985). In contrast, the same neurons become infected by PrV by the same route of administration (Martin and Dolivo, 1983). Similar results are obtained after intranasal inoculation: PrV infects the spinal trigeminal nucleus but rabies does not (Babic *et al.*, 1994; Lafay *et al.*, 1991). The reason for this failure to infect particular classes of neurons is unknown.

However, most neurons are infected by these viruses and this property could be exploited to identify neural connections which are not revealed using conventional neuroanatomical tracers. For example, the neurons of the internal plexiform layer of the olfactory bulb become infected within 48 hr after rabies virus is instilled in the nasal cavity, along with mitral and periglomerular cells, suggesting that internal plexiform layer neurons are also connected to olfactory neurons (Astic *et al.*, 1993).

It is certainly striking that two viruses as different as rabies and pseudorabies, in terms of morphology and molecular biology, have a very similar neurotropism: they infect most categories of peripheral neurons, although with a low efficiency, and propagate in most categories of connected central neurons.

The efficiency of transneuronal transfer is high and the number of infected neurons consistently increases at each successive cycle of viral multiplication.

V. Infection of Glial Cells

During the early stages of rabies infection, neurons containing viral material can be easily identified and no cytopathic effect occurs. Cell bodies, axons and dendrites containing Negri bodies (i.e., round inclusion bodies seen in the cytoplasm of neurons infected with rabies), can be intensively labeled with antibodies directed against rabies nucleocapsid. In the nervous system, rabies virus infects exclusively neurons (Fig. 3A), whereas PrV is capable of infecting both neurons and glial cells. By 48 hr postinfection, PrV-infected neurons are often surrounded by PrV-labeled glial cells (Fig. 3B). Whether replication of PrV in glial cells results in the production of infectious virus is uncertain (see Chapter 18 by Card).

VI. Are There Specific Receptors for Rabies and PrV?

Both rabies and PrV can readily enter peripheral neurons innervating any tissue and produce transneuronal infections in most neural systems. Therefore, the putative viral receptor(s) should be present on most nerve endings. In addition, during transneuronal transfer at synapses, the virus is delivered in close contact with the membrane of the next neuron to be infected (see Fig. 1), thereby eliminating the problem of finding (and attaching to) a suitable permissive cell. For a long time, the acetylcholine receptor which is mostly located on muscle cells has been thought to be the primary site of rabies binding (Lentz, 1985; Lentz *et al.*, 1984). This interaction could be important at the beginning of the infectious process, especially with viral strains which are able to replicate locally in muscle cells, a property which certainly increases their chance to find a nerve ending. However, the acetylcholine receptor cannot be the only rabies binding site since this virus can also bind to neurons which are not cholinergic such as trigeminal and dorsal root sensory neurons.

PrV, as well as HSV-1, interacts with heparin sulfate-like molecules on the cell surface (Mettenleiter *et al.*, 1990; Schnell *et al.*, 1994). This nonspecific attachment is followed by a more specific interaction with a host cell molecule, probably a protein, which has not yet been characterized (Karger and Mettenleiter, 1993). Nothing is known of the series of events that determines irreversible viral binding to neurons. From this point of view, it is interesting to note that glycoprotein gD is essential for penetration of PrV *in vitro* (Peeters *et al.*,

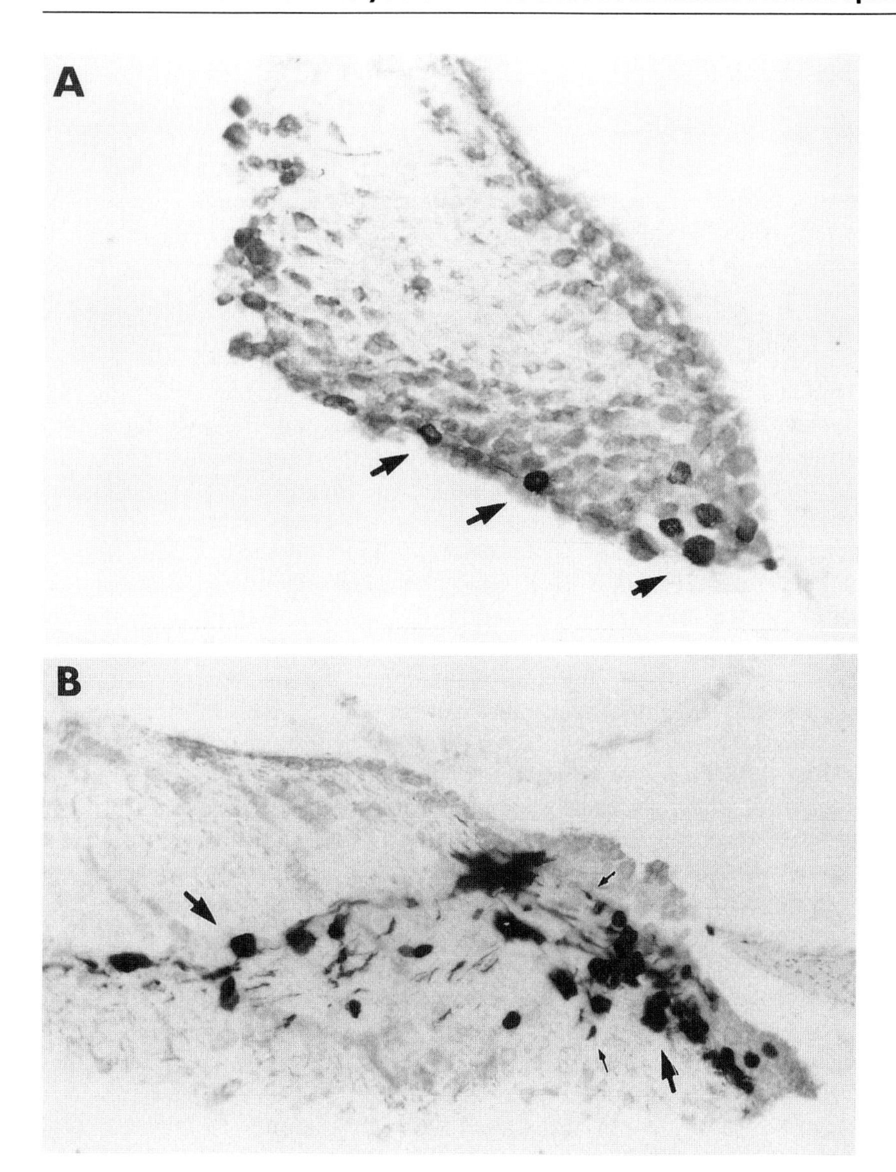

Figure 3 Infection of mouse trigeminal ganglia after intranasal inoculation of rabies or pseudorabies viruses. (A) Immunoperoxidase revelation of rabies virus infection 120 hr after injection of the CVS strain. Only the neurons (arrows) are infected. (B) X-gal reaction of a ganglion 52 hr after infection with the gG^{-} β-gal + mutant of PrV (Mettenleiter and Rauh, 1990). Beside the infection of large neurons (thick arrows), note the infection of small glial cells (thin arrows). Bars = 100 mm.

1992; Rauh and Mettenleiter, 1991) and in neurons at the periphery (Peeters *et al.*, 1993), but is not necessary for transneuronal infections to occur (Babic *et al.*, 1993; Heffner *et al.*, 1993).

After viral attachment, the next step in the infectious process is penetration into the host cell. Rabies virus enters cells through endocytosis, while PrV probably fuses directly at the cell surface. Even at this step, interactions with specific molecules could be necessary, but nothing is known on the subject, although efforts are being made to clarify these points in cell cultures.

Since both viruses enter almost all categories of neurons, we have to postulate that some ubiquitous molecule(s) must be present on every category of nerve ending which will permit viral penetration and, thus, we think that neurotransmitter receptors per se are not necessarily good candidates for this type of function. Interesting observations have come from studies directed at analyzing rabies virus neurotropism. Avirulent rabies mutants have been identified (Coulon *et al.*, 1982; Tuffereau *et al.*, 1989). When tested in animals, one of these mutants (strain AvO1), with a glutamine substituted for an arginine in position 333 of the viral glycoprotein, had a reduced ability to penetrate certain categories of peripheral neurons and was not transferred to certain classes of central neurons (Table 1). For example, it did not penetrate parasympathetic neurons after intraocular inoculation (Fig. 2) (Kucera *et al.*, 1985). After intranasal instillation, it infected olfactory neurons as efficient as wild-type rabies. However, it was not transmitted to mitral cells and to neurons of the anterior olfactory nucleus, although it propagated normally to other categories of second-order olfactory neurons (Table 2) (Lafay *et al.*, 1991). Since maturation of the mutant is not affected, as is generally the case for a glycoprotein mutant, we can postulate that the penetration into certain categories of neurons is difficult or impossible, probably because of lack of interaction between the mutated glycoprotein and molecules which still have to be identified. As a consequence, the virus would not be internalized and/or would not fuse. The fact that there are permissive and nonpermissive neurons for the mutant suggests that there will be at least two different receptor molecules involved in viral penetration.

Similar results have been obtained with PrV mutants. In this case, two categories of mutants have been investigated: mutants with deletions of two essential glycoproteins, gB and gD, are not able to penetrate any cell type including neurons. The gD^- mutant, but not the gB^- mutant, is able to propagate in cell culture and produce transneuronal CNS infections (Babic *et al.*, 1993; Heffner *et al.*, 1993). These mutants are not the most interesting because they behave similarly in nonnervous and nervous cells. Another category of mutants is affected in those glycoproteins which are not essential for multiplication in cell culture. Up to now, only gC^-, gE^-, and gI^- mutants have been

Table 2
Spread of CVS and AvO1 Rabies Viruses Via Olfactory Pathway

		Main olfactory bulb								
		Periglomerular + tufted		Mitral + internal plexiform		AON		HDB		
	Days	L	R	L	R	L	R	L	R	Raphe
CVS	2	0	6	0	0	0	0	0	3	0
	2.5	0	7	0	1	1	3	0	22	0
		0	7	0	0	0	1	0	2	0
	3	1	81	0	67	50	600	0	376	0
		13	32	14	49	31	330	14	385	0
		0	41	0	12	16	59	0	150	0
		0	56	0	33	0	43	0	100	0
		0	116	0	92	168	664	48	880	0
		0	135	0	117	81	405	15	852	0
		0	21	0	34	3	18	0	72	0
		0	6	0	6	0	15	0	135	0
AvO1	3	0	23	0	0	0	0	0	76	0
		0	23	0	0	0	0	0	110	0
		0	105	0	0	0	0	0	200	0
		20	40	0	0	0	0	16	180	0
		0	12	0	0	0	0	0	9	0
		0	12	0	0	0	0	0	192	0

Note. Each line represents one infected animal. Mice were instillated with 4×10^7 PFU of CVS or AvO1 viruses and sacrificed at various times after infection. Serial 30 μm frontal sections of the entire brains were made and the total number of infected neurons was determined after staining with fluorescein-conjugated anti-rabies nucleocapsid antibodies. The various categories of neurons correspond to second-order infected cells, connected to the neurons of the olfactory epithelium, which are the primary targets of CVS and AvO1 infection. Abbreviations: AON, anterior olfactory nucleus; HDB, horizontal limb of the diagonal band; L, left; R, right. Adapted from Lafay *et al.* (1991).

studied after intraocular injection in adult rats. Two of them (gE^- and gI^-) are less neurovirulent than wild-type virus (Card *et al.*, 1992). When injected intraocularly, gE^- and gI^- mutants are able to infect retinal ganglion cells but produce anterograde transneuronal infections only in specific subsets of second-order neurons in visual relay centers, while wild-type PrV (Becker strain) is transported to all known CNS sites that receive input from retinal ganglion cells (Whealy *et al.*, 1993). By this route of inoculation, gC^- mutant behaves like wild-type PrV (Card *et al.*, 1992). These results suggest that interaction between viral glycoprotein(s) gE and/or gI and molecules present at nerve endings is essential for penetration into specific categories of neurons.

VII. Neurotropic Viruses as Potential Tools for Genetic Manipulation of the Nervous System

Our studies, as well as others described in this monograph, show that viruses are unique tools for the study of the nervous system. One of the most topical issues of modern neurovirology is whether they can be used for gene transfer. Not all neurotropic viruses could be considered as suitable vectors. For instance, the fear of rabies will certainly discourage the use of this virus for gene therapy in humans. However, if rabies would be used for gene transfer under laboratory conditions, the feasibility of introducing foreign genes into this virus should be discussed. The rabies genome, like other nonsegmented RNA genomes, is not infectious. The RNA molecule needs to be adequately wrapped into the nucleocapsid protein and transcribed by two viral proteins, P and L, in order to initiate an infectious cycle. Until recently, it was impossible to reconstitute an infectious particle from the transcription of a full-length DNA copy of the genome. This has now been achieved (Schnell *et al.*, 1994), which will allow genetic manipulation of the virus. Introduction of foreign sequences in the rabies genome is then theoretically possible, although it might be impossible to consistently increase the size of the genome due to structural constraints of the virion. In any case, the gene of interest would be transcribed in the cytoplasm.

In the case of PrV, the problem is much simpler; it is well known that many viral genes are nonessential in cell culture and could easily be replaced by a gene of interest without affecting the viability of the virus. A gene carried by an herpes virus will be transcribed in the nucleus like the other viral genes, whether it is expressed as an immediate-early, early, or late gene, and will not be integrated into the cellular genome, which might be an advantage.

The problem of vector pathogenicity also needs to be considered. Rabies virus probably impairs the viability of the neurons (Matsumoto, 1975), although which function(s) is affected is not clear at the moment. After 2 or 3 days of infection, first-order motoneurons infected with rabies mutant AvO1 show signs of degeneration and die (Coulon *et al.*, 1989). At that time, first-order CVS-infected motoneurons are surrounded by more recently infected neurons and consequently are less easily visualized. Rabies virus also has a slow but measurable cytopathic effect in cell culture. Infected neurons are the target of the immune system in a way which is not totally clear but most certainly needs to be considered. Animals probably survive inoculation with avirulent mutants because infected neurons are eliminated by the immune system before the infection becomes too widespread. In the case of PrV, infected neurons degenerate much more rapidly than with rabies (Babic *et al.*, 1993) before the rise of circulating antibodies. Whether the virus kills neurons by inhibiting cellular metabolism or inducing apoptosis is not known, but it is an interesting problem that needs resolution. A picture of an infected oligodendrocyte with characteristic fragmented chromatin may suggest that PrV infection induces apoptosis (Martin and Dolivo, 1983). In any case, PrV would have to be genetically manipulated in order to lower its pathogenicity. Similarly, appropriate mutants of rabies with less cytopathic effect need to be selected and analyzed. In addition, the ideal virus should not stimulate the immune system because the latter will recognize and eliminate infected neurons. It might prove to be impossible to obtain a viral strain which would not have any residual cytopathic effect while keeping at least some ability to multiply and propagate from neuron to neuron. In addition, the presence of viral proteins (or virus-derived peptides) at the surface of infected cells, which would be the target of the immune system, seems to be a potential problem that has not yet been explored.

If fully infectious virus is used, it will propagate and finally invade the totality of the nervous system. To reduce the extent of the infection, specific mutants with reduced tropism for certain categories of neurons could be used, provided that the problem of the survival of infected cells is solved.

Another possibility would be to use defective viruses capable of going through only one cycle of replication in first-order infected cells, like the complemented PrV gB^- mutant. This is feasible with herpes viruses, although the problem of long-term survival of infected neurons is the same as that previously described. If specific genes are to be transferred into peripheral neurons, injection of the vector should be done at the periphery, with the difficulty of a low efficiency of penetration. For certain categories of neurons, direct injection into a particular field of innervation, tissue, or ganglion can be envisaged. This form of restricted infection might increase the chance of success. For transfer to the central nervous system, stereotaxic injections would have to be carried out, but we have no idea of the probability of infecting a sufficient number of relevant neurons.

Finally, hybrid defective vectors can theoretically be engineered with a gene of interest being inserted into a specific sequence, ensuring its transcription and eventually its replication. This artificial genome would have to be wrapped into suitable proteins in order to ensure its penetration and transport from the periphery to the cell body. Of course, for obvious reasons this kind of vector should be neurospecific, but not infectious, which means that it should not be transmitted from neuron to neuron. In other words, one should avoid building "new" infectious entities with unknown pathogenic properties. The rabies glycoprotein, whether wild type or mutant, could theoretically be used to address a particle to neurons. Here again, the problem of penetration efficiency may be crucial, especially if the vector is not replicating because very few, if any, particles will succeed in entering a given nerve terminal. Engineering an "artificial" vector requires an advanced knowledge of the interactions involved in building a viral capsid (or nucleocapsid) wrapped into a functional envelope. Unexpected solutions to this problem may be provided by fundamental research on viral families which currently do not seem promising for gene transfer. Thus, significant progress in gene therapy could result from research on viral tropism, protein–protein and protein–nucleic acid interactions, even in viral families other than adeno-, retro-, and herpes viruses.

Acknowledgments

The authors thank Dr. G. Ugolini for careful reading of the manuscript. Their work is supported by the Centre National de la Recherche Scientifique through the UPR 02431 and by the European Economic Community through contracts CHRX-CT92-0029 and SC1-CT92-0803.

References

Astic, L., Saucier, D., Coulon, P., Lafay, F., and Flamand, A. (1993). The CVS strain of rabies virus as transneuronal tracer in the olfactory system of mice. *Brain Res.* **619,** 146–156.

Babic, N., Mettenleiter, T. C., Flamand, A., and Ugolini, G. (1993). Role of essential glycoproteins gII and gp50 in transneuronal transfer of pseudorabies virus from the hypoglossal nerves of mice. *J. Virol.* **67,** 4421–4426.

Babic, N., Mettenleiter, T. C., Ugolini, G., Flamand, A., and Coulon, P. (1994). Propagation of pseudorabies virus in the nervous system of the mouse after intranasal inoculation. *Virology* **204,** 616–625.

Card, J. P., Whealy, M. E., Robbins, A. K., and Enquist, L. W. (1992). Pseudorabies virus envelope glycoprotein-gI influences both neurotropism and virulence during infection of the rat visual system. *J. Virol.* **66,** 3032–3041.

Charlton, K. M., and Casey, G. A. (1979). Experimental rabies in skunks. Immunofluorescence, light and electron microscopic studies. *Lab. Invest.* **41,** 36–44.

Coulon, P., Derbin, C., Kucera, P., Lafay, F., Préhaud, C., and Flamand, A. (1989). Invasion of the peripheral nervous system of adult mice by the CVS strain of rabies virus and its avirulent derivative AvO1. *J. Virol.* **63,** 3550–3554.

Coulon, P., Rollin, P., Aubert, M., and Flamand, A. (1982). Molecular basis of rabies virus virulence. I. Selection of avirulent mutants of the CVS strain with anti-G monoclonal antibodies. *J. Gen. Virol.* **61,** 97–100.

Heffner, S., Kovacs, F., Klupp, B. G., and Mettenleiter, T. C. (1993). Glycoprotein gp50-negative pseudorabies virus—A novel approach toward a nonspreading live herpesvirus vaccine. *J. Virol.* **67,** 1529–1537.

Karger, A., and Mettenleiter, T. C. (1993). Glycoprotein-gIII and glycoprotein-gp50 play dominant roles in the biphasic attachment of pseudorabies virus. *Virology* **194,** 654–664.

Kucera, P., Dolivo, M., Coulon, P., and Flamand, A. (1985). Pathways of the early propagation of virulent and avirulent rabies virus strains from the eye to the brain. *J. Virol.* **55,** 158–162.

Lafay, F., Coulon, P., Astic, L., *et al.* (1991). Spread of the CVS strain of rabies virus and of the avirulent mutant AvO1 along the olfactory pathways of the mouse after intranasal inoculation. *Virology* **183,** 320–330.

Lentz, T. L. (1985). Rabies virus receptors. *Trends Neurosci.* **8,** 360–364.

Lentz, T. L., Wilson, P. T., Hawrot, E., and Speicher, D. W. (1984). Amino acid sequence similarity between rabies virus glycoprotein and snake venom curaremimetic neurotoxins. *Science* **226**, 847–848.

Martin, X., and Dolivo, M. (1983). Neuronal and transneuronal tracing in the trigeminal system of the rat using the herpes virus suis. *Brain Res.* **273**, 253–276.

Matsumoto, S. (1975). Electron microscopy of central nervous system infection. In "The Natural History of Rabies" (G. M. Baer, Ed.), pp. 217–233. Academic Press, New York.

Mettenleiter, T. C., and Rauh, I. (1990). A glycoprotein gX$^-$$\beta$-galactosidase fusion gene as insertional marker for rapid identification of pseudorabies virus mutants. *J. Virol. Methods* **30**, 55–66.

Mettenleiter, T. C., Zsak, L., Zuckermann, F., Sugg, N., Kern, H., and Ben-Porat, T. (1990). Interaction of glycoprotein gIII with a cellular heparinlike substance mediates adsorption of pseudorabies virus. *J. Virol.* **64**, 278–286.

Murphy, F. A. (1977). Rabies pathogenesis. *Arch. Virol.* **54**, 279–297.

Murphy, F. A., Bauer, S. P., Harrison, A. K., and Winn, W. C., Jr. (1973). Comparative pathogenesis of rabies and rabies-like viruses: Viral infection and transit from inoculation site to the central nervous system. *Lab. Invest.* **28**, 361–376.

Peeters, B., de Wind, N., Hooisma, M., Wagenaar, F., Gielkens, A., and Moormann, R. (1992). Pseudorabies virus envelope glycoproteins gp50 and gII are essential for virus penetration, but only gII is involved in membrane fusion. *J. Virol.* **66**, 894–905.

Peeters, B., Pol, J., Gielkens, A., and Moormann, R. (1993). Envelope glycoprotein-gp50 of pseudorabies virus is essential for virus entry but is not required for viral spread in mice. *J. Virol.* **67**, 170–177.

Rauh, I., and Mettenleiter, T. C. (1991). Pseudorabies virus glycoproteins-gII and gp50 are essential for virus penetration. *J. Virol.* **65**, 5348–5356.

Rauh, I., Weiland, F., Fehler, F., Keil, G. M., and Mettenleiter, T. C. (1991). Pseudorabies mutants lacking the essential glycoprotein gII can be complemented by glycoprotein gI of bovine herpesvirus 1. *J. Virol.* **65**, 621–631.

Roizman, B., and Sears, A. E. (1990). Herpes simplex viruses and their replication. In "Virology" (B. N. Fields and D. M. Knipe, Eds.), pp. 1795–1841. Raven Press, New York.

Schnell, M. J., Mebatsion, T., and Conzelmann, K. K. (1994). Infectious rabies viruses from cloned cDNA. *Embo. J.* **13**, 4195–4203.

Shieh, M. T., WuDunn, D., Montgomery, R. I., Esko, J., and Spears, P. G. (1992). Cell surface receptors for herpes simplex virus are heparin sulfate proteoglycans. *J. Cell Biol.* **116**, 1273–1281.

Strack, A. M., and Loewy, A. D. (1990). Pseudorabies virus: A highly specific transneuronal cell body marker in the sympathetic nervous system. *J. Neurosci.* **10**, 2139–2147.

Tsiang, H., Derer, M., and Taxi, J. (1983). An in vivo and in vitro study of rabies virus infection of the rat superior cervical ganglia. *Arch. Virol.* **76**, 231–243.

Tuffereau, C., Leblois, H., Bénéjean, J., Coulon, P., Lafay, F., and Flamand, A. (1989). Arginine or lysine in position 333 of ERA and CVS glycoprotein is necessary for rabies virulence in mice. *Virology* **172**, 206–212.

Ugolini, G. (1994). Specificity of rabies virus as a transneuronal tracer of motor networks: Transfer from hypoglosomal motoneurons to connected second order and third order CNS cell groups. Submitted for publication.

Wagner, R. R. (1990). Rhabdoviridae and their replication. In "Virology" (B. N. Fields and D. M. Knipe, Eds.), pp. 867–881. Raven Press, New York.

Whealy, M. E., Card, J. P., Robbins, A. K., Dubin, J. R., Rziha, H-J., and Enquist, L. W. (1993). Specific pseudorabies virus infection of the rat visual system requires both gI and gp63 glycoproteins. *J. Virol.* **67**, 3786–3797.

CHAPTER 22

The Use of Retroviral Vectors in the Study of Cell Lineage and Migration during the Development of the Mammalian Central Nervous System

Kieran W. McDermott
Department of Anatomy
University College
Cork, Ireland

Marla B. Luskin
Department of Anatomy and Cell Biology
Emory University School of Medicine
Atlanta, Georgia

I. Introduction

Over the past decade or so the continuing emergence of molecular biological tools has proven of enormous benefit to the study of the development of the nervous system. Not the least among these newly introduced tools has been the modification of the genome of a retrovirus, as a means of permanently labeling clonally related populations of cells. This technology has proven invaluable in resolving certain questions in developmental neurobiology about how a highly complex interplay of genetic versus microenvironmental factors generate the vast array of cell types that exist in the immature and mature central nervous system (CNS). This chapter will review the application of retroviral-mediated gene transfer to studies concerned with cell lineage, commitment, migration, and differentiation in the CNS.

A. The Retroviral Life Cycle

Naturally occurring retroviruses contain single-stranded RNA which is encapsulated in a virally encoded protein shell and surrounded by a lipid bilayer derived from the host cell membrane (for general review of retroviruses see Levy, 1992). Retroviruses capable of infecting mammals, birds, reptiles, and fish have been described and all share a similar life cycle. Specific viral coat glycoproteins, encoded by the viral *env* gene, enable the retrovirus to interact

VIRAL VECTORS

with a receptor on its host cell membrane (White, 1990). This interaction culminates in the delivery of the viral particle into the cell's cytoplasm, thereby initiating the next phase of its life cycle. Once in the cell, uncoating takes place by poorly understood mechanisms and the unique retroviral enzyme, reverse transcriptase, encoded by the *pol* gene, begins transcribing the newly released viral RNA genome into double-stranded DNA (Baltimore, 1970; Temin and Mizutani, 1970). Viral enzymes then process the newly synthesized viral DNA and integrate it into the host cell DNA. Although it appears that most sites in the host genome are used for insertion, there may be certain sequences which are preferred targets for integration. Furthermore, synthesis and integration of viral DNA occurs much more efficiently in actively dividing cells than in nondividing cells. This precise and stable integration of the entire viral genome into the host cell DNA is of great importance in adapting retroviruses for use as gene vectors in biological manipulations. Once integrated, the provirus, as the inserted retroviral double-stranded DNA genome is now called, mainly utilizes the host cell transcription and translation apparatus to express the retroviral structural genes. The newly synthesized viral proteins, gag (capsid protein), pol (reverse transcriptase), and env (envelope glycoprotein), subsequently associate with viral genomic RNA via a special packaging sequence, called psi. The packaged proteins and RNA then interact with the host cell membrane resulting in the encapsidation and release, by budding off from the cell surface, of mature infectious virions.

B. Retroviral Vectors

As the wild-type retrovirus has the ability to convey genes efficiently and precisely into the genome of vertebrate cells which can subsequently express these genes at high levels, it is an obvious candidate for adaptation as a tool for experimentally introducing foreign genes into cells (for review of retroviral vector construction, production, and application see Stoker, 1993). This strategy has enormous potential in the study of normal vertebrate development and as a way to perform gene therapy in humans to treat inherited diseases.

The use of retroviruses as vectors to transduce genes into cells was made possible once techniques became available that enabled proviruses to be cloned from host cells. The cloned viral genome could then be manipulated as a bacterial plasmid. This technology has proceeded to yield a variety of useful engineered viral constructs. The usefulness of these constructs is that nonviral genes, selected according to requirements of the particular biological investigation, can be used to replace certain elements of the viral genome, usually including but not limited to the genes for the protein envelope. In replication-competent vectors the construct can retain certain essential *cis*-acting sequences such as the psi packaging signal, reverse transcription signal, and other transcriptional elements necessary for the viral life cycle. However, demand existed also, mostly from researchers studying early ontogenesis in the nervous system, for a gene-transfer system in which the retroviral vector was incapable of replication in the infected cell. This would restrict the transduced gene to the infected cell and its descendants—a heritable marker ideally suited for lineage studies.

In replication-incompetent vectors the genes required for the assembly of new viral particles, the genes *gag, pol,* and *env,* are deleted, leaving the nonviral gene(s), the packaging signal, and other promoter regions in the long terminal repeats which flank the genome intact. To produce infectious viral particles tailor-made "packaging" cell lines are used in which the necessary viral proteins are supplied in *trans* orientation from another transfected retroviral construct which expresses the viral proteins from RNA lacking the psi packaging signal. When the DNA construct of a retroviral vector containing the transgene of interest adjoining the packaging sequence is also transfected into the packaging cell line, the viral proteins can interact with the packaging signal allied with the vector and assemble new infectious vector\virus particles. Crucially, these viruses can infect cells within their host range and integrate into the host DNA just like wild-type virus but they lack the genes *gag, pol,* and *env* and so are incapable of reassembling new virus particles, thereby trapping the transgene in the infected cell. For stable integration and expression of the foreign gene the cell must undergo at least one cycle of division. The virally transduced genes are inherited, transcribed, and translated just like the host genes.

The wild-type viruses which have been most successfully used to produce vectors for lineage tracing in the nervous system include the murine leukemia viruses (Sanes *et al.,* 1986; Price *et al.,* 1987) and the avian sarcoma and leukosis viruses (Gray *et al.,* 1988). Newer replication-defective recombinant retroviruses continue to be introduced (Bonnerot *et al.,* 1987; Galileo *et al.,* 1990), often with applications to a broader range of species (Lin *et al.,* 1994). At the present time though, for studies of cell lineage in the nervous system, replication-incompetent vectors derived from the Moloney murine leukemia virus, which only infect rodent cells (ecotropic host range) and which use the enzymes *Escherichia coli* ß-galactosidase (*lac*Z) and alkaline phosphatase as markers, are the most widely used. Of these the BAG vector which encodes *E. coli lac*Z, the gene for ß-gal (Price *et al.,* 1987), and the DAP vector, encoding P ALP-1, the gene for placental alkaline phosphatase (Fields-Berry *et al.,* 1992), are the most commonly used at the present time. Both of these enzymes are either absent or present only at very low background levels in rodent nervous tissue and are easily detected in the progeny of infected cells using straightforward histochemical reactions or by immunocytochemistry.

C. Retroviral Vectors and Lineage Tracing

During ontogeny a vast amount of cell proliferation, migration, intracellular signaling, and a multitude of other cellular events are taking place which are governed by the switching on and off of genes. These events appear to follow distinct spatial and temporal gradients. Analysis of the control of ontogenetic development involves trying to separate out individual aspects of the overall developmental program for a given tissue or organism. One such aspect which may play a particularly prominent role is cell lineage. Lineage by definition imposes a genetically predetermined differentiation program on sets of lineally related cells. However, the ability to study lineage *in vivo* has long been impaired by the lack of suitable ways of marking precursor cells so that their descendants, many cells divisions further along in ontogeny, can be distinguished from their neighbors. In the past, tritiated thymidine and fluorescent dyes have been used for this purpose but their attendant disadvantages of label dilution with every division and fluorescence fading greatly restricted their use in experiments of longer duration. The development of retroviral vectors has essentially solved many of the problems associated with earlier tracers (although it is not without its own) and it is nowadays the technique of choice by many investigators involved in studies of cell lineage.

A number of features of replication-defective retroviruses make them ideally suited as markers to elucidate cell lineage in vertebrate tissues. First, they can efficiently transfer genetic material into dividing vertebrate cells. The integration of the viral genome into cellular DNA appears quite stable and strong expression of the transgene is commonly achieved. Furthermore, there are few detectable side effects in the infected cells provided an innocuous gene is chosen for transduction. A wide range of species and tissues can be infected and successful incorporation has been achieved *in vivo* and *in vitro* including in embryonic tissues. Perhaps of greatest importance for lineage marking is the fact that, once the transduced gene is incorporated into the host genome, the vector is incapable of replication, thus ensuring that no other cells except descendants of infected cells can express the reporter gene.

A typical strategy in cell lineage tracing experiments *in vivo,* for example, in the developing nervous system, is to introduce a small quantity of retroviral vectors into an area where the proliferating cells of interest are located. The retroviral vectors are obtained by harvesting the media containing secreted viral particles from the retroviral packaging lines described earlier. Among the more commonly used are the psi 2 and CRE fibroblast lines which package the BAG (lacZ expressing) vector (Price *et al.,* 1987; Danos and Mulligan, 1988). The DAP vector expressing the alkaline phosphatase gene is also packaged in psi 2 (Fields-Berry *et al.,* 1992). The retrovirus-containing supernatants can be further concentrated by ultracentrifugation and stored below -180°C until required. In most instances when studying lineage *in vivo* it is desirable to use a low retroviral titer so as to ensure that the number of precursor cells in a tissue which are infected is relatively small. By keeping the "hit rate" low, the chances of different clones of marked progeny, all descendants of the infected precursors, intermixing as development proceeds is minimized, although it is not entirely eliminated (Luskin *et al.*, 1993). This type of lineage analysis is essentially retrospective and there has been much discussion about whether or not the marked groups observed are indeed complete genetic clones each derived from a single infection event (vide infra).

Finally, once sufficient time for migration and differentiation of the progeny of infected cells has elapsed, the tissue can be removed, sectioned, and histochemically or immunocytochemically stained for the marker enzyme. Usually the end point in cell lineage experiments is the identification of the position and phenotype of all the marked progeny of a proportion of infected cells. Clonal analysis of this kind can provide direct evidence as to whether precursor cells are committed to a particular differentiation pathway (for review see Luskin, 1994). For example, if all clonally related cells bear the same phenotype it suggests that the precursor from which they originated was committed to production of that type of cell exclusively. This is especially valid if phenotypically different clones coexist in what appears to be a similar micoenvironment. Here we have chosen to focus on the CNS for the remainder of the review. However, retroviruses are also being used to elucidate the lineage of cells elsewhere in the nervous system to considerable advantage [e.g., in the olfactory epithelium (Caggiano *et al.*, 1994; Schwob *et al.*, 1994 and in the sympathetic nervous system (Hall and Landis, 1991)]. Unfortunately, space does not permit a fully comprehensive review of all lineage studies that have used retroviruses.

II. Cell Lineage in the Central Nervous System

The CNS is an example of a complex tissue system in which a diverse array of cell types act in concert to produce the integrated functions of the tissue. Understanding how the diverse cell types in the CNS are generated has been a long-standing goal in developmental neurobiology. It has been difficult, however, to unravel the mechanisms that control the generation and differentiation of these different cell types. This is largely because of the huge regional heterogeneity in the CNS and the considerable displacement of cells during embryogenesis from their birthplaces in specialized germinal zones to their final locations in the mature CNS.

One structure which has attracted much attention, not the least because its highly organized cytoachitecture is amenable to analysis, is the mammalian cerebral cortex. Numerous studies have elucidated the orderly "inside first–outside last" pattern of cell migration that eventually gives rise to the laminated cerebral cortex (McConnell, 1988; Rakic, 1988). Essentially, during precise periods of prenatal development, cells that will eventually form the cerebral cortex arise from a germinal zone surrounding the lateral ventricles and migrate toward the pial surface using a radial network of specialized glial cells, the so-called radial glia, as a means of guidance for at least part of the route to their final destinations. The postmitotic neurons complete their differentiation in the growing cortical plate. Gradually, the cortical plate becomes organized into layers wherein neurons born earlier lie deep to those generated later in development. Consequently, the latest-born neurons occupy the most superficial layer.

Cortical neurons are phenotypically distinguishable by their morphology, laminar position, physiological functions, and neurotransmitter profile. Nevertheless, despite a basic understanding of the mature cell types as well as their spatial and temporal patterns of proliferation and migration which characterize cortical histogenesis, our knowledge of how cortical neurons acquire their ultimate identities has until very recently, been sketchy. Likewise, the same questions can be asked about the generation of neuronal diversity in other parts of the CNS.

Besides the neurons, the other major cell types of neuroectodermal origin in the CNS are the glia (astrocytes and oligodendrocytes) which early in ontogeny must share ancestors with neurons, but which also must diverge from the neuronal lineage at some point. Progenitor cells in the germinal zone and their migrating progeny presumably must make a series of "decisions" which progressively restricts the fate of the cells (Luskin, 1994). Are these coordinated decisions regulated by interaction with environmental cues, such as contact with extracellular matrix molecules, other cells, and neurotrophic molecules, or solely by inherited factors, or are elements of both strategies required to achieve the complex array of cell types in the CNS? The design and availability of the retroviral vectors described previously has made it possible to begin to address these questions and recent investigations have utilized retroviral-mediated marker gene transfer into neural progenitor cells in the avian and mammalian brain both *in vivo* and *in vitro*. The primary aim of many of these studies has been to ascertain if and when separate committed progenitor cells for the major neural subtypes first appear in different parts of the developing CNS. Moreover, in many instances much information has also been obtained about the migration patterns of neural precursors and their descendants in the different parts of the CNS.

A. Tracing Cell Lineage in the Central Nervous System Using Retroviral Vectors

Sanes *et al.* (1986) first utilized recombinant retroviruses which encoded *lacZ* driven by the SV40 early promoter to investigate cell lineage in E7–E11 mouse embryos. The retrovirus was delivered through the uterine wall into the uterine cavity, and a few days later labeled cells were detected in whole mounts of embryonic tissue stained by X-gal. Due to the short survival time precluding large-scale cell migration, small clusters of labeled cells, representing the clonal progeny of infected progenitor cells, were observed. These were seen in the yolk sac, skin and amnion, and in the brain following intracranial injections at E12 and E13. Outside the CNS clones were always composed of cells of a single primary germ layer with different cell types from a particular layer sometimes found together within a clone. It was noticed that clone complexity decreased at later injection times. This latter observation is important as it confirmed the usefulness of this system in elucidating the timing of crucial events in the developing embryo such as the gradual lineage restrictions limiting cell fate. The study showed that it is also possible to label cells in the brain but that in the CNS identification of cell types poses special problems, a recurring theme in lineage studies in the nervous system.

Lineage analysis in the postnatal rat retina was facilitated by the highly ordered laminar arrangement of cells in the retina, their restricted migration, and the morphologically distinguishable cells types found in the different layers. A systematic analysis of lineage relationships among cells in the retina was undertaken by Cepko and co-workers who demonstrated, using the BAG recombinant retrovirus encoding *lacZ* driven by the LTR promoter, that retrovirally marked neural progenitors in the retina *in vivo* can generate clones of mixed phenotype. For example, numerous two-cell clones were identified which contained a Muller glial cell and a rod photoreceptor (Price *et al.*, 1987; Turner and Cepko, 1987). Moreover, a mixture of retinal cell types, with both neuronal and glial identities, was commonly generated (Turner *et al.*, 1990) following infection of retinal progenitor cells in the embryonic brain. These findings suggest that lineage appears to play, at most, a limited role in the generation of cell diversity in the retina. Instead, multipotential progenitor cells seem to predominate during retinal histogenesis. What finally determines cell fate in the retina remains unclear, but presumably position-dependent cues or cell–cell interactions around about the time of final mitosis or at a cell's final postmigratory location must play a significant role.

These observations naturally encouraged the view that the role of lineage was not central to cell determination in the CNS. Multipotential progenitor cells have also been found in other regions of the CNS using retroviral lineage marking. Mixed clones of several types of neurons, astrocytes, and radial glia were observed in the chick optic tectum following embryonic injections of retroviral marker into the tectal ventricle (Gray *et al.*, 1988; Galileo *et al.*, 1990; Gray *et al.*, 1990; Gray and Sanes, 1991, 1992). These workers concluded that a common multipotential progenitor or stem cell can produce neurons, astrocytes, and radial glia and that radial glia which may in fact act as stem cells may also be involved in guiding the migration of their clonal relatives (Gray and Sanes, 1992). The question of migration of clonal progeny is dealt with later in this chapter but the findings of these studies if universally true for other laminated regions of the CNS begin to throw light on the ontogenetic mechanisms which underscore their unique layered cytoarchitecture.

Evidence for lineage playing a more important role has, however, been obtained from retroviral lineage marking studies in other parts of the CNS. In order to study lineage in the developing murine cerebral cortex, Luskin *et al.* (1988) performed retroviral injections into the telencephalic vesicles of E12–E14 mice and analyzed marked cells up to 4 weeks later. Identification of cells was based on the light microscopic morphology of lacZ-positive cells and in a parallel *in vitro* study on immunocytochemical staining. Contrary to the findings for the retina and the chick optic tectum, in this study nearly all clones contained either neurons or glia, but not a mixture of both. This implied that lineage restrictions are being activated early in cortical development thereby specifying the fate of these progenitor cells and their progeny. An alternative, though perhaps less satisfactory, explanation could be that during development microenvironmental molecular cues commit the descendants of multipotential progenitors to particular phenotypic pathways; descendants moving off in different directions after being produced in the ventricular zone could be induced to differentiate along a number of different pathways. However, many neuronal clones containing similar cells seen by Luskin *et al.* (1988) were radially dispersed across different cortical laminae and, as such laminae presumably present different microenvironments, it is hard to reconcile this with the latter hypothesis. In similar studies in developing rat cerebral cortex (Price and Thurlow, 1988) and hippocampus (Grove *et al.*, 1992), discrete clusters, considered clones, were observed which were nearly always composed of phenotypically identical cells. These clusters either contained neurons or else they contained glia with the exception of a small number of clones which contained both neurons and white matter cells which were presumably glia. Exclusively glial clones could be further subdivided into gray matter astrocytes or white matter cells. Although these results generally support the notion of lineally committed progenitors being present in the cortical germinal zone, information on cell identity is often no more than inferred, thus increasing the possibility that clones may not be homogeneous.

The unequivocal identification of retrovirally marked cells has been unexpectedly troublesome. Initially, only cell morphology as demonstrated by X-gal histochemistry was used but it was soon realized that incomplete staining of cell processes and other sources of ambiguity posed a significant problem (Cepko, 1988). Attention was thus focused on methods of authenticating cell identities within clones as otherwise retroviral labeling as a research tool in the study of neural cell lineage is of relatively limited value.

Further clonal analyses in the developing cortex have been carried out in which identification of cells has not been confined to the light microscopic morphology of the lacZ-positive cells visualized histochemically. Instead, there has been an increasing reliance on more definitive techniques, often involving a combination of approaches including electron microscopy and intracellular dye injection. Furthermore, several investigators have examined with immunohistochemical markers the phenotype of the progeny of progenitor cells from the telencephalon infected with retrovirus *in vitro*.

In cultures of E16 rat telencephalon which were retrovirally infected and subsequently immunostained with neuronal and glial specific antibodies, Williams *et al.* (1991) found mostly clones composed of a single cell type, confirming earlier *in vivo* studies (Luskin *et al.*, 1988). In a similar analysis of glial lineage in cultures prepared from P1 striatum mixed clones of both astrocytes and oligodendrocytes were exceedingly rare (Vaysse and Goldman, 1990), suggesting that the main separate lineages exist for astrocytes and oligodendrocytes. However, Williams *et al.* (1991) also identified a very small number of clones which contained both neurons and oligodendrocytes, a highly unexpected pairing. Such clones were only seen when cells were dissociated and cultured prior to the onset of neurogenesis. If cultures and infections were made after E18 no mixed clones were seen. Even so, the rarity of these clones tends to argue against this being a significant neural lineage and perhaps relects the increased propensity for cluster\clone overlap which dense cultures allow or artefactual progenitor plasticity induced by culture conditions. Indirect evidence for such plasticity may also be discerned from *in vitro* retroviral lineage tracing experiments in which isolated glial progenitors are switched from medium containing low (1%) serum to medium with higher (10%) serum concentration (Lubetzki *et al.*, 1992). In low serum clones were never mixed and were composed solely of either immunocytochemically characterized astrocytes or oligodendrocytes, but when cultures were switched from low to high serum concentrations a small number of heterogeneous clones emerged. The longer the cells were maintained in low serum before the switch the fewer the mixed clones that appeared. These results can be viewed (as Lubetzki *et al.* do) as suggesting a strong microenvironmental influence in the determination of glial phenotype, but as the manipulation tested in the experiment does not attempt to mimic any particular *in vivo* microenvinomental features likely to be encountered by progenitors, artificially induced plasticity could equally well explain the observations.

Similar clones to those seen by Williams *et al.* (1991) containing both neurons and white matter glia have also been detected *in vivo* in lineage tracing studies in the developing forebrain. In embryos injected at E16 and analyzed at either P14 or P28/29 using lacZ staining and intracellular injections of Lucifer Yellow to confirm the identify of some neurons approximately 6% of clones were reported to contain a mixture of neurons and white matter glia (Grove *et al.*, 1993). It has been proposed previously that such mixed neuronal–glial clones may be a function of the age at which infections are made. Presumably, if infections are made early enough, before neuronal and glia lineages diverge, then such mixed clones could theoretically be encountered. One would not expect them after most neurons have been produced when neuronal and glial lineages have separated and, indeed, Williams *et al.* (1991)

did not identify such clones after E18 in the rat, when neurogenesis has largely ceased. Therefore, it was very surprising when similar clones were reported in a study in which the developmental fates of *postnatal* subventricular zone cells were analyzed in rats injected postnatally into the subventricular zone and allowed to develop for a further 14 days. In this study a very small proportion (1%) of the resultant clones was deemed to contain a mixture of neurons and glia (Levison and Goldman, 1993). Both types of glia, astrocytes and oligodendrocytes, were found with neurons in these clones and although identification of neurons was based solely on lacZ staining, immunocytochemistry of GFAP and CAII was used to identify astrocytes and oligodendrocytes, respectively. Of the other clones identified in this study about 15% contained mixtures of astrocytes and oligodendrocytes. Collectively, from these results, bi- or multipotential progenitor cells in the neuroepithelium of the pre- and postnatal telencephalon cannot be dismissed.

In another series of studies we have adopted a slightly different approach in tackling this vexed question of neuronal and glia fate determination, lineage divergence, and progenitor cell committment. These studies have looked at neuron and glial lineage in the cerebral cortex by tagging cells with a retrovirus lineage tracer both in the pre- and postnatal brain. However, so as to ensure rigorous identification of every cell in a clone, the tissues were processed so that all cells comprising clones could be analyzed at both light and electron microscopic levels and identified according to ultrastructural criteria. This has allowed cell-type determination to be made more definitively. The X-gal histochemical reaction product which appears blue at the LM level is also electron dense and easily detectable in sections which have been lightly osmicated and counterstained (Fig. 1) (Luskin *et al.*, 1993). Analysis of mature brains from rats injected with retrovirus at E16 and allowed to survive for 1 to 3 months postnatally revealed that by E16, close to the onset of neurogenesis, neuronal and glial lineages have diverged (Luskin *et al.*, 1993). In these brains closely spaced clusters of lacZ-positive cells were designated as clones when they were separated from any other stained cell by at least 500 μm [for critique of the criteria used for clonal designation see Luskin (1993) and Luskin *et al.* (1993)]. Ultrathin sections were prepared from each cell in a clone permitting all constituent cells to be classified according to their ultrastructural appearance (Fig. 1). In most cases it would not have been possible to identify the cells by their light microscopic morphology alone. This level of analysis revealed that all clones, bar one, were homogeneously composed of astrocytes, oligodendrocytes, or neurons, indicating that virtually all the progeny of an individual progenitor cell infected in the ventricular zone were of the same type (Luskin, 1993; Luskin *et al.*, 1993).

The approach has also been used to further explore lineage relationships among different subtypes of neurons in the CNS, such as between the morphologically and neurochemically distinct pyramidal neuron and interneuron (nonpyramidal neuron) populations of the cerebral cortex (Parnavelas *et al.,* 1991). Clones of retrovirally marked neurons were generated by embryonic intraventricular injections of the lacZ-encoded retrovirus lineage tracer. Relying on unequivocal ultrastructural evidence, such as the type of axosomatic synapses a cell received which differ for pyramidal and nonpyramidal neurons (Fig. 1), all cells were identified as either pyramidal or nonpyramidal neurons. A further indication that these were clonally related cells was that, typically, all the cells of a clone presented a very similar density and intracellular distribution of reaction product, whereas these parameters varied considerably between obvious clusters (Luskin *et al.,* 1993). Parnavelas *et al.* (1991) found that these clones were composed of either all pyramidal neurons or all nonpyramidal neurons with only one exception. Subsequent immunocytochemical staining with antibodies against glutamate and GABA, the neurotransmitters used by pyramidal and nonpyramidal neurons respectively, has substantiated these findings (Mione *et al.,* 1994). This further supports the notion that there are separate progenitor cells for pyramidal and nonpyramidal neurons, and that their divergence occurs by the onset of neurogenesis. Furthermore, considering the numerical composition of clones, this commitment must occur a number of divisions prior to the final division which produces at most two postmitotic neurons.

Decisions about where a newborn cortical neuron will eventually reside in the emerging layers of neocortex, i.e., its laminar fate, appear to be made shortly before the mitotic division which precedes it becoming postmitotic (McConnell and Kaznowski, 1991). The decision to become a neuronal or a glial cell or a pyramidal cell or nonpyramidal cell, as has been seen, precedes the decision about laminar fate. So it would appear that while a progenitor cell is still in the ventricular zone decisions are made in series about its progeny's ultimate fate. This decision-making mechanism is crucial to the question of the role of inherited factors versus environment in fate determination as each decision seems to progressively restrict the fate of these cells. Once taken, these decisions are seemingly irreversible and the progenitor cell is committed to generating progeny of a specified phenotype, as has been elegantly demonstrated for how laminar position is decided. Once the period during which this decision is made has elapsed, ventricular zone cells, if transplanted heterochronously into another ventricular layer, will still acquire the characteristic phenotype of the layer to which they have already been committed (McConnell and Kaznowski, 1991). Moreover, the retroviral lineage tracing studies show that earlier decisions result in the generation of homogeneous clones of progeny. This suggests that, at the very least, no environmental factor acted, after the decision was made, to divert any cell from the pathway of differentiation being followed by its siblings.

An environmental influence, acting on cells in the ventricular zone, on decisions which restrict lineages cannot be ruled out, but it has not been

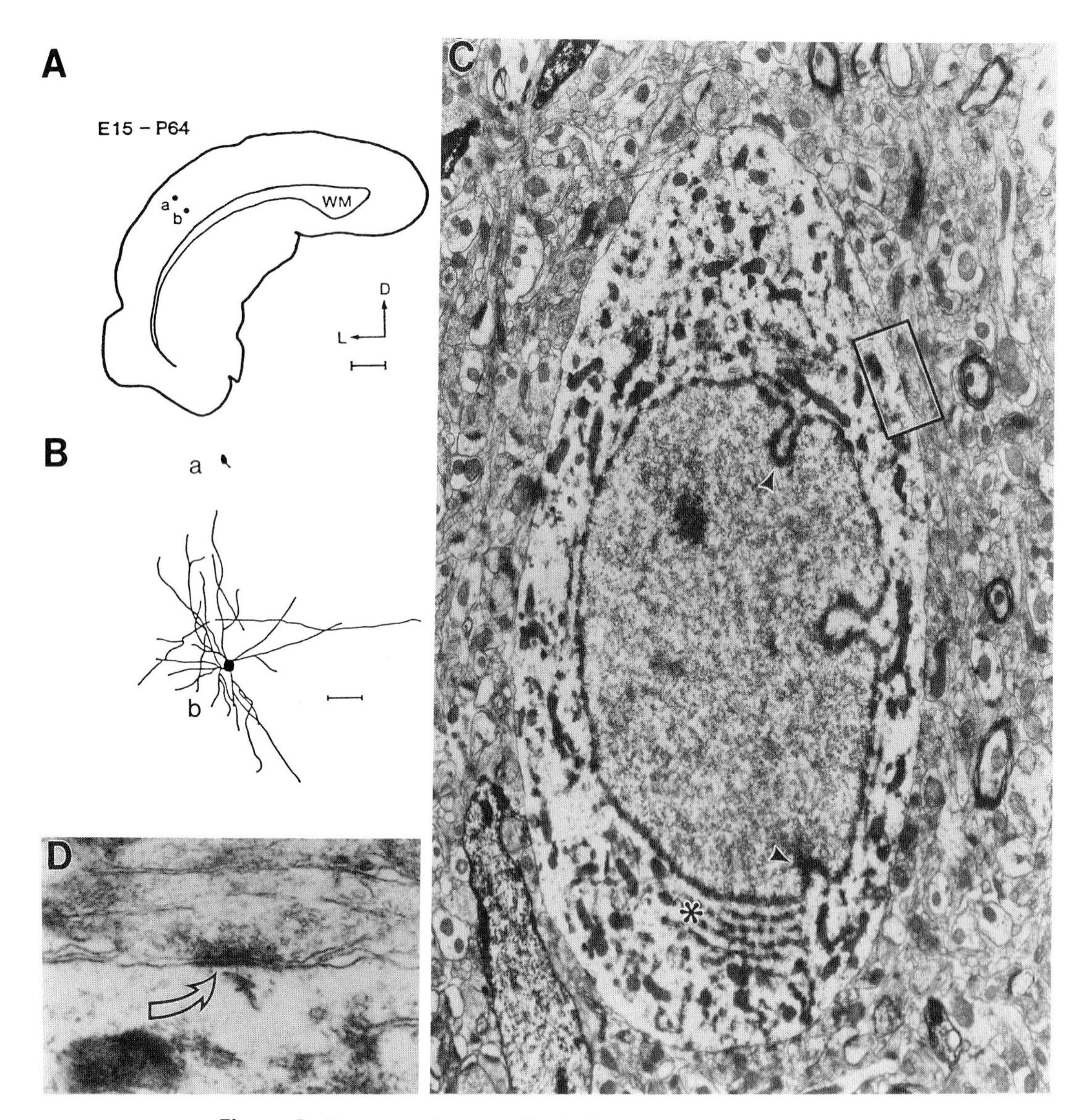

Figure 1 Representative example of a homogeneous nonpyramidal neuronal clone in the visual cortex of a rat brain injected with retrovirus near the onset of neurogenesis (E15) and examined at maturity (Postnatal Day 64). (A) This clone contained two closely spaced lacZ-positive cells (a and b) within three consecutive 100-μm thick sections. Abbreviations: D, dorsal; L, lateral; WM, white matter. Scale bar = 1 mm. (B) Camera lucida drawings of the two clonally related neurons to show the morphology revealed by the X-gal histochemical staining. Cell b revealed a very extensive dendritic field characteristic of a multipolar nonpyramidal neuron, while the phenotype of cell a could not be established by its light

conclusively demonstrated. The evidence from the laminar fate studies in fact suggests that microenvironmental modification of a progenitor cell's potential is likely to be an integral part of its commitment process and is probably highly organized and targeted on cells during precise periods of division and differentiation. Similarly, during migration of postmitotic neurons the environmental conditions they encounter may further modify differentiation by signaling the acquisition of certain neuronal features.

Taken together, these retroviral lineage tracing experiments clearly show that by E16, the onset of neurogenesis, a number of lineage restrictions have occurred in the developing cerebral cortex such that separate progenitor cells now exist within the ventricular zone for pyramidal neurons, nonpyramidal neurons, atrocytes, and oligodendrocytes. This conclusion can be drawn from the almost complete absence of any clones which contain mixtures of these cell types. The embryonic ventricular zone during the peak period of neurogenesis would appear to be composed of a mosaic of different, albeit morphologically similar, progenitor cells each committed to a particular lineage.

B. Lineage Analysis during Postnatal Gliogenesis

Although the results of the experiments by Luskin *et al.* (1993), in which retrovirus was injected at E16, demonstrated that by the onset of neurogenesis in the rat the lineage for astrocytes and oligodendrocytes had diverged, a further analysis was performed to investigate glial lineage during the early postnatal period when larger numbers of glia are being generated and most neurogenesis has ceased. This is also a time period in which previous *in vitro* studies of optic nerves had proposed that a glial progenitor cell capable of generating oligodendrocytes and certain astrocytes played a key role in gliogenesis (Raff *et al.*, 1983). Other studies subsequently identified this bipotential glial progenitor, now called the O2A progenitor, in the cultures of the cerebellum (Levi *et al.*, 1986) and cerebrum (Ingraham and McCarthy, 1989). On the other hand, the existence of a bipotential O2A glial progenitor in any part of the CNS *in vivo* has not been conclusively demonstrated.

To investigate glial lineage restrictions in the cerebrum of newborn rats *in vivo*, a retroviral lineage tracer was injected into the subventricular zone

---

microscopic appearance. Scale bar = 50 μm. (C and D) Electron micrographs of cell a showing the electron-dense ß-galactosidase histochemical reaction product preferentially associated with the nuclear membrane and endoplasmic reticulum (asterisk). Ultrastructural examination of a cell revealed that it was a nonpyramidal neuron with an indented nucleus and a number of both symmetrical and asymmetrical synapses; asymmetric axosomatic synapses (curved arrow in D) are an exclusive feature of nonpyramidal neurons in the cerebral cortex. The area within the box in C is shown at higher magnification in D. Magnification: C, ×8915; D, ×42050. (Modified with permission from Parnavelas *et al.*, 1991).

(Luskin and McDermott, 1994), the site of most postnatal gliogenesis. Following a survival period of 2 weeks, clones of lacZ-positive cells, the progeny of infected progenitors, were identified and ultrastructurally classified using established criteria (Parnavelas *et al.*, 1982). The results of this analysis demonstrated that all of the clones contained cells of the same phenotype and could be divided into four distinct groups: immature cell clones situated in the subependymal zone surrounding the lateral ventricle, oligodendrocytes clones, white matter astrocyte clones, and gray matter astrocyte clones (Fig. 2). Within some of these clones cells displaying both the mature phenotype and a less differentiated phenotype were found together, which indicates that lineally related cells can be generated over an extended period of time. Most importantly, clones containing both macroglial subtypes or clones containing neurons were not encountered. These results, which were based on rigorous ultrastructural analysis of each cell within each clone, cast considerable doubt on the existence postnatally of a bipotential progenitor in the cerebrum *in vivo*. Nonetheless, Levison and Goldman (1993), in a similar study which used light microscopic immmunocytochemical staining to classify cells types, reported that approximately 15% of clones (also derived from neonatal injections into the subventricular zone) contained both astrocytes and oligodendrocytes. This approach allows greater numbers of clones to be analyzed but difficulties can arise with the degree of certainty with which cells can be identified as the expression of cell-specific antigens at a level detectable by immunocytochemistry is not always achieved (Luskin and McDermott, 1994).

Another approach to reconcile these differences has been pursued in which O2A-like progenitors have been isolated in culture and then put back into the cerebrum. The transplanted cells only gave rise to oligodendrocytes (Espinosa del los Monteros *et al.*, 1993), which is further evidence suggesting that observed *in vitro* plasticity of these cells does not occur in normal circumstances in the brain. Retroviral lineage analysis has not, however, been performed *in vivo* in the optic nerve where O2A cells were originally identified. Generally, however, our studies and other recent *in vitro* studies (Lubetzki *et al.*, 1992; Vaysse and Goldman, 1990) suggest that bipotential O2A-like progenitor behavior is, at best, uncommon during the entire period of gliogenesis in the cerebrum.

C. Analysis of the Migration of Neural Progenitors Using Retroviral Tracers

Retroviral vectors have also advanced our understanding of the migratory mechanisms involved in the development of the cerebral hemispheres. For instance, there has been much discussion in the literature about whether or not the functional cortical columns which have been described (Hubel and Weisel, 1962; Mountcastle, 1979) derive ontogenetically from "radial units" composed of clonally related immature neurons (Rakic, 1988). The neurons destined for these columns supposedly migrate along the fibers of restricted cohorts of radial glial cells which are well known to act as a guidance system for migrating postmitotic neurons (Rakic, 1972; Hatten, 1990).

If lineally related neurons, arising from specified progenitors in fixed positions in the ventricular zone, do use distinct yet closely related sets of radial glial fibers as a means of constructing cortical columns, retroviral lineage markers are ideally suited to test the hypothesis. A number of retroviral studies have now looked directly at this issue. Based on analyses of clonal dispersion of migrating cells it appears that the notion that clonally related neurons maintain a simple one to one relationship with certain radial fibers designated by their positions in the ventricular zone is far from universally applicable. From a number of studies it has emerged that not all clonally related cells follow this straightforward pattern and variable degrees of dispersion tangential to the radial axis in the proliferative zone and the intermediate zone have been observed (Walsh and Cepko, 1988; Price and Thurlow, 1988; Austin and Cepko, 1990; Misson *et al.*, 1991; Walsh and Cepko, 1993). In some instances tangential dispersion of the labeled cells was of the order of several hundred micrometers, which begins to complicate the task of assigning clonal boundaries.

That the constituent cells of a clone could be distributed over such distances was conclusively shown in an interesting adaptation of the retroviral marking technique devised by Walsh and Cepko (1992). These investigators prepared a library of genetically different retroviruses each carrying unique nucleotide sequences as well as the *lac*Z gene and then injected this mixture of vectors bearing different tags into the lateral ventricles of embryonic rats in order to infect a small number of progenitor cells in the ventricular zone. Subsequently, Walsh and Cepko (1992) mapped the distribution of lacZ-positive cells in each brain, microdissected out the labeled cells, and used the polymerase chain reaction to amplify and determine the specific nucleotide sequence in each lacZ-positive cell. As this procedure was designed to minimize the possibility of duplicate viral constructs being injected it was assumed that the presence of the same viral construct in two or more cells predicated clonality regardless of their relative location in the brain. Conversely, cells with different constructs were assumed to have arisen from different progenitor cells. They found that, most frequently, cells bearing identical inserts occurred in clusters but also that some cells possessing identical DNA inserts were separated by many millimeters. While these experiments confirmed that the majority of discrete clusters with closely spaced lacZ-positive cells do arise from single progenitors and are therefore clonally related, some clones do disseminate widely in the tangential plane of cerebral cortex. This exceptional pattern of migration of cortical precursors has been documented by other studies using different methodologies (O'Rourke *et al.*, 1992; Tan and Breen, 1993).

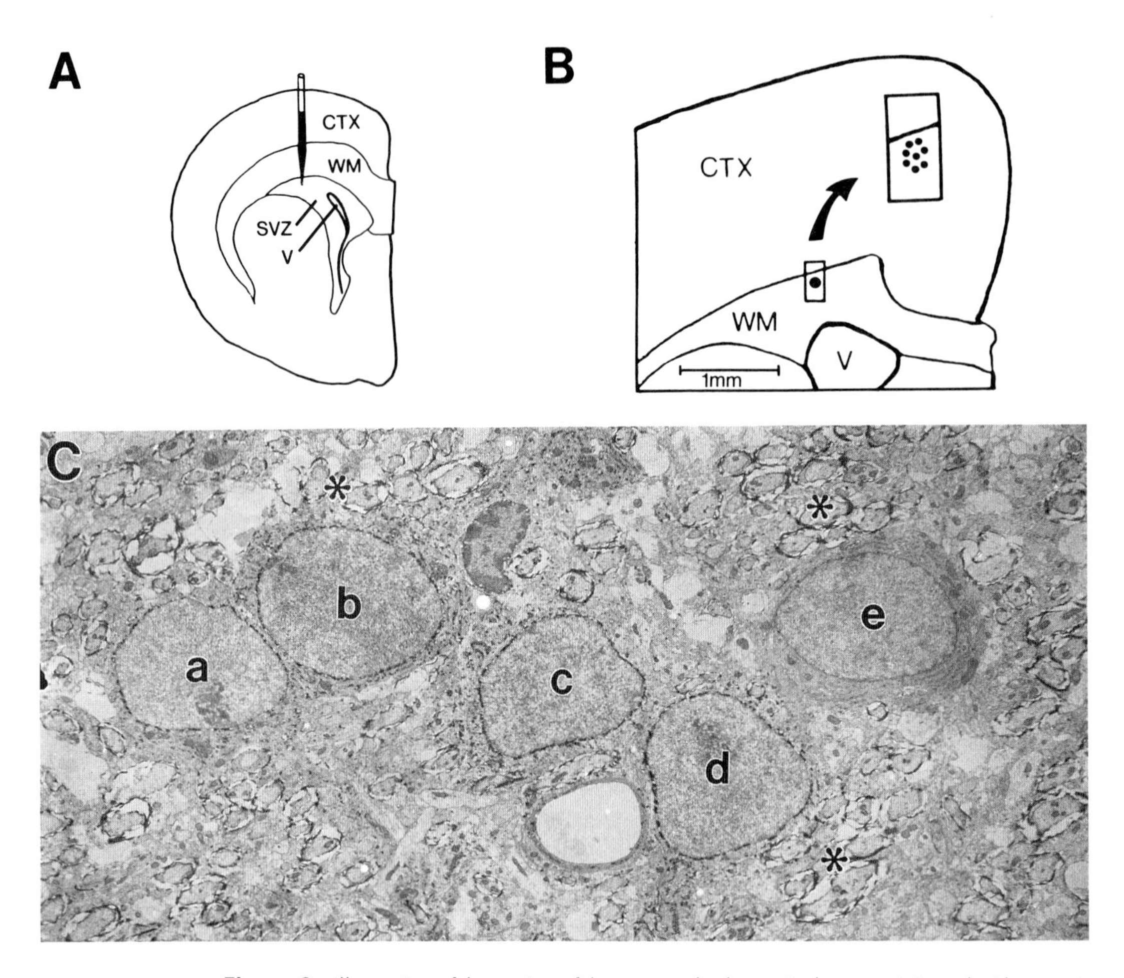

Figure 2 Illustration of the region of the neonatal subventricular zone injected with retrovirus to infect glial progenitor cells and a representative example of a homogeneous clone of oligodendrocytes. (A) Diagrammatic representation of the method of infecting progenitor cells of the neonatal subventricular zone with retrovirus. The desired injection site was the region of the subventricular zone surrounding the anterior and lateral part of the lateral ventricle, which is indicated by the tip of the pipette. (B and C) Distribution and ultrastructural appearance of a closely spaced clone of lacZ-positive oligodendrocytes situated in the white matter of the corpus callosum. B shows a camera lucida reconstruction of the position (large black dot) of an eight-cell clone of lacZ-positive cells in the P15 rat brain resulting from an intracerebral injection of retrovirus at P2. The positional relationships of the individual cells (small black dots) within the clone are depicted in the inset, indicated by the arrow. C shows four closely apposed lacZ-positive oligodendrocytes from the eight-cell clone. The cells labeled a–d all display perinuclear reaction product and have features typical of light oligodendrocytes, including dispersed chromatin and electron-dense cytoplasm. The remaining cell (e) is not labeled and not part of the clone. Asterisks indicate regions of

When retroviral lineage tracing was combined with antibody staining of radial glial elements in the developing mouse cerebral cortex, an even clearer picture of how clonally related cells are distributed began to emerge (Misson *et al.*, 1991). Following retroviral injection at E13 and subsequent lacZ histochemistry at E16, 17, and 18 to identify clones, brains were immunostained with monoclonal antibody RC2 which recognizes mouse radial glia cells and fibers. This method enabled the alignment of radial glial fibers and apparent patterns of alignment of lacZ-positive cells to be reconciled. Radial glial fibers showed substantial regional variation in alignment patterns but in all regions labeled migrating cells were aligned in parallel with adjacent fibers. In some regions, such as dorsal and medial parts of the cerebral hemisphere, the alignment of fibers was radial, whereas in lateral regions fiber alignment was nonradial. In lateral regions fibers leaving the ventricular zone initially converge but begin to diverge on leaving the intermediate zone. As single fibers fan out from the convergent fascicles, migrating neurons may then take different routes on the last leg of their journey to the cortical plate (Misson *et al.*, 1991). Thus, clonally related neurons, which in ventricular and intermediate zones have a radial and closely spaced alignment, may tend to diverge considerably in the lateral cortical regions due to the differential distribution of radial glial fibers in these regions. This seems a very plausible explanation for the tangential dispersion of clonally related cells seen in other studies (*vide supra*).

The idea that the embryonic ventricular zone contains a "protomap" composed of fixed proliferative units which specify strictly bounded cortical columns is not well supported by most retroviral studies. However, there are examples which demonstrate that mosaicism in the germinal neuroepithelium is present (Arimatsu *et al.*, 1993; Tan and Breen, 1993). Using retroviral marking it has recently been clearly demonstrated that the germinal zone for olfactory bulb interneurons is located in a discrete part of the postnatal subventricular zone (Luskin, 1993; Luskin and Boone, 1994). This unexpected and striking finding illustrates the usefulness of retroviral lineage tracers to reveal the phenotype and destination of the progeny of spatially restricted groups of progenitor cells rather than the lineage of individual cells. When retroviral injections were targeted specifically to the anterior part of the neonatal subventicular zone (SVZa) of the forebrain large numbers of labeled cells were detected (depending on the duration between injection of retrovirus and analysis) forming a stream of cells en route to the olfactory bulb or already in the bulb itself (Fig. 3). Cells undergoing this migration to the olfactory bulb, the longest of any cells in the CNS, travel along a highly circumscribed pathway

myelinated axons in the vicinity of the lacZ-positive oligodendrocytes. Abbreviations: CTX, cerebral cortex; SVZ, subventricular zone; V, lateral ventricle; WM, white matter. Magnification: C, ×3900. (Modified with permission from Luskin and McDermott, 1994).

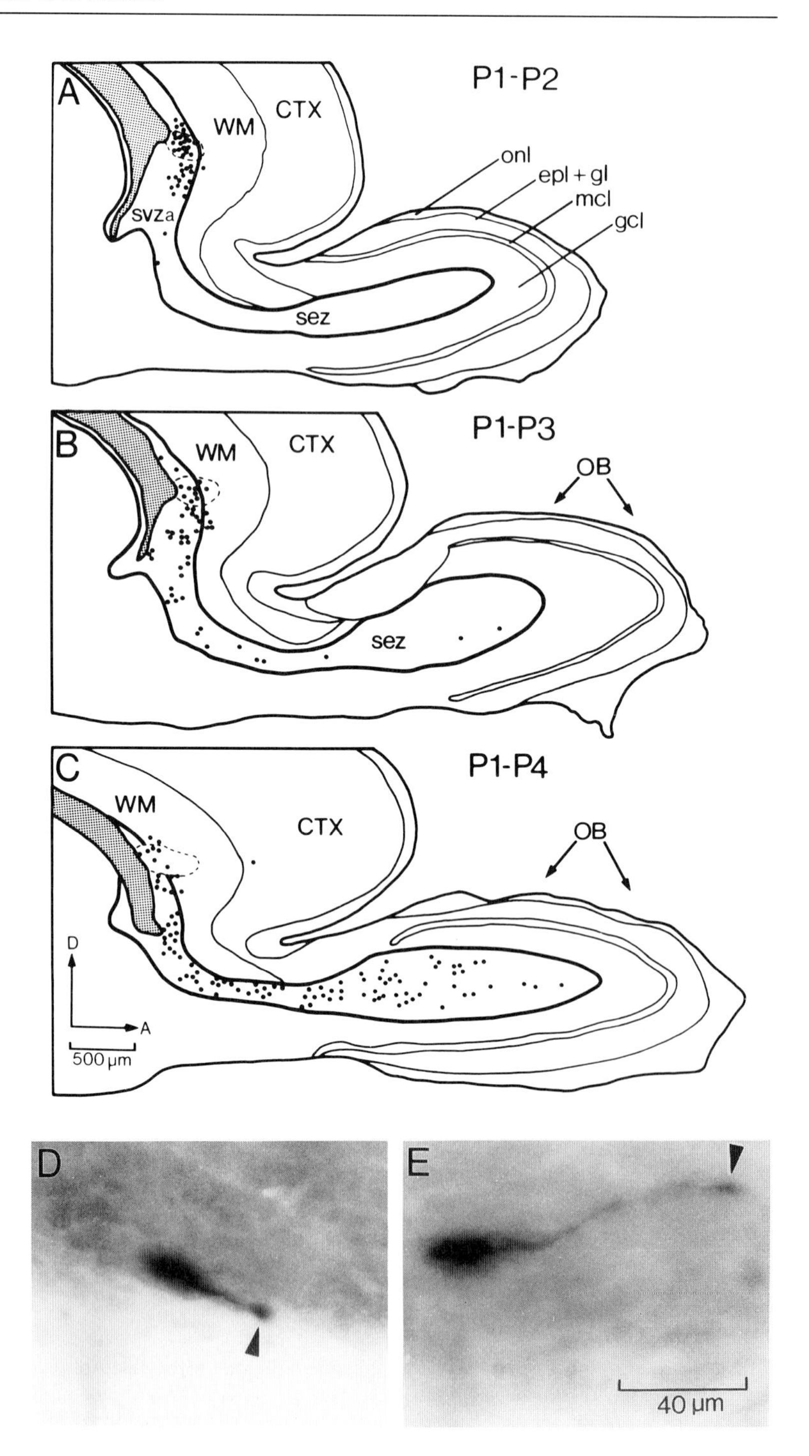
A
P1-P2
WM
CTX
svza
onl
epl + gl
mcl
gcl
sez
B
P1-P3
WM
CTX
OB
sez
C
P1-P4
WM
CTX
OB
D
A
500 μm
D
E
40 μm

which appears to run orthogonal to the general alignment of radial glial fibers used by cortical neurons. The SVZa-generated cells in the olfactory bulb were confirmed as neurons by antibody staining for neuron-specific class III β-tubulin (Menezes *et al.*, 1994). They become either granule cells residing mainly in the granule layer or periglomerular cells of the glomerular layer. None of the SVZa-derived cells had the appearance of glia or of mitral cells, the prenatally generated projection neurons of the olfactory bulb. If injections of retrovirus were made in any other part of the neonatal sunventricular zone, only glial clones were obtained. Therefore, the cells in the SVZa constitute a distinct set of phenotypically committed progenitor cells which reside alongside other progenitors committed to an entirely separate lineage. Whether this is a unique area in the subventricular zone or whether other specialized zones exist in the subventricular zone remains to be determined.

III. Conclusion

In this chapter we have reviewed a broad range of studies including a number carried out by ourselves which have applied the recently developed retroviral lineage tracing technology to address some of the many intractable issues in developmental neurobiology. It is clear that some of these studies have gone a long way in helping to further our understanding of the extremely complex mechanisms which subserve the development of our nervous system. They also show how important it is to continue to develop novel techniques

Figure 3 Spatio-temporal changes in the distribution of SVZa-derived cells along the migratory pathway leading to the olfactory bulb and their morphological appearance. (A–C) Camera lucida drawings of representative 20-μm sagittal sections of the forebrain illustrate the distribution of lacZ-positive cells (black dots) resulting from an injection of retrovirus into the SVZa at P1. The position of the injection site is circumscribed by dashed lines in A–C. Highlighted are the successive change in the distribution of lacZ-positive cells after progressively longer postinjection survival times. One day postinjection (A), the majority of lacZ-positive cells are situated near the injection site bordering the lateral ventricle (shaded region). Two days following an injection (B) labeled cells have migrated varying distances toward the olfactory bulb. Three days after an injection (C) there is a continuous stream of cells from the injection site to the center of the olfactory bulb, all of which are restricted to a well-defined migratory pathway. (D and E) The lacZ-positive cells in the migratoy pathway at P7 from a brain which received an injection of retrovirus into the SVZa at P2 show morphological features typical of migrating neurons. They have an elongated cell body and a leading process that ends in a growth cone (arrowheads). Abbreviations: A, anterior, CTX, cerebral cortex; D, dorsal; epl, external plexiform layer; gcl, granule cell layer; gl, glomerular layer; mcl, mitral cell layer; OB, olfactory bulb; onl, olfactory nerve layer; sez, subependymal zone; SVZa, anterior part of the subventricular zone; WM, white matter. The calibration bar in E also applies to D. (Modified with permission from Luskin, 1993).

in order to underpin new lines of research into nervous system ontogenesis. The use of retroviruses, as shown here, has, in only a few years, added greatly to our understanding of lineage and migratory strategies in the generation of cell diversity and the construction of the different cytoarchitectonic areas in the CNS. In other research not reviewed here retroviruses are also being used as gene vectors for producing immortalized cultures of different neural cell types (Snyder *et al.,* 1992). These lines of investigation are proving beneficial in studies exploring fundamental questions of neural cell biology and may also provide material for therapeutic use in certain neurodegenerative disorders.

Acknowledgments

We thank Kasey Nelson for his assistance in preparing the manuscript. The authors' work is supported by a grant from the Multiple Sclerosis Society of Ireland and the Health Research Board of Ireland (K.McD.) and from the National Institutes of Health (M.B.L.) and awards from the Pew Charitable Trusts and the March of Dimes Birth Defects Foundation (M.B.L.). Part of the work was carried out by K.McD. while in receipt of a Wellcome Trust Travel Grant.

References

Arimatsu, Y., Miyamoto, M., Nihonmatsu, I., Hirata, K., Uratani, Y., Hatanaka, Y., and Takiguchi-Hayashi, K. (1993). Early specification for a molecular phenotype in the rat neocortex. *Proc. Natl. Acad. Sci. USA* **89,** 8879–8883.

Austin, C. P., and Cepko, C. L. (1990). Cellular migration patterns in the developing mouse cerebral cortex. *Development* **110,** 713–732.

Baltimore, D. (1970). RNA-dependent DNA polymerase in virions of RNA tumour viruses. *Nature* **226,** 1209–1211.

Bonnerot, C., Rocancourt, D., Briand, P., Grimber, G., and Nicolas, J.-F. (1987). A ß-galactosidase hybrid protein targeted to nuclei as a marker for developmental studies. *Proc. Natl. Acad. Sci. USA* **84,** 6795–6799.

Caggiano, M., Kauer, J. S., and Hunter, D. D. (1994). Globose basal cells are neuronal progenitors in the olfactory epithelium: A lineage analysis using replication- incompetent retrovirus. *Neuron* **13,** 339–352.

Cepko, C. (1988). Retrovirus vectors and their applications in neurobiology. *Neuron* **1,** 345–353.

Danos, O., and Mulligan, R. C. (1988). Safe and efficient generation of recombinant retroviruses with amphotropic and ecotropic host ranges. *Proc. Natl. Acad. Sci. USA* **85,** 6460–6464.

Espinosa del los Monteros, A., Zhang, M. S., and De Vellis, J. (1993). O2A progenitor cells transplanted into the neonatal rat brain develop into oligodendrocytes but not astrocytes. *Proc. Natl. Acad. Sci. USA* **90,** 50–54.

Fields-Berry, S. C., Halliday, A. L., and Cepko, C. L. (1992). A recombinant retrovirus encoding alkaline phosphatase confirms clonal boundary assignment lineage analysis of murine retina. *Proc. Natl. Acad. USA* **89,** 693–697.

Galileo, D. S., Gray, G. E., Owens, G. C., Majors, J., and Sanes, J. R. (1990). Neurons and glia arise from a common progenitor in the chicken optic tectum: Demonstration with two retroviruses and cell type-specific antibodies. *Proc. Natl. Acad. Sci. USA* **87,** 458–462.

Gray, G. E., Glover, J. C., Majors, J., and Sanes, J. R. (1988). Radial arrangement of clonally related cells in the chicken optic tectum: Lineage analysis with a recombinant retrovirus. *Proc. Natl. Acad. Sci. USA* **85**, 7356–7360.

Gray, G. E., Leber, S. M., and Sanes, J. R. (1990). Migratory patterns of clonally related cells in the developing central nervous system. *Experientia* **46**, 929–940.

Gray, G. E., and Sanes, J. R. (1991). Migratory paths and phenotypic choices of clonally related cells in the avian optic tectum. *Neuron* **6**, 211–225.

Gray, G. E., and Sanes, J. R. (1992). Lineage of radial glia in the chicken optic tectum. *Development* **114**, 271–282.

Grove, E. A., Kirkwood, T. B. L., and Price, J. (1992). Neuronal precursor cells in the rat hippocampus formation contribute to more than one cytoarchitectonic area. *Neuron* **8**, 217–229.

Grove, E. A., Williams, B. P., Li, D.-A., Hajihosseini, M., Freidrich, A., and Price, J. (1993). Multiple restricted lineages in the embryonic rat cerebral cortex. *Development* **117**, 553–561.

Hall, A.K, and Landis, S. C. (1991). Early commitment of precursor cells from the rat superior cervical ganglion to neuronal or nonneuronal fates. *Neuron* **6**, 741–752.

Hatten, M. E. (1990). Riding the glial monorail: A common mechanism for glial guided neuronal migration in different regions of the developing mammalian brain. *TINS* **13**, 179–184.

Hubel, D. H., and Weisel, T. N. (1962). Receptive fields, binocular interaction and functional architecture in the cat's visual cortex. *J. Physiol. (London)* **160**, 106–154.

Ingraham, C. A., and McCarthy, K. D. (1989). Plasticity of process bearing glial cell cultures from the neonatal rat cerebral cortical tissue. *J. Neurosci.* **9**, 63–69.

Levi, G., Gallo, V., and Ciotti, M. T. (1986). Bipotential precursors of putative fibrous astrocytes and oligodendrocytes in the rat cerebellar cultures express aminobutyric acid transport. *Proc. Natl. Acad. Sci. USA* **83**, 1504–1508.

Levison, S. W., and Goldman, J. E. (1993). Both oligodendrocytes and astrocytes develop from progenitors in the subventricular zone of the postnatal rat forebrain. *Neuron* **10**, 201–212.

Levy, J. A. (1992). "The Retroviridae," Vol. 1. Plenum Press, New York.

Lin, S., Gaiano, N., Culp, P., Burns, J. C., Friedmann, T., Yee, J.-K., and Hopkins, N. (1994) Integration and germ-line transmission of a pseudotyped retroviral vector in zebrafish. *Science* **265**, 666–669.

Lubetzki, C., Goujet-Zalc, C., Demerens, C., Danos, O., and Zalc, B. (1992). Clonal segregation of oligodendrocytes and astrocytes during in vitro differentiation of glial progenitor cells. *Glia* **6**, 289–300.

Luskin, M. B., Pearlman, A. L., and Sanes, J. R. (1988). Cell lineage in the cerebral cortex of the mouse studied in vivo and in vitro with a recombinant retrovirus. *Neuron* **1**, 635–647.

Luskin, M. B., Parnavelas, J. G., and Barfield, J. A. (1993). Neurons, astrocytes and oligodendrocytes of the rat cerebral cortex originate from separate progenitor cells: An ultrastructural analysis of clonally related cells. *J. Neurosci.* **13**, 1730–1750.

Luskin, M. B. (1993). Restricted proliferation and migration of postnatally generated neurons derived from the forebrain subventricular zone. *Neuron* **11**, 173–189.

Luskin, M. B. (1994). Neuronal cell lineage in the vertebrate central nervous system. *FASEB J.* **8**, 722–730.

Luskin, M. B., and Boone, M. S. (1994). Rate and pattern of migration of lineally related olfactory bulb interneurons generated postnatally in the subventricular zone of the rat. *Chem. Senses* **19**, 695–714.

Luskin, M. B., and McDermott, K. (1994). Divergent lineages for oligodendrocytes and astrocytes originating in the neonatal forebrain subventricular zone. *Glia* **11**, 211–226.

McConnell, S. K. (1988). Development and decision-making in the mammalian cerebral cortex. *Brain Res. Rev.* **13**, 1–23.

McConnell, S. K., and Kaznowski, C. E. (1991). Cell cycle dependence of laminar determination in developing neocortex. *Science* **254**, 282–285.

Menezes, J. R. L., Nelson, K., and Luskin, M. B. (1994). Simultaneous migration and proliferation of neuroblasts originating in the anterior subventricular zone in the neonatal rat forebrain. *Neurosci. Abstr.* **20**, 1671.

Mione, M. C., Danevic, C., Boardman, P., Harris, B., and Parnevelas, J. G. (1994). Lineage analysis reveals neurotransmitter (GABA or glutamate) but not calcium-binding protein homogeneity in clonally related cortical neurons. *J. Neurosci.* **14**, 107–123.

Misson, J.-P., Austin, C. P., Takahashi, T., Cepko, C. L., and Caviness, V. S. (1991). The alignment of migrating neural cells in relation to the murine neopallial radial glial fiber system. *Cereb. Cortex* **1**, 221–229.

Mountcastle, V. B. (1979). An organizing principle for cerebral function: The unit module and the distributed system. In "The Neurosciences: Fourth Study Program" (F. O. Schmitt and F. G. Worden, Eds.), pp. 21–42. MIT Press, Cambridge.

O'Rourke, N. A., Dailey, M. E., Smith, S. J., and McConnell, S. K. (1992). Diverse migratory pathways in the developing cerebral cortex. *Science* **258**, 299–302.

Parnavelas, J. G., Luder, R., Pollard, S. G., Sullivan, K., and Lieberman, A. R. (1983). A qualitative and quantitative ultrastructural study of glial cells in the developing visual cortex of the rat. *Phil. Trans. R. Soc. London (Biol.)* **301**, 55–84.

Parnavelas, J. G., Barfield, J. A., Franke, E., and Luskin, M. B. (1991). Separate progenitor cells give rise to pyramidal and non-pyramidal neurons in the rat telencephalon. *Cereb. Cortex* **1**, 463–468.

Price, J., Turner, D., and Cepko, C. (1987). Lineage analysis in the vertebrate nervous system by retrovirus-mediated gene transfer. *Proc. Natl. Acad. Sci. USA* **84**, 154–160.

Price, J., and Thurlow, L. (1988). Cell lineage in the rat cerebral cortex: A study using retrovira-mediated gene transfer. *Development* **104**, 473–482.

Raff, M. C., Miller, R. H., and Noble, M. (1983). A glial progenitor that develops in vitro into an astrocyte or an oligodendrocyte depending on the culture medium. *Nature* **303**, 390–396.

Rakic, P. (1972). Mode of cell migration to the superficial layers of fetal monkey neocortex. *J. Comp. Neurol.* **145**, 61–84.

Rakic, P. (1988). Specification of cerebral cortical areas. *Science* **241**, 170–176.

Sanes, J. R., Rubenstein, J. L. R., and Nicolas, J.-F. (1986). Use of a recombinant retrovirus to study post-implantation cell lineage in mouse embryos. *EMBO J.* **5**, 3133–3142.

Schwob, J. E., Huard, J. M. T., Luskin, M. B., and Youngentob, S. L. (1994). Retroviral lineage studies of the rat olfactory epithelium. *Chem. Senses* **19**, 671–682.

Snyder, E. Y., Deitcher, D. L., Walsh, C., Arnold-Aldea, S., Hartwieg, E. A., and Cepko, C. L. (1992). Multipotent neural cell lines can engraft and participate in development of mouse cerebellum. *Cell* **68**, 33–51.

Stoker, A. W. (1993). Retroviral vectors. In "Molecular Virology. A Practical Approach" (A. J. Davidson and R. M. Elliot, Eds.), pp. 171–197. Oxford Univ. Press, Oxford.

Tan, S.-S., and Breen, S. (1993). Radial mosaicism and tangential cell dispersion both contribute to mouse neocortical development. *Nature* **362**, 638–640.

Temin, H. M., and Baltimore, D. (1972). RNA-directed DNA synthesis and RNA tumor viruses. *Adv. Virus Res.* **17**, 129–186.

Turner, D., and Cepko, C. L. (1987). A common progenitor for neurons and glia persists in rat retina late in development. *Nature* **328**, 131–136.

Turner, D. L., Snyder, E. Y., and Cepko, C. L. (1990). Lineage independent determination of cell type in the embryonic mouse retina. *Neuron* **4**, 833–845.

Vaysse, P. J.-J., and Goldman, J. E. (1990). A clonal analysis of glial lineages in neonatal forebrain development *in vitro*. *Neuron* **5**, 227–235.

Walsh, C., and Cepko, C. L. (1988). Clonally related cortical cells show several migration patterns. *Science* **241**, 1342–1345.

Walsh, C., and Cepko, C. L. (1992). Widespread dispersion of neuronal clones across functional regions of the cerebral cortex. *Science* **255**, 434–440.
Walsh, C., and Cepko, C. L. (1993). Clonal dispersion in proliferative layers of the developing cerebral cortex. *Nature* **362**, 626–638.
White, J. M. (1990). Viral and cellular membrane fusion proteins. *Annu. Rev. Physiol.* **52**, 675–697.
Williams, B. P., Read, J., and Price, J. (1991). The generation of neurons and oligodendrocytes from a common precursor cell. *Neuron* **7**, 685–693.

CHAPTER 23

Retroviral Vectors for the Study of Neuroembryology: Immortalization of Neural Cells

Evan Y. Snyder
Departments of Neurology & Pediatrics
Harvard Medical School
Children's Hospital
Boston, Massachusetts

I. Introduction

The generation, analysis, and transplantation of immortalized neural cells has received attention for both research and clinical applications (Bjorklund, 1993; Gage, 1992; Gage *et al.*, 1995; Snyder, 1994). For neuroscientists the expectation has been that such neural cell lines might lend insight into the cellular and molecular processes underlying development of the mammalian nervous system. For clinicians, the hope has been that such lines might play pivotal roles in gene therapy and repair of the central nervous system (CNS). The ability of immortalized neural progenitors to integrate normally into the nervous system following transplantation offers a potential therapeutic approach to certain neurodegenerative diseases. It also suggests that immortalization may be a legitimate neuroembryologic tool.

This field is still in its infancy. While there are a few seminal papers that form its foundation (Snyder *et al.*, 1992, 1995a; Renfranz *et al.*, 1991; Onnifer *et al.*, 1993a; Groves *et al.*, 1993; Anton *et al.*, 1994), investigations in this area are progressing so rapidly that citing even promising work still in abstract form at this writing may not be sufficient to keep this review current.

This chapter is focused on a detailed discussion of retrovirally immortalized primary neural progenitors which give rise to neurons, particularly from the standpoint of transplantation into the mammalian CNS.[1]

II. Why Immortalize Neural Cells?

Developmental neurobiologists have long fantasized about having a library of cell lines. Each line might represent cells plucked from defined developmental times and specific locations. Each line could provide an unlimited source of homogeneous material from which factors could be purified and against which reagents could be assayed. In some nonneural systems (e.g., hematopoetic, myogenic) cell lines have, in fact, been useful in isolating gene products that control commitment and differentiation. In neurobiology as well, cell lines have historically played a prominent role: PC12 (Greene and Tischler, 1976), a cell line derived from a mouse neural crest tumor (pheochromocytoma), has been instrumental for two decades in studying neurotrophin action. Although cell lines have been derived from CNS tumors, they have an ill-defined history, do not constitute a developmental spectrum of cells from defined locations and stages, have lost many of the properties and developmental programs of the region from which they originated, and are usually tumorigenic, which precludes their use for transplantation experiments. Other cell lines derived from embryonic carcinoma cells, e.g., P19 (McBurney *et al.*, 1988), differentiate into neurons in response to retinoic acid, but also are not engraftable and have an uncertain relationship to normal development. A neural cell line derived from a human teratocarcinoma (NT-2 cells) appears to differentiate into engraftable neurons following retinoic acid treatment (Trojanowski *et al.*, 1993); however, the degree to which the oncogenic potential has been abrogated and the extent to which the line faithfully models primary neurons remain to be established.

[1] Therefore, a number of intriguing topics in neural cell line biology must unfortunately be excluded. While their similarity to retrovirally immortalized progenitors will be mentioned, lines perpetuated by chronic neurotrophin exposure will not be detailed [e.g., (Gage *et al.*, 1994; Geller and Dubois-Dalcq, 1988; Kilpatrick and Bartlett, 1993; Ray *et al.*, 1993; Reynolds and Weiss, 1992; Reynolds *et al.*, 1992)]. Immortalized glial lines (Groves *et al.*, 1993) [including CG9 (Louis *et al.*, 1992)] will not be discussed. Also excluded will be lines derived from "spontaneous" tumors [e.g., hNT (Trojanowski *et al.*, 1993; Kleppner *et al.*, 1995), P19 (McBurney *et al.*, 1988)], and EC cells (Wojcik *et al.*, 1988), "induced" tumors [e.g., in transgenic mice (Hammang *et al.*, 1990; Largent *et al.*, 1993; Mellon *et al.*, 1990; Suri *et al.*, 1993)], "fusions" to tumors [e.g., somatic cell hybrids (Blusztajn *et al.*, 1992; Crawford *et al.*, 1992)], or other aberrant conditions [e.g., megencephaly: HCN-1 (Poltorak *et al.*, 1992)]. For retrovirally immortalized lines, this chapter will not detail technique, but rather explore the biologic insights and therapeutic advantage one might gain through such lines. For methods, the reader is referred to some of the original descriptions [e.g., (Bartlett *et al.*, 1988; Bernard *et al.*, 1992a,b; Birren *et al.*, 1992; Cepko, 1991; Evrard *et al.*, 1990; Frederickson *et al.*, 1988; Geller and Dubois-Dalcq, 1988; Giordano *et al.*, 1993; Ryder *et al.*, 1990; White and Whittemore, 1992; Whittemore *et al.*, 1995)].

Because mature end-differentiated neurons do not, in normal circumstances, divide either in the brain or in culture, the key appeared to be to derive normal neurons from normal neural precursors. However, progenitors removed from the brain also do not normally remain in a constant proliferative state *in vitro* (Davis and Temple, 1994; Temple, 1989; Temple and Davis, 1994): after a limited number of mitoses, if any, they cease dividing and differentate. To maintain them in a proliferative state requires an intervention. These limitations may be circumvented by (a) transduction of immortalizing genes into neural progenitors (Anton *et al.*, 1994; Bernard *et al.*, 1992a; Birren and Anderson, 1990; Evrard *et al.*, 1990; Frederickson *et al.*, 1988; Geller and Dubois-Dalcq, 1988; Ryder *et al.*, 1990; Snyder *et al.*, 1992; White and Whittemore, 1992), (b) chronic exposure of progenitors to mitogenic cytokines (Catteneo and McKay, 1990; Gensburger *et al.*, 1987; Groves *et al.*, 1993; Kilpatrick and Bartlett, 1993; Ray *et al.*, 1993; Reynolds and Weiss, 1992; Reynolds *et al.*, 1992), or (c) coculture of progenitors with astroglial membrane homogenates (Temple and Davis, 1994). In this chapter we will focus on the techniques and utility of the first approach.

In this first strategy, replication-incompetent retroviral vectors encoding an immortalizing gene are typically introduced into freshly plated dissociated primary cultures of neural tissue (Fig. 1). Because a retrovirus stably integrates its genome into only mitotic host cells, the immortalizing gene is transduced into progenitors present in the culture and is passed to their progeny in a Mendalian fashion.[2] First, isolating and expanding infected colonies, and then characterizing early passages with antibodies to neural antigens, identifies those lines that are derived from immortalized neural cells. A retrovirus, for practical purposes, integrates its genome randomly into the host chromosome. Therefore, the clonal relationship between lines can be affirmed by the presence of a single and identical site of viral insertion in each cellular population (see Chapter 5).

The neural cell types one might expect from a given line probably depend on the potential of the precursors normally present in a given region at the developmental stage from which the primary tissue was obtained. For example, primary cultures of postnatal rodent neocortex, in which neurogenesis has normally ceased but gliogenesis persists, contain predominantly active glial precursors. On the other hand, neonatal cerebellum (Snyder *et al.*, 1992; Snyder, 1992), where a subpopulation of neurons are still born, or embryonic CNS tissue (Anton *et al.*, 1994; Ray *et al.*, 1993; Whittemore and White, 1993; Renfranz *et al.*, 1991; Reynolds *et al.*, 1992), where a wide range of

[2] Some investigators, such as Horellou and colleagues (Snyder, 1995b), have attempted to take advantage of the fact that astroglia in primary dissociated cultures of embryonic neural tissue divide in the presence of fetal calf serum or growth factors (e.g., bFGF) in order to potentiate their infection *in vitro* by recombinant retroviruses. The cells most efficiently infected, however, are probably astroglial progenitors.

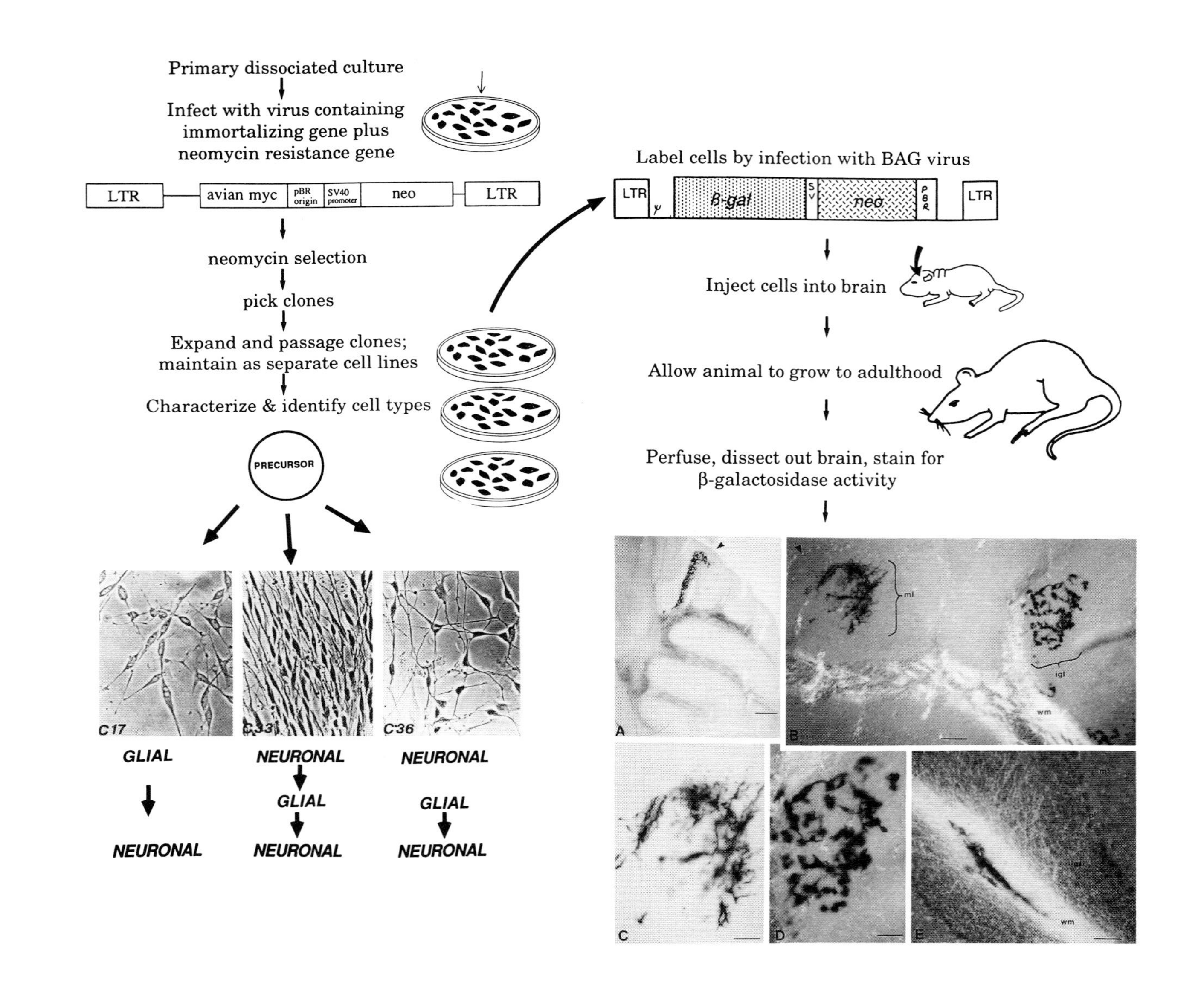

Primary dissociated culture
Infect with virus containing immortalizing gene plus neomycin resistance gene
LTR
avian myc
pBR origin
SV40 promoter
neo
LTR
neomycin selection
pick clones
Expand and passage clones; maintain as separate cell lines
Characterize & identify cell types
PRECURSOR
C17
C33
C36
GLIAL
NEURONAL
NEURONAL
GLIAL
NEURONAL
NEURONAL
GLIAL
NEURONAL
Label cells by infection with BAG virus
LTR
ψ
β-gal
SV
neo
pBR
LTR
Inject cells into brain
Allow animal to grow to adulthood
Perfuse, dissect out brain, stain for β-galactosidase activity
A
B
ml
igl
wm
C
D
E
wm

neurons and glia are being elaborated, contain precursors with broader potential. Even precursors from the autonomic nervous system have been immortalized. Birren and Anderson (1990) immortalized embryonic sympathoadrenal progenitors and generated a cell line (MAH cells) that, in its responsiveness to and dependence on trophins, mimics many of the properties of those primary progenitors that generate sympathetic neurons and adrenal chromaffin cells.

Immortalization might actually broaden the phenotypic range beyond what one would naturally see from the corresponding "normal," endogenous progenitor, [see Gao and Hatten (1994) and discussion under Section III,B,5]. On the other hand, most immortalized cell lines behave remarkably similar to endogenous stem cells. In fact, such lines have been termed "stem-*like* cells." Further study may elucidate whether immortalization subverts differentiation

---

Figure 1 Schematic for the generation, characterization, and transplantation of immortalized neural progenitors. The progenitors illustrated were derived from primary cultures of neonatal mouse cerebellum and were immortalized with avian *myc* (v-*myc*) transduced by a retroviral vector derived from Maloney Murine Leukemia Virus (MoMLV). The immortalizing gene is transcribed from the LTR, while the selection gene (neomycin resistance) is transcribed from an internal SV40 promoter. Infected, neomycin-resistant colonies were isolated, expanded, propagated, and characterized as independent lines (Ryder *et al.*, 1990; Snyder *et al.*, 1992; Snyder, 1992). Clonality of a given line was confirmed by the presence of a single retroviral insertion site. Because, for practical purposes, a retrovirus integrated its genome randomly into the host genome, the presence of a single and identical site of retroviral integration in a given line and its subpopulations suggests that they were all derived from the infection and immortalization of a single progenitor. The phenotypic range of a given clonal line was established by staining the progeny cells with antibodies to cell type-specific antigens. Surprisingly, it was established that a single progenitor could give rise to progeny of diverse cell types, both glial and neuronal. To determine whether this multipotency *in vitro* reflected the true potential of progenitors *in vivo,* clonal lines were stably labeled by infecting them with a retroviral vector encoding the *lacZ* reporter gene transcribed from the LTR (BAG vector) and were transplanted into newborn mouse cerebellum, the region and developmental period from which the lines were originally derived (Snyder *et al.*, 1992; Snyder, 1992). This labeling allowed donor-derived, *lacZ*-expressing cells to stain blue when processed with the X-gal histochemical reaction. At adulthood, blue donor-derived cells were seen to have integrated into the host cerebellum in a nontumorigenic, cytoarchitecturally appropriate manner. Importantly, cells from the same clonal line gave rise to neurons (e.g., granule cells in the internal granule layer) and glia (astrocytes in granular and molecular layers; oligodendrocytes in white tracts) appropriate to the site of engraftment, as if responding to particular microenvironmental cues. The multipotency seen *in vitro* was therefore recapitulated *in vivo.* Implicit in these studies was the fact that a foreign gene (*lacZ*) has been stably expressed for sustained periods within donor cells which had become integral members of the host's brain parenchyma. These two observations—integration of appropriately differentiated cells and foreign gene expression—suggested that the transplantation of immortalized neural cells lines may be useful for repair of and gene transfer into the mammlian brain. (Photomicrographs reproduced with permission from Ryder *et al.*, 1990 and Snyder *et al.*, 1992).

programs or simply suspends the ordinarily rapid narrowing of phenotypic possibilities. Maintaining progenitors longer than usual in their plastic state may allow investigators to discern the full potential of a progenitor. In other words, immortalization experiments (which may preserve phenotypic options) when combined with lineage mapping studies [see Chapter 22], might help neuroembryologists make the often difficult distinction between a progenitor's "fate" as opposed to its "potential."

The genes which have been most successful at immortalization have been avian *myc* (v-*myc*) (Birren and Anderson, 1990; La Rocca *et al.*, 1989; Ryder *et al.*, 1990; Palmieri *et al.*, 1983; Snyder *et al.*, 1992) and SV40 large T antigen [particularly the temperature-sensitive (ts) mutant] (Anton *et al.*, 1994; Eves *et al.*, 1992; Evrard *et al.*, 1990; Giordano *et al.*, 1993; Miller *et al.*, 1993; Whittemore and White, 1993; Renfranz *et al.*, 1991). These genes induce unlimited mitotic activity *in vitro* without promoting oncogenic qualities. Lines immortalized by either gene behave similarly (though ts lines, because they are grown at 33°C, divide more slowly in culture). Both types of lines become contact inhibited. Upon transplantation, both undergo approximately two divisions before becoming postmitotic. Tumors are never seen. Therefore, it is not appropriate to describe these immortalized lines as "transformed."

A variation on retroviral transduction of an immortalizing gene is to use a strain of transgenic mice which harbours ts SV40 large T antigen under the transcriptional control of an inducible promoter (Jat *et al.*, 1991). Immortalization of a variety of cell types, including those from the CNS, can be achieved by maintaining cells derived from these animals under permissive conditions.

Chronic neurotrophin exposure—e.g., to epidermal growth factor (EGF), basic fibroblast growth factor (bFGF), and/or platelet-derived growth factor (PDGF)—represents another approach to the immortalization of at least a subset of neural progenitors (i.e., those with appropriate receptors). These progenitors appear to maintain or reacquire proliferative potential in the presence of these factors and withdraw from the cell cycle upon their removal (Bogler *et al.*, 1990; Kilpatrick and Bartlett, 1993; Raff *et al.*, 1988; Richardson *et al.*, 1988; Ray *et al.*, 1993; Reynolds and Weiss, 1992; Reynolds *et al.*, 1992).

Based on similarities between variously generated progenitor lines, some investigators believe the diverse methods of immortalization may feed into a common process which suspends differentiation (Davis and Temple, 1994; Gage *et al.*, 1995; Kilpatrick and Bartlett, 1993; Snyder, 1994; Temple and Davis, 1994; Vescovi *et al.*, 1993). In their ability to renew themselves, even when obtained from adult brain, and to give rise to multiple cell types, immortalization appears to have captured these cells in, or caused them to revert to, a stem cell-like state. When primary neural progenitors are immortalized, they share properties with normal stem cells, not only from the CNS but from other organs as well (Kitchens *et al.*, 1994). Such observations have, in fact, forced a reexamination of the definitions of "stem cell," "precursor," and "progenitor" as they pertain to the nervous system. Immortalized progenitors have been termed "stem-like cells" by some investigators (Gage *et al.*, 1995; Snyder 1994, 1995a; Snyder and Flax, 1995; Snyder *et al.*, 1995b).

Some stem-like cells not only differentiate into neurons *in vitro*—spontaneously (Anton *et al.*, 1994; Giordano *et al.*, 1993; Onnifer *et al.*, 1993a; Ray *et al.*, 1993; Ryder *et al.*, 1990; Snyder *et al.*, 1992; Whittemore and White, 1993) or following stimulation (Eves *et al.*, 1992; Ip *et al.*, 1994; Vescovi *et al.*, 1993)—but do so when transplanted into the rodent CNS, without forming tumors or disrupting normal cytoarchitecture. They are also capable of being transduced with, and stably expressing, foreign genes *in vivo*. To date, these *in vivo* abilities have been explored most extensively in the genetically immortalized lines,[3] though the engraftment and genetic manipulation of neurotrophin-immortalized lines is also now being studied (Gage *et al.*, 1995; Ray *et al.*, 1994; Svendsen *et al.*, 1994).[4]

The source of neural progenitors for transplant studies has been varied: neonatal cerebellum (Frederickson *et al.*, 1988; Kitchens *et al.*, 1994; Snyder *et al.*, 1992, 1993), embryonic (Martinez-Serrano *et al.*, 1994; Ray *et al.*, 1994; Renfranz *et al.*, 1991) and adult (Gage *et al.*, 1994) hippocampus, embryonic (Anton *et al.*, 1994; Reynolds *et al.*, 1992; Svendsen *et al.*, 1994) and adult (Reynolds and Weiss, 1992) striatum, and embryonic raphe nucleus (Onnifer *et al.*, 1993a; Shihabuddin *et al.*, 1993). Nevertheless, the similarities in behavior of some of these progenitor-derived lines is a bit remarkable. Often they give rise *in vitro* and *in vivo* to both neurons and glia (astrocytes and oligodendrocytes). They are mitogenic in response to EGF and/or bFGF (Kitchens *et al.*, 1994). They express nestin (Kitchens *et al.*, 1994; Martinez-Serrano *et al.*, 1994; Renfranz *et al.*, 1991; Snyder *et al.*, 1993; Snyder, 1995a). They are often quite migratory (Flax *et al.*, 1994; Martinez-Serrano *et al.*, 1994; Snyder *et al.*, 1993, 1995a). They also seem to be able to engraft in multiple regions (Macklis *et al.*, 1994; Martinez-Serrano *et al.*, 1994; Onnifer *et al.*, 1993a; Renfranz *et al.*, 1991; Snyder *et al.*, 1993, 1995a; Snyder, 1995a) and at multiple developmental stages (Flax *et al.*, 1994; Macklis *et al.*, 1994; Snyder *et al.*, 1993). Even when their cell-type phenotypic range is more restricted—for example, raphe-derived serotonergic lines that give rise only to neurons (Whittemore *et al.*, 1995)—they seem to be able to assume a neuronal phenotype in more than one region within the adult CNS (Onnifer *et al.*, 1993a; Shihabuddin *et al.*, 1993).

[3] First detailed contemporaneously in Snyder *et al.*, 1992 and Renfranz *et al.*, 1991, these results have been subsequently supported by the following reports: Anton *et al.*, 1994; Doering and Henderson, 1994; Flax *et al.*, 1994; Groves *et al.*, 1993; Lacorraza *et al.*, 1995; Macklis *et al.*, 1994; Martinez-Serrano *et al.*, 1994; Onnifer *et al.*, 1993a; Ray *et al.*, 1994; Shihabuddin *et al.*, 1993; Snyder, 1992, 1995a; Snyder *et al.*, 1993, 1994, 1995a; Snyder and Flax, 1995.

[4] While lines perpetuated by chronic neurotrophin exposure have been harder to engineer *ex vivo* using retroviruses, adenovirus vectors have been effective (Gage *et al.*, 1994).

III. Immortalized Neural Cells as Models for Development

A. *In Vitro* Studies

Using *primary* cell culture to observe progeny from a single progenitor [e.g., single cells isolated in a "miniwell" (Davis and Temple, 1994; Temple, 1989; Temple and Davis, 1994)] has been useful for defining the potential of some CNS progenitors. Immortalization and cloning of individual progenitors (i.e., the generation of clonal neural cell lines) represents an elaboration on, and a complement to, this approach because it overcomes some of the limitations of primary culture for studying certain properties of precursors (Snyder, 1992). For instance, when differences in behavior are observed for different progenitors isolated in different miniwells, one cannot often discern whether these simply represent stochastic differences or result from subtle differences in microenvironmental signals or reflect a fundamental heterogeneity of the progenitor population. The availability of an unlimited number of cells derived from a single precursor can facilitate the observation of progeny of the same progenitor for prolonged periods, over many generations, and in environments that can be systematically manipulated to alter their fates. Furthermore, maintaining a progenitor perpetually mitotic in culture helps overcome the risk that simply isolating a primary progenitor *in vitro* (even when subsequently reimplanted *in vivo*) changes it because its phenotypic range narrows as it is removed from its germinal zone and begins to withdraw from the cell cycle (McConnell and Kaznowski, 1991).

An additional hope was that immortalization of individual progenitors from defined developmental times and locations might "freeze" a progenitor in a particular state (Snyder, 1992). The availability of many such clonal lines, each with homogeneous cells in abundance, could then facilitate the isolation of developmentally important factors and provide systems for the expression of developmentally interesting genes. This attempt to create "freeze frames" of development—probably naive—has been only partially successful. Progenitors immortalized from a variety of structures and developmental periods, while homogeneous in terms of ancestry, have generally *not* been homogeneous in phenotype. A single progenitor *in vitro* could give rise to a variety of both glia and neurons. Some investigators have even found phenotypic instability within clonal lines wherein subpopulations of the same clone express various neuronal and/or glial characteristics to waxing and waning degrees over time and passages (Ryder *et al.*, 1990; Snyder *et al.*, 1992; Snyder, 1992). A lack of phenotypic homogeneity has, in fact, been a common observation by many investigators immortalizing progenitors derived from such diverse structures as cerebellum (Ryder *et al.*, 1990; Snyder *et al.*, 1992; Snyder, 1992), hippocampus

(Eves *et al.*, 1992; Gage *et al.*, 1994; Ray *et al.*, 1994; Renfranz *et al.*, 1991), and striatum (Evrard *et al.*, 1990; Reynolds and Weiss, 1992; Reynolds *et al.*, 1992; Vescovi *et al.*, 1993), even at fairly late stages in development. The failure to create developmental "snapshots" probably enunciates a more fundamental principle for the mammalian CNS: the great multipotency of some progenitors. Nevertheless, some lines have been generated that do appear to have a single cell type fate. Whittemore and colleagues (1994) have generated serotonergic lines from embryonic raphe nucleus that give rise to only neurons. Therefore, there is a point in the differentiation continuum where the phenotypic options of progenitors become restricted such that immortalization will not preserve pluripotency. That precise point still remains elusive and difficult to predict reliably or to control.

A growing knowledge of the factors that direct multipotent progenitors down various phenotypic pathways may, however, ultimately afford the experimenter the control necessary to address these issues. Even in these early stages of the field, the approach of using neural cell lines *in vitro* has been used to good advantage to elucidate, for example, the various trophic requirements of progenitors for proliferation and differentiation (Eves *et al.*, 1992; Ip *et al.*, 1994; Kilpatrick and Bartlett, 1993; Kitchens *et al.*, 1994; Ray *et al.*, 1993; Vescovi *et al.*, 1993). EGF and bFGF in serum-free media appear to be as pivotal for the maintanence of neural precursors as they are for stem cells from other organ systems (Bogler *et al.*, 1990; Catteneo and McKay, 1990; Kilpatrick and Bartlett, 1993; Kitchens *et al.*, 1994; Ray *et al.*, 1994; Vescovi *et al.*, 1993). In fact, responsiveness to such factors may prove to be a fundamental, defining characteristic of neural stem-like cells. In both a CNS-derived ts-large T-immortalized progenitor line (White *et al.*, 1994) and a v-*myc*-immortalized embryonic sympathoadrenal progenitor line (MAH cells) (Birren *et al.*, 1992), depolarization by KCl was necessary to enhance neurotrophin responsiveness (e.g., BDNF in the former, bFGF in the latter). Neurotrophins employing distinct intracellular signaling pathways may need to work in concert to drive progenitors toward a particular terminally differentiated phenotype. FGF has been established to work synergistically with NGF (Catteneo and McKay, 1990), EGF (Vescovi *et al.*, 1993; Kitchens *et al.*, 1994), and PDGF (Campbell *et al.*, 1994; Raff *et al.*, 1988; Richardson *et al.*, 1988) in the maintenance and/or differentiation of neuronal and glial CNS progenitors perpetuated in culture. BDNF enhances the arborization of neurons derived from EGF-perpetuated CNS progenitors, though it does not appear to prolong neuronal survival *in vitro* nor increase the number of neurons differentiating from such progenitors (Ahmed *et al.*, 1994). Exposure of MAH cells to bFGF (Birren and Anderson, 1990), particularly in combination with depolarization (Birren *et al.*, 1992) or with CNTF (Ip *et al.*, 1994), induces an NGF-dependence similar to that of primary, postmitotic sympathetic neurons and to that of primary chromaffin cells during their transdifferentiation into sympathetic neurons.

CNTF (which is distantly related to a subfamily of hematopoietic cytokines) and FGF and NGF (which activate receptor tyrosine kinases) utilize completely different receptor and intracellular signaling systems.

In related studies, an immortalized serotonergic neuronal line was used to determine that neurotransmitter synthesis and neurofilament synthesis are independently regulated developmental processes which are influenced by different trophic agents (White *et al.*, 1994b).

Some multipotent neural stem cells may acquire their differentiated cell type identity as they withdraw from the cell cycle (McConnell and Kaznowski, 1991; Snyder, 1995a; Snyder and Flax, 1995). By manipulating the microenvironment these stem-like cells encounter, investigators are identifying factors that might influence differentiation. In serum-free medium, that phenotype can be experimentally altered. For example, IGF-2 seems to potentiate neuronal and/or oligodendroglial differentiation for some progenitors. In the absence of instructive signals, however (e.g., in unsupplemented, mitogen-free, serum-free medium), the cells pursue a default pathway which, for at least some multipotent stem-like cells, appears to be neuronal.

In addition to their responsiveness to cytokines (Kitchens *et al.*, 1994), immortalized neural progenitors share many other *in vitro* properties with primary stem cells (Table 1), e.g., they are nestin$^+$ (Snyder *et al.*, 1993), they express homeodomain proteins (Snyder *et al.*, 1992), neural cell adhesion molecules (Weinstein *et al.*, 1990), precursor-specific transcription factors (Snyder *et al.*, 1993), and developmentally regulated proteoglycans (Snyder *et al.*, 1993), and they can promote the differentiation of other immature neural cells (Weinstein *et al.*, 1990). Therefore, by virtue of their genetic homogeneity and abundance, they may serve as valid models for other neurobiologic inquiries, including assaying neurotoxicity and isolating neural-specific genes, transcription factors, receptors, etc.

B. *In Vivo* Transplant Studies

1. Returning Cells to a Normal Environment

Both primary culture and immortalized cell lines suffer from the same inherent limitation: removal of a cell from its *in vivo* context. Transplantation of neural tissue has been a classic tool for studying the behavior of donor cells in a normal environment (e.g., McConnell, 1988). The most compelling evidence for authenticity of progenitor lines would be their incorporation into the normal cytoarchitectonics of the appropriate brain region at the appropriate host age in a functionally meaningful manner. Conclusions regarding stem-like cell behavior *in vitro* might then have relevance to the *in vivo* condition (Snyder, 1992).

Stem-like cells do indeed integrate into the host CNS in a cytoarchitecturally appropriate, nontumorgenic fashion (Onnifer *et al.*, 1993a; Ray *et al.*,

Table 1
Some Characteristics of Immortalized Neural Progenitors in Culture

Expression of neural "stem cell" markers (e.g., Nestin$^+$)	
Gives rise to diverse neural cell types expressing various differentiated markers: e.g.,	
Neurofilament (NF)$^+$	Neuron-specific enolase (NSE)
Microtubule associated protein (MAP)$^+$	GAP 43$^+$
Glutamate +	GAD$^+$
Glial fibrillary acidic protein (GFAP)$^+$	A2B5$^+$
Galactocerebroside C (GalC)$^+$	CNPase$^+$
Myelin basic protein (MBP)$^+$	
Possess ions channels	
Lack characteristics of tumor cells (e.g., contact inhibited; no growth in soft agar)	
Capable of being maintained in serum-free medium (N_2)	
Mitogenic response to bFGF and/or EGF	
Differentiate upon exit from cell cycle (e.g., in response to pharmacologic mitotic inhibition, coculture, serum-free medium without mitogens, etc.)	
?Enriched for neurons upon abrupt withdrawal from cell cycle in serum containing medium (e.g., with mitotic inhibition)	
?Enhanced differentiation into neurons and/or oligodendrocytes with IGF II	
Differentiate in response to inhibition of protein kinase	
Differentiate when cocultured with primary neural culture	
Promote neurite outgrowth of cocultured neurons	
Inhibit proliferation and induce differentiation of glioma cell lines when cocultured	
Express various cell cycle-specific proteins (e.g., cyclin D_2 homologue +)	
Express various patterning and differentiation-promoting gene products e.g., *engrailed* 1$^+$ and 2$^+$, *notch* homologue$^+$, *disheveled* homologue$^+$	
Express various neural-specific extracellular matrix molecules e.g., NCAM$^+$, TAG$^+$, Neural-specific syndecan$^+$	
Express neuronal-specific transcription factors (e.g., MEF-2 C$^+$)	
Express various lysosomal enzymes e.g., β-glucocerebrosidase$^+$, β-glucuronidase$^+$, hexosaminidase A$^+$ and B$^+$	

1994; Renfranz *et al.*, 1991; Shihabuddin *et al.*, 1993; Snyder *et al.*, 1992, 1993, 1995a; Snyder, 1992). Progenitor clones which were multipotent *in vitro* often recapitulated that multipotency *in vivo*. They differentiated into neurons or glia in a manner appropriate to their site of engraftment. Some transplant-derived neurons received appropriate synapses and possessed appropriate ion channels (Macklis *et al.*, 1994; Snyder *et al.*, 1992, 1993; Snyder, 1992). In some cases, the full expression of a mature, differentiated phenotype (e.g., some electrophysiologic properties and cytoskeletal molecules) was not attained by these stem-like cells until they were returned, through transplantation, into their "natural" *in vivo* environment. The observation that immortalized progenitors could participate in normal development and differentiate into appropriate cells *in vivo* further validated these cell lines as models for studying commitment and differentiation of endogenous CNS progenitors or stem cells.

2. Tracing the Lineage of Neural Cell Types

In situ lineage analyses, using retroviral vectors and tracer dyes to track the progeny of individual endogenous progenitors in certain parts of the vertebrate CNS (e.g., retina, tectum, spinal cord), indicated that some progenitors give rise to multiple neural cell types, both neuronal and glial (Cepko *et al.*, 1993; Luskin, 1994; Snyder, 1992) [see Chapter 22]. Though the degree to which this is true remains controversial, this assessment was enabled by the fact that clonally related cells in these particular regions remained as identifiable clusters into adulthood (Turner *et al.*, 1990). However, the ability to extend this technique and its conclusions to other regions of the mammalian CNS was confounded by the extensive migration of some clonally related cells, e.g., in neocortex, striatum, and cerebellum (Fishell *et al.*, 1993; Lois and Alvarez-Buylla, 1994; O'Rourke *et al.*, 1992; Snyder *et al.*, 1992; Walsh and Cepko, 1993). Thus, there were limited interpretable data on the potency of an individual progenitor in such areas. Although techniques have been devised for circumventing this limitation [e.g., tagging each clone with a unique retroviral insert (Walsh and Cepko, 1992)], immortalization and cloning of individual progenitors, followed by their transplantation into the developing CNS represented one alternative approach to the question of potency. That the multipotency observed *in vitro* was recapitulated by clonal progenitor lines *in vivo* suggested that this characteristic may be a universal property of neural progenitors and stem cells (Snyder, 1992, 1994).

3. How Influential Are Environmental Cues?

Transplantation has been a time-honored technique for determining whether grafted cells follow autonomous developmental programs or accommodate to their new surroundings (McConnell, 1988; Sotelo and Alvarado-Mallat, 1987a,b). The use of primary fetal tissue for such investigations, while appealing in its direct connection to normal development, is complicated by the lack of homogeneity of its progenitor population, a limitation similar to that discussed for primary cultures (Section III,A).

The degree to which some immortalized progenitor clones can accommodate to their environments has actually been a bit surprising. For example, the v-*myc*-immortalized C17-2 clonal progenitor line (Ryder *et al.*, 1990; Snyder *et al.*, 1992) was originally derived from neonatal mouse cerebellum (Fig. 1). One would presume that, by this relatively late stage in CNS development, the phenotypic choices for even pluripotent progenitors would be fairly restricted. However, when examined at maturity, following transplantation into various germinal zones at various developmental time points, these same immortalized, postnatally derived cerebellar progenitors engrafted extensively and participated in development of multiple regions along the neuraxis and at multiple stages from embryo to adult. They differentiated into multiple cell types in these various CNS loci, apparently in response to microenvironmental signals. Donor progenitors intermingled nondisruptively with local endogenous progenitors and responded to the same spatial and temporal cues in the same manner as host progenitors. They differentiated into the types of neuron and glia expected in the respective region at the particular developmental stage of the transplant (e.g., Fig. 2). The engrafted cells resided on the "brain side" of the blood–brain barrier (BBB) and developed into integral members of the CNS cytoarchitecture, e.g., donor-derived neurons received appropriate synapses; the BBB remained intact where donor-derived astroglia put foot processes onto cerebral vasculature; donor-derived oligodendroglia myelinated neuronal processes. Transplanted mice exhibited no indications of neurologic dysfunction. Thus, the structures that received contributions from donor cells appeared to have developed in a functionally normal way (Snyder *et al.*, 1993, 1995a).

Though the number and dispersion of cells often seen at maturity suggested a component of mitosis and migration post-transplant and prior to end differentiation—donor progenitors may, in fact, need to be plastic and mitotic at the time of initial engraftment—no brain tumors were ever seen. The immortalization process did not subvert the ability of these progenitors to respond to normal cues (e.g., withdraw from the cell cycle, differentiate, interact with host cells). Stable integration and/or differentiation may, in fact, be accompanied by a diminution in v-*myc* immunoreactivity, suggesting downregulation or degradation of the immortalizing gene product.

To help rule out the possibility that there was mere selection by a particular host environment for particular donor cells rather than accommodation by progenitors to various host environments, the following manipulation was performed. Stably engrafted transplant-derived cells were recultured from the forebrain of an adult mouse who, as a fetus, had received a C17-2 cerebellar progenitor line implant into the rostral ventricular zone (VZ). These "retrieved" cells, derived from progenitors which had been stably integrated, were then propagated and reimplanted back into the cerebellum of a newborn mouse. The cells successfully engrafted and differentiated into the cerebellar cell types previously reported (Snyder *et al.*, 1992). Being able to retrieve donor cells, long stably integrated in one region of the brain, and reimplant them into an entirely different region, at an entirely different time, would seem to rule out selection by a particular host environment for particular donor cells, reinforcing the notion of plasticity of such progenitors. Furthermore, while the role of the immortalizing gene cannot be discounted, this observation is intriguingly consistent with recent findings that the adult brain may either harbor plastic progenitors in a quiescent state or that differentiated neural tissue can, in the presence of appropriate stimulatory factors, experience a reactivation of

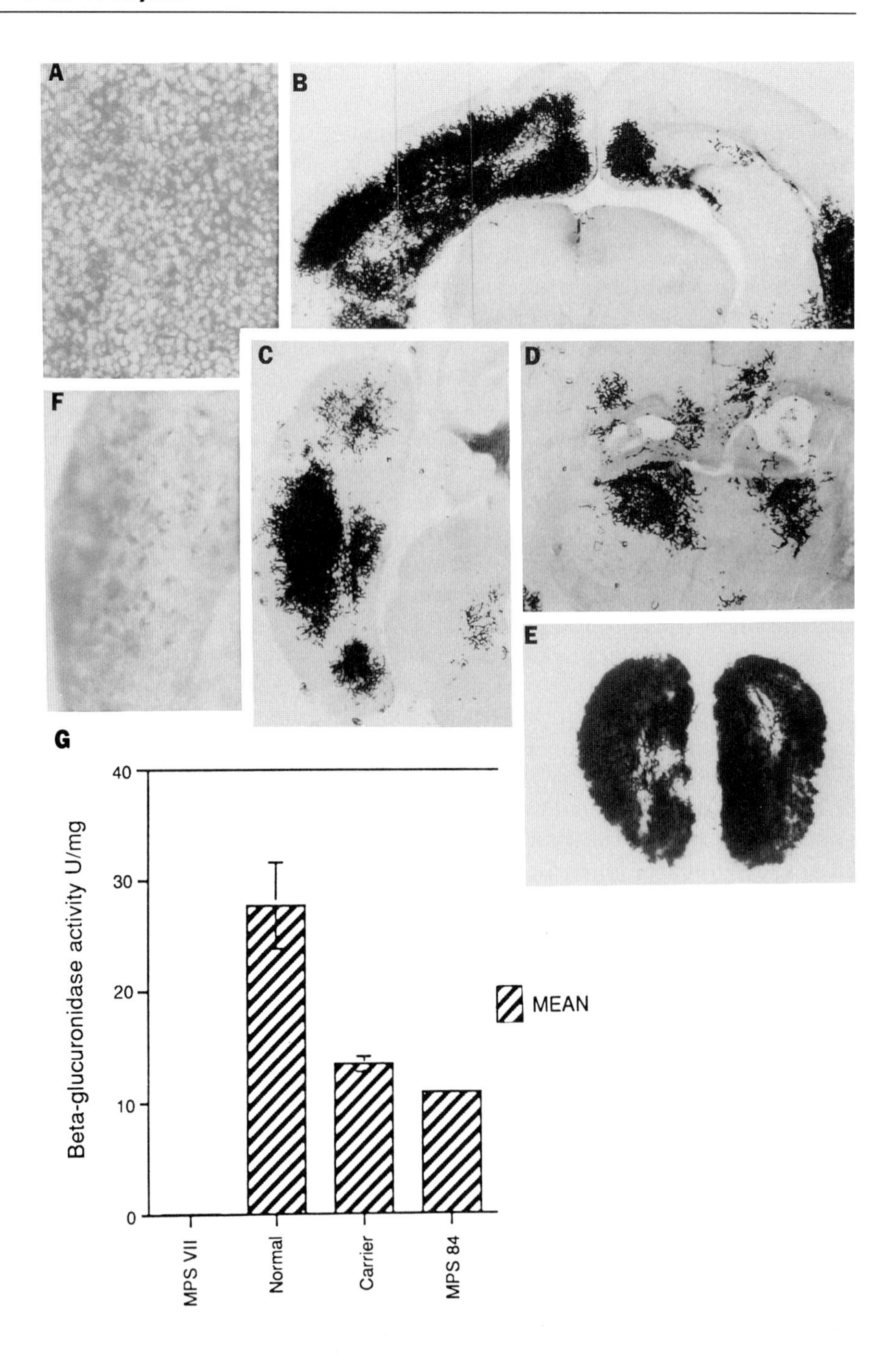
A
B
C
D
E
F
G
40
30
20
10
0
Beta-glucuronidase activity U/mg
MEAN
MPS VII
Normal
Carrier
MPS 84

dormant programs (e.g., Gage *et al.*, 1995; Reynolds and Weiss, 1992; Morshead *et al.*, 1994; Ray *et al.*, 1993).[5]

The ability of immortalized progenitors derived from one region to engraft in another region has also been demonstrated for embryonic hippocampal-derived lines (Renfranz *et al.*, 1991; Martinez-Serrano *et al.*, 1994) and for embryonic raphe-derived lines (Onnifer *et al.*, 1993a; Shihabuddin *et al.*, 1993).

These observations suggest that a single CNS progenitor (even that obtained from a postnatal animal) possesses a plasticity sufficient to give rise to progeny with multiple fates. These data also suggest that immortalized lines may not be committed/restricted to their region of origin, but may represent CNS progenitors with broader potential.

4. Can Progenitors Switch in Response to Altered Cues?

Examining the differentiation of progeny from one progenitor within host environments where normal local cellular and molecular signals have been altered in defined ways—e.g., in mutants, lesion paradigms, gene "knockouts"—may be another way in which immortalized clonal cell lines may help in the study of the interaction between internal and external cues in neural precursor development.

[5] While such findings have provocative developmental implications, they have heuristic value as well. Not all progenitor lines engraft. Furthermore, not all passages or subclones of even a competent line engraft with equal efficiency. Lines "retrieved" and studied may be one way to discern the properties of subclones whose "ancestors" were proven to be particularly good "engrafters" and "gene expressors." In fact, ensuring efficient engraftment and safety of recipients will entail, in general, identifying the variables which direct successful engraftment and understanding the attributes and fate of cells that *do* engraft and express their transgene [Sections IV,C and V,A].

Figure 2 Brain of a mature homozygote mouse with mucopolysaccharidosis (MPS) VII (Sly's disease; β-glucuronidase deficiency) who, as a newborn, received an intracerebral transplant diffusely of the C17-2 neural progenitor cell line expressing β-glucuronidase (GUSB). (A) GUS staining of cells in culture prior to transplantation. (B–E) Transplant-derived C17-2 cells, identified via the X-gal histochemical reaction for the *lacZ* reporter gene, can be seen to have integrated normally throughout the cytoarchitecture of the recipient brain. The representative structures pictured are (B) telencephelon at the level of the hippocampus (including parietooccipital cortex and hippocampus), (C) posterior telencephelon (including occipital cortex) and midbrain, (D) telencephaleon at the level of the striatum (including striatum, corpus callosum, and frontoparietal cortex), and (E) olfactory bulbs (magnified). (F) Close up of engrafted *lacZ*$^{+}$ cells from region shown in (C), now stained for GUSB. (G) Assay of GUSB activity in these regions of this engrafted brain is orders of magnitude greater than in untransplanted MPS VII mice and at nearly the level present in heterozygotes. Levels as low as 2% of normal are probably sufficient to decrease storage based on observations in liver and spleen (Wolfe *et al.*, 1990, 1992). Untransplanted MPS VII mice show no GUSB activity biochemically [as in (G)] or histochemically (not shown). (Portions reproduced with permission from Snyder *et al.*, 1995a).

This premise is illustrated in the following study in which the differentiation of a progenitor cell line was examined in a developmentally perturbed environment (Macklis *et al.*, 1994). "Targeted photolytic cell death" is a technique for experimentally eliminating a particular subpopulation of neurons (Macklis, 1993; Sheen *et al.*, 1992; Sheen and Macklis, 1994). It can provide a highly controllable model of neuronal degeneration even in the adult CNS. Briefly, when photoactive nanospheres are microinjected unilaterally into adult mouse neocortex, they are incorporated by axon terminals and retrogradely transported to the cell bodies of pyramidal neurons of the contralateral neocortex whose axons project across the corpus callosum. These nanospheres carry a chromophore which produces toxic singlet oxygen following photoactivation. A highly selective, geographically defined, and cell population-specific degeneration can be produced noninvasively by photoactivation of these retrogradely transported nanospheres within targeted neurons by using deeply penetrating, near-infrared, laser illumination with optics that limit the beam to lamina II/III. There is no injury to nontargeted, intermixed neurons, glia, axons, or connective tissue. Targeted host neurons, however, undergo apoptotic degeneration (Sheen *et al.*, 1992; Sheen and Macklis, 1994). Apoptosis not only plays a role in various neurodegenerative processes, but is also pivotal in normal CNS development (Johnson and Deckworth, 1993; Loo *et al.*, 1993). It was hypothesized, therefore, that apoptotic degeneration of a targeted adult neuronal population may create a microenvironment not normally available in adulthood, and that this process might recapitulate signals mediated by apoptosis during normal development to which neural progenitors could respond.

Cells from the C17-2 multipotent progenitor line (Ryder *et al.*, 1990; Snyder *et al.*, 1992, 1995a; Snyder, 1992) were transplanted into the neocortex of either intact adult mice or adult mice rendered selectively deficient of layer II/III callosally projecting pyramidal neurons via targeted photolysis (Macklis *et al.*, 1994). In intact neocortex, donor progenitors consistently differentiated into only glia, not an unexpected finding given that gliogenesis is the predominant process normally ongoing in adult neocortex. However, in adult recipients with photolytic degeneration, the donor progenitors, within regions of selective neuronal death, differentiated into pyramidal neurons, extending axons and dendrites and establishing afferent synaptic contacts, as if replacing the degenerated neuronal population. Engrafted cells within even closely approximated normal structures (e.g., thalamus, corpus callosum, even nearby neocortex) displayed glial phenotypes resembling those from the control group.

The differentiation of progenitors within these regions of neuronal death in adult cortex suggested that this form of degeneration created a microenvironment permissive or instructive for neuronal differentiation, perhaps through reactivation of signals ordinarily available only during embryonic corticogenesis. The use of clonal, multipotent cell lines allowed the inference that some progenitors are sufficiently plastic that their progeny can actually shift their fates in response to new signals, including those that accompany some forms of degeneration. Furthermore, by virtue of their abundance and homogeneity, these lines might facilitate the isolation of those factors responsible for neuronal differentiation. Finally, because there was a suggestion that these engrafted progenitors were replacing, at least in part, the degenerated neuronal population, this study also established a paradigm for neural progenitor transplantation as a possible cell replacement therapy, even in the adult CNS, for some degenerative, developmental, and acquired diseases of neocortex and other CNS structures (see Section IV,C,1).

5. How Representative Are Immortalized Progenitors?

Whether these stem-like cells are representative of the majority of mammalian CNS progenitors, or merely an unusual subtype, in uncertain. This uncertainty, in fact, represents one of the limitations in the use of progenitor lines (immortalized by any means) for addressing questions of normal development. Gao and Hatten (1994) voice caution in assuming that immortalized progenitors accurately reflect the behavior of endogenous progenitors *in vivo*. They determined that, while immortalized progenitors derived from the external granular layer (EGL) could, upon transplantation, give rise to a wide range of interneurons and glia in adult cerebellum (as described in Renfranz *et al.*, 1991; Snyder *et al.*, 1992), cultures of primary EGL cells, similarly implanted, yielded only granule cells. Immortalization, therefore, while not promoting abnormal behavior by progenitors, may artificially suspend their commitment and differentiation. The phenotypic range of endogenous progenitors might become narrowed more quickly than data from immortalized cells would predict. This speculation is consistent with findings by some investigators who, after tracking the progeny of some endogeneous progenitors *in vivo*, report that the divergence of lineages for various neuronal and glial cell types in various CNS regions may be quite early in embryogenesis (reviewed in Luskin, 1994).

That it is not the immortalizing gene per se that subverts differentiation programs is suggested by recent observations that neural cell lines perpetuated by chronic neurotrophin exposure (e.g., to bFGF or EGF), when engrafted, may also give rise to multiple cell types (Gage *et al.*, 1994, 1995; Ray *et al.*, 1994; Svendson *et al.*, 1994). Therefore, it is likely that keeping progenitors proliferative and from exiting the cell cycle preserves their phenotypic options. Furthermore, the extensive multipotency often seen in immortalized progenitors is consistent with that observed in primary neural stem cells. Such stem cells have recently been reported to constitute 7% of the progenitors in primary cultures of embryonic rodent VZ (Davis and Temple, 1994). It remains unclear how VZ cells become restricted from multipotency to oligo- or unipotency, leaving a minor residual stem cell population among a larger population of more restricted progenitors. In forestalling phenotypic restriction, immortalization may simply maintain progenitors in their normal antecedent "stem cell-like" state—self renewing and multipotent. Immortalized neural cell lines may, therefore, be valid models for primary neural stem cells and might aid developmental biologists in making the often difficult distinction between a progenitor's fate (as revealed through *in vivo* tracing of endogenous precursors) and its potential (as elucidated through the experimental manipulation of stem-like cells).

IV. Transplanting Immortalized Neural Cell Lines for Gene Therapy and Repair

A. Rationale

In many neurologic conditions, the absence of an identified enzyme, cofactor, neurotransmitter, or trophin has been implicated as part of the disease process (Breakefield *et al.,* 1993; Friedman, 1994; Suhr and Gage, 1993). Pharmacologic agents administered systemically for such conditions not only often have erratic or transient efficacy, but frequently produce undesireable side effects. For various inherited metabolic diseases bone marrow transplantation (BMT) and enzyme replacement have been successful in addressing peripheral manifestations but have been disappointing in reversing or forestalling damage to the CNS, presumably because of restrictions imposed by the BBB to entry of therapeutic molecules supplied peripherally. Also, BMT entails conditioning irradiation which is inimical to developing CNS. The delivery of gene products directly to the CNS might circumvent these problems.

In other types of neurologic disorders, specific neural cell types or circuits degenerate. These losses may be due to processes intrinsic to the dying cell, but may also result from an insufficiency of various trophins or from the presence of certain toxins in the milieu.

In some conditions (e.g., ischemia, trauma), multiple pathologic processes may be at work, each exacerbating the other.

"Gene therapy" is defined most broadly as providing a gene whose product can alleviate the consequences of a defective gene (Kay and Woo, 1994; Mulligan, 1993; Snyder, 1995b). The minimal requirements for such therapy are simply absence of undesirable side effects and sustained production of the gene product. Gene transfer into the CNS may be achieved by the direct delivery of genetic material to the host's own neural tissue. The vectors currently available, however, are difficult to target *in situ* to the specific neural cell types and regions most in need of correction (Fisher and Ray, 1994) (see Section I). Retroviral vectors infect only mitotic cells, which are less prevalent in postnatal CNS and often not the cells needing therapy. The safety and efficacy of herpes and adenovirus vectors for postmitotic neural tissue *in vivo* remains to be established. Alternatively, genes may be imported into the host CNS by the implantation of genetically modified donor cells that can reside within the CNS and make exogenous gene products accessible to the host's neural tissue. Donor cells may be chosen for their ability to provide a source of exogenous substances that can diffuse to appropriate targets, to become integral members of the host cytoarchitecture and circuitry, or to do both (Snyder, 1994, 1995a,b; Snyder and Flax, 1995; Snyder *et al.*, 1995a).

Neurons derived from the CNS would seem to be the ideal graft material. However, there are restrictions on the types and ages of neurons that successfully survive implantation in a functionally meaningful way for prolonged periods (Fisher and Gage, 1993; Freed, 1993). Also, primary neurons, which have limited mitotic capacity, cannot be efficiently transduced with foreign genes by standard retrovirus-mediated infection. Therefore, they have limited usefulness as vehicles for gene transfer (although herpes and adenovirus vectors may ultimately make postmitotic neurons more easily genetically manipulated *ex vivo* (Fisher and Ray, 1994; Suhr and Gage, 1993). Primary fetal neuronal tissue has historically proven to be the most successful donor tissue for CNS grafting and has shown promise recently for the amelioration of certain neurologic conditions (e.g., Parkinsonism) (Freed, 1993). However, the routine use of such tissue raises significant concerns, both biologic and ethical (Gage, 1993): e.g., availability of suitable material, ensuring survival of desired cells in tissue which is typically heterogeneous and contains nonneural cells, and augmenting the amount of desired factors produced by such cells (primary fetal tissue is not easily genetically engineered).

Alternative sources to neuronal grafts have, therefore, been sought that might provide exogenous therapeutic gene products and/or effect repair of damaged host brain as integral members of the cytoarchitecture (Bjorklund, 1993; Fisher and Gage, 1993; Gage, 1992). Although genetically engineered nonneural cells are excellent vehicles for passive, localized delivery of discrete molecules to the CNS, they lack the ability to incorporate into host cytoarchitecture in a functional manner following implantation (Snyder, 1995b). For that reason, not only may essential circuits not be reformed, but the regulated release of certain substances through feedback loops may be missing. For some substances, unregulated, inappropriate, excessive, or ectopic release may actually be harmful to the host. Furthermore, the loss of foreign gene expression may leave engineered nonneural cells incapacitated, whereas donor tissue originating from brain may intrinsically produce various CNS factors, allowing correction to proceed despite inactivation of the introduced gene. In fact, CNS-derived tissue may provide as yet unrecognized endogenous neural-specific substances that are beneficial to the host.

Therefore, the demonstration that immortalized neural progenitors could integrate appropriately into the CNS and stably express a foreign gene made this strategy an attractive alternative for CNS gene therapy and repair (Table 2). It represented a unique example of integration by exogenous mammalian CNS tissue which was neither of tumor or primary fetal origin. (First reported contemporaneously in Renfranz *et al.* (1991) and Snyder *et al.* (1992), these findings have been subsequently supported in the following references: Anton *et al.* (1994), Flax *et al.* (1994), Gage *et al.* (1994), Macklis *et al.* (1994), Lacorraza *et al.* (1995), Martinez-Serrano *et al.* (1994), Onnifer *et al.* (1993a), Ray *et al.* (1994), Shihabuddin *et al.* (1993), Snyder *et al.* (1993, 1995a,b), Snyder (1995a), Snyder and Flax (1995), and Svendsen *et al.* (1994).

B. Strategies

Once it had been established that immortalized neural cell lines could engraft and participate in the normal development of a wide range of structures

Table 2
Advantages of Immortalized Neural Progenitors over Other Potential Sources of Transplant Material

Advantages over primary fetal tissue
Homogeneous
Abdundant
Predictably accessible
Characteristics and quality of material known and controllable
Growth and differentiation systematically manipulatable
Advantages over primary neurons
Homogenous
Abundant
Predictably accessible
Characteristics and quality of material known and controllable
Growth and differentiation systematically manipulatable
Mitotic
Easily maintained for long periods and/or multiple passages *in vitro*
Easily genetically modified by retrovirus
Cells successfully expressing a transgene can be easily selected
Plastic and multipotent (greater range of implantation sites, precise targeting less crucial, accommodate to graft site)
Advantages over nonneural cells (primary or cell lines)
Often constitutively produce neural-specific enzymes
Often contain neural-specific promoters
Able to integrate functionally (e.g., able to receive and make synapses)
Might respond to normal neural signals (e.g., receive feedback, release gene product in a regulated fashion)
Nondisruptive, nontraumatic, and, vs nonneuronal cell lines, nontumorigenic
Plastic and multipotent (greater range of implantation sites, precise targeting less crucial, accommodate to graft site)
Capable of addressing developmental questions (e.g., neural lineage, differentiation, and potential)

along the neuraxis and at multiple stages (Snyder *et al.*, 1993), the next step was to test their ability to effect repair by repopulating the cytoarchitecture of lesioned or mutant animals. (Restitution of lost function in deficient hosts would, in fact, be an additional test of the authenticity of these cells.) The availability of mouse models with specific cell type, structural, and/or enzyme deficiencies (some of which also typified human neurodegenerative conditions) made feasible such complementation experiments in animals. Correction of neuropathology by a given progenitor cell line might be accomplished by gene product (enzyme) replacement, cell replacement, or, conceiveably, both simultaneously.

To these ends, many of the inherent biologic properties of immortalized progenitors and stem-like cells could be exploited (Table 3): for example, (a) mitosis, enabling efficient *ex vivo* genetic modification via retrovirus vectors of this ready and unlimited supply of well-characterized, uniform cells; (b) pluripotency, allowing the cells to assume an array of phenotypic and regional CNS fates within various environments and obviating concerns for precise targeting or the need for tissue-specific vectors; (c) facile engraftability, particularly into germinal zones; and (d) capacity to migrate, particularly if implanted within migratory germinal zones (Fishell *et al.*, 1993; Flax *et al.*, 1994; Lois and Alverez-Buylla, 1993; O'Rourke *et al.*, 1992; Walsh and Cepko, 1993). There is a growing recognition that progenitors and germinal zones persist throughout development, including within adult CNS (Daadi *et al.*, 1994; Gage *et al.*, 1994a; Lois and Alvarez-Buylla, 1993; Morshead *et al.*, 1994; Reynolds and Weiss, 1992; Richards *et al.*, 1992). Therefore, a strategy employing immortalized neural progenitors might be invoked at any stage of life. The introduction of these cells into a large, migratory germinal zone (from fetus to adult) provides a means for their diffuse engraftment and delivery of gene products throughout the CNS as integral cytoarchitectural components (Flax *et al.*, 1994; Snyder *et al.*, 1993, 1995a). [At particular developmental stages, this approach may actually entail no more than injecting progenitors into a recipient animal's cerebral ventricles (Snyder *et al.*, 1993, 1995a) (Fig. 2) (see Sections IV,C,2 and IV,D,2)]. However, this method of delivery does not preclude being able to transport gene products into the cytoarchitecture of circumscribed regions in order to effect selective manipulations (e.g., Anton *et al.*, 1994; Gage *et al.*, 1994a; Macklis *et al.*, 1994; Martinez-Serrano *et al.*, 1994; Onnifer *et al.*, 1993a; Renfranz *et al.*, 1991; Snyder *et al.*, 1992) and avoid extensive genetic alteration (often inherent in other methods of gene delivery).

Therefore, a strategy of progenitor-mediated gene transfer is also very adaptable.

One scenario might have normal donor progenitors participate early on in the development of abnormal hosts. Cells degenerated by mechanisms *intrinsic* to the host cell, or in response to a formerly inhospitable but now rectified host environment, might be replaced. However, even cells that had died by

Table 3
"Virtues" of Immortalized Neural Progenitors that Make Them Well Suited for Transplantation as Vehicles for Gene Transfer Directly to the CNS and as Tools for Neural Cell Replacement

- Mitotic in culture
  - Facilitates efficient *ex vivo* genetic manipulation via retroviruses
    - Genes of therapeutic interest
    - Genes of developmental interest
    - Genes which might promote differentiation of both host & donor cells
- Plasticity
  - Allows for an array of phenotypic and regional fates
  - Less concern over precise targeting or target selection
- Facile engraftability and integration
  - Particularly into germinal zones (which persists even in adulthood)
    - Utility, therefore, at multiple points within life span
- Migration
  - Cells inherently migratory
  - Cells may integrate into germinal zones which are also migratory
- Adaptable to variations in delivery technique
  - Introduction of cells into large and migratory *germinal zones* promotes *diffuse* engraftment and gene product delivery
  - Introduction of cells into *circumscribed* neural regions allows *selective, constrained* manipulation
  - Both techniques, however, still restricted to CNS—avoids systemic side effects from peripheral delivery of some pharmacologic agents
- Incorporate appropriately into host cytoarchitecture
  - Might reform essential *circuitry* in functional manner
  - Might reconstitute feedback loops for *regulated* release of substances
- Might, at times, circumvent loss of foreign gene expression
  - Because cells are neural derived, they may endogenously produce certain neural gene products of interest, allowing correction to proceed despite inactivation of introduced gene
- Might provide as-yet unrecognized neural-derived substances
  - Such endogenous substances may be equally beneficial to host

etiologies *extrinsic* to the host cell (e.g. a refractory enzyme deficiency or toxin) might conceivably be replaced by progenitors engineered to be *resistant* to ongoing metabolic insult. Or, if the lethal agent were operative only within a particular spatial or temporal window in the development of a host structure, normal progenitors could be implanted "downstream" of its action, thus evading destruction and successfully replacing degenerated host cells. Finally, such implanted donor cells might provide factors, "bridges," and/or cell–cell contact signals that might induce the injured host to reconstitute its *own* lost cells and connections.

Examples of some of these therapeutic paradigms are summarized below. It's instructive to note that many of these examples actually employ the same cell line (C17-2), reinforcing the breadth of a multipotent progenitor line's potential.

C. Cell Lines for Gene Transfer and Enzyme Replacement

Data suggested that gene products of therapeutic interest might be transfered and expressed—as was the *lacZ* reporter gene—by exogenous progenitors becoming integral members of host CNS cytoarchitecture. To test this hypothesis, transplants were performed into well-established rodent models in which defects in single genes or factors had been defined.

1. Transplantation for Focal Enzyme Replacement

In the first studies employing immortalized neural cell engraftment, investigators demonstrated the feasibility of transporting foreign—often therapeutic—genes into the CNS by transplanting specifically into circumscribed neural regions [e.g., cerebellum (Snyder *et al.*, 1992; Renfranz *et al.*, 1991), cortex (Macklis *et al.*, 1994; Onnifer *et al.*, 1993a), hippocampus (Renfranz *et al.*, 1991), striatum (Anton *et al.*, 1994), septum (Martinez-Serrano *et al.*, 1994)]. Anton *et al.* (1994), for example, showed that tyrosine hydroxylase expressed in this way within the striatum of Parkinsonian rats and monkeys may improve motor performance. Martinez-Serrano *et al.* (1994) showed that nerve growth factor expressed in this fashion near fimbria–fornix lesions in rats appeared to salvage severed septal-hippocampal cholinergic fibers and aid performance on memory tasks.

2. Transplantation for Widespread CNS Enzyme Replacement

Many of the mouse models, however, in which gene products are deleted or defective are exemplars of inherited metabolic neurodegenerative lysosomal storage diseases. Humans with these conditions are mentally retarded. The mice are characterized by widespread neuropathology that would not be addressed by local CNS grafts.

Recently, a rapid intraventricular injection technique was devised for diffuse engraftment of immortal progenitors in order to effect sustained, direct delivery of a therapeutic gene product throughout the recipient brain and to treat widespread neuropathology resulting from neurogenetic metabolic lesions (Snyder *et al.*, 1995a) (Fig. 2). Injecting the progenitors into the lateral ventricles presumably allows them to gain access to most of the subventricular germinal zone (Altman, 1969; Levison and Goldman, 1993; Lois and Alvarez-Buylla, 1993, 1994; Luskin and McDermott, 1994; Smart, 1961), including that of IIIrd and IVth ventricles, as well as to networks of cerebral vasculature, along the surface of which they might also migrate. Recipients survive this rapid, simple procedure without cerebrospinal fluid obstruction or other morbidity. This approach works equally well in the fetus in which donor progenitors gain access to the VZ (Snyder *et al.*, 1993). Donor cells migrate into the parenchyma from the ventricles within 24 hr of transplantation and take up residence on the brain side of the blood–brain barrier as integral members of the CNS cytoarchitecture (see Section III,B,3). This engraftment technique, therefore, exploits many of the inherent properties of stem-like cells to become components of normal structures throughout the host brain, permitting missing gene products to be delivered in a sustained, direct, and perhaps regulated fashion, without disturbing other neurobiological processes. [As discussed later (Section IV,D,2), this administration technique might be extended to other types of diffuse neuropathologies, including even those requiring widespread cell replacement].

a. Mucopolysaccharidosis Type VII (MPS VII) Mouse The MPS VII mouse, a prototype for genetic neurovisceral lysosomal storage diseases, was an ideal model recipient for testing these hypotheses. Mice homozygous for a deletion mutation of the β-glucuronidase (GUSB) gene are devoid of that secreted enzyme. They experience lysosomal accumulation of glycosaminoglycans in brain and other tissues causing a fatal progressive degenerative disorder, including mental retardation, which mimics the inherited human MPS VII condition (Sly's disease) (Sly *et al.*, 1973; Vogler *et al.*, 1990). Disease progression in the CNS begins at birth and reaches its full extent by 3 weeks of age. Treatments are designed to provide a source of normal enzyme for uptake by diseased cells. The presence of a receptor for GUSB on all cells in the MPS VII mouse (including within CNS) makes cross-correction by GUSB-expressing cells possible. However, for this syndrome, as for many genetic metabolic diseases, while BMT, somatic cell therapy, and/or peripheral enzyme replacement have met with success in reversing hematologic, skeletal, hepatic, and other visceral manifestations, they have been largely unsuccessful in reversing or forestalling damage to CNS, presumably because of restrictions imposed by the BBB to the sustained entrance of corrective GUSB levels (Birkenmeier *et al.*, 1991; Moullier *et al.*, 1993; Sauds *et al.*, 1993, 1994; Wolfe *et al.*, 1990, 1992a,b).

The C17-2 immortalized mouse neural progenitor line (Snyder *et al.*, 1992, 1993) constitutively expresses GUSB. The line was additionally exposed *ex vivo* to a retroviral vector encoding human GUSB typically enabling infected cells to release five-fold higher levels of GUSB (Wolfe *et al.*, 1995). By injecting GUSB-expressing C17-2 cells into the cerebral ventricles of newborn MPS VII mice, the donor cells engrafted throughout the neuraxis of 94% of transplanted mutants (Snyder *et al.*, 1995a). At maturity, donor-derived cells were present as normal constituents of diverse brain regions (Fig. 2). Intense GUSB activity colocalized with integrated donor cells. All of the engrafted recipients had evidence of corrective GUSB activity throughout the brain with nearly heterozygote levels in some regions (Fig. 2). In fact, approximately half the engrafted brain regions expressed more GUSB than is actually necessary to reverse *prexisting* storage in other tissues (Wolfe *et al.,* 1990, 1992a). This diffuse GUSB expression along the neuraxis resulted in widespread cross-correction of lysosomal storage in both mutant host neurons and glia in all engrafted areas examined—including cerebral cortex—compared to age-matched untreated MPS VII controls (Fig. 3). Engraftment, expression, and neuropathologic rescue could be detected at least 8 months post-transplant (typical life span of an MPS mouse). While histochemical stains and assays for GUSB are unable to distinguish between the mouse product (endogenously produced by donor cells) and human GUSB (introduced via retrovirus), PCR for proviral sequences indicated the long-term presence of the latter in engrafted brains. No tumors were ever seen and the CNS cytoarchitecture of recipients was never disrupted. Observation of cage behavior suggested that MPS recipients might be more active and alert than age-matched controls.

In short, the neonatal transplantation of GUSB-producing neural progenitors produced long-term improvements in the MPS VII mouse brain, as neonatal BMT and peripheral enzyme replacement did for the skeletal and visceral disease but were unable to do for the CNS. These observations, therefore, suggested a strategy for "CNS gene therapy" of a class of neurogenetic diseases which, heretofore, had not been adequately treated by extant techniques. Therapy instituted early in life might arrest CNS disease progression in many inherited metabolic diseases. This may be important for treating conditions causing mental retardation because it is unclear whether permanent alternations in brain function occur early that may not be reversible even if normal metabolism is later restored.

Though most interventions to date have been instituted as early as possible (newborns, fetuses), recent findings suggest that similar treatment of adult mutants might be worth attempting. C17-2 cells, if injected into the ventricles of normal adult mice, integrate into the subependymal germinal zone (SEZ) and migrate long distances—for example, to the olfactory bulb where they differentiate into interneurons and, occasionally, into subcortical parenchyma where they become glia (Flax *et al.,* 1994). This behavior is identical to that of endogenous SEZ progenitors in adult rodents (Lois and Alvarez-Buylla, 1994). Because reporter genes are successfully expressed by C17-2 in these distant locations, they might also be capable of expressing therapeutic enzymes which might cross-correct host cells even in *adult* mutant brains.

b. β-Hexosaminidase A (HexA) Progenitor cell-mediated gene transfer is being extended to the replacement of other enzymes whose absence produces metabolic disease extensively involving the CNS.

One such example is HexA, which catalyzes the hydrolysis of GM_2-ganglioside to GM_3-ganglioside. The HexA isoenzyme is a heterodimer composed of an α-subunit (Hex-α) and a β-subunit (Hex-β). Mutations in either of the subunit genes lead to a deficiency in HexA activity, leading to the accumulation of GM_2-ganglioside in lysosomes and resulting in severe neurodegeneration. Deficiencies in hexosaminidase activity are seen in Tay–Sachs disease, Sandhoff's disease, and other gangliosidoses. The most frequent mutation affects Hex-α. The C17-2 neural progenitor line (Snyder *et al.*, 1992, 1993) constitutively produces HexA. When transduced with a retroviral vector encoding the human HexA α-chain, the line produces 6- to 15-fold higher levels of HexA compared with nontransduced subclones (Lacorraza *et al.*, 1995). Transduced lines engrafted extensively following transplantation into fetal and neonatal mouse brain and expressed substantial HexA *in vivo* for at least 2 months as integral cytoarchitectural components throughout recipient brains. Engraftment into normal mice boosted whole brain HexA levels by ~25%.

Figure 3 Decreased lysosomal storage throughout the brains of MPS VII recipients of C17-2 transplants. As per Fig. 2, GUSB activity was distributed extensively throughout recipient brains. Some regions in engrafted MPS VII mice reached nearly heterozygous levels and most regions of all mice had at least 2 or 3% of normal activity, a level sufficient to correct lysosomal storage in other organs (Wolfe *et al.,* 1990, 1992). Lysosomal storage was examined in all regions by toluidine blue staining of X-gal-processed, plastic-embedded sections and found to be absent or diminished compared to analogous regions of untransplanted, age-matched MPS VII controls. Representative regions are shown, but the other regions of treated brains showed similar decreases in storage. (A–D) Donor cell GUSB activity and reduction of storage in the hippocampus. (A) Hippocampus of an *untransplanted* 3 week old MPS VII mouse was completely negative for GUSB histochemical reaction product and B showed abundant white cytoplasmic vacuoles representing distended lysosomes (arrowheads). (C) The same region from a 3 week old MPS VII mutant transplanted neonatally with C17-2 cells. The dense stain represents extensively distributed normal GUSB expression by engrafted cells (as seen in Fig. 2). (D) A toluidine blue-stained, plastic-embedded section of the hippocampus from this mouse, adjacent to the GUSB-positive section shown in C and analogous to the untreated control section in B, showing the absence of white cytoplasmic vacuoles in neurons and glia where GUSB is secreted. (The large white holes are blood vessels.) The section had previously been processed with X-gal histochemistry. (E–H) Decreased lysosomal storage in the neocortex of the same MPS VII mouse. (E and F) Toluidine blue-stained cortex and subcortical area of an *untransplanted* 3 week old MPS VII mouse showing abundant white cytoplasmic vacuoles representing distended lysosomes (arrowheads). (Affected neurons can be seen particularly well in F.) (G and H) Toluidine blue-stained, X-gal-processed, plastic-embedded sections from a transplanted MPS VII mouse. These sections are from an analogous region to those shown in the control, untreated MPS mouse (E and F). Little if any lysosomal storage is seen in either neurons or glia, which are probably cross-corrected cells of host origin because they are X-gal-negative and because many of the corrected cells are neurons, a cell type into which progenitors in a postnatal neocortex would not differentiate. Scale bars represent 320 μm in A and C, 31 μm in B, D, G, and H, and 25 μm in E and F. (Reproduced with permission from Snyder *et al.,* 1995a).

Such results lend further support to the conclusion that progenitor-mediated gene transfer may be a useful strategy for diseases that are amenable to intercellular cross-correction by cells providing recombinant enzymes. This approach is being tested in the recently generated Tay–Sachs ("HexA-knock-out") mouse model (Yamanaka *et al.*, 1994).

In summary, transplanting immortalized neural precursors intrinsically secreting missing gene products, or genetically engineered to do so, may provide a strategy (in combination with other systemic or somatic cell therapies) for

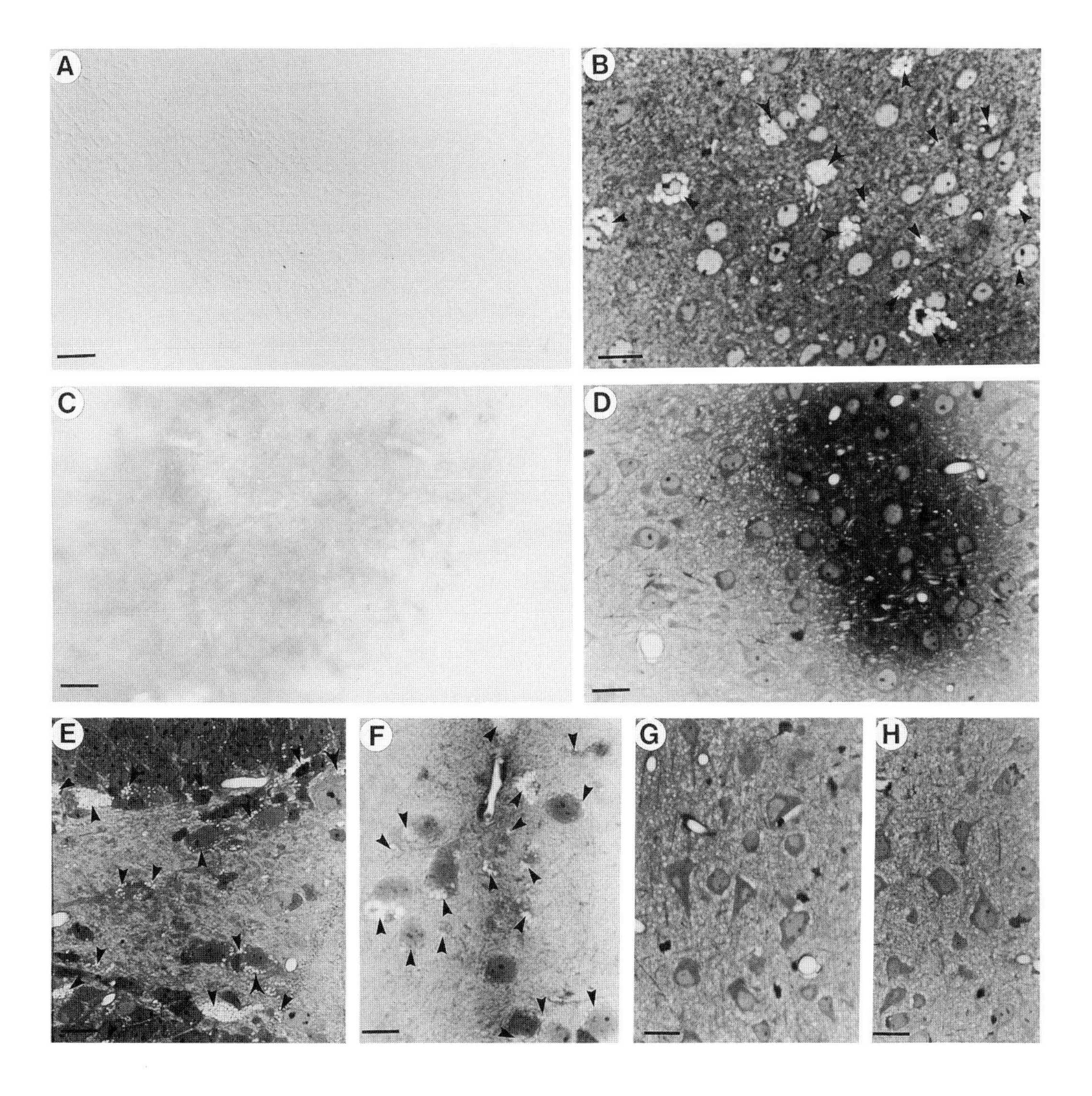

the treatment of the CNS manifestations of a number of neurogenetic diseases. It may, in fact, serve as a paradigm for using such cells for the transfer of other types of factors of therapeutic or developmental interest into the CNS. Although many gene-transfer methods are under study for the brain, exploiting the properties of neural progenitors to become integral members of normal structures throughout the host CNS may allow foreign gene products to be delivered in a sustained, direct, regulated fashion without disturbing other neurobiological processes.

D. Cell Lines for Therapeutic Neural Cell Replacement

1. Replacing Neurons?

a. In Models of Neuronal Degeneration The experiments described under Section III,B,4 suggested the possibility that transplanted immortalized neural progenitors might be capable of replacing certain degenerated neurons, even in the adult CNS. To review briefly, when cells from the C17-2 progenitor line were transplanted into adult mouse cerebral cortex undergoing experimentally induced apoptosis of a subclass of pyramidal neurons (Macklis, 1993; Sheen *et al.*, 1992; Sheen and Macklis, 1994), the donor progenitors integrated within the regions of neuronal degeneration and differentiated into that type of degenerating neuron, partially replacing the lost neuronal population (Macklis *et al.*, 1994). Apoptosis is not only pivotal in normal CNS development (Johnson and Deckwerth, 1993) but is postulated to play a role in various neurodegenerative processes (Johnson and Deckwerth, 1993; Loo *et al.*, 1993). These findings hinted that neural progenitors or stem cells may be capable of partially reconstituting a degenerated neural system. In fact, because these data actually represented the first "repopulation" by stem-like cells of *any* solid organ, they further suggested that this strategy might be useful to investigators studying the therapeutic potential of progenitor or stem cells from nonneural organ systems as well.

b. In Models of Abnormal Neuronal Development Many of the naturally occurring mouse mutants are prototypes for failure of specific types of neurons to develop properly.

The *meander tail* (*mea*) mutation causes a neural developmental defect largely restricted to the anterior lobe (AL) of the cerebellum, featuring a cell-poor EGL that generates a markedly reduced number of granule cell (GC) interneurons within the adult internal granular layer (IGL) (Ross *et al.*, 1990). The C17-2 progenitor line was originally derived from the immortalization of normal neonatal mouse EGL cells (Ryder *et al.*, 1990; Snyder *et al.*, 1992). As previously reported, C17-2 is capable of integrating orthotopically following transplantation into the cerebellum of normal newborn mice and differentiating into various neuronal and glial cell types (Snyder *et al.*, 1992). The cerebellums of *mea* pups were implanted at birth with C17-2 cells (Snyder *et al.*, 1995b).

When cerebellar development was complete, the brains were processed to characterize donor-derived cells. The posterior lobe (PL) for each transplanted mutant served as an internal control for its abnormal AL. Robust engraftment was evidence throughout the mutant cerebellum, giving rise to neurons and glia as per normal mice (Snyder *et al.,* 1992). The deficient AL was crowded with donor-derived cells. A significant number of engrafted progenitors differentiated into GCs. The more progenitors injected, the more donor-derived GCs were found. Some transplant-derived GCs received appropriate synaptic input. The percentage of donor-derived cells which had become GCs in AL was the same as that in PL. Therefore, there was the suggestion that the deficient IGL of the *mea* AL had, indeed, been partially reconstituted by the transplantation of immortalized cerebellar progenitors.

Repopulating deficient regions of brain in models of selective cell-type maldevelopment reinforces the therapeutic utility of transplanting immortal neural progenitors for the replacement of diseased cells in certain conditions. In naturally occurring mutants, such an approach may also help to test hypotheses regarding the pathophysiology underlying a particular mutation (Gao *et al.,* 1992; Gao and Hatten, 1993). For instance, the experiment described above demonstrated the use of a neural cell line to help determine whether a given cell-type degeneration was due to pathology *intrinsic* to the cell or extrinsic to it. That exogeneous normal cerebellar progenitors could survive transplantation into the effected region of the newborn mutant's cerebellum, and differentiate into the deficient neural cell type, suggested that the microenvironment was *not* inimical to normal GC development—or at least not at the developmental stage (i.e., postnatal) or in the region of implantation (i.e., the EGL) of these experiments. Such data suggested that the *mea* gene exerts its deleterious action within the *mea* EGL cell. Alternatively, it is possible that *mea* action is extrinsic to the EGL cell but that its disruptive impact on GC development is restricted to only prenatal events. Immortalized neural progenitor transplantation may, therefore, be therapeutic for cell replacement in disease models if (a) the defect is intrinsic to the host cell to be replaced, (b) the donor cells are "resistant" to a defect extrinsic to the degenerating host cell, or (c) the donor cells can be implanted "downstream" (temporally and/or spatially) of a cell-extrinsic defect.

2. Replacing Glia?

Attempts are ongoing by various investigators to immortalize glial cells via retroviral gene transduction (see footnote 2). Interestingly, however, if it is multipotent, an immortalized neural progenitor line may be used not only for neuronal replacement but also for glial replacement. This point is illustrated by experiments in which the C17-2 multipotent line (Ryder *et al.,* 1990; Snyder *et al.,* 1992) was transplanted into a "glial mutant," the *shiverer* (*shi*) mouse

(Snyder *et al.*, 1995b). C17-2 was the same line employed in the enzyme-deficient and neuron-deficient models described under Sections IV,C,2 and IV,D,1.

The homozygote *shi* mouse suffers from extensive white matter disease (dysmyelination/hypomyelination) because of a deletion of five of seven exons of the gene-encoding myelin basic protein (MBP) which is essential for proper myelination by oligodendroglia. [Many traumatic, ischemic, neurogenetic, metabolic, immunologic, and infectious neurologic conditions are characterized by white matter disease (Fisher and Gage, 1993)]. The *shi* phenotype can be rescued genetically by introducing the wild-type MPB gene into the germ line of *shi* mice (Readhead *et al.*, 1987) or, by cellular methods, in a more modest fashion (e.g., small, discrete foci of spinal cord) by injecting a fragment of primary CNS tissue containing normal, mature, MBP^+ oligodendrocytes (Gout *et al.*, 1988) or immature oligodendrocytes (Groves *et al.*, 1993). Therefore, the possibility seemed to exist that normal neural stem-like cells might be useful for diffuse cellular correction. It was known that, in culture, a number of multipotent progenitor lines (including C17-2) could differentiate into cells which express MBP and, upon transplantation, differentiate into oligodendroglia (Ryder *et al.*, 1990; Snyder *et al.*, 1992; Snyder, 1992). Transplantation of C17-2 cells into *shi* newborns yielded extensive engraftment throughout the mutant brains at maturity, including within white tracts (Snyder *et al.*, 1995b). The intraventricular implantation technique, described under Section IV,C,2 for the widespread engraftment of progenitors, ensured donor cell distribution throughout the mutant's effected brain, an important requirement for the treatment of widespread white matter diseases. MBP was repleted diffusely by donor cells throughout the MBP-deficient *shi* brain, and many donor progenitors differentiated into mature oligodendroglia, some of which myelinated neuronal processes and laid down healthier, more compacted myelin.

Similar experiments with progenitors perpetuated by EGF exposure are being attempted for the focal remyelination of small segments of spinal cord in the myelin-deficient rat (Duncan *et al.*, 1993).

V. Technical Aspects of Grafting Immortalized Neural Cells

A. Labeling Donor Cells for Engraftment

Often the most challenging aspect to the grafting of immortalized neural cells is reliably recognizing donor cells within the recipient brain. With nonneural grafts, the donor cells often form an easily discernable mass. However, grafts of immortalized neural cells will ideally intermingle with host neural tissue making their identification all the more difficult. While donor cells can be labeled prior to transplantation with DiI, Hoechst blue, or PKH, over extended periods of time (sometimes as short as a week), dyes tend to diffuse (mislabeling host cells) or to dissipate (precluding their use for analysis following long intervals post-transplant). Some investigators (Anton *et al.*, 1994; Martinez-Serrano *et al.*, 1994; Renfranz *et al.*, 1991) have used mitotic markers (e.g., thymidine) to label cell lines in culture prior to transplantation. This technique, though labor intensive, when combined with other histologic stains for visualizing cells upon which grains can be seen to be superimposed, can be quite convincing.

If cells from one species are transplanted into hosts of another species, then antibodies to species-specific cell markers may be used to identify donor cells. The mouse-specific M6 cell surface antibody marker has been used to identify engrafted mouse cells in rat host brain (Campbell *et al.*, 1994). Antibodies to human neurofilament have been used to trace axons from engrafted human fetal dorsal root ganglion entering adult rat host spinal cord (Kozlova *et al.*, 1994). The concern with this approach is that cross-species transplantation may not be as robust or reliable as transplants within the same species; graft rejection may compromise the experiments.

A histochemical or immunocytochemical reaction for a reporter gene product is the most efficient way for distinguishing donor cells. The *lacZ* gene product, which can be recognized by the X-gal histochemical reaction, has been used to good advantage by some investigators (Breakefield *et al.*, 1993; Gage *et al.*, 1994a; Macklis *et al.*, 1994; Onnifer *et al.*, 1993a; Snyder *et al.*, 1992, 1993, 1995a,b; Snyder, 1992). In some cases the blue X-gal reaction product fills donor cells so exuberantly that it confers on them almost a Golgi stain quality, making cell type identification quite easy (Snyder *et al.*, 1992) (Fig. 1). The X-gal reaction product is also electron dense (Luskin *et al.*, 1993; Snyder *et al.*, 1992, 1993; Snyder, 1992) permitting ultrastructural analysis not only for purposes of cell type identification but also for ascertaining synapse formation, myelination, and cytoskeletal structures. Immunocytochemical detection of *lacZ* (particularly when enhanced with nickel cobalt) often delineates the distal reaches of neural processes better than does X-gal histochemistry under light microscopy (Flax *et al.*, 1994; Onnifer *et al.*, 1993a,b; Shihaubuddin *et al.*, 1993).

A reporter gene may be easily introduced into lines via retroviral vector transduction (Breakefield *et al.*, 1993; Onnifer *et al.*, 1993a; Renfranz *et al.*, 1991; Snyder, *et al.*, 1992) (Fig. 1). Alternatively, transgenic mice, in which a reporter gene, transcribed from a cell type-specific promoter, is constitutively expressed within the neural cell type of interest, may be used to provide the primary neural tissue from which immortalized lines can subsequently be generated. Immortalization may be performed by either transduction of an immortalizing gene or by chronic neurotrophin exposure (Hammang *et al.*, 1990).

If identification of engrafted donor cells is to be by virtue of a retrovirally transduced reporter gene, which is probably the best labeling technique, a few potential pitfalls must nevertheless be anticipated and avoided. First, it must be remembered that engrafted cells can only be identified if they express their transgene. Marker gene "non-expression" can be misinterpreted as "non-engraftment," leading to an erroneously lower apparent grafting efficiency. (The actual efficiency of transgene expression in neural cell line transplantation is still unknown.) Second, tests for "helper" virus, i.e., the recrudescence of replication-competent virus by recombination, must be ruled out prior to transplantation (Ryder *et al.*, 1990) (see Chapter 5). The presence of helper virus might mean that the retroviral genome encoding the reporter gene could escape and infect host cells, leading to their misidentification as donor cells. The "helper test" can be performed by simply demonstrating that the supernatant from a neural cell line is incapable of transmitting the reporter gene to naive cells (Ryder *et al.*, 1990; Snyder *et al.*, 1992). Some studies have gone as far as demonstrating that X-gal$^+$ cells within a recipient brain actually shared the same unique retroviral insertion site as the donor clonal cell line in the culture dish (Snyder *et al.*, 1992). This method offered unambiguous proof that blue cells were of donor origin. Furthermore, the integration site could serve as a tag for donor-derived cells which was independent of gene expression and at no risk for dilution, diffusion, or phagocytosis.

B. Immune Rejection

An additional observation of note is the apparent ability of immortalized neural cell lines to engraft in multiple mouse strains, even at adulthood (Macklis *et al.*, 1994; Snyder *et al.*, 1992, 1993, 1995a). There is evidence as well that some mouse-derived neural progenitor cell lines (e.g., C17-2) can engraft in newborns of other rodent species (e.g., rats) without immune suppression. Recent data suggest that this capability may be rooted in fundamental neural progenitor biology: once differentiated, immortalized CNS lines apparently downregulate cell surface molecules necessary for recognition and lysis by cytotoxic T lymphocytes (White *et al.*, 1994). This tolerance may expand the usefulness of a relatively small number of competent neural cell lines. In cases in which immune rejection is problematic, however, cyclosporin immunosuppression of hosts is efficacious.

VI. Future Challenges

Future challenges in the transplantation of immortalized neural cell lines into the CNS include the following: to better understand the basic biology of the cells that have been immortalized, to determine the parameters that optimize engraftment, to discern the triggers that direct the phenotypic fate of progenitors in the brain, and to identify the mechanisms that dictate efficiency of foreign gene expression by engrafted neural cells. Generating such lines from human neural tissue—the ultimate clinical goal—will entail being able to prospectively identify progenitors that will prove amenable to immortalization and transplantation. This, in turn, will require a better understanding of fundamental neural progenitor cell biology. Are the neural cell lines that exist truly representative of the majority of progenitors in the CNS or only an unusual minor subgroup (albeit a subgroup with potential clinical import)? The extent to which such lines are representative of the true *in vivo* condition will require constant reality checks—that is, investigators working with lines comparing their data with those obtained by groups studying primary progenitors and neural cells *in vivo* and *in vitro*.

VII. Summary and Conclusions

Two emerging areas of inquiry into CNS dysfunction have converged: *neural progenitor and stem cell biology* with *gene therapy and repair of the CNS via transplantation*. Immortalized neural progenitors, when transplanted into the CNS, seem to intermingle nondisruptively with endogenous progenitors and respond to the same spatial and temporal developmental signals in the same manner as host progenitors. Immortalization does not appear to subvert their ability to respond to normal cues (Section III,B). Their successful engraftment represents integration by exogenous central neural tissue that is neither tumorigenic nor of tumor or primary fetal origin. Clonal neural cell lines have many advantages over primary fetal tissue as graft material for both developmental (Section III,B) and therapeutic (Section IV) purposes (Table 2). These cells can express foreign genes in an exuberant, stable fashion within CNS parenchyma for prolonged periods. These observations suggest that using immortalized multipotent progenitors as transduction agents for exogenous factors (e.g., therapeutic gene products) (Section IV,C) and/or as integral members of CNS cytoarchitecture (e.g., repopulating missing neural cell types) (Section IV,D) may be feasible for clinical and research applications. Experiments affirming these abilities in animal models are progressing. The grafting of immortalized progenitors may, in fact, be a flexible, broadly applicable, and parsimonious strategy for simultaneously addressing a number of the pathophysiologies that often characterize CNS dysfunction. First, transplanted progenitors may literally differentiate into and replace damaged neurons. Cells degenerated by either cell-intrinsic or cell-extrinsic processes might be so replaced (Section IV,D,1,b). Second, implanted cells, being of CNS origin, may intrinsically provide bridges, growth factors, and/or cell–cell contact signals that might allow the injured host to regenerate its own lost cells and reform its own connections. Third, the cells might constitutively synthesize, or be genetically engineered *ex vivo* before transplantation to become factories for, proteins known to promote regeneration or to forestall degeneration resulting from insufficiency of a trophin or enzyme in the milieu (Section IV,C). These cells make it possible to deliver foreign gene products either to wide regions of the brain or to discrete areas, depending on the age of the host and route of transplantation (e.g., see Section IV,C,2). This flexibility may be useful in basic neurobiology to study the effect of introduced genes on brain function. Because immortal progenitors may incorporate into host cytoarchitecture in a functional manner, they may prove more than vehicles for "passive" delivery of substances: the regulated release of certain substances through feedback loops may be reconstituted as might the reformation of essential circuits. Intriguingly, by virtue of the basic biology of some progenitors, one cell line could, under certain conditions, perform many of these above-described tasks (Table 3).

That such cell lines may participate in normal development and differentiate into appropriate cell types *in vivo* offers reason to suspect that some observations may be applicable to endogenous CNS neurons and progenitors (Section III). These lines, therefore, because of their homogeneity and abundance, might prove useful as *in vitro* models for neurotoxicity, for isolation of neural-specific or developmentally important genes and transcription factors, for assaying the effects of trophins, etc. (Section III,A).

A cautionary note must be sounded, however. The process of immortalization may create a selection bias and/or actually forestall the restriction of phenotypic range that would otherwise take place in the unmanipulated CNS (Section III,B,5). On the other hand, maintaining progenitors in a proliferative state *in vitro* and transiently suspending their differentiation might be the only way to help distinguish experimentally between a progenitor's "fate" and its "potential." The fate of a cell does not always reflect its potential, nor does its potential always portend its fate. Immortalization may maintain progenitors in a more stem cell-like state. (A knowledge of its potential may permit its fate to be altered if necessary.) Ultimately, however, observations from multiple methods of analysis (e.g., *in vivo* lineage mapping of endogenous progenitors, transplantation of immortal and primary progenitors, *in vitro* analysis of isolated primary and immortal progenitors) will need to be compared and assimilated in order to provide a complete picture of normal mammalian CNS development. In the interim, the use of immortalized or perpetual neural progenitors may allow restorative neurologists to exploit the inherent biologic attributes of progenitors and stem-like cells for therapeutic purposes (Snyder, 1992, 1994).

References

Ahmed, S., Reynolds, B. A., and Weiss, S. (1994). BDNF enhances differentiation but not survival of neurons and neuronal precursors derived from EGF-responsive CNS stem cells. *Soc. Neurosci. Abstr.* 20, 458.

Altman, J. (1969). Autoradiographic & histologic studies of postnatal neurogenesis. IV. Cell proliferation & migration in thc antcrior forcbrain, with special reference to persisting neurogenesis in the olfactory bulb. *J. Comp. Neurol.* **137**, 433–458.

Anton, R., Kordower, J. H., Maidment, N. T., Manaster, J. S., Kane, D. J., Rabizadeh, S., Schueller, S. B., Yang, J., Rabizadeh, S., Edwards, R. H., Markham, C. H., and Bredesen, D. E. (1994). Neural-targeted gene therapy for rodent and primate hemiparkinsonism. *Exp. Neurol.* **127**, 207–218.

Bartlett, P. F., Reid, H. H., Bailey, K. A., and Bernard, O. (1988). Immortalization of mouse neural precursor cells by the c-myc oncogene. *Proc. Natl. Acad. Sci. USA* **85**, 3255–3259.

Bernard, O., Reid, H. H., and Bartlett, P. F. (1992a). Role of the c-myc and the N-myc proto-oncogenes in the immortalization of neural precursors. *J. Neurosci. Res.* **24**, 9–20.

Bernard, O., Drago, J., and Sheng, H. (1992b). L-*myc* & N-*myc* influence lineage determination in the CNS. *Neuron* **9**, 1217–1224.

Birkenmeier, E. H., Barker, C. A., Vogler, C. A., Kyle, J. W., Sly, W. S., Gwynn, B., Levy, B., and Pegors, C. (1991). Increased life span and correction of metabolic defects in murine mucopolysaccharidosis type VII after syngeneic bone marrow transplantation. *Blood* **78**, 3081–3092.

Birren, S. J., and Anderson, D. J. (1990). A v-myc-immortalized sympathoadrenal progenitor cell line in which neuronal differentiation is initiated by FGF but not NGF. *Neuron* **4**, 189–201.

Birren, S. J., Verdi, J. M., and Anderson, D. J. (1992). Membrane depolarization induces p140trk and NGF responsiveness, but not p75LNGFR, in MAH cells. *Science* **257**, 395–397.

Bjorklund, A. (1993). Better cells for brain repair. *Nature* **362**, 414–415.

Blusztajn, J. K., Benturini, A., Jackson, D. A., Lee, H. J., and Wainer, B. H. (1992). Acetylcholine synthesis and release is enhanced by dibuturyl cyclic AMP in neuronal cell line derived from mouse septum. *J. Neurosci.* **12**, 793–799.

Bogler, O., Wren, D., Barnett, S. C., Land, H., and Noble, M. (1990). Cooperation between 2 growth factors promotes extended self-renewal & inhibits differentiation of O2A progenitor cells. *Proc. Natl. Acad. Sci. USA* **87**, 6368–6372.

Breakefield, X. O., *et al.* (1993). *Nature Genet* **3**, 187–189.

Campbell, K., Olsson, M., and Bjorklund, A. (1994). Restricted fates of neuronal precursors derived from the mouse lateral ganglionic eminence after transplantation into the embryonic rat forebrain. *Soc. Neurosci. Abstr.* **20**, 205–210.

Catteneo, E., and McKay, R. (1990). Nerve growth factor regulates proliferation and differentiation of neuronal stem cells. *Nature* **347**, 762–765.

Cepko, C., Turner, D., Price, J., Ryder, E., and Snyder, E. (1987). Retrovirus-mediated gene transfer & expression in the nervous system. In *Current Communications in Molecular Biology* (J. H. Miller and M. P. Calos, Eds.), pp. 15–17. Cold Spring Harbor Laboratory Press, Cold Spring Harbor, NY.

Cepko, C. L., Ryder, E. F., Austin, C. P., Walsh, C., and Fekete, D. M. (1993). Lineage analysis using retrovirus vectors. *Methods Enzymol.* **225**, 933–960.

Cepko, C. L. (1991). Transduction of genes using retrovirus vectors. In *Current Protocols in Molecular Biology* (F. M. Ausubel, R. Brent, R. E. Kingston, D. D. Moore, J. G. Seidman, J. A. Smith, and K. Truhl, Eds.), Vol. 1, pp. 9.10.1–9.14.3. Wiley, New York.

Corotto, F. S., Henegar, J. A., and Maruniak, J. A. (1993). Neurogenesis persists in the subependymal layer of the adult mouse brain. *Neurosci. Lett.* **149**, 111–114.

Crawford, G. D., Le, W. D., Smith, R. G., Xie, W-J., Stefani, E., and Appel, S. H. (1992). A novel Ni18Tg × mesencephalic cell hybrid expresses properties that suggest a dopaminergic cell line of substantia nigra origin. *J. Neurosci.* **12**, 3392–3398.

Daadi, M., Hewson, J., Wheatley, M., Reynolds, B. A., and Weiss, S. (1994). The EGF-responsive cell of the adult murine striatum is a multipotent stem cell residing in the subependynmal cell layer. *Soc. Neurosci. Abstr.* **20**, 458.

Davis, A. A., and Temple, S. (1994). A self-renewing multipotential stem cell in embryonic rat cerebral cortex. *Nature* **372**, 263–266.

Doering, L. C., and Henderson, J. T. (1994). C6 cells modified to secrete CNTF promote the survival of MAH cell co-grafts in the brain. *Soc. Neurosci. Abstr.* **20,** 472.

Duncan, I. D., Archer, D. R., and Hammang, J. P. (1993). EGF-responsive neural stem cells isolated from rat and mouse brain are capable of differentiating into oligodendrocytes and of forming myelin following transplantation into the myelin deficient rat. *Soc. Neurosci. Abstr.* **19,** 689.

Eves, E. M., Tucker, M. S., Roback, J. D., Downen, M., Rosner, M. R., and Wainer, B. H. (1992). Immortal rat hippocampal cell lines exhibit neuronal and glial lineages and neurotrophin gene expression. *Proc. Natl. Acad. Sci. USA* **89,** 4373–4377.

Evrard, C., Borde, Marin, P., Galiana, B. E., Premont, J., Gros, F., and Rouget, P. (1990). Immortalization of bipotential & plastic glio-neuronal precursor cells. *Proc. Natl. Acad. Sci. USA* **87,** 3062–3066.

Fishell, G., Mason, C. A., and Hatten, M. E. (1993). Dispersion of neural progenitors within the germinal zones of the forebrain. *Nature* **362,** 636–638.

Fisher, L. J., and Gage, F. H. (1993). Grafting in the mammalian central nervous system. *Physiol. Rev.* **73,** 583–616.

Fisher, L. J., and Ray, J. (1994). In vivo and ex vivo gene transfer to the brain. *Curr. Opin. Neurobiol.* **4,** 735–741.

Flax, J. D., Villa-Komaroff, L., and Snyder, E. Y. (1994). Multipotent immortalized neural progenitors differentiate into glia and neurons following engraftment into the subependymal germinal zone of adult mice. *Soc. Neurosci. Abstr.* **20,** 1672.

Frederickson, K., Jat, P. S., Valtz, N., Levy, D., and McKay, R. (1988). Immortalization of precursor cells from the mammalian central nervous system. *Neuron* **1,** 439–448.

Frederickson, K., and McKay, R. D. G. (1988). Proliferation & differentiation of rat neuroepithelial precursor cells in vivo. *J. Neurosci.* **8,** 1144–1151.

Freed, W. J. (1993). Neural transplantation: Prospects for clinical use. *Cell Tranplant.* **2,** 13–31.

Friedmann, T. (1994). Gene therapy for neurological disorders. *Trends Genet.* **10,** 210–214.

Gage, F. H. (1993). Fetal implants put to the test. *Nature* **361,** 405–406.

Gage, F. H. (1992). Repopulating the mortal brain—With immortal cells. *Curr. Biol.* **2,** 232–234.

Gage, F. H., Coates, P. W., Ray, J., Peterson, D. A., Suhr, S. T., Fisher, L. J., Kuhn, H. G., and Palmer, T. D. (1994). Cells from the adult hippocampus survive in vitro and following grafting to the adult brain. *Soc. Neurosci. Abstr.* **20,** 670.

Gage, F. H., Ray, J., and Fisher, L. J. (1995). Isolation, characterization and use of stem cells from the CNS. *Annu. Rev. Neurosci.* **18,** 159–192.

Gao, W.-Q., Liu, X.-L., and Hatten, M. E. (1992). The weaver gene encodes a nonautonomous signal for CNS differentiation. *Cell* **68,** 841–854.

Gao, W.-Q., and Hatten, M. E. (1993). Neuronal differentiation rescued by implantation of weaver granule cell precursors into wild-type cerebellar cortex. *Science* **260,** 367–369.

Gao, W.-Q., and Hatten, M. E. (1994). Immortalizing oncogenes subvert the establishment of granule cell Identity in developing cerebellum. *Development* **120,** 1059–1070.

Geller, H. M., and Dubois-Dalcq, M. (1988). Antigenic & functional characterization of a rat CNS-derived cell line immortalized by a retroviral vector. *J. Cell Biol.* **107,** 1977–1986.

Gensburger, C., Laboaurdette, G., and Sensenbrenner, M. (1987). Brain basic fibroblast growth factor stimulates the proliferation of rat neuronal precursor cells in vitro. *FEBS Lett.* **217,** 1–5.

Giordano, M., Takashima, H., Herranz, A., Poltorak, M., Geller, H. M., Marone, M., and Freed, W. J. (1993). Immortalized GABAergic cell lines derived from rat striatum using a temperature-sensitive allele of the SV40 large antigen. *Exp. Neurol.* **124,** 395–400.

Greene, L. A., and Tischler, A. S. (1976). Establishment of a noradrenergic clonal line of rat adrenal pheochromocytoma cells which respond to nerve growth factor. *Proc. Natl. Acad. Sci. USA* **73,** 2424–2428.

Gout, O., Gansmuller, A., Baumann, N., and Gumpel, M. (1988). Remyelination by transplanted oligodendrocytes of a demyelinated lesion in the spinal cord of the adult shiverer mouse. *Neurosci. Lett.* **87**, 195–199.

Groves, A. K., Barnett, S. C., Franklin, R. J. M., Crang, A. J., Mayer, M., Blakemore, W. F., and Noble, M. (1993). Repair of demyelinated lesions by transplantation of purified O2A progenitor cells. *Nature* **362**, 453–455.

Hammang, J. P., Baetge, E. E., Behringer, R. R., Brinster, R. L., Palmiter, R. D., and Messing, A. (1990). Immortalized retinal neurons derived from SV40-t-antigen-induced tumors in transgenic mice. *Neuron* **4**, 775–782.

Ip, N. Y., Boulton, T. G., Li, Y., Verdi, J. M., Birren, S. J., Anderson, D. J., and Yancopoulos, G. D. (1994). CNTF, FGF, and NGF collaborate to drive the terminal differentiation of MAH cells into postmitotic neurons. *Neuron* **13**, 443–455.

Jat, P. S., Nobel, M. D., Ataliotis, P., Tanaka, Y., Yannoutous, N., Larssen, L., and Kioussis, D. (1991). Direct derivation of conditionally immortal cell lines from an H-2K^{b}tsA58 transgenic mouse. *Proc. Natl. Acad. Sci. USA* **88**, 5096–5100.

Johnson, E. M., and Deckwerth, T. L. (1993). Molecular mechanisms of developmental neuronal death. *Annu. Rev. Neurosci.* **16**, 31–46.

Kay, M. A., and Woo, S. L. C. (1994). Gene therapy for metabolic disorders. *Trends Genet.* **10**, 253–257.

Kilpatrick, T., and Bartlett, P. F. (1993). Cloning and growth of multipotential neural precursors: Requirements for proliferation & differentiation. *Neuron* **10**, 255–265.

Kitchens, D. L., Snyder, E. Y., and Gottlieb, D. I. (1994). bFGF & EGF are mitogens for immortalized neural progenitors. *J. Neurobiol.* **25**, 797–807.

Kleppner, S. R., Robinson, K. A., Trojanowski, J. Q., and Lee, V. M-Y. (1995). Transplanted human neurons derived from a teratocarcinoma cell line (NTera-2) mature, integrate, and survive for over one year in the nude mouse brain. *J. Comp. Neurol.*, (in press).

Kozlova, E. N., Rosario, C. M., Stomberg, I., Carlstedt, T., Bygdeman, M., Sidman, R. L., and Aldskogius, H. (1994). Axons from peripherally grafted human fetal dorsal root ganglion cells grow into adult rat spinal cord. *Soc. Neurosci. Abstr.* **20**.

Lacorraza, H. D., Flax, J. D., Snyder, E. Y., and Jendoubi, M. (1995). In vivo gene transfer and expression of human β-hexosaminidase α-subunit into mouse brain. *J. Neurochem.* **64**, suppl, 59. [Abstract]

Largent, B. L., Sosnowski, R. G., and Reed, R. R. (1993). Directed expression of an oncogene to the olfactory neuronal lineage in transgenic mice. *J. Neurosci.* **13**, 300–312.

La Rocca, S. A., Grossi, M., Falcone, G., Alema, S., and Tato, R. (1989). Interaction with normal cells suppresses the transformed phenotype of v-myc-transformed quail muscle cells. *Cell* **58**, 123–131.

Levison, S. W., and Goldman, J. E. (1993). Both oligodendrocytes & astrocytes develop from progenitors in the subventricular zone of postnatal rat forebrain. *Neuron* **10**, 201–212.

Lois, C., and Alvarez-Buylla, A. (1994). Long distance neuronal migration in the adult mammalian brain. *Science* **264**, 1145–1148.

Lois, C., and Alvarez-Buylla, A. (1993). Proliferating subventricular zone cells in the adult mammalian forebrain can differentiate into neurons & glia. *Proc. Natl. Acad. Sci. USA* **90**, 2074–2077.

Loo, D. T., Copani, A., Pike, C. J., Whittemore, E. R., Walencewicz, A. J., and Cotman, C. W. (1993). *Proc. Natl. Acad. Sci. USA* **90**, 7951–7955.

Louis, J. C., Magal, E., Muir, D., Manthrope, M., and Varon, S. (1992). CG-4, a new bipotential glial cell line from rat brain, is capable of differentiating in vitro into either mature oligodendrocytes or type-2 astrocytes. *J. Neurosci. Res.* **31**, 193–204.

Luskin, M. B. (1994). Neuronal cell lineage in the vertebrate central nervous system. *FASEB J.* **8**, 722–730.

Luskin, M. B. (1993). Restricted proliferation & migration of postnatally generated neurons derived from the forebrain subventricular zone. *Neuron* **11**, 173–189.
Luskin, M. B., and McDermott, K. (1994). Divergent lineages for oligodendrocytes & astrocytes originating in the neonatal forebrain subventricular zone. *Glia* **11**, 211–226.
Luskin, M. B., Parnavelas, J. G., and Barfield, J. A. (1993). Neurons, astrocytes, & oligodendrocytes of the rat cerebral cortex originate from separate progenitor cells: An ultrastructural analysis of clonally related cells. *J. Neurosci.* 13(4), 1730–1750.
Macklis, J. D. (1993). Transplanted neocortical neurons migrate selectively into regions of neuronal degeneration produced by chromophore-targeted laser photolysis. *J. Neurosci.* **13**(9), 3848–3863.
Macklis, J. D., Yoon, C. H., and Snyder, E. Y. (1994). Immortalized neural progenitors differentiate toward repletion of a neuronal population selectively eliminated from adult mouse neocortex by targeted photolysis. *Exp. Neurol.* **127**, 9. [Abstract]
Martinez-Serrano, A., Fischer, W., Nikkah, G., Lundberg, C., McKay, R. D. G., and Bjorklund, A. (1994). Nerve growth factor gene transfer to the adult and aged rat brain using cns derived neural progenitor cells cellular effects and behavioral recovery. *Soc. Neurosci. Abstr.* **20**, 1099.
McBurney, M. W., Reuhl, K. R., Ally, A. I., Soma, N., Bell, J. C., and Craig, J. (1988). Differentiation & maturation of embryonal carcinoma-derived neurons in cell culture. *J. Neurosci.* 8(3), 1063–1073.
McConnell, S. K., and Kaznowski, C. E. (1991). Cell cycle dependence of laminar determination in developing neocortex. *Science* **257**, 282–285.
McConnell, S. K. (1988). Fates of visual cortical neurons in the ferret after isochronic & heterochronic transplantation. *J. Neurosci.* **8**, 945–974.
McDermott, K. W., and Lantos, P. L. (1991). Distribution & fine structural analysis of undifferentiated cells in the primate subependymal layer. *J. Anat.* **178**, 45–63.
Mellon, P. L., Windle, J. J., Goldsmith, P. C., Padula, C. A., Roberts, J. L., and Weiner, R. I. (1990). Immortalization of hypothalamic GnRH neurons by genetically targeted tumorigenesis. *Neuron* **5**, 1–10.
Miller, G. M., Silverman, A-J.,Roberts, J. L., Dong, K. W., and Gibson, M. J. (1993). Functional assessment of intrahypothalamic implants of immortalized gonadotropin-releasing hormone-secreting cells in female hypogonadal mice. *Cell Transplant.* **2**, 251–257.
Miller, M. W., and Nowakowski, R. S. (1988). Use of BRDU-immunohistochemistry to examine the proliferation, migration, & time of origin of cells in the CNS. *Brain Res* **457**, 44–52.
Morshead, C. M., Reynolds, B. A., Craig, C. G., McBurney, M. W., Staines, W. A., Morassutti, D., Weiss, S., and van der Kooy, D. (1994). Neural stem cell is the adult mammalian forebrain: A relatively quiescent subpopulation of subependymal cells. *Neuron* **13**, 1071–1082.
Moullier, P., Bohl, D., Heard, J.-M., and Danos, A. (1993). Correction of lysosomal storage in the liver & spleen of MPS VII mice by implantation of genetically modified skin fibroblasts. *Nature Genet.* **4**, 154–159.
Mulligan, R. C. (1993). The basic science of gene therapy. *Science* **260**, 926–932.
Onnifer, S. M., Whittemore, S. R., and Holets, V. R. (1993a). Variable morphological differentiation of a raphe-derived neuronal cell line following transplantation into the adult rat CNS. *Exp. Neurol.* **122**, 130–142.
Onnifer, S. M., White, L. A., Whittemore, S. R., and Holets, V. R. (1993b). In vitro labeling strategies for identifying primary neural tissue & a neuronal cell line after transplantation in the CNS. *Cell Transplant.* **2**, 131–149.
O'Rourke, N. A., Dailey, M. E., Smith, S. J., and McConnell, S. K. (1992). Diverse migratory pathways in the developing cortex. *Science* **258**, 299–302.
Palmieri, S., Kahn, P., and Graf, T. (1983). Quail embryo fibroblasts transformed by v-myc containing virus isolates show enhanced proliferation but are nontumorigenic. *EMBO J.* **2**, 2385–2389.

Poltorak, M., Isono, M., Freed, W. J., Ronnett, G. V., and Snyder, S. H. (1992). Human cortical neuronal cell line (HCN-1): Further in vitro characterization and suitability for brain transplantation. *Cell Trans.* **1,** 3–15.

Raff, M. C., Lillien, L. E., Richardson, W. D., Burnes, J. F., and Noble, M. D. (1988). Platelet-derived growth factor from astrocytes drives the clock that times oligodendrocyte development in culture. *Nature* **333,** 562–565.

Ray, J., Peterson, D. A., Schinstine, M., and Gage, F. H. (1993). Proliferation, differentiation, & long term culture of primary hippocampal neurons. *Proc. Natl. Acad. Sci. USA* **90,** 3602–3606.

Ray, J., Fisher, L. J., Kuhn, H. G., Peterson, D. A., Tuszynski, M., and Gage, F. H. (1994). Neuroblasts cultured from embryonic hippocampus and spinal cord can survive and express neural markers after grafting in the adult CNS. *Soc. Neurosci. Abstr.* **20,** 670.

Readhead, C., Popko, B., Takahashi, N., Shine, H. D., Saavedra, R. A., Sidman, R. L., and Hood, L. (1987). Expression of a myelin basic protein gene in transgenic shiverer mice: Correction of the dysmyelinating phenotype. *Cell* **48,** 703–712.

Renfranz, P. J., Cunningham, M. G., and McKay, R. D. G. (1991). Region-specific differentiation of the hippocampal stem cell line HiB5 upon implantation into the developing mammalian brain. *Cell* **66,** 713–719.

Reynolds, B. A., and Weiss, S. (1992). Generation of neurons & astrocytes from isolated cells of the adult mammalian central nervous system. *Science* **27,** 1707–1710.

Reynolds, B., Tetzlaff, A., and Weiss, S. (1992). A multipotent EGF-responsive striatal embryonic progenitor cell produces neurons & astrocytes. *J. Neurosci.* **12,** 4565–4574.

Richards, L. J., Kilpatrick, T. J., and Bartlett, P. F. (1992). De novo generation of neuronal cells from the adult mouse brain. *Proc. Natl. Acad. Sci. USA* **89,** 8591–8595.

Richardson, W. D., Prtingle, N., Mosley, M., Westernmark, M., and Dubois-Dalcq, M. (1988). A role for platelet-derived growth factor in normal gliogenesis in the central nervous system. *Cell* **53,** 309–319.

Ross, M. E., Fletcher, C., Mason, C. A., and Hatten, M. E. (1990). Meander tail reveals a discrete developmental unit in the mouse cerebellum. *Proc. Natl. Acad. Sci. USA* **87,** 4189–4192.

Ryder, E. F., Snyder, E. Y., and Cepko, C. L. (1990). Establishment & characterization of multipotent neural cell lines using retrovirus vector-mediated oncogene transfer. *J. Neurobiol.* **21,** 356–375.

Sands, M. S., Barker, J. E., Vogler, C. A., Levy, B., Gwynn, N., Galvin, N., Sly, W. S., and Birkenmeier, E. H. (1993). Treatment of murine MPS VII by syngeneic bone marrow transplantation in neonates. *Lab Invest.* **68,** 676–686.

Sands, M. S., Vogler, C., Kyle, J. W., Grubb, J. H., Levy, B., Galvin, N., Sly, W. S., and Birkenmeier, E. H. (1994). Enzyme replacement therapy for murine mucopolysaccharidosis Type VII. *J. Clin. Invest.* **93**(6), 2324–2331.

Sheen, V. L., Dreyer, E. B., and Macklis, J. D. (1992). Calcium-mediated neuronal degeneration follows singlet oxygen production by chromophore-targeted photolysis. *Neuroreport* **3,** 705–708.

Sheen, V. L., and Macklis, J. D. (1994). Apoptotic mechanisms in targeted photolytic neuronal cell death by chromophore-activated photolysis. *Exp. Neurol.* **130,** 67–81.

Shihabuddin, L. S., Whittemore, S. R., and Holets, V. R. (1993). The adult CNS retains the potential to direct specific differentiation of a transplanted neuronal precursor cell line. *Soc. Neurosci. Abstr.* **19,** 1319.

Sly, W. S., Quinton, B. A., McAlister, W. H., and Rimoin, D. L. (1973). β-Glucuronidase deficiency: Report of clinical, radiologic and biochemical features of a new mucopolysaccharidosis. *J. Pediatr.* **82,** 249–257.

Smart, I. (1961). The subependymal layer of the mouse brain & its cell production as shown by radiography after ^{3}H-thymidine injection. *J. Comp. Neurol.* **116,** 325–347.

Snyder, E. Y., Deitcher, D. L., Walsh, C., Arnold-Aldea, S., Hartwieg, E. A., and Cepko, C. L. (1992). Multipotent neural cell lines can engraft & participate in development of mouse cerebellum. *Cell* **68,** 33–51.

Snyder, E. Y., Yandava, B. D., Pan, Z-H., Yoon, C., and Macklis, J. D. (1993). Immortalized postnatally-derived cerebellar progenitors can engraft & participate in development of multiple structures at multiple stages along mouse neuraxis. *Soc. Neurosci Abstr.* **19**, 613.

Snyder, E. Y., Taylor, R. M., and Wolfe, J. H. (1995a). Transplantation of b-glucuronisase-expressing immortalized neural progenitors for gene transfer into the central nervous system of the Mucopolysaccharidosis VII mouse, a prototype of a genetic neurovisceral lysosomal storage disease. *Nature* **374**, 367–370.

Snyder, E. Y., Yandava, B. D., Rosario, C. M., Kosaras, B., and Sidman, R. L. (1995b). Use of immortalized neural progenitors or stem-like cells for specific cell type replacement in neurologic mutants: Granule cell interneurons in *meander tail;* oligodendroglia in *shiverer. Exp. Neurol.* (in press). [Abstract]

Snyder, E. Y., and Flax, J. D. (1995). Transplantation of neural progenitors and stem-like cells as a strategy for gene therapy and repair of neurodegenerative diseases. *Mental Retardation Dev. Disabilities Res. Rev.* (in press).

Snyder, E. Y. (1992). Neural transplantation: An approach to cellular plasticity in the developing CNS. *Semin. Perinatol.* **16**, 106–121.

Snyder, E. Y. (1994). Grafting immortalized neuronal cell lines to the CNS. *Curr. Opin. Neurobiol.* **4**, 742–751.

Snyder, E. Y. (1995a). Immortalized neural stem cells: Insights into development; prospects for gene therapy & repair. *Proc. Assoc. Am. Physicians* (in press).

Snyder, E. Y. (1995b). Use of non-neuronal cells for gene delivery. *NeuroReport* (in press).

Sotelo, C., and Alvarado-Mallat, R. M. (1987a). Reconstruction of the defective cerebellar circuitry in adult PCD mutant mice by Purkinje cell replacement through transplantation of embryonic implants. *Neuroscience* **20**, 1–22.

Sotelo, C., and Alvarado-Mallat, R. M. (1987b). Embryonic & adult neurons interact to allow Purkinje cell replacement in mutant cerebellum. *Nature* **327**, 421–423.

Suhr, S., and Gage, F. H. (1993). Gene therapy for neurologic disease. *Arch. Neurol.* **50**, 1252–1268.

Suri, C., Fung, B. P., Tischler, A. S., and Chikaraishi, D. M. (1993). Catecholaminergic cell lines from the brain and adrenal glands of tyrosine hydroxylase-SV40-T-antigen transgenic mice. *J. Neurosci.* **13**, 1280–1291.

Svendsen, C. N., Clarke, D. J., Fawcett, J. W., Haque, N., and Dunnett, S. B. (1994). Grafting of cultured human and rat EGF-responsive progenitor cells into the striatum of 6-OHDA lesioned rats. *Soc. Neurosci. Abstr.* **20**, 471.

Temple, S. (1989). Division & differentiation of isolated CNS blast cells in microculture. *Nature* **340**, 471–473.

Temple, S., and Davis, A. A. (1994). Isolated rat cortical progenitor cells are maintained in division in vitro by membrane-associated factors. *Development* **120**, 999–1008.

Trojanowski, J. Q., Mantione, J. R., Lee, J. H., Seid, D. P., You, T., Inge, L., and Lee, V. M. Y. (1993). Neurons derived from a human teratocarcinoma cell line establish molecular & structural polarity following transplantation into the rodent brain. *Exp. Neurol.* **122**, 283–284.

Turner, D. L., Snyder, E. Y., and Cepko, C. L. (1990). Lineage-independent determination of cell type in the embryonic mouse retina. *Neuron* **4**, 833–845.

Vescovi, A. L., Reynolds, B. A., Fraser, D. D., and Weiss, S. (1993). bFGF regulates the proliferative fate of unipotent (neuronal) & bipotent (neuronal/astroglial) EGF-generated CNS progenitor cells. *Neuron* **11**, 951–966.

Vogler, C., Birkenmeier, E. H., Sly, W. S., Levy, B., Pegors, C., Kyle, J. W., and Beamer, W. G. (1990). A murine model of MPS VII: Gross & microscopic findings in β-glucuronidase deficient mice. *Am. J. Pathol.* **136**, 207–217.

Vogler, C., Sands, M., Higgins, A., Levy, B., Grubb, J., Birkensmeier, E. H., and Sly, W. S. (1993). Enzyme replacement with recombinant β-glucuronidase in the newborn MPS Type VII mouse. *Pediatr. Res.* **34,** 837–840.

Walsh, C., and Cepko, C. L. (1992). Widespread dispersion of neuronal clones across functional region of the cerebral cortex. *Science* **255,** 434–440.

Walsh, C., and Cepko, C. L. (1993). Dispersion of neural progenitors within the germinal zones of the forebrain. *Nature* **362,** 632–635.

Weinstein, D. E., Shelanski, M. L., and Liem, R. K. (1990). C17, a retrovirally immortalized neuronal cell line, inhibits the proliferation of astrocytes & astrocytoma cells by a contact-mediated mechanism. *Glia* **3,** 130–139.

White, L. A., and Whittemore, S. R. (1992). Immortalization of raphe neurons: An approach to neuronal function in vitro and in vivo. *J. Chem. Neuroanat.* **5,** 327–330.

White, L. A., Keane, R. W., and Whittemore, S. R. (1994). Differentiation of immortalized CNS neuronal cell lines decreases their susceptibility to cytotoxic T lymphocyte cell lysis in vitro. *J. Neuroimmunol.* **49,** 135–143.

White, L. A., Eaton, M. J., Castro, M. C., Klose, K. J., Globus, M. Y-T., Shaw, G., and Whittemore, S. R. (1995). Distinct regulatory pathways control neurofilament expression and neurotransmitter synthesis in immortalized serotonergic neurons. *J. Neurosci.* **14(11),** 6744–6753.

Whittemore, S. R., and White, L. A. (1993). Target regulation of neuronal differentiation in a temperature-sensitive cell line derived from medullary raphe. *Brain Res.* **615,** 27–40.

Whittemore, S. R., White, L. A., Shihabuddin, L. S., and Eaton, M. J. (1995). Phenotypic diversity in neuronal cell lines derived from raphe nucleus by retroviral transduction. In *NeuroProtocols, 4: Neuronal Cell Lines: Applications in Neurobiology* (A. Russo and S. Green, Eds.). Academic Press, San Diego (in press).

Wojcik, B. E., Nothias, F., Lazar, M., Jouin, H., Nicolas, J-F., and Peshanski, M. (1993). Catecholaminergic neurons result from intracerebral implantation of embryonal carcinoma cells. *Proc. Natl. Acad. Sci. USA* **90,** 1305–1309.

Wolfe, J. H., Schuchman, E. H., Stramm, L. E., Concaugh, E. A., Haskings, M. E., Aguiree, G. D., Patterson, D. F., Desnick, R. J., and Gilboa, E. (1990). Restoration of normal lysosomal function in mucopolysaccharidosis type VII cells by retroviral vector-mediated gene transfer. *Proc. Natl. Acad. Sci. USA* **87,** 2877–2881.

Wolfe, J. H., Sands, M. S., Barker, J. E., Gwynn, B., Rowe, L. B., Vogler, C. A., and Birkenmeier, E. H. (1992a). Reversal of pathology in murine MPS type VII by somatic cell gene transfer. *Nature* **360,** 749–753.

Wolfe, J. H., Deshmane, S. L., and Fraser, N. W. (1992b). Herpesvirus vector gene transfer & expression of β-glucuronidase in the CNS of MPS VII mice. *Nature Genet.* **1,** 379–384.

Wolfe, J. H., Kyle, J. W., Sands, M. S., Sly, W. S., Markowitz, J. G., and Parente, M. K. (1995). High level expression and export of β-glucuronidase from murine mucopolysaccharidosis VII cells corrected by a double copy retrovirus vector. *Gene Ther.* **2,** 70–78.

Yamanaka, S., Johnson, M. D., Grinberg, A., Westphal, H., Crawley, J. N., Taniike, M., Suzuki, K., and Proia, R. L. (1994). Targeted disruption of the HexA gene results in mice with biochemical and pathologic features of Tay-Sachs disease. *Proc. Natl. Acad. Sci. USA* **91 21,** 9975–9979.

Index